Estuaries of South Africa

Estuaries of South Africa presents an authoritative and comprehensive review of the current status of the country's estuarine research and management. Information is provided on a wide range of topics, including geological, physical and chemical processes, diversity and productivity of plant and animal communities, interactions between estuarine organisms, and system properties, ecological modelling and current management issues. This broad scope is complemented by a comparative perspective, resulting in a volume which provides a unique contribution to the subject of estuarine ecology, relevant to all those working in this field throughout the world.

BRIAN ALLANSON was a full Professor in and Head of the Department of Zoology and Entomology, Rhodes University, South Africa from 1963 until retirement in 1988. He is co-author of *Inland Waters of Southern Africa* (1990) and editor of *Lake Sibaya* (1979).

DAN BAIRD is a full Professor in and Head of the Department of Zoology, University of Port Elizabeth, South Africa.

Estuaries of South Africa

Edited by
Brian Allanson
Dan Baird

CAMBRIDGE UNIVERSITY PRESS
Cambridge, New York, Melbourne, Madrid, Cape Town, Singapore, São Paulo, Delhi

Cambridge University Press
The Edinburgh Building, Cambridge CB2 8RU, UK

Published in the United States of America by Cambridge University Press, New York

www.cambridge.org
Information on this title: www.cambridge.org/9780521584104

First published 1999
This digitally printed version 2008

A catalogue record for this publication is available from the British Library

Library of Congress Cataloguing in Publication data
Estuaries of South Africa/edited by B.R. Allanson, D. Baird.
    p.   cm.
  ISBN 0 521 58410 8 (hb)
  1. Estuarine ecology – South Africa.   2. Ecosystem management-South
  Africa.   I. Allanson, Brian R.   II. Baird, D. (Daniel), 1944–   ·
  QH195.S6E88   1999
  577.7′86′0968 – dc21   98-17406   CIP

ISBN 978-0-521-58410-4 hardback
ISBN 978-0-521-08776-6 paperback

This review is dedicated to all who go down to the muddy shores of the estuary to enquire of its mysteries

Contents

Contributors

Dr. J.B. Adams Department of Botany, University of Port Elizabeth, P.O. Box 1600, Port Elizabeth 6000.

Prof. B.R. Allanson Institute for Water Research, Rhodes University *and* Knysna Basin Project, P.O. Box 1186, Knysna 6570.

Prof. D. Baird Zoology Department, University of Port Elizabeth, P.O. Box 1600, Port Elizabeth 6000.

Prof. G.C. Bate Department of Botany, University of Port Elizabeth, P.O. Box 1600, Port Elizabeth 6000.

Prof. J.A.G. Cooper School of Environmental Studies, University of Ulster, Coleraine, BT52 1SA, Northern Ireland, UK.

Dr C.J. de Villiers Department of Zoology, University of the Western Cape, Private Bag X17, Bellville 7335.

Prof. A.T. Forbes Department of Biology, University of Natal, King George V Avenue, Durban 4001.

Prof. A.E. Heydorn WWF (SA), P.O. Box 456, Stellenbosch 7599.

Prof. P.A.R. Hockey Percy FitzPatrick Institute of African Ornithology, University of Cape Town, Rondebosch 7700.

Prof. A.N. Hodgson Department of Zoology and Entomology, Rhodes University, Grahamstown 6140.

Dr J.L. Largier Department of Oceanography, University of Cape Town, Private Bag, Rondebosch 7700 *and* Scripps Institute of Oceanography, University of California, San Diego, CA 92093-0209, USA.

Prof. J.F.K. Marais Zoology Department, University of Port Elizabeth, P.O. Box 1600, Port Elizabeth 6000.

Dr T.R. Mason The Planetarium, Armagh, Northern Ireland, UK.

Dr P.D. Morant ENVIRONMENTEK, CSIR, P.O. Box 320, Stellenbosch 7599.

Dr N.W. Quinn Institute for Natural Resources, University of Natal, Private Bag X01, Scottsville 3209.

Dr M. O'Callaghan National Botanical Institute, Kirstenbosch, Private Bag X7, Claremont 7735.

Dr E.H. Schumann Department of Geology, University of Port Elizabeth, P.O. Box 1600, Port Elizabeth 6000.

Dr J.H. Slinger ENVIRONMENTEK, CSIR, P.O. Box 320, Stellenbosch 7599.

Prof. T.D. Steinke Violet Cottage, Morgan's Bay 5292.

Dr J.K. Turpie Percy Fitzpatrick Institute of African Ornithology, University of Cape Town, Rondebosch 7700.

Dr A.K. Whitfield J.L.B. Smith Institute of Ichthyology, Private Bag 1015, Grahamstown 6140.

Dr D. Winter Zoology Department, University of Port Elizabeth, P.O. Box 1600, Port Elizabeth 6000.

Prof. T. Wooldridge Zoology Department, University of Port Elizabeth, P.O. Box 1600, Port Elizabeth 6000.

Dr C.I. Wright Joint Geological Survey/University of Natal Marine Geoscience Unit, King George V Ave, Durban 4001.

Acknowledgements

Our sincere thanks are due to Fiona Thomson of Cambridge University Press, Cape Town who had sufficient faith to support our proposal; and to a host of reviewers around the world who gave so willingly of their time to assess the merit of our contributions.

Susan Allanson provided gentle, but firm syntactical skill and prepared the final copy for presentation to the publishers. Jane Bulleid's care as our copy editor is without equal, thank you! Susan Allanson prepared the indices and the cooperation afforded us by Dr Maria Murphy of Cambridge University Press is likewise much appreciated.

The financial assistance of the University of Port Elizabeth towards costs is warmly acknowledged as is the contribution of the Water Research Commission towards the costs of the colour plates for Chapter 5. Prof. Tris Wooldridge supplied the photograph of the Mpako estuary, Eastern Cape, used on the cover.

1   Fifteen years on! Or one hundred and fifty years on!

Brian Allanson and Dan Baird

The Knysna marshes are even more interesting than the forests, for there live the vast families of waders and other water birds which find food and shelter in the reed covered swamps and marshy inlets of the estuary. Immense flocks of curlews darken the mud banks at low water and fill the air with their doleful cries. Lines of snow white pelicans and at times the ruddy crimson of the flamingo extend along the shores or rest on the water covered rocks of the Haven during

The Knysna embayment and estuary

the winter months. Terns and ibises, whimbrels and wild geese flock thither in countless myriads, while herons, bitterns, cranes, storks, divers, and the thousand varieties of sandpiper. snipe, dotterel and wild duck make the marshes resound with their clamour . . . (Extract from *A Cape Traveller's Diary, 1856* by Robert Wilmot.)

These images have gone forever! The modern traveller may wonder at what has been lost but recognises the inevitable press of human demands upon ecosystems which were not designed to withstand them. The birds are the first to go, submitting to our unrelenting over-utilisation of the natural resources of the estuaries – the most ephemeral of the coastal scenery.

Fortunately with an increasing environmental sensitivity among the community, interest in estuaries has increased enormously largely because of their focus as nodes of urban development. During the 1980s and 1990s the demands being made on the estuarine environment, both directly and indirectly, by anthropogenic activities were considerable. Within the littoral and river catchments these demands were in danger of modifying river and tidal flow to such an extent that the estuarine habitat along the coast could disappear, leaving in its stead salty or brackish water lagoons, exhibiting all the signs of crippling hypersalinity or clogging enrichment.

Hand in hand with this assessment went the need to recognise that present and future management of tidal estuaries, in particular, depends upon a good understanding of how these systems function. What are the essential physical, chemical and biological processes required to sustain an acceptable estuarine environment against the certainty that they are short-lived in geological terms? And because of this, their facies change rapidly if the tidal and river pulses which reset the ecological (or environmental) clocks are not sustained. If these rhythms are disturbed for long periods of time the consequences are often severe and irreversible. Fortunately the accumu-

lation of biophysical information which has grown from the early work at the University of Cape Town has contributed significantly to this understanding and the amelioration of human impact.

While there is no historical account of the personae involved in the South African estuarine work, the originator of serious work was J.H. Day who, during his tenure of the chair of Zoology at the University of Cape Town (1946–74), gathered a team of young researchers around him who contributed to our knowledge of the biological structure of a wide diversity of estuaries. These early studies were in many ways equivalent to the pioneering work by T.A. Stephenson on the rocky shores of the subcontinent. The fruits of this extraordinary labour were synthesised by John Day and his co-authors in *Estuarine Ecology*, published in 1981, which has provided the sure foundation upon which this new volume rests.

As time passed more and more university and museum groups and the South African Council for Scientific and Industrial Research (CSIR) took up the challenge of expanding the work of the Cape Town school, and with the substantial financial and administrative support of the South African National Council for Oceanographic Research (SANCOR), many of the research groups flourished. To judge by the overall spread in the primary peer reviewed literature, this was a particularly rich and rewarding period of investigation. The bibliographic assessment of the SANCOR literature pertinent to Southern Africa by Allanson (1992), and the later more exhaustive assembling of estuarine literature by Whitfield (1995), are testimony to the productivity of this period, 1980–96.

It will be obvious from a survey of the cited literature for each chapter that considerable use has been made of the so called 'grey' literature. It is the nature of much modern work that investigations are required for agencies demanding a description of an estuarine system, or of some structural feature or process. It is the reports arising from such work which constitute such literature. And while subsequent peer review publication may occur, it is not invariably so! This means, in effect, that often new and important results are not readily accessible. The advantage of a review such as this is that it allows the editors to sift material, and bring the more significant to the attention of the international audience.

There existed at last a framework of sensibly connected elements of estuarine properties and function which could be used in the verification of, for example, Decision Support Models (Slinger 1996, and Chapters 3 and 11). These models are directed towards management of tidal ecosystems even in the face of persistent demands that, in a country with one of the lowest mean annual precipitation:mean annual runoff (MAP:MAR) conversion ratios in the world, and with only two natural lakes, river water must be conserved at all costs, meaning in many instances the demise of an estuary. Fortunately, recognition of the intrinsic requirements of the river and its valley and its interface with the sea through the ever widening debate between ecologists, engineers, politicians and the informed public, has generated a national policy embodied in a new Water Act of 1998 which will prevent this loss and save our estuaries!

Recent quantitative studies have moved away from the purely descriptive ecology of plants and animals of earlier years and provided more information about what is happening in the water column, the surface of the sediments and the role the intertidal wetlands play in the transfer and sequestration of nutrients and particles.

These studies depended initially upon an ecophysiological approach in the resolution of the biology of single species in the ecosystem. Laboratory and mesocosm experiments in which the community of plants and animals is relatively undisturbed, provided findings which, when coupled with quantitative measurements of losses and gains in particulate and dissolved carbon, nitrogen and phosphorus, provided, in large measure, reasonable estimates of the magnitude of sinks and export for nutrients. The holistic interpretation of systems' responses is a logical outcome of this experimental era.

The objective of this volume is to present comprehensive and authoritative reviews of research conducted in South African estuaries on various aspects of the geological, physical, chemical and biological processes typical of these systems, on anthropogenic influences, system modelling and management implications. The volume provides in 13 chapters, a diverse, yet cohesive overview of our current knowledge and understanding as it developed since the publication of Day's *Estuarine Ecology*. The authors and editors have been at some considerable pains to avoid the highly technical or abstruse in their respective treatments of material. In adopting this approach we hope that those responsible for formulating policy at local, provincial and national levels will find this review helpful in directing attention to what is relevant in the effective management of estuaries.

Each chapter has been reviewed, as far as was possible, by at least two reviewers. Their singular contribution to this volume is warmly acknowledged – the errors are collectively ours! Nevertheless, we hope that our findings and interpretations may have relevance to similar systems elsewhere in the coastal regions of Earth.

A note about geographical nomenclature and climate

With the emergence of post-apartheid South Africa, the internal political boundaries of the Republic were redrawn. This has resulted in the coastal regions being divided between four Provinces: the Northern Cape, Western Cape, Eastern Cape and KwaZulu-Natal (Figure 1.1). Two erstwhile geopolitical areas are likely to cause some confusion as authors may have inadvertently used the names Ciskei and Transkei largely from habit. Each area possessed a significant segment of the southeast coast of the Republic (Figure 1.1), but both are now part of the Eastern Cape Province, and are not always referred to by these names. A somewhat similar situation pertains with respect to the names of rivers: the combination of African languages, notably Zulu and Xhosa, Afrikaans and English has resulted in considerable variance in the spelling of names, depending upon which language is being used. A case in point is the spelling of the name Krom or Kromme River. In Afrikaans the correct spelling would be 'Kromrivier', although the use of Krom or Kromme (a variant of Krom) River when writing in English is correct. We have endeavoured to standardise upon spellings which are in current use, but readers will, we hope, forgive lapses which may have slipped through.

The subdivision of the coast into three climatological regions by de Villiers and Hodgson (Chapter 8) has provided a frame of reference within which the estuarine features are placed. They are (1) the **subtropical** to the Mbashe River and (2) the **warm temperate** from the Mbashe River to Cape Point, under the influence of the warm Agulhas current; (3) a **cool temperate** region incorporating the west coasts of the Western and Northern Provinces under the influence of the Benguela current.

Subtropical control is effected by a northward movement of the South Indian Anticyclone, and a tropical easterly flow. KwaZulu-Natal is predominantly a summer-

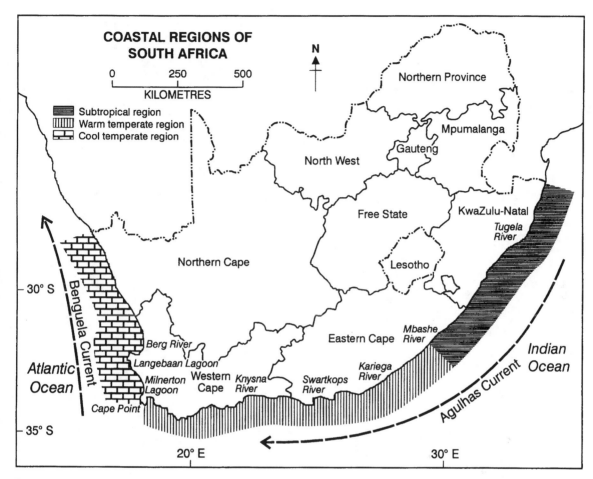

Figure 1.1. **The geopolitical structure of South Africa and the division of the coast into three geographical regions. (After de Villiers & Hodgson, Chapter 8.)**

rainfall Province because of this flow. Temperate control is brought about by westerly troughs which sweep along the coast bringing rain to the southwest areas of the Western Cape Province in winter and creating a marked mediterranean climate. The Eastern Cape coast, while affected by westerly fronts, does come under the influence of the south-erly meridional flow in early summer and autumn, so that the coastal regions experience a bimodal rainfall pattern – a pronounced peak in spring or summer and a smaller peak in autumn. A more detailed treatment of the atmospheric circulation and weather over southern Africa is given by Preston-Whyte and Tyson (1988), and in Chapter 3.

References

Allanson, B.R. (1992). *An assessment of the SANCOR Estuaries Research Programme from 1980 to 1989.* Committee for Marine Science Occasional Report No. 1. Pretoria: Foundation for Research Development.

Day, J.H. (ed.) (1981). *Estuarine ecology with particular reference to southern Africa.* Amsterdam: A.A. Balkema.

Preston-Whyte, R.A. & Tyson, P.D. (1988). *The atmosphere and weather of Southern Africa.* Cape Town: Oxford University Press.

Slinger, J. (1996). *A co-ordinated research programme on decision support for the conservation and management of estuaries.* WRC Project No. K5/577/0/1. Final Report of the Predictive Capability sub-project. Pretoria: Water Research Commission.

Whitfield, A. (1995). *Available scientific information on individual South African estuarine systems.* WRC Report No. 577/1/95. Pretoria: Water Research Commission.

Wilmot, R. (1856). *A Cape Traveller's Diary.* Friends of the Library Publication No.2. University of the Witwatersrand. Johannesburg: A.D. Donkin Publisher.

2 Geomorphology and sedimentology

Andrew Cooper, Ian Wright and Tom Mason

Mkweni River estuary

Introduction

The South African coast is particularly well endowed with estuaries which comprise a dominant component of its coastal geomorphological elements. Heydorn and Tinley (1980) list some 166 estuaries between the Orange River and the Great Kei. A further 60 have been listed in the former Transkei, and Begg (1978) describes the largest 73 of the 239 independent river outlets that occur on the KwaZulu-Natal coastline. In this context it is perhaps noteworthy that the volume *Estuarine ecology with particular reference to Southern Africa* (Day 1981) contained no section on estuarine geomorphology. Such a shortcoming was, however, entirely consistent with the state of knowledge and the dominantly ecological focus of research on South African estuaries at the time. The only existing research of significance on South African estuarine geomorphology at the time was the work of Orme (1973, 1974, 1975) in KwaZulu-Natal. Noting this lack of knowledge and perhaps recognising the fundamental role of geomorphology in estuarine ecology, the South African National Committee for Oceanographic Research (SANCOR) Estuaries Committee initiated estuarine geomorphological research through the Universities of Port Elizabeth and Natal in the early to mid-1980s (Bowmaker *et al.* 1986). This was complemented by geomorphological research by the Council for Scientific and Industrial Research (CSIR) on several Cape and Natal systems (van Heerden 1985, 1986; van Heerden & Swart 1986).

The results of that research initiative, and associated and subsequent geomorphological research, have greatly increased the understanding of South African estuarine geomorphology and have illustrated the direct relevance of geomorphological research in the interpretation of ecological observations. Indeed, it has been possible to challenge previous interpretations, based almost entirely on ecolog-

ical findings, in the light of this geomorphological research. Although much remains to be studied in the field of South African estuarine geomorphology, results to date have not only elucidated the manner in which some of South Africa's estuaries function on a temporal scale and vary spatially, but have contributed to the global understanding of estuaries (Boyd *et al.* 1992; Dalrymple *et al.* 1992; Cooper 1994*a*), since the estuaries of South Africa include among them several types that are not widely distributed elsewhere.

A second phase of estuarine geomorphological research is now warranted to fill the gaps (particularly concerning the numerous smaller estuaries) and to implement estuarine geomorphological research in the context of estuarine management. Research on the latter is under way and a geomorphological classification forms the basis of the Estuarine Health Index (Cooper *et al.* 1995) which is currently being applied to all South African estuaries.

In this section the essential characteristics of estuaries in general and South African estuaries in particular are outlined. (For a review of the current state of knowledge regarding estuarine geomorphology on a global scale, the reader is referred to Perillo 1995). The factors which contribute to variation in morphology are then considered and a broad division of estuaries into four zones is proposed on the basis of spatial variability in these controlling factors throughout the country. The diagnostic elements of estuaries in each zone are then discussed and variation within each zone is assessed by reference to previously compiled locally-based classification schemes. Temporal variation in estuarine morphology is driven largely by river discharge variations on a variety of time scales. The impacts of this variability are, however, spatially controlled and estuaries in different zones respond differently to discharge variation. The chapter concludes with an assessment of human impacts on estuarine geomorphology.

Essential elements of estuaries

Estuaries are the meeting place of terrestrial drainage systems with the coast. As such they are highly variable environments in both space and time. A concise geomorphological definition of an estuary is difficult to achieve as estuaries are transitional to both deltas and lagoons and the boundaries between the types are not clear cut. For the purposes of this chapter the term estuary will be used in a generic sense to apply to all transitional fluvio-marine systems in South Africa.

South African estuaries have almost all originated in formerly incised bedrock valleys cut during lowered sea level periods of the Pliocene and Pleistocene epochs. Some valleys may even have earlier origins extending back to Cretaceous times, soon after the South African coastline was initiated. Of most relevance to their present morphology is the Holocene rise in sea level, during which sea level rose from −130 m about 13000 years ago to within a few metres of the present during the last 5000–6000 years (Cooper 1991a; Ramsay 1995). This period of rapid rise in sea level caused a landward retreat of the shoreline and modified the lower reaches of rivers as it did so. It is important to realise that not all estuaries were affected in the same way. It is commonly held that contemporary estuarine morphology arose from drowning of former river valleys and their subsequent infilling to a greater or lesser extent. However it is possible to demonstrate that there are marked spatial controls on the nature of the infilling and that it is the balance between rate of sea level rise and rate of sedimentation that influences the morphology of an estuary. Broadly speaking this defines deficit, balance and surplus sediment supply conditions (Nichols 1989; Cooper 1994a).

Estuaries receive sediment from three main sources: the sea, inflowing rivers, and material generated within the estuary itself. The last category consists largely of organic material while the other categories comprise dominantly inorganic sediment. Marine sediment input to estuaries occurs via (i) barrier overwash forming overwash fans (Cooper & Mason 1986), (ii) tidal inflow to form flood-tidal deltas (Reddering 1983), and (iii) wind action whereby dunes advance into estuarine channels (Illenberger 1988). The relative contribution from each of these sources varies from one estuary to the next. Aeolian inputs are particularly important along the Eastern Cape coast (e.g. Sundays) and overwash dominates marine sediment input in most closed estuaries (e.g. Mhlanga) although there are cases (e.g. Bulura) where flood-tidal deposition occurs during the open phases of modally closed lagoons. Tidal currents deposit flood-tidal deltas in permanently open estuaries

with large tidal prisms (e.g. Kosi, Keurbooms, Kariega) and these frequently provide extensive intertidal sandflats. To a large extent estuarine morphology is a function of the relative importance of two groups of factors, tidal currents and fluvial currents. It is important to establish the nature of these forces as controls on estuarine morphology. River discharge is an essentially independent variable in this regard in as much as channel and floodplain morphology respond to variations in discharge. Thus, in a spatial sense, rivers with a given discharge may be braided or meandering according to the dominant fluvial sediment texture. River discharge also exerts a temporal control; river channels may be scoured or infilled by deltaic sedimentation following river floods, or they may diminish in area during prolonged low discharge conditions. Tidal influence is, on the other hand, an entirely dependent variable for the amount of sea water that can enter a coastal embayment on a tidal cycle (the tidal prism) is governed by the morphology of that embayment. Changes in the volume of the embayment thus govern the extent of tidal influence. This link is embodied in the linear relationship between tidal inlet dimensions and tidal prism (Bruun & Gerritson 1960). This fundamental difference between tidal and fluvial influences, which is stressed by Pethick (1984), is of great importance in understanding estuarine geomorphology and in managing estuaries.

No two estuaries are identical, but Figure 2.1 shows a hypothetical estuary in which all the essential elements are included. The extent to which each of these environments is represented within a given estuary is discussed in the following section.

Barriers are a near-ubiquitous element of South African estuaries. These range from a few metres in length to several tens of kilometres (e.g. St Lucia) but are typically in the range 100–1000 m long. Barriers on South African estuaries are mainly sandy and may or may not be topped by a coastal dune system. The origins of South African barriers have not been studied in detail but it is clear in most cases that they have been derived from marine sands carried landward during the Holocene marine transgression. In a few instances where river sediment supply has been high, contemporary barriers are augmented by or even largely composed of fluviatile sediments. Although it appears that most barriers accumulated via landward sediment transfer, modern-day processes on many barriers are controlled by a combination of longshore and cross-shore sediment transport. These processes typically act in a cyclic fashion, as in the case of the Keurbooms estuary (Reddering 1981) or as an adjustment to episodic impacts such as river floods (Reddering & Esterhuysen 1987a; Cooper et al. 1988, 1990). In both instances they operate so as to restore a dynamic equilibrium between barrier morphology and coastal dynamics.

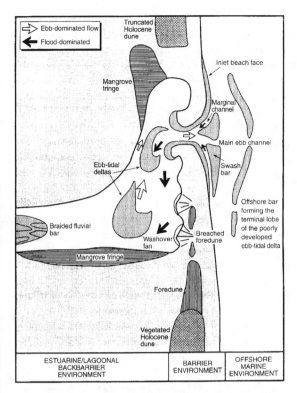

Figure 2.1. **An idealised model of a typical estuary, showing the major morphological features. Not all features may be present at any single estuary and individual components may be developed to varying extents.**

Estuaries may lack a barrier if insufficient sediment is available, a characteristic of many cliffed coastlines. Relatively few South African estuaries fall into this category but examples include the Storms river mouth on the Tsitsikamma coast, the Steenbras in False Bay as well as the much larger Knysna system on the south coast of the Western Cape (Reddering & Esterhuysen 1987b).

Spatial variation in estuarine morphology

A number of factors may be identified which influence the extent to which an estuary resembles the hypothetical archetype outlined above. A number of these (recent tectonic history, rate of sea level rise, tidal range and Quaternary sea level history) exhibit such limited spatial variation in a regional South African context that they may be considered constants on the basis of our present understanding. The major controls on estuary morphology may be considered under the following headings: climate, hinterland topography, wave energy, sediment supply and coastal lithology.

Climate

In terms of the geomorphology of estuaries, climatic controls are important in that they influence river discharge

and evaporation rates as well as the nature of estuarine vegetation. Evaporation also plays an important role in estuarine geomorphology for if it exceeds fluvial discharge a dry or hypersaline basin will ultimately result. Most of the estuaries of the arid to semi-arid Northern Cape coast are thus affected and are essentially little different from inland pans or dry river beds. They lack typical estuarine circulation, depositional patterns, sediment type and vegetation and although they provide important wetland habitats most are not truly estuaries.

The estuaries of much of the remaining warm temperate Western Cape coast experience seasonally variable rainfall. Some receive winter rainfall, some summer rainfall and a few areas have year-round or bimodal rainfall (Heydorn & Tinley 1980), but essentially the river discharge exceeds evaporation in most cases. Instances have been recorded where, during drought, certain estuaries in this zone have become hypersaline (Whitfield & Bruton 1989) and certain small catchment systems dry up in the dry season, but such systems are relatively infrequent along the coast. The seasonal discharge regime does give rise to dry-season closure of many systems which are unable to maintain a connection to the sea during low flow periods. Such systems are, however, relatively more common in the subtropical zone. Seasonal patterns in river flow commonly give rise to prolonged low salinity periods followed by prolonged high salinity periods in Western and Eastern Cape estuaries such as the Breede, Great Berg and Great Kei. Wooldridge (1992) draws attention to the crucial role of mouth formation and persistence in the functioning of South African estuaries, so that while the term 'estuary' in South Africa may include those systems that are temporarily open to the sea (Day 1981) there is a range of frequencies of opening. Reddering and Rust (1990) suggested that only 13% of river mouths in South Africa are permanently open. Most of the remaining 87% have mouths that open occasionally. Within this range there are river mouths that open seasonally and those that open only under extreme floods, or some that open only when the river that feeds them flows.

In the temperate zone climatic conditions favour the growth of saltmarsh on higher intertidal and low supratidal areas. Such vegetation gives cohesion to otherwise unconsolidated sands and muds and may exert a strong control on geomorphology by increasing the resistance of the sediment to erosion.

In the subtropical zone that stretches from central Eastern Cape to the Mozambique border, rivers are perennial and subject to a distinct summer rainfall peak. Here a seasonal geomorphological signal may be imparted by the ability of rivers to force a free connection with the sea only after strong river flow during which ephemeral deltas may form (Cooper 1990). Thus many estuaries may be unconnected to the sea for much of the year. Although not widespread, climatic conditions in this zone favour the growth of mangroves which take the place of the salt-marshes of the temperate zone. The mangrove stands provide cohesion to estuarine sediments and thus influence channel morphology within estuaries. This factor is particularly important in sandy estuaries such as Kosi Bay where the stability provided by mangroves has allowed for the establishment of a number of small islands.

Hinterland topography
Hinterland topography is an important control on estuarine morphology for it controls a number of parameters, including drainage basin size and shape, and influences other factors such as fluvial sediment yield. Across much of the country, the hinterland topography is relatively gentle and rivers are relatively long so that gentle river gradients result. In such instances river courses have matured by the time they reach the coast, and discharge extremes have relatively broad peaks and troughs. For example, the Orange River flood of May 1988 lasted for about two weeks and could be predicted several days in advance of its arrival at the river mouth (Bremner *et al.* 1990). This reflects the buffering capacity of low gradient rivers on river flows. Steep and narrow hinterlands characterise much of southern KwaZulu-Natal and impart high river gradients. Such rivers are susceptible to shortlived but often extreme peaks and troughs in discharge. The September 1987 floods in KwaZulu-Natal experienced a shorter peak which lasted overall for about five days but which contained a maximum sub-peak lasting a few hours (Kovacs 1988). The velocity of such steep rivers enables the rapid and efficient transfer of sediment to the coast, but the availability of such sediment in the catchment and the propensity for a river to flood are both strongly influenced by climate. Areas such as the Tsitsikamma coastal stretch are steeper than the regional trend while areas such as the northern KwaZulu-Natal coastal plain have exceedingly low gradients.

River gradient also determines the extent to which the tidal prism may extend upstream in an estuary. Water cannot flow uphill and so only those estuaries whose water surface lies below high tide level are capable of experiencing tidal inflow. In steep river channels such areas are much shorter than in rivers with more gentle gradients. Consequently the estuaries of southern KwaZulu-Natal are less likely to have large tidal prisms. In the low gradient systems of the Western Cape (e.g. Great Berg, Breede) tidal prisms are large and extend for tens of kilometres upstream.

Coastal dynamics
Of importance to estuarine geomorphology in this category is the nature of the coastal wave regime. The impor-

tance of waves in overwashing sediment, water and organisms from the sea into estuaries, even when closed, has been acknowledged by many authors (Begg 1978; Cooper & Mason 1986). Begg (1984) even cited a case in the Mnamfu when overwashing waves cut a channel into the barrier and breached the enclosed lagoon. Waves have also been acknowledged (sometimes wrongly) as important in inducing longshore sediment transport and the longshore growth of sand spits across the mouths of estuaries. In many instances in KwaZulu-Natal these appear to be opposed to the dominant drift direction, but this situation arises through formation of a river outlet in the lowest wave energy zone which is typically immediately updrift of the downdrift cell boundary (typically a rock headland). Such a situation is exemplified by the Zinkwasi, Mvoti and Mgobeseleni where the estuary mouth is located on the southern margin of an embayment in the lee of a rock headland. This gives the incorrect impression of a southerly directed spit. The importance of waves on estuarine geomorphology is, however, best exemplified through their influence on beach state. This concept, introduced by Short and Wright and their co-workers (Wright & Short 1984) recognised six microtidal beach states on a continuum from dissipative to reflective on the basis of wave breaker height, wave period and sediment hydraulic grain size.

The dissipative beach state which is associated with high swell waves and fine sand is a low gradient, wide beach with a well-defined surf zone which acts to reduce wave energy and enhance the stability of such beaches. This state which is of low gradient and low elevation permits frequent connection between back-barrier areas and the sea via shallow (typically between <1 and 5 cm water depth) surface flow, often without formation of a distinct channel. This may be landward (driven by wave overwashing) or seaward (effectively overspilling of estuarine waters), depending upon the relative elevations of sea and estuary. Such conditions are typical along much of the Western and Eastern Cape coast.

The KwaZulu-Natal coast is characterised by intermediate to reflective beach states which arise partly as a result of coarser grained beach sediment and reduced wave energy. Such beaches typically have a relatively high berm which reduces the ability of waves to overwash on a regular basis. The berm elevation is of importance in mouth hydrodynamics in that it may permit much seaward discharge to be effected via seepage and percolation without surface flow. However, during prolonged river flow, when seepage and evaporation are unable to keep pace with freshwater inflow, the barriers may be overtopped by estuary waters. The great vertical elevation difference between the berm and sea enhances downcutting and the barrier may fail catastrophically. Begg (1978) describes the breaching of the

Mdloti lagoon under such conditions. Such systems usually revert relatively rapidly to their former morphology as river flow is then unable to maintain the outlet against wave-induced sediment transport.

Such a fundamental difference in outlet/inlet conditions merits closer study, particularly in terms of its ecological implications (Harrison 1993) and from a management perspective.

Sediment supply

Availability of sediment and the ability of hydrodynamic forces to transport it provides the material for the construction of estuary barriers and the infilling of estuarine channels. Climate plays a role in the weathering rates in the hinterland, and thus influences sediment availability from that source. Two main sediment sources are evident: fluvial and marine; while a third source (intra-estuarine sediment, which is mainly organic) typically contributes only a minor proportion of the total sediment retained in South African estuaries. Reddering & Esterhuysen (1984) noted that the mud content of fluvially-derived sediment in Eastern Cape estuaries exerted a strong control on channel morphology, the sandy sediments of mud-depleted rivers being characterised by wide intertidal flats (principally the flood-tidal delta sediments, unlike Natal where sand-rich sediments produce braiding in the river-dominated estuaries) whereas mud-rich rivers produced confined channels with limited intertidal areas. The extent of flood-tidal deposition (admittedly potentially controlled by channel width) is the true determinant of intertidal exposure in the Eastern Cape. Several estuaries (e.g. Gamtoos, Sundays) have large flood-tidal deltas within laterally confined channels and thus do contain fairly extensive intertidal areas. Cooper (1991a) noted that in the river-dominated systems of Natal, sediment type exerts a strong control on channel form, and that mud-rich systems are more likely to be deep and sinuous or meandering while sandy systems have shallow braided or anastomosing channels. Sediment availability determines whether an estuary has a barrier or not. Those on the Tsitsikamma coast which lack barriers, do so because of a lack of available sediment. In some cases barriers do not reach their full development because of this limitation and are reduced to intertidally exposed flood-tidal deltas which are fully submerged on high tides. Most South African estuaries have sandy barriers and only in a few cases (e.g. Kranshoek, Crooks in the Western Cape) is a significant gravel or pebble component present.

It is worth stating at this stage that most South African estuaries have a barrier that is composed of marine sediment. Only in a few localised instances outside KwaZulu-Natal is this not the case. In KwaZulu-Natal, the sediment-filled nature of estuaries and their distinctive

flood response (see below) yields a significant fluvial component of sand-sized sediment. Only the Tugela and Great Berg supply sufficient sediment to have formed appreciable deltas in historical times, which because of the strong wave regime and associated longshore drift at their mouths are offset from the estuary mouth and form elongate beachridge plains downdrift of the river outlet. In the case of the Tugela, accretion has reached average rates of up to 5 m per year since 1937 (Cooper 1991b).

Coastal lithology

The coastal geology, influenced morphologically by past sea level history, provides the bedrock framework within which estuaries subsequently evolve. If a common Holocene sea level history is assumed for the South African coast, then lithology is the major control. The deeply entrenched rock channels beneath many KwaZulu-Natal estuaries indicate appreciable downcutting during rejuvenation caused by several Tertiary marine regressions (Orme 1974). Boreholes failed to reach bedrock 40 m below sea level in the Mfolozi Valley some 16 km from the present coast (Hobday 1979). This downcutting was followed by erratic Pleistocene and Holocene sedimentation, although the laterally confined estuaries of South Africa commonly record only the latest phase of infilling in their bedrock valleys. The sediment-filled rock channels subsequently anchor the estuary mouths by restricting lateral movement of the estuaries. In general it can be verified empirically that estuaries formed in soft lithologies (shales, mudstones and unconsolidated sediments) are much wider than those associated with rivers of equivalent discharge characteristics in more resistant lithologies (sandstone, granitic and metamorphic lithologies) (Cooper 1991a). While it may be difficult to see immediately how this influences contemporary geomorphology, it is the dimensions of this bedrock valley that set the limits within which estuaries subsequently evolved. Those which are constricted (e.g. Mgeni) are, for example, typically more prone to flood-scour than adjacent systems (e.g. Mvoti) which are formed in wider valleys. Those with wide bedrock valleys (e.g. Mvoti: Cooper 1993b) are more likely to have extensive floodplain or marginal tidal depositional environments (saltmarshes, intertidal flats) than those which have steep-sided valleys (e.g. Mtamvuna: Cooper 1993a).

Temporal variation in estuarine geomorphology

Variation in the morphology of estuaries occurs on several time scales. The longest of these time scales that is relevant to this discussion is the millennium-long time scale over which most estuaries have evolved to their present morphology. On this time scale the main temporal controls on estuarine geomorphology are the rates of sea level change and sediment supply as described above in the context of estuarine evolution. Such changes are commonly studied through stratigraphic analysis of estuarine sediments and their texture, chemical composition and contained microfossils (e.g. Martin 1962).

A second category of temporal changes (often termed mesoscale change) takes place on a time scale of centuries to decades. These changes are those which may be recorded during a lifetime or shorter period and which are best studied by reference to historical records (maps, air photographs, historical accounts). Changes in this category may be progressive (e.g. estuarine infilling), or cyclic whereby erosion and accretion are balanced and little net overall morphological change occurs on this time scale. The latter category includes the cyclic patterns of sedimentation and scour recorded in Eastern Cape and KwaZulu-Natal estuaries (Reddering & Esterhuysen 1987a; Cooper 1989, 1994a, c; Cooper et al. 1990) and the inlet migration noted by Reddering (1983) at the Keurbooms estuary. Although the patterns of erosion and accretion in each of these locations is different, the cycle of flood and fair weather periods assists in maintaining the geomorphological equilibrium of the system.

Several instances of apparently progressive estuarine change have been documented, e.g. Onrus, Mcantsi, Siyai. The first two examples have been related to enhanced siltation on a decadal time scale. One of the best documented instances is the Siyai where progressive change may be linked to progradation of the coastline and the progressive elongation of the Siyai river channel. This has prompted a series of actions designed to 'restore' the estuary. Several instances of apparently progressive infilling in KwaZulu-Natal (Begg 1984) have subsequently had to be reinterpreted as part of a long-term cyclicity in geomorphology. The floods of 1987 scoured all the sediment that had accumulated in the Mgeni estuary since 1917 (the last flood of such magnitude) and sedimentation since 1987 has followed the pattern recorded in aerial photographs dating from the early 1930s (Cooper 1994a). Such a pattern, which is clearly cyclic in the mesoscale, might be legitimately regarded as progressive in the short term.

Mesoscale changes are best interpreted in terms of climatic variation (which influences marine process intensity and freshwater discharge), historical variation in sea level, variation in sediment supply rates and human activities.

Episodic events on this time scale can be responsible for dramatic morphological change but can also influence sediment yield to the coast (McCormick et al. 1992). Bremner et al. (1990), for example, calculated that during the 1988 Orange River flood an amount equivalent to the mean annual sediment load was carried in a three month period. An amount equal to about 5% of the total load was deposited as bedload offshore from the river mouth and suggests that during such events in the Orange River, bedload comprises only about 5% of the total load.

Also in a mesoscale context, the effect of periods of drought on estuarine geomorphology become important (Reddering 1988b; Whitfield & Bruton 1989). Drought may cause estuaries to become hypersaline for a period and exposed former estuary beds may become desiccated or colonised by terrestrial vegetation. Reddering (1988b) assessed the role of drought or artificially diminished freshwater inflow to Eastern Cape estuaries and concluded that under such conditions flood-tidal deltas could grow to unusually large size and muddy formerly submerged sediments could be exposed and attain greater cohesion. The net effect of such conditions is that when floods do occur, the scouring and sediment removal would not be as great as predicted and a much greater vertical rise in water level would be produced as the flood arrived at the estuary than for equivalent floods in previous periods. In addition, the channel dimensions of non-flooding rivers would be greatly reduced.

A third category of changes concerns those which take place at time scales of less than one year (often termed microscale changes). These span the temporal boundary between the disciplines of process sedimentology and geomorphology. They may be related either to predictable seasonal climatic, tidal and ecological variability or to changes on a minute to second time scale in, for example, flow velocities, which alter sediment transport rates in the short term. Seasonally variable rainfall may prompt the breaching of barriers which isolate estuaries from the sea during the dry season as in the case of the Mhlanga estuary (Cooper 1989). This seasonal variation may be mimicked by distinctive ecological cycles (Harrison 1993) in which ecological cycles (recruitment, emigration, etc.) are adapted to geomorphological changes.

An understanding of estuarine geomorphology on these time scales but in particular in the mesoscale is crucial to estuarine management if observed changes are to be interpreted in proper context. In the case of the Mgeni estuary, for example, perceived siltation since the earliest photographs of 1931 was attributed to bad land use practices (Begg 1984). The flood of 1987 showed these changes to be part of a cyclic pattern of instantaneous erosion followed by a long period of deposition and redistribution of sediment within the estuarine channel.

Coastal geomorphological zones

Considering the role played by each of the above parameters on the morphology of estuaries it is possible to recognise four distinct zones on a broad scale. These largely incorporate the parameters outlined above and although there is significant within-group variation the estuaries within each zone may be characterised in general terms. Previous attempts to categorise estuaries in South Africa include the work of Reddering (1988a) in the Eastern Cape, who recognised five main controls on estuarine morphology: state of the tidal inlet, volume of the tidal prism, cohesion of the sediment bed, wave climate, and morphology of the adjacent coastline. This classification sought to assess variability in terms of observed morphological parameters of estuaries but was confined to the Eastern Cape. Cooper (1991a) classified Natal's estuaries on the basis of variation in mouth persistence, floodplain width, catchment geology, barrier length and catchment area in an attempt to identify natural groupings on the basis of these genetic variables. Deviation from the group morphology could then be interpreted in terms of human alteration of a system. Morphometric measurements of estuaries have been made in an attempt to classify all South African estuaries in the context of application of the estuarine health index (Cooper et al. 1995). Day (1981) and Whitfield (1992) have attempted to distinguish estuarine types at a South African scale. Day (1981) discussed several estuarine types in each of three biogeographic regions (the west coast, the temperate southern and southeast coast, and the subtropical northeast coast). Whitfield (1992) identified five types of estuary on the South African coast: estuarine bay; permanently open estuary; river mouth; estuarine lake; and temporarily closed estuary. While such an approach is valid it is clear that such distinct variation in morphology exists even within these categories, particularly in terms of inlet type and persistence (see below), that further subdivision is necessary. The following discussion takes a combined geomorphological and biogeographic/climatic approach to a discussion of essentially regional variation in estuarine types. Four zones are identified; the Western Cape north coast, the Western Cape south and the southeast coast of the Eastern Cape, the southern KwaZulu-Natal coast, and the northern KwaZulu-Natal coast. A number of subzones also occur and several estuarine types are found in more than one zone but these are discussed in the relevant sections.

The Western Cape north coast

This area, which is arid, shows spatial variation in rainfall, with aridity increasing northward. Much of the coast north of the Olifants River receives less than 200 mm per year while the area between this and St Helena Bay receives between 200 and 300 mm. This area can thus be classified as a desert or semi-desert (Day 1981). The coastal water bodies of deserts and semi-deserts have been little studied globally. There is therefore little against which to compare

these systems. Southward, rainfall increases to an annual average of 400–700 mm in coastal areas of the southern Western Cape (Heydorn & Tinley 1980). Here there is a marked seasonal fluctuation in precipitation and most rain falls in the winter.

Evaporation, which is controlled mainly by temperature, also shows marked variation in this area. Recorded evaporation varies from over 2400 mm per year at St Helena Bay to 1800 mm per year between Saldanha Bay and Hermanus (Day 1981). High evaporation tends to counter runoff and may produce dry or seasonally dry river channels.

No appreciable difference occurs in tidal range on the west coast. Heydorn and Tinley (1980) list the following mean tidal ranges: Hermanus 1.4 m; Simon's Town 1.5 m; Table Bay 1.5 m; Saldanha Bay 1.5 m; Port Nolloth 1.6 m. Although a slight northward reduction in peak wave heights has been noted (Day 1981), wave energy and direction of approach does not vary appreciably along the Atlantic west coast. A dominant littoral drift toward the north has been identified.

The rivers which enter the sea on the Western Cape's north coast have been little studied from a geomorphological viewpoint, but with the exceptions of the Orange, Great Berg and Olifants, none maintains a near-permanent tidal exchange with the sea. Estuaries in this zone are typically confined within bedrock valleys and most of the systems are subject to periodic discharge regimes and may not sustain surface flow for long periods. The Holgat, for example, last flowed in 1925 (Keyser 1972). A provisional assessment of the geomorphology of these systems (J.A.G.C., unpublished data) indicates that most may be regarded as ephemeral coastal pans in which there is a range of conditions which vary from brackish through hypersaline water bodies to dry river beds (Bickerton 1981). Any spatial or temporal link between these types has yet to be established, although there is a broad northward increase in aridity. This is not completely matched by river bed type because of the important control exerted by groundwater in these systems. As a rule these systems receive inputs from the sea via barrier overwash and discharge water seaward via percolation through the barrier. A free surface water connection with the sea forms very infrequently and only during rare rainfall events. Some such as the Buffels, Papkuils, Wadrif and Jakkals tend to contain a high proportion of freshwater and therefore have less tendency to become hypersaline. They also tend to have relatively long barriers that may promote overwash and hence maintain hyposalinity. Other ephemeral coastal pan systems (Spoeg, Holgat, Groen, Swartlintjies, Bitter) appear to be more commonly either dry or hypersaline. The reasons for (and precise nature of) this variation are not clear although groundwater levels undoubtedly exert a strong control on surface water persistence in this area.

The large estuarine systems of the west coast vary considerably in their geomorphology, although they each discharge to the sea through a near-permanent surface channel. Each drains a large hinterland that extends beyond the arid coastal zone and thus may maintain surface flow throughout the year, although there is marked seasonal variation. The Orange River which drains an area of about 520 000 km² is the largest river in the sub-continent. At the coast its mainly sandy channel is characterised by a shallow, braided pattern (van Heerden 1986) featuring vegetated (chiefly grasses and reeds) braid bars and shallow intervening channels. Considering the importance of the Orange, surprisingly little research has been conducted on it.

The estuary is located in the arid and cold temperate climatic region and is fed by one of the few perennial rivers in the region. For this reason it seldom dries up and thus offers one of the few areas of perennial, calm water in the region. Mean annual rainfall at the estuary exceeds 100 mm and evaporation is over 2000 mm (Heydorn & Tinley 1980). Salinity measured in the estuary in September 1993 varied from 14 at the surface and 29 at the bottom near the barrier to 3 at the surface and 16.5 approximately 4 km upstream. At this time the mouth was closed but the salinity variation showed a typically estuarine pattern, indicative of a recent open phase. The coastal barrier at the mouth of the Orange is about 4500 m long and about 1500 m of this is backed by open water. The remainder is backed by a degraded coastal saltmarsh. Near the coast the river floodplain is typically 3000 m wide and the channel is typically about 1600 m wide.

The Orange has been described variously as a delta (van Heerden 1986), an estuary (Bremner et al. 1990), and lacking 'a real estuary' (Day 1981). Although it has a shallow sandy channel that extends right to the barrier it does receive tidal inflow and experiences tidal rise and fall over a distance of about 4 km. The estuary is best termed a river-dominated estuary *sensu* Cooper (1993b, 1994a) in that it does not have a subaerial delta extending seaward of the coastline. Although it receives a considerable volume of sediment (which has decreased following upstream impoundments), the sediment is dispersed offshore via wave and tidal action and is not retained onshore as a deltaic accumulation.

The Olifants River, which drains an area of about 45 600 km², has the second largest catchment on the west coast and one of the largest in the subcontinent. It has a channel that is bounded by steep banks and generally lacks emergent sandbars and intertidal areas. The channel contains more muddy sediment than the Orange (Day 1981) and this may account for the difference in channel

form. Clean sands near the mouth are probably deposited as flood-tidal deltas. While distinct seasonal changes in salinity have been noted (Day 1981), during sampling in September 1993, salinity varied from 35.9 at the mouth (surface and bottom) to 1.9 at the surface and 7 at the bottom some 4 km upstream. Salinities of over 30 were recorded at this upstream location in September 1973 (Day 1981). The coastal barrier at the mouth of the Olifants is about 3000 m long and about 300 m of this is backed by open water. The remainder is backed by a coastal salt-marsh which is periodically inundated. Near the coast the river floodplain is 1800 m wide and the channel is typically 250 m wide.

The Olifants has all the characteristics of a wave-dominated estuary, in that it has a well-defined flood-tidal delta, deep muddy middle reaches and a sandy upper section. It experiences variation in freshwater runoff from a maximum in winter to virtually no flow in summer (Day 1981). The estuary is up to 15 km long in the dry period. Tidal flow is responsible for maintaining the inlet and the island at the inlet probably represents a now stabilised flood-tidal delta.

The Great Berg River, which drains an area of about 7700 km² (Heydorn & Tinley 1980) or 4000 km² (Day 1981), is the third largest perennial river in the area and one of the largest in the western and southern Cape. The larger estimate seems more likely. It has a channel that is bounded by steep banks and generally lacks emergent sandbars and intertidal areas. The meandering channel contains more muddy sediment than the Orange (Day 1981) and this, coupled with its very low gradient accounts for the difference in channel form. Clean sands near the mouth are probably deposited as flood-tidal deltas. While distinct seasonal changes in salinity have been noted (Day 1981) (the estuary is reported to be tidal for up to 70 km during low river flow in summer), at the time of sampling in September 1993, salinity varied from 32.5 (surface) and 34.4 (bottom) at the mouth to 5.5 at the surface and 5.2 at the bottom, 12.5 km upstream. The estuary of the Great Berg is very much longer than that of the Orange or Olifants although its hydrology is artificially manipulated at present, which probably reduces the length of the estuary (Day 1981).

The coastal barrier (sandbar) at the mouth of the Great Berg is about 1800 m long and about 1500 m of this is backed by open water in a coast-parallel channel extension. The outlet is now artificially fixed by a pair of piers. The remainder is backed by a coastal saltmarsh which is periodically inundated. Near the coast the river floodplain is typically 1600 m wide of which the meandering channel typically occupies 200 m. Day (1981) considered the Great Berg a true estuary and Whitfield (1992) considered it the most important permanently open estuary on the west coast. North of its mouth is an elongate beachridge plain which appears to consist of reworked fluvial sediment, although no detailed formal study of the geomorphology of this system has been conducted.

South coast of Western Cape and Eastern Cape coast

The south coast of the Western Cape and the Eastern Cape coast forms the next main geomorphological subdivision. Here arid conditions are replaced by more humid and temperate conditions in which seasonal rainfall regimes predominate. Rainfall reaches an annual average of 400–700 mm in coastal areas of the south Western Cape (Heydorn & Tinley 1980). Wave energy and direction of approach does not vary appreciably along the south Western Cape coast where the continental shelf reaches its greatest extent around the South African coast. Those rivers that drain only the coastal area will be influenced by rainfall in that area; however, other rivers that drain areas further inland may be affected by a range of rainfall scenarios. A major change in rainfall pattern (winter to all seasons rainfall) is present at Cape Agulhas, and a second subzone of higher rainfall is present on the Tsitsikamma coast. The coastal belt west of Cape St Francis generally experiences between 400 and 500 mm of rainfall per year while the Tsitsikamma coastal sector receives between 700 and 1000 mm. Analysis of mean annual runoff and river catchment area data suggests an approximately tenfold increase in runoff for a given catchment area in the Tsitsikamma area compared with areas of lower rainfall east of Cape St Francis. Although numerous small rivers in the southeastern Cape area drain only the coastal areas, the relatively high rainfall throughout the year means that they seldom dry up although some have been reported to become hypersaline. Between Cape St Francis and Great Fish Point annual rainfall is between 450 and 600 mm. There is a marked increase between Great Fish Point and East London where annual rainfall amounts to 800 mm.

Estuaries in this zone are typically confined within bedrock valleys and many are influenced by aeolian sediment deposition as many dunes along the coast are unvegetated. The estuaries of this region may be divided into two broad types, tidal estuaries and non-tidal lagoons, although the latter may become tidal for a period after breaching. The tidal estuaries of the Western and Eastern Cape coasts are numerous and several, including for example the Keurbooms, Kromme and Sundays, have been studied from a geomorphological perspective (Reddering 1981; Reddering & Esterhuysen 1981, 1983). In general, these estuaries comprise three distinct geomorphological zones; a sandy barrier in which a constricted tidal inlet is formed and landward of which is an extensive flood-tidal delta. The flood-tidal delta is deposited as a result of the

flood dominance of tidal currents at constricted inlets. Dunes are typical of many barriers, particularly on the Eastern Cape coast between Port Elizabeth and East London where extensive unvegetated dunes migrate into estuary mouths. Ebb-tidal deltas are typically absent or limited in extent owing to high wave energy.

Landward of the barrier-associated environments is a deeper water area typified by fine sediment deposition, in many instances enhanced by flocculation of suspended clays in the saline estuarine water. These estuarine channels may be flanked by intertidal sandflats and salt-marshes, both of which form important habitats. At their upstream limit such estuaries typically exhibit a fluvial delta where bottom sediments coarsen and water depths are reduced. Such a tripartite arrangement of sedimentary facies is typical of microtidal, wave-dominated estuaries (Roy 1984; Dalrymple et al. 1993).

A comprehensive set of surveys on the Nahoon, Kwelera and Gqunube estuaries following river floods in 1985 (Reddering & Esterhuysen 1987a) provided a rare insight into the role of extreme events on such systems. It was found that floods produced significant scour of the uncohesive sandy sediments of the flood-tidal deltas in the lower reaches of these systems, whereas the upstream estuarine channels were largely unaffected. Such a pattern points to a mesoscale cyclicity between progressive flood-tidal deposition and rapid erosion following floods. Reddering & Esterhuysen (1987a) also found that most of the eroded sediment (385×10^3 m^3) was removed from the subtidal zone compared with only 23×10^3 m^3 from the intertidal zone. Calculations indicated that the sediment removed by the flood of 1985 would be replaced by flood-tidal deposition within 125 years assuming that no further floods occurred in the interim.

A variation on this morphological theme is afforded by several estuaries in this zone (Gwaing, Maalgate, Bloukrans, Lottering) which, through a limited marine sediment supply, do not have subaerially exposed barriers and are blocked by flood-tidal deltas that are completely submerged at high tide. In extreme cases (Storms, Steenbras) no barrier is present at all, but even in such instances wave energy is markedly reduced by interaction with the seabed topography and the estuarine area remains a zone of low energy. The back-barrier areas of these estuaries may also be typified by bedrock, boulder and gravel bed rivers in cases where the supply of fine-grained sediment from fluvial sources is also limited.

Although they are more common in KwaZulu-Natal there are a few river-dominated estuaries in the Eastern Cape region, e.g. Great Fish (Reddering & Esterhuysen 1982), Great Kei, which maintain an outlet in spite of a low subtidal and intertidal volume. In such instances sustained strong fluvial discharge plays a dominant role in maintaining the outlet and tidal influences, although not absent, are areally restricted by small estuarine volumes. The river channels in such systems are typically shallow and dominated by fluvial sediments; flood-tidal deposition is limited.

The non-tidal lagoons of this region exhibit much variation in their morphology. None is able to maintain a tidal outlet permanently as a result of small subtidal volumes and consequently small tidal prisms, and instead, periods of high fluvial discharge occasionally promote breaching of the barrier. Much of the morphological variation may be ascribed to the size of the inflowing stream and the surrounding topography (cliffed banks, floodplain development, etc.) but an important source of morphological variation lies in the nature of the sand barrier and the extent of water exchange between land and sea. Exchange may take place (i) via barrier overwash, whereby waves introduce seawater into the estuary across the barrier either as sheet flow or channelised flow; (ii) by overspilling, whereby lagoon water flows seaward across the barrier without development of a deep channel; (iii) by seepage, in either a landward or seaward direction according to the relative levels of the estuary and the sea; and (iv) by channelised seaward flow promoted by seaward flow of estuary water from an estuary whose water level is high above a given tidal level. These closed systems maintain various and variable water levels according to the combination of discharge variability and barrier morphology. Most lack flood-tidal deltas although a few contain evidence of flood-tidal deposition for brief periods when open. Many such systems, however, have distinct fluvial deltas at their upper reaches which appear to be graded to the modal water level in the system rather than to a tidally-controlled level as in open estuaries.

Delta progradation gradually reduces the volume of water in such systems, which may ultimately be transformed into river channels whose dimensions are controlled by freshwater discharge in the way that river-dominated systems are influenced.

Although only a few of these lagoons have been studied in detail from a geomorphological perspective, for example the Bot (van Heerden 1985) and Seekoei (Esterhuysen 1982), it is clear from synoptic observations and measurements that a series of estuarine types exists that vary according to the relative importance of each of these processes. The low, dissipative beaches of the Eastern Cape coast are more prone to overwashing than those of KwaZulu-Natal and consequently marine influence may be much greater in such systems (e.g. Ngculura, Kiwane). Several, however, are perched above the normal high tidal levels and act mainly as stream outlets (e.g. Blind at East London). The potential must exist for a seasonal pattern to the dynamics of such systems with overwash being more common during winter

storms and freshwater discharge being greater in the rainy season. The relative elevation of seawater and estuarine water exerts a strong influence on the geomorphology of these systems. Those that are perched may breach and drain following freshwater floods, while those that are near high tide level may accomplish the necessary discharge by shallow surface flow. Those that are deeper may become tidal after breaching but the tidal prism is too small to maintain the inlet for long periods, whereas those that are shallower or whose outlet is controlled by rock outcrop or existing water levels may not experience tidal inflow.

Thus while a number of observations of mouth condition have been made, it is important that the mouth dynamics be studied in much greater detail and attention be given to the mode of exchange of marine and freshwater in such systems. Further research on this aspect is necessary in order to understand adequately the dynamics of such systems which cannot simply be regarded as being similar in terms of geomorphological process. The classification of South African estuaries which is incorporated in ongoing estuarine health index research (Cooper et al. 1995) will address the classification of estuaries on a national scale and identify estuaries of several types which will aid the selection of systems for further investigation.

Southern KwaZulu-Natal and Transkei coast, Eastern Cape Province

This zone extends from the Tugela River to the Bashee and is characterised by bedrock-confined river valleys subjected to moderate to high wave energy under a subtropical climatic regime. Most of the 73 systems described by Begg (1978) are located south of the Tugela River in this region although the estuaries of Transkei have been little studied from a geomorphological perspective. The essential characteristics of this environmental setting are a steep hinterland, deep weathering profiles in soils (Orme 1973; Partridge & Maud 1987), subtropical climate and seasonal rainfall which combine to produce rapid fluvial discharge and high sediment yields. The presence of mangroves in several systems throughout this zone introduces a potential geomorphological control which has yet to be investigated in a South African context.

Estimates of denudation rates in the hinterland, based on sedimentation rates and other environmental factors, have been published by Rooseboom (1975, 1978). Fleming and Hay (1984) consider Rooseboom's work to be the most reliable source on sediment yields although work by Cooper (1991a) and McCormick et al. (1992) suggests that the data may be an overestimation for KwaZulu-Natal's smaller estuaries. The average sediment loss from the whole KwaZulu-Natal hinterland is approximately 323 tonnes per square kilometre per year (Martin 1987), which is considerably higher than the figure of 80 per square kilo-metre per year considered by Milliman and Meade (1983) as representative of east African rivers. The average sediment yield per square kilometre for KwaZulu-Natal is only exceeded in China, southeast Asia, northwest South America, Central America and glacial areas of Alaska (Milliman & Meade 1983).

Present KwaZulu-Natal coastal morphology has been divided into several types (Cooper 1991b, c, 1994b), of which the most important are headland-embayment, linear clastic and prograding beachridge coasts. Much of the KwaZulu-Natal coastline (including northern KwaZulu-Natal) appears to be in equilibrium with the contemporary wave field. Only the area around Port Durnford exhibits progressive long-term erosion and the area north of the Tugela River exhibits long-term accretion (Cooper 1991b).

Four basic estuary types were identified in southern KwaZulu-Natal by Cooper (1991a). These were (i) drowned river-valley estuaries which are similar to the tidal estuaries of the south and southeastern Cape (e.g. Mtamvuna: Cooper 1993a); (ii) river-dominated estuaries (e.g. Mgeni: Cooper 1994a); (iii) river-dominated coastal lagoons (e.g. Mhlanga: Cooper 1989); and (iv) non-tidal river mouths (e.g. Mvoti: Cooper 1993b). The differences in evolutionary processes are illustrated diagrammatically in Figure 2.2. The major difference between river-dominated estuaries and non-tidal river mouths is in the lack of tidal inflow to the latter in spite of an open mouth. This characteristic does not, however, necessarily remove all the estuarine characteristics of such a system; seawater may enter via barrier overwash and marine organisms may enter such systems even against high current velocities (Harrison & Cooper 1991).

In the estuaries of this region high wave energy and a steep nearshore gradient preclude the formation of ebb-tidal deltas or offshore barriers and it follows that long-shore extension of river courses into extensive lagoons is generally restricted. Deltas deposited after major floods are rapidly reworked into barriers or are distributed along and offshore by wave action (Cooper 1990). Furthermore, the lack of a coastal plain and a rocky coastal framework means that the lower reaches of rivers are laterally confined. High sediment yields from the steep catchments mean that most of these estuaries (with the exception of the Mtamvuna) may be regarded as river-dominated (Cooper 1994a) and do not conform to the typical model for microtidal estuaries proposed by Roy (1984).

River-dominated estuaries and non-tidal river mouths
High fluvial sediment supply is basically responsible for the sediment-filled nature of Natal estuaries, although as discussed below this does not necessarily imply a human role in accelerated sedimentation. The river-dominated

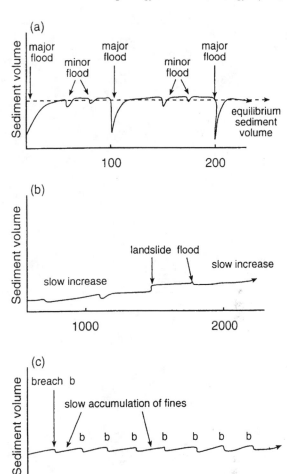

Figure 2.2. **Conceptual view of temporal change in estuarine sedimentation in (a) river-dominated estuaries, (b) tidal estuaries and (c) temporarily open estuaries/lagoons.**

estuaries south of the Tugela generally have channels that are filled with fluvial sediment and are in equilibrium with fluvial discharge. In most instances this fluvial sediment extends right to the mouth which is maintained not by tidal currents but by fluvial discharge. This low estuarine intertidal volume means that flood-tidal deltas are restricted or non-existent in these systems (Cooper 1988). Consequently most marine-derived sediment enters these systems via barrier overwash (Cooper & Mason 1986). Limited flood-tidal deposition may produce distinctive assemblages of foraminifera in KwaZulu-Natal's estuaries (Cooper & McMillan 1987; Wright et al. 1990), with planktonic forms being carried further upstream in suspension than the benthic forms which form part of the bedload.

The best studied examples of river-dominated estuaries are the Mgeni (Cooper 1994a, Cooper et al. 1990) and Mvoti

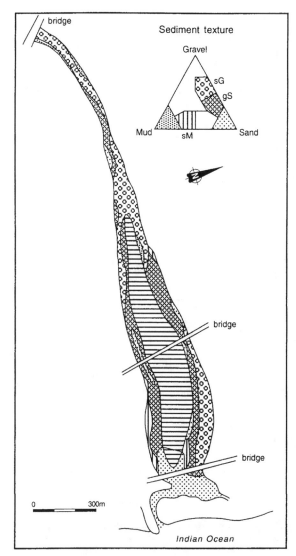

Figure 2.3. **The major geomorphological features of the Mgeni estuary. Note the central island (horizontal shading) separating two subtidal channels.**

(Cooper 1993*b*). In the period 1931–86, the Mgeni River was transformed from a wide, shallow, braided estuary to one with two well-defined anastomosing channels with a vegetated central island (Figure 2.3). The change occurred through coalescence and emergence of braid bars and the incorporation of side-attached bars into the bank. At the same time island formation allowed downcutting of the flanking channels and maintenance of relatively deep channels, so the subtidal volume increased. This permitted deposition of a flood-tidal delta which was previously impossible due to the elevated bed levels of the estuary.

From 1937 to the present, the back-barrier lagoon north of the tidal inlet was transformed from a shallow,

muddy lagoon with unvegetated intertidal mudflats to a high intertidal mangrove swamp and supratidal saltmarsh drained by a narrow channel. A major feature of the long-term development of the Mgeni estuary area is the progressive erosion of the barrier north of the inlet (Cooper 1994*a*). Erosion was concentrated in the 1200 m north of the groyne and led to the formation of a concavity there. Comparison of the shoreline positions indicated that most of the erosion occurred up to 1973, after which the coast stabilised. In 1983 and 1985 erosion culminated in severe overwash which deposited sand in the back-barrier mangrove swamp and filled a former channel (Cooper & Mason 1986). The fact that erosion is restricted mainly to the 1200 m north of the inlet is direct evidence that the groyne is the cause of it. Extremely high rainfall during the September 1987 cut-off low flood event caused up to 800 mm of rain to fall over five days, from 26 September to 1 October. During the associated floods, water levels in the Mgeni River rose to 5 m above normal high tide level, inundating the surrounding low-lying land and the mangrove swamp. The constricted outlet was enlarged and the mid-channel island completely eroded. A total of 1.8×10^6 m^3 of sediment was removed from the mouth.

The post-flood reconstruction of the barrier was rapid, and took place approximately 20 days after the flood peak. Response to major floods varies greatly within KwaZulu-Natal's estuaries. In the Mvoti estuary, which has a wide floodplain, downward scour was reduced in comparison to the other systems during the 1987 floods (Cooper 1993*b*) and fine-grained sand was deposited across the floodplain as the flood waned. Under high discharges river-dominated estuaries generally show a combination of the following responses (Figure 2.4).

- The gradient is increased by the river taking a shorter course through the barrier.
- The gradient is increased by scouring the downstream end.
- Channels expand by vertical erosion.
- Channels expand by lateral erosion.
- Sediment is eroded until the remaining lag grain sizes exceed the transport capacity of the outflowing river.
- Sediment is eroded until bedrock is reached.
- Overbank flow accommodates increasing discharge without associated erosion.

River-dominated lagoons

The river-dominated lagoons of Natal share many common characteristics, chief among which is the propensity for seasonal breaching after summer rainfall events, although

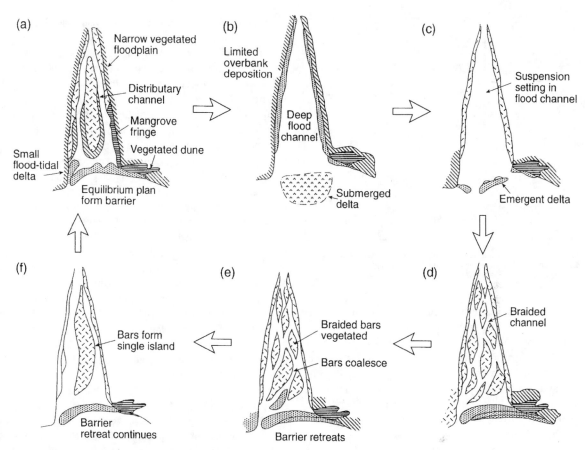

Figure 2.4. **Schematic diagram to illustrate cyclic sedimentary processes in KwaZulu-Natal estuaries (Cooper 1991a).**

breaching is not solely limited to these periods. Breaching has also been ascribed to overwash channel formation. The high barriers and distinct berms of these steep beaches mean that lagoons typically breach and drain, exposing most of the lagoon floor (Grobbler 1987; Cooper 1989). The Mhlanga, Mdloti and uMgababa are the best studied examples of this type of system (Grobbler 1987; Grobbler *et al.* 1987; Cooper 1989; Cooper *et al.* 1990).

The channel of the Mhlanga is incised into a 500 metre wide alluvial floodplain. The channel follows a sinuous course across the floodplain, but where it approaches the coast it is diverted southward into an elongate coast-parallel back-barrier area, approximately 1 km long and 150 m wide (Figure 2.5). The southern end of the lagoon consists of a narrow backwater channel draining *Phragmites* reedswamps and is protected from the Indian Ocean by a high, vegetated, Holocene dune barrier up to 30 m high. Freshwater runoff is normally insufficient to maintain an outlet against littoral drift and strong wave action, and therefore no connection exists between the lagoon and the ocean for about 90% of the year (Begg 1984). The range of sediment grain sizes in the Mhlanga Lagoon is

restricted because of the limited size range of grains available in the catchment and the low fluvial flow velocities (under normal flow conditions) which are ineffective in transporting coarse-grained sediments. The only sand that is transported into the estuary is deposited by washover fans and wind action off the beach.

Under certain conditions seepage through the barrier is insufficient to cope with the discharge and a mouth forms when the rising water level overtops the lowest point on the barrier. The overflowing water erodes the non-cohesive sand of the barrier and forms a constricted outlet. In KwaZulu-Natal, artificial breaching of the beach barrier is illegal, but outlet formation may arise under three sets of natural circumstances.

1   Following fluvial floods associated with heavy rain in the catchment.
2   After continuous rainfall over a prolonged period, which causes the lagoon water level to rise slowly, ultimately overtopping and breaching the barrier.
3   After the formation of marine overwash channels which intersect the lagoon water level.

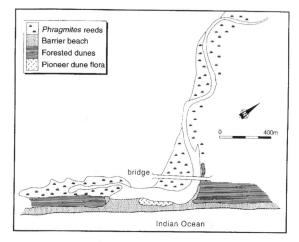

Figure 2.5. **The major geomorphological features of the Mhlanga Lagoon. Note the barrier environment being breached. The lagoon remains closed for c. 90% of the year.**

The short period in which the lagoon drains ensures that the ephemeral mouth bars deposited opposite the reaches are rapidly reworked landward, closing the mouth.

Flood response in the Mhlanga results in the scour of only fine-grained sediments from the back-barrier, without significant depth increase. A similar response was noted in the uMgababa (Grobbler *et al*; 1987, Cooper *et al.* 1988). This is attributed to a lower peak discharge in small catchment rivers which consequently have less erosive capability and which have a short flood peak during the period that the lagoon drains, after which only a limited section of the exposed bed is subject to flood effects.

Northern KwaZulu-Natal

Although the smallest of the coastal geomorphological sectors, this area is distinctive in that it has a wide coastal plain composed of unconsolidated Tertiary to Pleistocene sediments with a thin Holocene veneer. The zone extends northward into Mozambique. A morphologically similar zone occurs around the Wilderness Lakes in the south Western Cape. Estuaries in these settings are fundamentally different from those elsewhere on the coast which are confined in bedrock valleys, in that their margins are erodible on a short time scale. Although most systems on the northern KwaZulu-Natal coast are associated with former incised river valleys, lateral erosion of unconsolidated surface sediments means that most estuaries in this zone have relatively large water areas and are comparatively shallow. They include St Lucia (which formerly shared a common mouth with the Mfolozi), Richard's Bay, Nhlabane, Mgobeseleni, Sibaya and Kosi, each of which has developed a distinctive morphology. In this area too, low river gradients and the sandy substratum reduce surface flow velocities and impart great

importance to groundwater as a control on estuarine processes and form. Unfortunately few of these systems have been studied from a geomorphological perspective. Each is quite unique even on a local scale and it is difficult to identify more than a few generic similarities.

The lack of resistant outcrop along the northern KwaZulu-Natal beaches allows their mouth positions to change drastically. Prior to human interference with the St Lucia estuary mouth, starting in the 1930s, the combined Mfolozi–St Lucia mouth used to migrate up to 2 km northward (Wright & Mason 1990). Of particular importance to the barrier structure on the KwaZulu-Natal coast, is the formation of beachrock which may form the core of barriers across the mouths of several rivers (Cooper 1991*d*). The presence of beachrock tends to stabilise the mouth position. In northern KwaZulu-Natal, beachrock forms the only restriction to lateral mouth movement. The Kosi, Mgoboseleni and Mfolozi River mouths all have beachrocks that prevent a southern migration of the mouth.

The long, flat, sandy coastal plain forms a shallow gradient enabling floodplains to form. The transitional fluvio-marine environment is complex as the flat nature of the surrounding topography allows lateral migration of the estuary mouths. The high wave energy and shallow continental shelf do not allow the formation of an offshore barrier complex. The dominant southeasterly swells ensure the development of longshore extensions of estuary mouths and this, combined with the lack of a rocky coastal framework, means that these systems are very dynamic. The evolutionary path of these systems is one of progressive infilling (Hobday 1979; Cooper 1994*b*).

Over the last 60 years the Mfolozi/St Lucia estuarine complex has suffered excessive sedimentation due to bad catchment management. The cyclical climatic pattern, described by Tyson (1986), compounds the problem as drought periods are invariably broken by floods. Since the catchment areas have been heavily affected by droughts there is little or no vegetation to hinder soil erosion, and this impacts directly on the estuaries.

Wright & Mason (1993) describe marine sand encroachment up the St Lucia estuary via flood-tidal deposition at an approximate rate of 1200 m per year. Floods caused by Cyclone Domoina in 1984, and the September 1987 cut-off low flood, caused the Mfolozi–St Lucia estuarine system to change dramatically (van Heerden & Swart 1986; Wright & Mason 1990, 1993; Wright 1995). Both the St Lucia estuary mouth and the Mfolozi estuary mouth were scoured wide open and the main channel scoured to depths in excess of 15 m.

The St Lucia–Mfolozi system has changed drastically during the last 50 years due to human interference. Figure 2.6 illustrates Crofts' (1905) survey of the system, showing it in its natural state before poor farming tech-

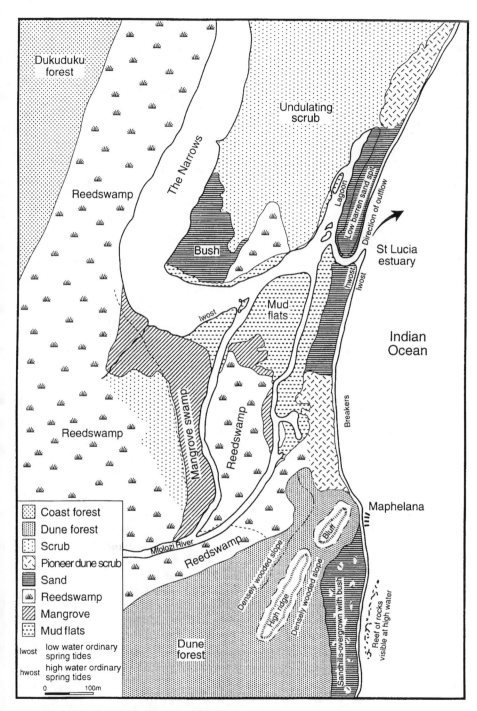

Figure 2.6. **Crofts' 1905 survey of the combined Mfolozi/St Lucia estuarine complex showing conditions before human degradation of the catchment.**

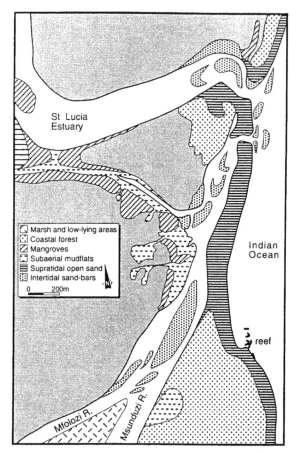

Figure 2.7. **The major geomorphological features of the Mfolozi/St Lucia estuarine complex. Compare with Figure 2.6 to see the vast changes the system has undergone in the last 60 years.**

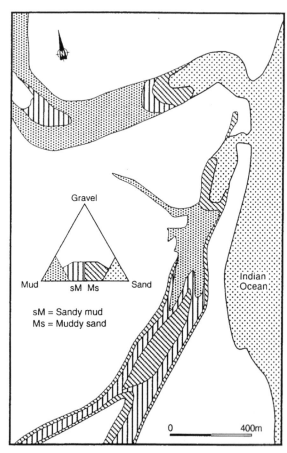

Figure 2.8. **The sedimentary texture of the Mfolozi/St Lucia estuarine complex (1988).**

niques and uncoordinated management changed the system. Due to increased farming within the Mfolozi River catchment, and channelling of the Mfolozi River floodplain, large amounts of sediment were dumped into the lower Narrows area of St Lucia, causing the St Lucia estuary mouth to close from April 1951 to April 1956 when it was artificially opened. The Mfolozi River was artificially diverted to a new mouth position north of Mapelane. Other human-induced changes include a large network of canals to supply water from the Mfolozi River to Lake St Lucia via the Narrows. A large groyne complex was completed in 1969, in an attempt to stabilise the mouth and keep it open by tidal flushing action. The groyne complex was only partially successful as a dredger still had to be operated to keep the mouth open. The whole groyne complex was destroyed in 1984 by Cyclone Domoina, being either swept out to sea or undercut and buried. Since 1984 the Mfolozi and St Lucia estuary mouths have been free of man-made structures except for the present dredging operation (Figure 2.7).

The Mfolozi River has an extremely erratic flow with

average winter flows being as little as 3 m^3 per second and average summer flows 60 m^3 s^{-1} although flood events such as Cyclone Domoina in 1984 are capable of delivering in excess of 16 000 m^3 s^{-1}.

During a 5-year period in the 1980s two major floods have occurred in the St Lucia–Mfolozi catchment. During Cyclone Domoina in February 1984, a maximum flood peak of 16 730 m^3 s^{-1} was recorded in the Mfolozi River (Wright 1990, 1995). Widespread heavy rainfall for five days from 25 September 1987 resulted in major flood by the afternoon of 29 September. Although the September 1987 flood discharged only 4440 m^3 s^{-1} in the Mfolozi River, both floods were extreme events.

Although the Mfolozi and St Lucia systems are presently separated by a small sandy beach (Figure 2.8), their physical dynamics are very similar (Lindsay *et al.* 1996a, b). Both systems are driven by flood events and the sequence of events can be summarised as follows.

1   The flood waters scour the mouth.

2   The mouth position widens as the St Lucia mouth

shifts northward, and the Mfolozi mouth south, taking the shortest route to the sea.

3   After the floodwaters abate a complex barrier forms and slowly migrates up the estuary.

4   The ebb-tidal delta re-forms offshore of the mouth.

5   The ebb-tidal delta refracts the dominant southeasterly swell to form a local reversal in longshore drift.

6   The current reversal forms a southerly protrusion from the north bank of the St Lucia estuary mouth.

7   The ebb-tidal flow takes on a meandering pattern eroding the spits and areas along the north bank.

8   Washover fans overtop the barrier depositing sand into the estuary.

9   The St Lucia estuary mouth occupies a more southerly position to that cut by the flood and the Mfolozi mouth a more northerly position.

10 An increased volume of marine sediment is brought in on the flood tide and is responsible for constriction and shoaling at the estuary mouths which ultimately leads to the closure of the mouths, if a flood does not remove the accumulated sand.

Sediment in both estuaries is dominated by marine sand in the lower estuary, and catchment-derived mud and sand in the middle and upper estuary (Wright & Mason 1990). The marine sand slowly moves up the estuary. The middle and upper estuary undergoes siltation by catchment-derived mud settling out, and sediment trapping is enhanced by the numerous mangrove root systems and filter-feeding organisms. The lower estuary is very dynamic, and the estuary mouth often changes position.

Discussion

It is interesting to note the marked spatial variability among South African estuaries which precludes the application of generic models without careful consideration of the genetic controls on estuarine geomorphology, even in systems which appear physically similar. A consideration of the relative rates of sea level change and sedimentation (Figure 2.9) provides an indication of the situation. In this scenario a constant sea level change is superimposed on a variable volume increase in an estuary (in this case mediated by changing bedrock channel morphology but which could equally be associated with variably changing sea level rise). What is evident is that an estuary could move from a deficit to a surplus situation as a result of the interplay of these factors. Surplus estuaries could be equated with the river-dominated systems of

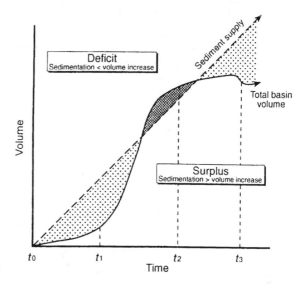

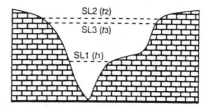

Bedrock valley cross-section

Figure 2.9. **Conceptual view of estuarine development under constant sediment supply and variable rate of change in volume via basin morphology and variable sea level. Deficit and surplus conditions are defined which broadly equate to tidal estuaries and river-dominated estuaries.**

Natal whereas those of the Cape which are dominated by tidal currents have not infilled to the same extent and hence retain a capacity for sediment deposition. If conditions are right it is conceivable that some systems may have remained shallow and river-dominated throughout the Holocene. This is believed to be the case in many Natal systems (Cooper 1991d). Others may have been drowned and are now undergoing infilling at variable rates as predicted by conceptual models of tidal estuaries on microtidal coasts. The ultimate state of an estuary at any given time may thus be interpreted in terms of sea level history and sediment supply rates. The mesoscale behaviour of river-dominated and tidal estuaries is contrasted in Figure 2.2 and compared with closed lagoons. This conceptual model is amenable to further testing and refinement by further studies on South African estuaries.

In a spatial context it is useful to view the variability among estuaries on a global scale in terms of the interplay between wave, tidal and fluvial influences following the approach of Galloway (1975) to deltas. Such an approach applied to estuaries and to lagoons identifies river, wave

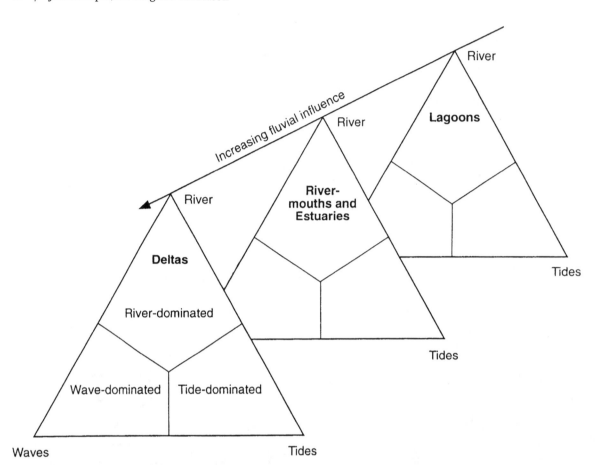

Figure 2.10. **Conceptual global view of estuarine morphology as a function of variable control of waves, tides and fluvial influences. The full range of transitional fluvio-marine environments can be envisaged if deltas as a group are considered to be more river-dominated than all estuaries which are in turn more river-dominated than all lagoons. South African systems equate with wave-** dominated deltas (Tugela, Groot Berg), river-dominated estuaries (Orange, Great Fish, Mvoti, Mgeni), wave-dominated estuaries (Kromme, Kariega, Sundays, Mtamvuna) and river-dominated lagoons (Mhlanga, uMgababa) and wave-dominated lagoons (Bot, Kiwane).

and tide-dominated end members. This can be integrated (Figure 2.10) into an holistic view of transitional fluvio-marine environments in which the relative influence of riverine inputs increases from lagoons to deltas. At a South African level this enables identification of estuaries as mainly wave-dominated but with river-dominated systems located mainly in the subtropical zone. Tidal range in South Africa is too small for tidally-dominated systems on a global scale to be recognised. Lagoons (or temporarily open/closed estuaries) may likewise be divided into river-dominated and wave-dominated systems according to the extent that fluvial influence extends to the coast. Tidal range is too small for tidally-dominated lagoons to be recognised in South Africa. Although no typical deltas occur on the South African coast, the beach ridge plains of the Tugela and Great Berg may be regarded as offset, wave-dominated deltas, although the source of the sediment as fluvial has yet to be established with confidence.

Future research

Geomorphological and sedimentological studies should underpin ecological research on estuaries. It is only by fully understanding the underlying physical processes that biological, chemical and human aspects of estuaries can be placed in context and utilised in a management framework. In this regard it is interesting to note the recent establishment by Harrison (1993) of a model of fish community structure in small KwaZulu-Natal estuaries that follows a cyclic pattern which corresponds to the pattern of sedimentary processes in such estuaries documented by Cooper (1991a).

South African estuaries are particularly well (perhaps uniquely) suited to the study of climatic and sea level rise controls on estuarine systems as a result of a good historical hydrological and climatic database and an enviable record of morphological change recorded in sequential aerial photography since the 1930s in most instances. In a longer term context the single cycle infills below most South African estuaries afford the opportunity via stratigraphical analysis to interpret environmental changes during the Holocene.

The incorporation of an underlying geomorphological component in the multidisciplinary Estuarine Health Index (Cooper et al. 1995), which is currently being used to monitor and assess the status of all South African estuaries, represents a new understanding of the underlying importance of geomorphological and sedimentological processes in estuarine management. This research will also be directed at gaining an understanding of the spatial controls on estuarine morphology across marked environmental gradients that characterise both the marine and terrestrial environments of the South African coast.

References

Begg, G.W. (1978). *The Estuaries of Natal*. Natal Town & Regional Planning Commission Report 41.

Begg, G.W. (1984). *The Estuaries of Natal, Part 2*. Natal Town & Regional Planning Commission Report 55.

Boyd, R., Dalrymple, R.W. & Zaitlin, B.A. (1992). Classification of clastic coastal depositional environments. *Sedimentary Geology*, **80**, 139–50.

Bickerton, I.B. (1981). *Estuaries of the Cape. Part 2. Synopses of available information on individual systems*, ed. A.E.F. Heydorn & J.P. Grindley. *Report No. 5. Holgat (CW2)*. CSIR Research Report 404. Stellenbosch: CSIR.

Bowmaker, A.P., Van der Zee, D. & Ridder, J.H. (eds.) (1986). *Marine Research in Natal*. South African National Scientific Programmes Report No. 139. Pretoria: CSIR.

Bremner, J.M., Rogers, J. & Willis, J.P. (1990). Sedimentological aspects of the 1988 Orange River floods. *Transactions of the Royal Society of South Africa*, **47**, 247–94.

Bruun, P. & Gerritson, J. (1960). *Stability of tidal inlets*. Amsterdam: North Holland Publishers.

Cooper, J.A.G. (1988). Sedimentary environments and facies of the subtropical Mgeni estuary, southeast Africa. *Geological Journal*, **23**, 59–73.

Cooper, J.A.G. (1989). Fairweather versus flood sedimentation in Mhlanga Lagoon, Natal: implications for environmental management. *South African Journal of Geology*, **92**, 279–94.

Cooper, J.A.G. (1990). Ephemeral stream-mouth bars at flood-breach river-mouths: comparison with ebb-tidal deltas at barrier inlets. *Marine Geology*, **95**, 57–70.

Cooper, J.A.G. (1991a). *Sedimentary models and geomorphological classification of river mouths on a subtropical, wave-dominated coast, Natal, South Africa*. PhD thesis, University of Natal, Durban.

Cooper, J.A.G. (1991b). *Shoreline changes on the Natal coast. Tugela River mouth to Cape St Lucia*. Natal Town & Regional Planning Report 76.

Cooper, J.A.G. (1991c). *Shoreline changes on the Natal coast. Mkomazi River mouth to Tugela River mouth*. Natal Town & Regional Planning Report 77.

Cooper, J.A.G. (1991d). Beachrock formation at low latitudes: implications for coastal evolutionary models. *Marine Geology*, **98**, 145–54.

Cooper, J.A.G. (1993a). Sedimentation in the cliff-bound Mtamvuna estuary, South Africa. *Marine Geology*, **112**, 237–56.

Cooper, J.A.G. (1993b). Sedimentary processes in the river-dominated Mvoti estuary, South Africa. *Geomorphology*, **9**, 271–300.

Cooper, J.A.G. (1994a). Sedimentation in a river-dominated estuary. *Sedimentology*, **40**, 979–1017.

Cooper, J.A.G. (1994b). *Shoreline changes on the Natal Coast. Mtamvuna River mouth to Mkomazi River mouth*. Natal Town & Regional Planning Report 79.

Cooper, J.A.G. (1994c). Lagoons and microtidal coasts. In *Coastal evolution*, ed. R.W.G. Carter & C. Woodroffe, pp. 219–65. Cambridge: Cambridge University Press.

Cooper, J.A.G. & Mason, T.R. (1986). Barrier washover fans in the Beachwood Mangrove area, Durban, South Africa: cause, morphology and environmental effect. *Journal of Shoreline Management*, **2**, 285–303.

Cooper, J.A.G. & McMillan, I.K. (1987). Foraminifera of the Mgeni estuary and their sedimentological significance. *South African Journal of Geology*, **90**, 489–98.

Cooper, J.A.G., Mason, T.R. & Grobbler, N.G. (1988). Flood effects in Natal estuaries: a sedimentological perspective. In *Floods in Perspective*. Pretoria: CSIR conference centre.

Cooper, J.A.G., Mason, T.R., Reddering, J.S.V. & Illenberger, W.K. (1990). Geomorphological effects of catastrophic flooding on a small subtropical estuary. *Earth Surface Processes and Landforms*, **15**, 25–41.

Cooper, J.A.G., Ramm, A.E.L. & Harrison, T.D. (1995). The Estuarine Health Index: a new approach to scientific information transfer. *Ocean and Shoreline Management*, **25**, 103–41.

Crofts, C.J. (1905). *Survey of mouth of Umvolosi River and entrance to St Lucia Lake*. Zululand Coast Surveys. Durban: Natal Harbour Department.

Dalrymple, R.W., Zaitlin, B.A. & Boyd, R. (1992). Estuarine facies models: conceptual basis and stratigraphic implications. *Journal of Sedimentary Petrology*, **62**, 1130–46.

Day, J.H. (ed.) (1981). *Estuarine ecology with particular reference to southern Africa*. Cape Town: A.A. Balkema.

Esterhuysen, K. (1982). *The Seekoei estuary: example of disrupted sedimentary ecology*. ROSIE report 5. Department of Geology, University of Port Elizabeth.

Fleming, B.W. & Hay, R. (1984). On the bulk density of South African marine sands. *Transactions of the Geological Society of South Africa*, **87**, 233–6.

Galloway, W.E. (1975). Process framework for describing the morphological and stratigraphic evolution of deltaic depositional systems. In *Deltas: models for exploration*, ed. M.L. Broussard, Houston Geological Society: 87–98.

Grobbler, N.G. (1987). *Sedimentary environments of Mdloti, uMgababa and Lovu Lagoons, Natal, South Africa*. MSc thesis, University of Natal, Durban.

Grobbler, N.G., Mason, T.R. & Cooper,

J.A.G. (1987). *Sedimentology of Mdloti Lagoon*. Sedimentation in Estuaries and Lagoons (S.E.A.L. Report no. 5). Durban: Department of Geology and Applied Geology, University of Natal.

Harrison, T.D. & Cooper, J.A.G. (1991). Recruitment of juvenile grey mullet (Teleostei: Mugilidae) into a small coastal lagoon against high current velocities. *South African Journal of Science*, **87**, 395–6.

Harrison, T.D. (1993). *Ecology of the ichthyofauna in three temporarily open/closed estuaries on the Natal coast.* MSc thesis, Rhodes University, Grahamstown.

Heydorn, A.E.F. & Tinley, K. (1980). *Estuaries of the Cape*. Part 1. *Synopsis of the Cape Coast*. Stellenbosch: CSIR.

Hobday, D.K. (1979). Geological evolution and geomorphology of the Zululand coastal plain. In *Lake Sibaya*, ed. B.R. Allanson, pp.1–16. Monographiae Biologicae 36. The Hague: Dr W. Junk.

Illenberger, W.K. (1988). The Holocene evolution of the Sundays estuary and adjacent coastal dunefields, Algoa Bay, South Africa. In *Geomorphological studies in Southern Africa*, ed. G.F. Dardis & B.P. Moon, pp. 389–408. Rotterdam: A.A. Balkema.

Keyser, U. (1972). The occurrence of diamonds along the coast between the Orange River estuary and the Port Nolloth Reserve. *Geological Survey of South Africa Bulletin*, **54**, 1–23.

Kovacs, Z.P. (1988). Preliminary hydrological assessment of the Natal flood. *The Civil Engineer in South Africa*, January 1988: 7–13.

Lindsay, P., Mason, T.R., Pillay, S. & Wright, C.I. (1996a). Sedimentology and dynamics of the Mfolozi estuary, north KwaZulu-Natal, South Africa. *South African Journal of Geology*, **99**.

Lindsay, P., Mason, T.R., Pillay, S. & Wright, C.I. (1996b). Suspended particulate matter and dynamics of the Mfolozi estuary, north KwaZulu-Natal, South Africa. *Environmental Geology*, **28**, 40–51.

Martin, A.K. (1987). Comparison of sedimentation rates in the Natal Valley, south-west Indian Ocean, with modern sediment yields in east coast rivers of Southern Africa. *South African Journal of Science*, **83**, 716–24.

Martin, A.R.H. (1962). Evidence relating to the Quaternary history of the Wilderness Lakes. *Transactions of the Geological Society of South Africa*, **65**, 19–42.

McCormick, S., Cooper, J. A. G. & Mason, T.R. (1992). Fluvial sediment yield to the Natal coast: a review. *South African Journal of Aquatic Sciences*, **18**, 74–88.

Milliman, J.D. & Meade, R.H. (1983). World-wide delivery of river sediments to the oceans. *Journal of Geology*, **91**, 1–21.

Nichols, M.M. (1989). Sediment accumulation rates and relative sea level rise in lagoons. *Marine Geology*, **88**, 201–19.

Orme, A.R. (1973). *Barrier and lagoonal systems along the Zululand coast, South Africa*. Office of Naval Research, Technical Report 1.

Orme, A.R. (1974). *Estuarine sedimentation along the Natal coast, South Africa*. Office of Naval Research, Technical Report 5.

Orme, A.R. (1975). Late Pleistocene channels and Flandrian sediments beneath Natal estuaries: a synthesis. *Annals of the South Africa Museum*, **71**, 77–85.

Partridge, T.C. & Maud, R.R. (1987). Geomorphic evolution of South Africa since the Mesozoic. *South African Journal of Geology*, **90**, 179–208.

Perillo, G.M.E. (ed.) (1995). *Geomorphology and sedimentology of estuaries*. Developments in Sedimentology, 53. Elsevier.

Pethick, J. (1984). *An introduction to coastal geomorphology*. London: Edward Arnold.

Ramsay, P.J. (1995). 9000 years of sea level change along the southern African coastline. *Quaternary International*, **31**, 71–5.

Reddering, J.S.V. (1981). *The sedimentology of the Keurbooms estuary*. M.Sc Thesis, University of Port Elizabeth.

Reddering, J.S.V. (1983). An inlet sequence produced by migration of a small microtidal inlet against longshore drift: the Keurbooms inlet, South Africa. *Sedimentology*, **30**, 201–18.

Reddering, J.S.V. (1988a). Coastal and catchment basin controls on estuary morphology of the south-eastern Cape coast. *South African Journal of Science*, **84**, 154–7.

Reddering, J.S.V. (1988b). Prediction of the effects of reduced river discharge on the estuaries of the south-eastern Cape province, South Africa. *South African Journal of Science*, **84**, 726–30.

Reddering, J.S.V. & Esterhuysen, K. (1981). *Sedimentation in the Sundays estuary*. ROSIE Report 4. Department of Geology, University of Port Elizabeth.

Reddering, J.S.V. & Esterhuysen, K. (1982).

Fluvially-dominated sedimentation in the Great Fish estuary. ROSIE Report 4. Department of Geology, University of Port Elizabeth.

Reddering, J.S.V. & Esterhuysen, K. (1983). *Sedimentation in the Kromme estuary*. ROSIE Report 4. Department of Geology, University of Port Elizabeth.

Reddering, J.S.V. & Esterhuysen, K. (1984). *Sedimentation in the Gamtoos estuary*. ROSIE Report 7. Department of Geology, University of Port Elizabeth.

Reddering, J.S.V. & Esterhuysen, K. (1987a). The effects of river floods on sediment dispersal in small estuaries: a case study from East London. *South African Journal of Geology*, **90**, 458–70.

Reddering, J.S.V. & Esterhuysen, K. (1987b). Sediment dispersal in the Knysna estuary: environmental management considerations. *South African Journal of Geology*, **90**, 448–57.

Reddering, J.S.V. & Rust, I.C. (1990). Historical changes and sedimentary characteristics of southern African estuaries. *South African Journal of Science*, **86**, 425–8.

Rooseboom, A. (1975). *Sedimentproduksiekart vir Suid Afrika*. Department of Water Affairs. Technical Report No. 61. Pretoria: Department of Water Affairs.

Rooseboom, A. (1978). Sedimentafvoer in Suider-Afrikaanse riviere. *Water SA*, **4**, 14–17.

Roy, P.S. (1984). New South Wales estuaries: their origin and evolution. In *Coastal geomorphology in Australia*, ed. B.G. Thom, pp. 99–121. Australia: Academic Press.

Tyson, P.D. (1986). *Climatic variability in Southern Africa*. Cape Town: Oxford University Press.

van Heerden, I.Ll. (1985). Barrier–estuarine processes: Bot River estuary – an interpretation of aerial photographs. *Transactions of the Royal Society of South Africa*, **45**, 239–51.

van Heerden, I.Ll. (1986). Fluvial sedimentation in the ebb-dominated Orange estuary. *South African Journal of Science*, **82**, 141–7.

van Heerden, I.Ll. & Swart, D.H. (1986). *An assessment of past and present geomorphological and sedimentary processes operative in the St Lucia estuary and environs*. Marine Geoscience and Sediment Dynamics Division, National Research Institute for Oceanology. CSIR Research Report No. 569. Stellenbosch: CSIR.

Whitfield, A.K. (1992). A characterization of Southern African estuarine systems. *Southern African Journal of Aquatic Science*, **18**, 89–103.

Whitfield, A.K. & Bruton, M.N. (1989). Some biological implications of reduced fresh water inflow into Eastern Cape estuaries: a preliminary assessment. *South African Journal of Science*, **85**, 691–4.

Wooldridge, T.H. (1992). *Biotic and abiotic exchange across estuarine tidal inlets, with particular reference to southern Africa*. Estuaries Joint Venture Programme Progress Report 1990–92. Institute of Natural Resources.

Wright, C.I. (1990). *The sediment dynamics of St Lucia estuary mouth, Zululand*. MSc thesis, University of Natal, Durban.

Wright, C.I. (1995). The sediment dynamics of St Lucia and Mfolozi estuary mouths, Zululand, South Africa. *Geological Survey of South Africa Bulletin*, **109**.

Wright, C.I., McMillan, I.K. & Mason, T.R. (1990). Foraminifera and sedimentation patterns in St Lucia estuary mouth, Zululand, South Africa. *South African Journal of Geology*, **94**.

Wright, C.I. & Mason, T.R. (1990). The sedimentary environments and facies of St Lucia estuary mouth, Zululand, South Africa. *Journal of African Earth Sciences*, **11**, 411–20.

Wright, C.I. & Mason, T.R. (1993). Management and sediment dynamics of St Lucia estuary mouth, Zululand, South Africa. *Environmental Geology*, **22**, 227–41.

Wright, L.D. & Short, A.D. (1984). Morphodynamic variability of surf zones and beaches: a synthesis. *Marine Geology*, **56**, 93–118.

3 Estuarine hydrodynamics

Eckart Schumann, John Largier and Jill Slinger

Fafa River estuary

Introduction

Estuaries lie at the interface between the ocean and the land, forming the meeting place of the saltwater regime of the sea and the freshwater flow of parent rivers. The oceanic input is driven primarily by the regular forcing of the tides, while the freshwater input is dependent on variable rainfall in the catchment areas of the rivers. It is the continual interaction between saltwater and freshwater that forms the basis of estuarine hydrodynamics, compounded by other influences such as channel structure and sediment movement, and the effects of wind, waves, insolation, anthropogenic inputs and biotic processes.

The objective of this chapter is to provide a description of the physical processes inherent in estuarine hydrodynamics. In the South African situation the dominant tidal influence is semidiurnal, i.e. there are two high tides and two low tides per day, though the diurnal tide can be important, and the spring–neap variation is also substantial. On much shorter time scales surface gravity waves, internal waves and turbulence must be considered, particularly when mixing processes are involved. On longer time scales, wind-generated surges and coastal trapped waves can markedly alter ocean–estuary exchanges on sections of the South African coast.

Seasonal changes are important because variations in freshwater runoff, winds, waves and insolation cause regular and substantial alterations in estuarine circulation and water column structure, while interannual variability and its consequences are still poorly understood. Episodic events, such as sustained strong winds, high waves, and extreme rainfall or drought represent major perturbations to an estuary, and can result in significantly altered systems; an obvious example is the periodic scouring of estuary mouths by floods after heavy rains. However,

longer term trends such as the reduced freshwater runoff due to the building of dams in the catchment areas of rivers, or the potential rise in sea level associated with climatic change, may have even more significant consequences. These effects need extended time series of good quality data for detailed analyses, and unfortunately such series are generally unavailable at present.

Hypersaline conditions occur in many South African estuaries in the upper reaches of tidal influence when freshwater inflow is limited. On occasion high evaporation rates lead to salinities of 100 or more, causing considerable stress to biotic communities. Lower salinities are restored once adequate rainfall occurs, though dam storage can delay flushing and exacerbate hypersaline conditions. There are now some estuaries with salinity values consistently around 40 in their upper reaches.

In developing an understanding of how an estuary functions physically, it is useful to model the system; such modelling generally involves analytical or numerical models with various degrees of complexity. It is important to realise that it is not possible to obtain an overall solution to the equations describing the system, and generally only specific aspects can be modelled. The advantage of such models is that they give quantitative results, and therefore have a predictive capability and ultimately allow an improved comparison of different possible management scenarios. Modelling also facilitates the transfer of understanding to other similar systems.

It is important to understand that estuaries are inherently variable, responding to various inputs over a broad range of periods. To consider time averages or instantaneous 'snap-shots' as being representative of an estuary can be highly misleading, and it is essential to consider

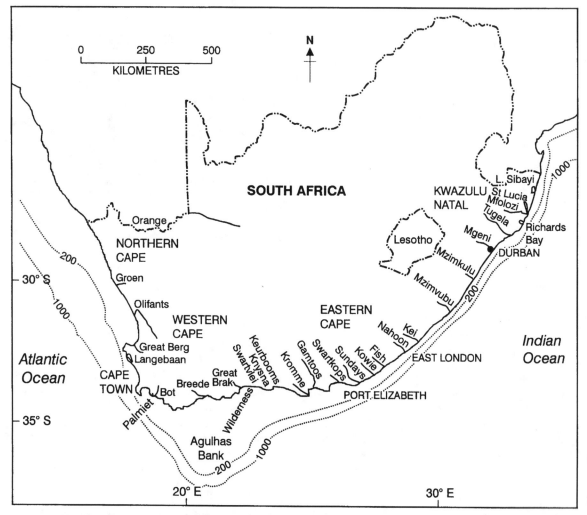

Figure 3.1. **Map of South Africa showing the positions of estuaries discussed in the text, as well as other regions and sites mentioned.** Isobaths are indicated with a dotted line, with depths given in metres.

measurements of selected parameters taken at short enough intervals and for long enough periods to assess the degree of variability (Wolfe & Kjerve 1986). Estuarine spatial variability also occurs on a variety of scales, with the shorter spatial scales generally associated with shorter time scales, e.g. turbulence and mixing processes, whereas the longer spatial scales extend over estuary basins and are generally more persistent.

All these estuarine processes are covered in this chapter to give a conceptual picture of the physical characteristics of South African estuaries. However, it is important to realise that there are marked variations between different estuaries, and it is the mix of different conditions and processes which determines individual estuarine characteristics. An assessment in terms of known variables, as well as additional site-specific measurements are required before an estuary can be classified in terms of all the different parameters.

Figure 3.1 shows the location of South African estuaries discussed in the text, while additional areas and sites mentioned are also indicated. It is clear that only a small percentage of all the estuaries are covered here, firstly because they are used to demonstrate specific characteristics, and secondly because little physical work has been done (and reported) on those remaining.

A wide range of estuarine types can be found in South Africa, reflecting substantially different physical environments. Rivers in the Northern Cape fall into a mediterranean to desert climate region, and flow only sporadically after good rains, e.g. about once every 5 years in the case of the Groen River (Bickerton 1981); the Orange River, with its vast catchment in the interior, is an obvious exception. In contrast, it is in the eastern section of South Africa where most rain falls, and it is there that most rivers rise: from minor streams to major rivers such as the Tugela, Mgeni,

Mzimkulu, Mzimvubu, Kei and Fish. However, the presence of all the fresh water means this is also where developments are taking place, and dams are being mooted and built to utilise this resource.

Climate and biogeographic regions

The characteristics of estuaries are dependent on where they are, and therefore on the climate of the region, including also the oceanographic environment. In comparison with many other parts of the world, the flow of freshwater into South African estuaries is limited. In many cases this flow averages less than $1 \text{ m}^3 \text{ s}^{-1}$, and consequently the eventual outflow into the sea does not appreciably affect conditions on the adjacent shelf regions except during floods. The average flow of the Tugela River varies from $74 \text{ m}^3 \text{ s}^{-1}$ during winter to $481 \text{ m}^3 \text{ s}^{-1}$ in summer (Begg 1978), while the Orange River averages $370 \text{ m}^3 \text{ s}^{-1}$ over a year, and reached an estimated peak value of $8300 \text{ m}^3 \text{ s}^{-1}$ during the record-breaking floods in March 1988 (Swart et al. 1990). To put these values into context, the average flow of the Mississippi is about $18\,400 \text{ m}^3 \text{ s}^{-1}$ (Milliman & Meade 1983), while the flow of the Amazon is an order of magnitude greater than that.

Climatic regions

Climates in the subtropical regions of the Earth are characterized by a high degree of intra- and inter-annual variability. In the context of estuaries, important climatic variables are the rainfall, winds, waves, insolation and evaporation, and a brief description of the relevant South African conditions is given below.

Rainfall

Rainfall over South Africa is particularly erratic, and the isohyets have a nearly north–south trend, except along the south coast (Figure 3.2a). The 40 cm isohyet divides South Africa into wetter eastern and drier western halves, with over 100 cm in some eastern sections, and almost complete aridity in places in the west.

Rainfall is also highly seasonal (Figure 3.2b). More than 80% of the rainfall in the northern regions occurs in summer between October and March, while in the south Western Cape the situation is reversed with over 80% of the rain falling in winter. In contrast, the narrow southern coastal belt and the adjacent interior regions receive rainfall more uniformly throughout the year. In all regions topography exerts a strong influence, the mountains enhancing precipitation, with the result that marked spatial anomalies can occur.

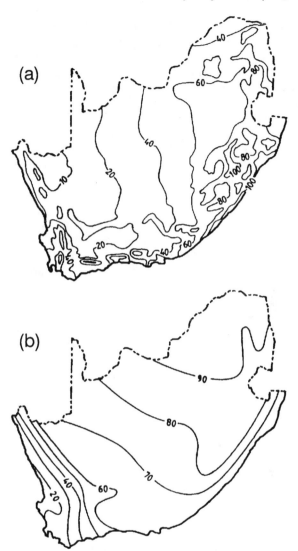

Figure 3.2. **(a) Mean annual rainfall for South Africa (isohyets given in cm). (b) Seasonal rainfall over South Africa, with the summer rainfall (October to March), given as a percentage of the annual rainfall (from Tyson 1986).**

Winds

Seasonal variations in the wind field over South Africa are associated with the latitudinal movement of the discontinuous high pressure belt that circles the southern hemisphere at about latitude 30° S (Preston-Whyte & Tyson 1988). The consequence is that in winter the cold fronts associated with mid-latitude cyclones further south sweep from west to east across the southern portion of the subcontinent, bringing with them northwesterly winds to the south Western Cape and a greater percentage of westerly winds on the south coast. In contrast, in summer the cyclones generally stay much further south, and the South Atlantic high pressure system ridging south of the subcontinent causes strong southeasterly winds to blow in the south Western Cape. To the east the westerly winds still dominate, but with a higher percentage of easterly winds in summer. The frontal systems often move offshore along the Transkei and KwaZulu-Natal coasts, particularly in summer, and the region comes under the influence of easterly winds associated with the Indian Ocean high pressure system. However, there are shallow low pressure systems – called **coastal lows** – which are associated with the frontal systems and propagate in an anticlockwise sense around southern Africa. These coastal lows can have a substantial influence on coastal weather: in particular, they are often preceded by warm, offshore *Berg* winds, to be followed by a rapid change to blustery westerly winds. These characteristics are shown in the seasonal wind variability depicted in Figure 3.3 for Cape Town, Port Elizabeth and Durban. The dominant wind directions in the three cases follow the orientation of the coastline, i.e. northwesterlies and southeasterlies at Cape Town, southwesterlies and north-easterlies at Port Elizabeth, and south-southwesterlies and north-northeasterlies at Durban. The curves show the strongest seasonal variation in the easterly component winds, in particular at Cape Town where they also tend to blow continuously for longer periods. In all cases strongest winds occur in October/November, with the calmest periods in June.

Waves

Most of the waves impinging on South Africa's coast come from the southwest, generated by storm winds in the vast reaches of the Southern Ocean. The increased influence of these weather systems during the winter months means that high wave conditions are generally more frequent in winter than in summer (Rossouw 1984). There is also a gradual reduction in wave height northwards on both the east and west coasts. On the east coast high waves also come from the east, and tropical cyclones can occasionally generate very high waves along the northern KwaZulu-Natal coast; the cyclone season extends from November to May, with an average frequency of 4 or 5 per annum in the western Indian Ocean.

The effect of waves on beaches, and therefore also on estuary mouths, depends on the wave heights and periods. Storm waves can be expected to erode beaches, while smoother, smaller waves will bring sediment back onshore and thus cause beach accretion; waves approaching a coastline obliquely also generate alongshore transport of sediment.

Incident radiation energy and evaporation

Incident solar radiation (insolation) is important for estuaries since it directly affects air and water temperatures.

Clouds reflect such insolation, and it means that the net input of surface radiation energy to an estuary is dependent on both insolation and atmospheric conditions. Water also loses or gains energy through back-radiation, conduction and evaporation/condensation, and it is the overall balance which will dictate whether temperatures increase or decrease.

The western regions of South Africa have minimal cloud cover, and consequently these regions receive the greatest average annual surface incident radiation energy; however, along the coast the occurrence of fog can cause a marked reduction. On the other hand, the eastern regions receive the most diffuse sky radiation (Schulze 1986).

South Africa falls into a mid-latitude region where there is a marked seasonal variation in insolation. Because of its mediterranean climate, the south Western Cape receives the greatest surface radiation input during summer when there is limited cloud cover, with a corresponding decrease eastwards. In contrast to summer, during winter there is a gradual increase of incident surface radiation northwards, but with a minimum in the south Western Cape.

The loss of water through evaporation from the surface and through evapotranspiration from aquatic macrophytes (e.g. mangroves) may be a significant factor in the hydrology budget for South African estuaries. Evaporation rates are highest where the air is hot and dry, and seaward winds enhance evaporation by bringing in further hot and dry air. High evaporation is thus anti-correlated with high rainfall, with the greatest evaporation found in the west of the country, particularly in the dry season and during prolonged droughts (Figure 3.4). Evaporation increases to about 1 cm per day during the dry summers in the Western Cape, and this loss of freshwater may exceed the inflow into estuaries, resulting in a net evaporative condition during the dry summer season. Examples are the Langebaan Lagoon (not strictly an estuary) on the Western Cape coast and the Kromme River estuary on the Eastern Cape coast, while the St Lucia estuary experiences multi-year warm dry periods when the hydrological balance is net evaporative (Begg 1978).

Biogeographic regions and estuarine classification

Whitfield (1992) has described biogeographic subdivisions of South African estuaries. These subdivisions range from subtropical on the southeast coast, warm temperate on the south coast to cool temperate on the west coast, and have been linked to ichthyofaunal diversity. They originate basically because of the different oceanic regions off the South African east and west coasts, i.e. the warm Agulhas Current in the east (Schumann 1987), and the cold Benguela system in the west (Shannon 1985). They also tie

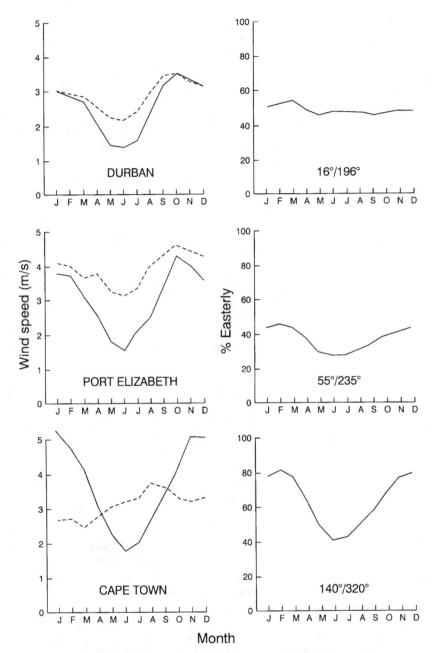

Figure 3.3. **Monthly variations of mean daily wind speeds measured at Durban, Port Elizabeth and Cape Town. The dominant directions are orientated approximately parallel to the local coastline, i.e. 140/320° T at Cape Town, 55/235° T at Port Elizabeth and 16/196° T at Durban; the winds blow *from* the given direction. In each case the** **solid line depicts the wind with an easterly component, and the dashed line the westerly component; percentage occurrence of the easterly winds is shown on the right, with the westerly component forming the complement to 100%. (Adapted from Schumann & Martin 1991.)**

in with seasonal rainfall, wave and wind distributions, and the availability of nutrients from the marine environment.

The physical classification of estuaries is not straightforward, in part because the term covers such a wide diversity of sizes and shapes. Fischer *et al.* (1979) describe hydrodynamic categories involving stratifica-

tion, geomorphological categories, and analytical methods; these influences are discussed in more detail later in this chapter. On the other hand, Whitfield and Wooldridge (1994) classified South African estuaries according to the state of the estuary mouths, and identified five types useful in ecological and management studies, namely:

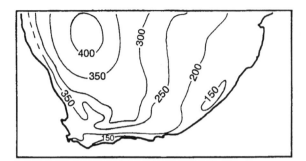

Figure 3.4. **Average annual evaporation (in cm) from Class 'A' pans over southern Africa (from Schulze 1986).**

- Permanently open estuaries
- Temporarily open/closed estuaries
- River mouths
- Estuarine lakes
- Estuarine bays

Tidal and longer-period water level fluctuations

In most South African estuaries tidal exchanges constitute the major regular forcing, and it is this input and removal of water through an estuary mouth that determines many of its characteristics. There are also longer-period fluctuations in sea level that force flow in and out of estuaries: these are generated by weather events, in particular wind. As a result they are not regular like the tides, but depend on fluctuating winds and occur sporadically and with varying amplitudes. Nonetheless, they can have important influences on estuarine dynamics.

The manner in which the exchanges take place through an estuary mouth is strongly dependent on the mouth configuration. Chapter 2 has already dealt with the nature of the sedimentation that can occur in such a mouth, in particular the buildup of flood-tidal deltas, the bar at the mouth, and the occasional scouring that takes place during flood events. Exchanges with the ocean can be severely restricted by the flood-tidal deltas or bars, the extreme case being that of an estuary closed completely by such sedimentation.

This section will investigate these tidal and longer-period exchanges, and present quantitative criteria to describe their occurrence and nature. At this stage homogeneous water bodies will be assumed, with the further influence of stratification addressed later.

Tidal amplitudes and periods

The dominant tide around the South African coast is that due to the lunar M_2 tidal constituent, i.e. it is **semidiurnal** with a period of 12.42 hours. Nonetheless, the diurnal components, with a period of about 24 h, are also important and lead to succeeding high tides of differing amplitudes.

Because of the relatively straight nature of much of the South African coastline there is little variation in tidal height, though there is some indication that the wide Agulhas Bank may cause slightly higher tides at sites along the south coast. Most models of the M_2 tide find an **amphidromic** point, i.e. a site of no tidal variation, in the Southern Ocean south of Africa, and tidal propagation then occurs as a **Kelvin wave** southwards along the west coast, and then from west to east; it means that the tide along the west coast precedes that along the south and east coast (for an explanation, see e.g. Pond and Pickard, 1986). Tidal records (SAN Hydrographer 1997) show that there is about a 15–30 minute delay between Cape Town and Port Elizabeth, but very little further delay along the east coast.

Tidal height is defined as the difference between sea level at high tide and that at low tide. Considerable variability occurs along the South African coast, with tidal heights of more than 2 m at spring tides and as little as 0.5 m at neap tides. On a world-wide classification, such tidal variations fall into the microtidal category (0–2 m). In South African waters spring low tides generally occur between 0900 or 1000 h, with the corresponding high tides just over 6 hours later. Approximately a week later the neap high tides occur in the morning, and the neap low tides in the afternoon.

Measurements of sea level are made on a continuous basis at all South African ports. These measurements are made relative to **Chart Datum**, which is now specified as −0.900 m relative to Land Levelling Datum, commonly called **Mean Sea Level** (MSL) by land surveyors (SAN Hydrographer 1997). Tidal constituents can be derived from such a sea level record by carrying out a harmonic analysis, and these can then be used in predicting future tidal conditions. The longer the original record, the more accurate these predictions are likely to be, though it is important to realise that meteorological effects can cause substantial deviations from predicted tidal sea levels.

Tidal asymmetry

In the open ocean the tidal variations follow smooth sinusoidal curves as a result of the regular forcing of the gravitational attractions of the Earth, moon and sun systems and their consequent orbital patterns. However, the shallow depths of coastal seas can alter these smooth variations, while constrictions in the mouths of estuaries will further distort the tidal signal resulting in asymmetric tidal exchanges. A flood dominant system is characterised by a longer ebb tide in the estuary, and a shorter, more intense flood tide, while conversely an ebb dominant system has a longer flood tide and a shorter ebb tide.

Aubrey and Friedrichs (1988) have analysed the conditions necessary for the occurrence of flood- or ebb-tidal estuaries with a negligible freshwater outflow. At present there is no finality about which processes dominate, and while various explanations have been proposed, each estuary needs to be considered in terms of its own configuration. While the nature of the tidal cycle and the overall morphology determine the flood or ebb dominance, the location and structure of specific channels will result in some channels being ebb dominant and others flood dominant. This is evident at the mouth of most bar-built estuaries, where the channel in line with the mouth is dominated by the inertia of the flood tide, while deeper, curved channels approaching the mouth from one or other side are dominated by the ebb tide.

One perspective considers tidal propagation along the coast as a wave, and the water moving into and out of an estuary is a continuation of that wave. Further propagation upstream can also occur as a non-reflected tidal wave, moving at the shallow water wave speed C, namely

$$C = \sqrt{gh} \qquad (1)$$

Here $g (= 9.81 \text{ m s}^{-2})$ is the gravitational constant, and h is the estuary depth. In shallow estuaries, this formula shows that tidal asymmetry can be accentuated upstream, since at the peak of the wave the depth h – and therefore also speed C – is greater than at the trough of the wave; the peak therefore tends to overtake the preceding trough. This formula will also enable an approximate determination to be made of the time delay from the estuary mouth to its head.

Water will flow from the ocean into an estuary if there is a head with a corresponding pressure gradient, i.e. the water level outside the estuary is higher than that inside the estuary, and similarly for water to flow out the level inside the estuary must be higher than that of the sea. The flood-tidal deltas effectively create a sill, or bar, at the mouth of the estuary, and in many cases the sea's low tide level is actually below that of the sill. The consequent water level change in the estuary and the development of the flood-tidal asymmetry can then be explained very simply with reference to Figure 3.5.

Thus the incoming tide in the ocean only establishes a head with respect to the water level in the estuary from point A, and then there is a rapid flood tidal inflow, though the estuary water level lags behind that of the sea because of constrictions to flow at the mouth. This head is maintained through the peak ocean tide at B, and it continues to the point C, when the two water levels cross again. Sea level continues to fall until point D, with a reverse pressure gradient now established from the estuary to the sea.

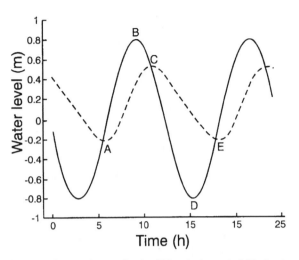

Figure 3.5. **Schematic water level variations in the sea (solid line) and adjacent estuary (broken line) depicting a South African flood-tidal system. (Adapted from MacKay & Schumann 1991.)**

Water flows out of the estuary, but as it does so the flow over the sill becomes shallower and shallower, increasing frictional effects and slowing the velocity. Consequently the water level in the estuary does not drop as quickly as the rise at the flood tide and, moreover, it clearly cannot fall below that of the sill. Sea level then starts to rise again, decreasing the head and reducing the outflow further, and when the levels cross at point E the whole sequence recommences.

Ebb dominance appears to require relatively deep channels and extensive intertidal water storage. If there are large areas of intertidal marshes and flats, low velocities in these areas tend to delay the turn to ebb, giving a longer flood tide. On the other hand, at the end of the ebb the flow is constrained to the relatively deep channels, giving stronger currents and a faster turn to the next flood tide.

Bays on the landward side of small inlets will also affect the occurrence of flood or ebb dominant systems, depending on the natural frequencies of the bays (DiLorenzo 1988). However, in South African estuaries there appear to be no such bays which could markedly affect the exchanges in this manner.

The M4 overtide

In mathematical terms, the asymmetric water level variation in estuaries can be modelled by adding in an additional tidal constituent, namely the M_4 overtide. This oscillation has a period of half that of the M_2 tide, i.e. 6.21 hours, and is assumed to be generated by non-linear effects at the mouth.

The water level amplitude A in the estuary can then be modelled very simply as

$$A = a_0 + a_2 \cos(\omega t - \theta_2) + a_4 \cos(2\omega t - \theta_4) \qquad (2)$$

where ω is the tidal frequency, a_0 is a constant, a_2 the amplitude of the M_2 tide, a_4 the amplitude of the M_4 over-tide, and θ_2 and θ_4 the corresponding phases. A direct measure of the non-linear distortion is given by the ratio of the amplitudes of the M_2 and M_4 tides, i.e.

$$\text{Distortion } S = \frac{a_4}{a_2} \qquad (3)$$

and an undistorted tide has $a_4 = 0$ and therefore $S = 0$.

Whether the resulting system is flood or ebb dominated depends on the respective phase angles, and the magnitude of the parameter P, where

$$P = 2\theta_2 - \theta_4 \qquad (4)$$

If P lies between $0°$ and $180°$ the system is flood dominated, while if P lies between $180°$ and $360°$ the system is ebb dominated.

Upstream variation

The changes in the duration of ebb and flood tides within an estuary have already been discussed. The major change takes place at the mouth where flow is strongly constricted, and as the tide propagates upstream further changes can be expected, again depending on the particular estuary being considered. Thus the existence of extensive intertidal flats will slow down the tidal propagation, while narrow, deep channels will allow much faster propagation; this will occur both because of vegetation and also because of the dependence on water depth (equation 1).

As the tide propagates upstream, changes in tidal amplitude can also be expected. Frictional forces, in particular at the edges of the estuary, will dissipate energy and therefore decrease the amplitude. On the other hand, if the estuary narrows then more water will be forced into a smaller area and the tidal amplitude will increase; further increases will result as shallower water decreases the speed of the tide wave and correspondngly its wavelength also decreases.

Each estuary therefore needs to be assessed individually to determine the manner in which the tidal amplitude and phase vary, and this can only be done by carrying out the necessary measurements of water level variation at different points in the estuary. This may also depend on spring and neap tides, and longer-period variations in sea level.

South African estuaries

These theories can be applied to South African estuaries, and most appear to be flood dominated, with the consequent establishment of extensive flood-tidal deltas. Figure

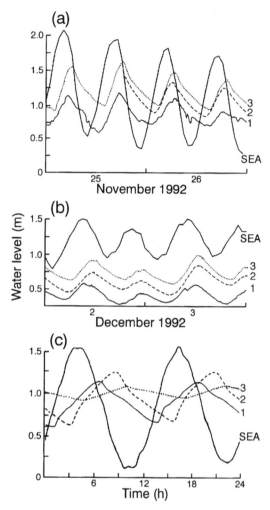

Figure 3.6. **Tidal water level variations in two selected South African estuaries. (a) The Gamtoos River estuary at spring tide, showing the water level in the sea (dark solid line), and at 3 sites in the estuary at distances approximately 1.8 km (1: light solid line), 8.3 km (2: broken line) and 18 km (3: dotted line) from the mouth. The water levels are not shown relative to an absolute reference level. (From Schumann & Pearce 1997. Reprinted by permission of the Estuarine Research Federation. © Estuarine Research Federation.) (b) The Gamtoos River estuary at neap tide; details as in (a). (c) The Great Berg River estuary, showing water level in the ocean (dark solid line) and at distances of 14.4 km (1: light solid line), 46 km (2: broken line) and 56 km (3: dotted line) from the mouth (from Slinger & Taljaard 1994).**

sea. It is apparent that large variations occur, and it is important to note the marked reduction in the amplitude of the dominant M_2 tidal variation in going from the ocean to the estuary. This can be ascribed directly to the limited exchange of water that can take place through the narrow mouth channels over the tidal period. However, the exchange can vary considerably in any one of the estuaries depending on the stage of development of the flood-tidal delta.

Berg Estuary the depth of the channel remains fairly constant from 14 to 45 kilometres upstream of the mouth but the width becomes smaller (Slinger & Taljaard 1994), while for the Gamtoos Estuary the width decreases, and the elevation increases with a consequent decrease in depth (Schumann & Pearce 1997).

The difference in the time between the occurrence of the peak and the following trough in the ocean, and that occurring in the two estuaries is also evident. In particular, in the estuaries there is a much longer delay from the peak to the trough than from the trough to the peak, leading to the typical asymmetrical flood-tidal character of the water level variations.

Tidal prism and tide discharge curves

The tidal prism (or intertidal volume) is the volume of water exchanged between the sea and an estuary over each tidal cycle. The exchanges with the ocean and the manner in which an estuary is flushed are dependent on the ratio of this volume relative to the remnant volume of water left in the estuary at low tide.

To determine the tidal prism from measurements means that the surface area of the estuary and water levels need to be known over the whole estuary, over the relevant tidal cycle: the tidal prism is then the volume calculated by multiplying the successive surface areas by the incremental changes in water levels. A rough estimate of the tidal prism can be made by multiplying the area of the estuary by the tidal height. Such a calculation is reasonable if the estuary channels are steep, but difficulties arise when the water surface area changes dramatically with water level, e.g. in the case of extensive tidal flats. Complications also arise in calculating the tidal prism when the tidal height varies with position in an estuary. Nonetheless, the method is useful and can yield reasonable estimates of the tidal prism if done carefully.

On the other hand, the tide discharge curves are a measure of the instantaneous volume flows over a tidal cycle. These require measurements of the currents and cross-sectional area at all stages of the tide, so that the time-dependent volume flows can be calculated. Practical problems involved in these determinations include the necessity of interpolations, boundary layer profiles, and the fact that measured velocities are seldom perpendicular to the cross-section alignment (Hume & Bell 1993). The tidal prism over that particular tidal cycle can then also be found by integration of these volume flows, although the ebb flows will also include freshwater inflow from the parent river, and this must be subtracted so that the tidal prism reflects only the exchanges with the ocean.

In many cases the tide discharge variations do not follow smooth curves with time because of variations in current structures due to estuary basin morphology. Thus flood velocities may be high initially when they are confined to the main channels, but then decrease owing to increased frictional resistance as the water spills out of the channels and over intertidal flats. Thereafter the flows will increase again as the water depth over the intertidal areas deepens. With all these variations it is apparent that the shape of the tide discharge curves, and therefore the tidal prism, can vary markedly in one estuary with changes in tidal amplitude over the spring–neap cycle. Moreover, any longer-period changes in sea level, for instance due to coastal trapped waves, can also result in changes to the parameters.

Numerical models, properly calibrated for particular estuaries, can be used directly to calculate both the tidal prism and tide discharge curves. Such models are convenient, since they allow calculation of the tidal prism over any tidal cycle or river flow condition.

Coastal trapped waves

Longer-period sea level variations also occur in the ocean, and can then be transmitted through the mouth and up an estuary. Of importance along the South African west, south and southeast coasts are the so-called **coastal trapped waves** (CTW). These are waves which are generated by the wind, and then propagate around the coast with their crests perpendicular to the coastline and also with maximum amplitudes at the coast. Substantial amplitudes have been measured, with peak-to-trough values of up to 1 m (Schumann & Brink 1990), and such CTWs can then substantially alter water levels in an estuary.

Because CTWs are generated by weather systems, it is not possible to predict their occurrence or amplitude; typical periods appear to lie in the band of about 4–10 days, but often they may not be present. If sea level records are available from a nearby port, then analyses can indicate their possible presence in retrospect. Such longer periods mean that water levels in an estuary have a longer time to equilibrate to those in the ocean, and consequently the effects of CTWs will be felt higher up an estuary than shorter-period fluctuations with the same amplitude.

The importance of CTWs for a particular estuary can be gauged from their amplitudes relative to the tidal amplitudes. For instance, at neap tide the M_2 tidal height is around 50 cm, and if the trough of a CTW with a similar height occurs at the same time, then the measured high tide will be no higher than low tide would have been; similarly at the peak of a CTW the high water will be much higher than expected.

Shorter-period trapped waves have also been reported, generated by travelling micropressure oscillations in the atmosphere (Shillington 1984). These are termed edge waves, with heights up to about 50 cm, but with periods of the order of 20–60 minutes. Because of the short periods, it is unlikely that substantial ocean–estuary exchanges will take place, though some penetration will occur.

Stratification, fronts and mixing processes

Freshwater input from rivers has a salinity around 0, whereas seawater has a salinity of about 35. Influent water, whether from the sea or river, may also have a temperature different from that in an estuary, and it is the combination of salinity and temperature that defines the resultant density of a mass of water (see e.g. Pond & Pickard 1986, for an equation relating density to salinity and tempera- ture). Generally less dense water will lie above more dense water, giving a stably stratified water column in which salinity increases with depth and temperature decreases with depth. Abrupt changes in temperature with depth are termed **thermoclines**, while abrupt changes in salin- ity are termed **haloclines**; an abrupt change in density, irrespective of its origin, is termed a **pycnocline**. The pro- cesses inherent in setting the structures and mixing between different water masses form a major component of estuarine hydrodynamics, and are discussed in this section.

As a comparison, a change in temperature from 5°C to 25 °C will change the density of water with a salinity of 35 by about 0.006 kg m^{-3}, while a change in salinity from 0 to 30 will change the density by around 0.027 kg m^{-3}. Because of the limited inflow of freshwater, in most South African estuaries salinity varies over the entire range from 0 to 35 and more, whereas temperature varies between about 5 °C and 30 °C, though seldom through a range of more than 10 °C at any one time. Consequently, salinity is generally the dominant factor in determining water density variations in estuaries.

Estuarine circulation

As already stated, the physical nature of estuaries is largely defined by the interaction between dense, saline, ocean water entering and leaving the system in response to tidal and longer-period forcing, and less dense freshwater enter- ing from parent rivers. The less dense river water will tend to spread out over the more dense seawater which, at the same time, will spread out along the bottom beneath the fresher water. This layered structure, called a **stratified system**, represents a low state of potential energy.

However, kinetic energy is available in the movement of estuarine waters, owing to tides, wind and density differences. This kinetic energy, and associated turbulent kinetic energy, can do work in lifting heavier water and in mixing the stratified water column – a mixed water

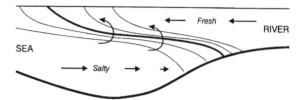

Figure 3.7. **A schematic representation of estuarine circulation. The dark solid line indicates the halocline, while the lighter lines depict additional isohalines.**

column representing a higher state of potential energy. The nature of the source waters and the degree of mixing will thus determine the stratification and the temperature and salinity of the water types in an estuary. At any one location, there may be one or more layers.

As a result of gravity, the interface between two water types of different density represents an energy surface, i.e. energy must be supplied to water in the one layer if it is to intrude vertically into the other layer. Thus these pycno- clines will be horizontal in the absence of motion. Any slope in the pycnocline will create a horizontal **baroclinic** pressure gradient force in the lower layer that will acceler- ate water beneath the pycnocline. Conversely, a slope in the water surface will give a pressure gradient force over the whole water column, resulting in **barotropic** flow.

Water in the lower reaches of a river will move seaward owing to a seaward slope in the water surface. As this fresh- water meets the seawater in the estuary, the water surface will still slope seaward and the barotropic pressure gradi- ent will continue to move near-surface water towards the sea. Beneath the landward sloping pycnocline, however, the landward baroclinic pressure gradient will be stronger than the seaward barotropic pressure gradient, resulting in landward movement of the dense near-bottom water. This simultaneous seaward movement of near-surface water and landward movement of near-bottom water is known as **estuarine circulation** (Figure 3.7).

In the presence of tidal motions, both the near-surface and near-bottom waters may flow in on the flood tide and out on the ebb tide, but the inflow will be stronger at depth, with the outflow stronger near the surface (Figure 3.8). Over a tidal cycle, the average velocity will be land- ward in the bottom layer and seaward in the surface layer. Implicit in this circulation, and prescribed by continuity, is an upward entrainment of saltier bottom water into the seaward moving surface layer (Figure 3.7). This conveyor belt circulation characterises estuaries and ensures a rapid exchange of waters between the ocean and the estuary as well as a rapid movement of river water through the system and out to sea.

With this understanding of how land runoff and sea- water interact to form layers, it is apparent that there will not be sudden along-estuary changes from river water to

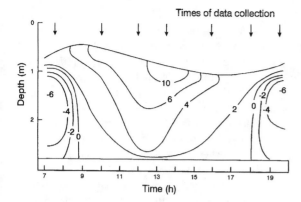

Figure 3.8. **Laterally-averaged water velocities (cm s⁻¹) over a high tide/low tide period in a section across the Palmiet River estuary (from Largier 1986). The arrows indicate times of data collection, the water surface level varies because of the tidal fluctuation, and the shaded portions (negative velocities) show inflow as opposed to positive values indicating outflow.**

seawater. In contrast to the sharp vertical gradients in salinity, the horizontal gradients are typically gradual, indicating that isopycnals (lines of equal density) are almost horizontal (Figure 3.7). As the low-salinity surface layer flows seaward, it wil entrain underlying salty water and the salinity of the surface layer will increase gradually towards the sea. Similarly, the intruding bottom salty layer may entrain fresher surface water and the salinity of the bottom layer will then decrease gradually towards the land. Under certain circumstances, sharp horizontal gradients (**fronts**) may form, and these are discussed in the next sections.

Types of estuarine stratification

Stratification in an estuary depends on the extent to which waters of different temperatures and salinities mix. At the molecular level, **diffusion** takes place across a gradient from higher to lower temperatures, or from more saline to less saline regions (Fischer *et al.* 1979). This molecular diffusion is a slow process, essentially important only in stationary fluids or in laminar flow conditions. However, fluid motions in the environment are almost always turbulent, and then laminar flow breaks down into a range of turbulent eddies, and **eddy diffusion** dominates any effect of molecular diffusion.

While turbulence causes mixing, density differences inhibit the vertical extent of particle excursions and thus limit mixing. The degree of stratification of the water column in an estuary then depends on the balance between the density difference of the source waters (sea and river) and the amount of energy available for turbulent mixing. The greater the supply of buoyancy, i.e. the density difference of source waters, the more likely it is that stratification will be strong. The greater the source of

mixing due to tides, winds and density-driven flows, the more likely it is that stratification will be broken down. In the case of wind, the energy is supplied as a stress at the surface boundary, while in the case of tides it is a stress at the bottom boundary. Thus the amount of mixing energy necessary to break down stratification also depends on the water depth. In the case of density-driven flows the stress is internal, i.e. on the interfaces between layers of different density.

The relationship between velocity structures and density structures remains a topic of active research. The two are intimately linked: velocities are constrained by density structures, which are in turn largely determined by velocity structures. If there is strong vertical mixing there will be no vertical stratification (layering) and both velocity and density will be roughly uniform in the vertical. In contrast, if vertical mixing is not strong relative to the supply of buoyancy, the surface and deeper currents will remain sheared. In this situation the less dense water of the surface current and the more dense water of the deeper current will serve to enhance or maintain the vertical stratification in spite of some limited mixing of the two layers.

The effects of three different mixing processes are considered below.

Internal

The balance between the mixing effect of velocity shear ($\partial u/\partial z$) across the pycnocline and the stabilizing effect of the density difference ($\partial \rho/\partial z$) across the pycnocline can be given by the (gradient) Richardson number R_i, where

$$R_i = \frac{g\,\partial\rho/\partial z}{\rho(\partial u/\partial z)^2} \tag{5}$$

and ρ is the average density of the water, u is the horizontal velocity and z is measured vertically. If $R_i < 0$ then the system is unstable and the density variations will enhance the turbulence, while if $R_i > 0$ they tend to reduce it. Theoretically, if $R_i > ¼$ then mixing will not take place across the pycnocline, but in practice there are variations about this critical value; note that it is also difficult to measure the vertical gradients appearing in equation (5).

Velocity shear on the interface can also result in internal waves on the density interface analogous to wind-generated waves on the sea surface. As the upper layer is generally faster flowing and more turbulent, these waves will break upward and water from the upper levels of the pycnocline can be entrained into the upper turbulent layer (analogous to spray being blown off the surface of the sea in a storm). This process is dominated by upward entrainment, though there may be limited downward transfer as well. The pycnocline is not necessarily broken down by this entrainment, but may in fact be sharpened and deepened.

Tidal

While equation (5) describes the balance between stratification and internal stress, Fischer *et al.* (1979) noted that tidal bottom stress is the primary reason for the breakdown of stratification in estuaries. They therefore suggested the use of a non-dimensional Estuarine Richardson Number, R, given by

$$R = \frac{\frac{\Delta\rho}{\rho}gQ}{wu_t^3} \qquad (6)$$

where $\Delta\rho$ is the difference in density between the river and ocean water, ρ is the density of the bottom ocean water in the estuary, Q is the input of fresh water, W the channel width and u_t is the root-mean-square tidal velocity. As for the general Richardson number (equation (5)), the buoyancy contribution appears in the numerator, with the mixing energy in the denominator. Large values of R (typically >0.8) indicate that the estuary is likely to be stratified, whereas small values (typically <0.08) indicate that the estuary is likely to be mixed.

Wind

The importance of wind stress in vertical mixing has been well studied in lakes and in the ocean, where it controls the weaker thermal stratification. In estuaries, stratification and mixing is usually dominated by tidal forcing, although this is not always true where tides are weak and winds are strong. Although not often used in estuaries, the Wedderburn number W is a derivation of a Richardson number for use in lakes (Imberger & Patterson 1990), and is a useful parameter in wind-dominated estuaries. It is given by

$$W = \frac{\frac{\Delta\rho}{\rho}gh_1^2}{\tau L} \qquad (7)$$

where h_1 is the depth of the pycnocline, τ is the surface wind stress and L is the wind fetch, i.e. the length of the basin in the direction of the wind. Large values of W (typically >1) indicate wholesale mixing of the upper and lower layers through both interfacial shear and upwelling, whereas moderate values of W indicate tilting of the pycnocline without interfacial breakdown, and small W (typically <0.1) indicates little wind effect on the stratification. The direct effect of wind-generated turbulence is considered to be important only with the shallowest of thermoclines.

Stratification

Three types of stratification are recognised in estuarine studies, namely **highly stratified**, **partially mixed** and **well mixed**. These stratification types may change in both space and time, for instance an estuary could be stratified during the rainy season and mixed during periods of low inflows, or it could be stratified during neap tides and mixed during spring tides, or stratified in the upper reaches and mixed nearer the mouth. Highly stratified estuaries have an intense, thin pycnocline and the vertical structure is approximately two-layered (Figure 3.9a). The density difference between surface and bottom is larger than the density difference between the seaward and landward ends in the surface layer, i.e. vertical density gradients dominate longitudinal density gradients.

In well-mixed estuaries (Figure 3.9b), there is a negligible density difference between surface and bottom along the whole estuary, and the water column approximates a single layer. The density difference between the seaward and landward ends of the estuary is much larger than that between the surface and bottom, i.e. longitudinal density gradients dominate vertical gradients. It is the vertical density difference between surface and bottom waters that characterises highly stratified estuaries, whereas well-mixed estuaries are characterised by a greater longitudinal difference between landward and seaward ends. Well-mixed estuaries have little or no estuarine circulation and often exhibit weak longitudinal dispersion, which leads to limited exchanges at the landward end of the estuary.

Partially mixed estuaries fall between these two extremes and do not have a sharp pycnocline anywhere in the longitudinal section, but they do exhibit a significant density difference between the surface and the bottom (Figure 3.10). The vertical density structure approximates linear stratification, and while there is no one barrier to vertical movement, vertical mixing is suppressed by the density differences.

In South Africa, the river inflow to most estuaries may be small, but the tidal range is also small and the full range of stratification types can be observed. However, a critical human impact can be appreciated by noting from equation (6) that a reduction in river flow Q, due to the construction of upstream reservoirs, is likely to lead to a more mixed estuary with decreased estuarine circulation and an increased residence time in the upper estuary. This can often lead to hypersalinity in the upper reaches, and these effects have been described in the Kromme River estuary by Wooldridge (Chapter 7, this volume).

The balance between the potential energy of stratification and the kinetic energy of mixing is not steady, and there are continual changes in the supply of buoyancy and in the availability of mixing energy. The most important changes occur on seasonal and tidal time scales. On a seasonal time scale, rivers flow strongly in the wet season and may run dry in the dry season. There are also large changes in the temperature of the river and seawater. In particular, on the west coast of South Africa the summers are characterised by coastal upwelling and a minimum in sea temperatures at the same time as the river water attains a

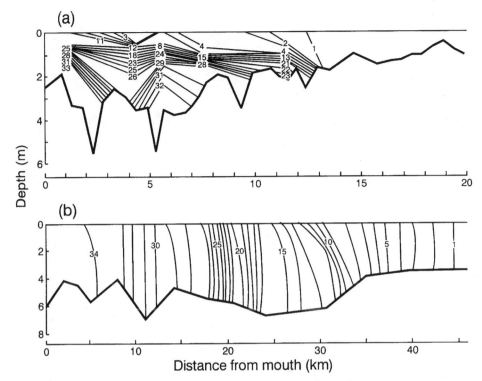

Figure 3.9. (a) A salinity section along the Gamtoos River estuary at ebb tide with an inflow of freshwater showing highly stratified conditions. (From Schumann & Pearce 1997. Reprinted by permission of the Estuarine Research Federation. © Estuarine Research Federation.) (b) A salinity section along the well mixed Great Berg River estuary at flood tide (from Slinger & Taljaard 1994).

maximum (Largier *et al.* 1997a). In winter, the temperature difference is reversed and much smaller; these seasonal effects are discussed further in the next sections.

Regarding mixing energy, it has been shown that wind forcing for South African estuaries also varies seasonally. However, in most estuaries it is the tidal energy that controls mixing. There are temporal variations in tidal current on diurnal and semi-diurnal time scales as well as on fortnightly spring–neap time scales. Tidally-modulated mixing may result in estuaries that are stratified on neap tides when there is less tidally-induced mixing, but mixed on spring tides when there is greater mixing (Nunes and Lennon 1987). As an example, stratification in the Sundays River estuary is shown in Figure 3.10, and similar conditions have been observed under constant river inflows in the upper Knysna Estuary and the Gamtoos Estuary (Schumann & Pearce 1997).

In many estuaries (or parts of estuaries) classed as vertically mixed, a brief period of stratification can be observed during slack tide as the longitudinal density gradient (the baroclinic pressure gradient) relaxes and the denser and less dense water spread under and over each other. This transient stratification and pulse of estuarine circulation occurs in all but the most well-mixed estuaries and it can be quite effective in bringing about longitudinal disper-sion in estuaries. By the same token, it is only the most highly stratified estuaries that do not experience limited mixing across the pycnocline at some stage of the tidal cycle, particularly towards the mouth.

Estuarine fronts

The above description of density structures in estuaries suggests a smooth variation in water type over horizontal distances in an estuary. This is often true, but is not always the case, and spatial changes in topography (e.g. depth or width) and temporal changes in inflow (e.g. tide or river) can result in regions where the water characteristics change abruptly over short distances. These high-gradient regions are known as **fronts** and they are usually accompanied by a density difference across the front. An intriguing variety of estuarine fronts can be collected together into three primary groups: intrusion fronts, shear fronts and mixing fronts (O'Donnell 1993).

Intrusion fronts occur at the surface where less dense water spreads over more dense water and at the bottom where denser water intrudes under less dense water. The maximum speed of propagation of the intruding water and front relative to the receiving water is set by the internal gravity wave speed $c = (g' h_i)^{1/2}$, where now h_i is the depth of the intruding layer and $g' = g \, \Delta\rho/\rho$ is the so-called

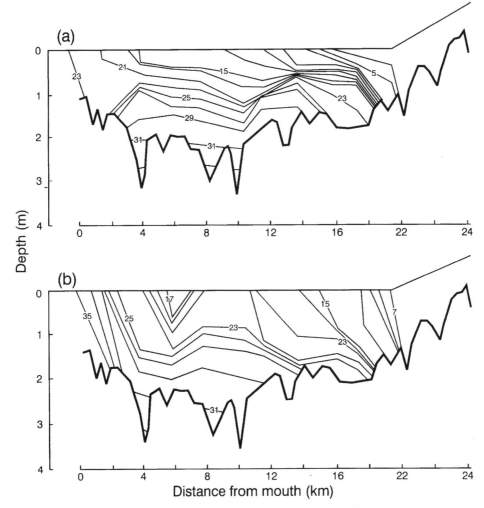

Figure 3.10. **Isohaline contours at neap tide in the partially mixed Sundays River estuary, showing the stratification at (a) low tide and** **(b) high tide, with distance from the mouth (from MacKay & Schumann 1990).**

reduced gravity (cf. equation 1). This implies that an intrusion of river water will back up until h_1 is thick enough to flow at a speed and with a cross-sectional area sufficient to remove the intruding water at the same rate as it is introduced. The front of this thickened intruding layer is an intrusion front, observed at the seaward edge of a freshwater outflow as it propagates seaward, e.g. the Orange River front in the 1988 floods (Shillington *et al.* 1990). Such intrusion fronts discharging into the sea are not very common off the South African coast as few rivers have large enough volume outflows, and because open coastal waters are characterised by a high level of wave energy that acts to break down the stratification in the plume. Transient intrusion fronts can be observed following flood flows in rivers (e.g. off the Tugela River).

Tidal intrusion fronts occur where inflowing seawater plunges beneath less dense estuarine water (Simpson &

Nunes 1981), and have been observed in South African estuaries (Largier 1992). Associated with these tidal intrusion fronts are fronts at the landward (forward) edge of bottom density currents. Both tidal intrusion fronts and bottom density current fronts have been directly observed in the Palmiet estuary (Largier & Taljaard 1991; Largier *et al.* 1992) and their presence has been visually confirmed or anticipated in other South African estuaries (Largier & Slinger 1991). Tidal intrusion fronts are visually obvious owing to a foam line on the surface where the intruding seawater plunges beneath the less dense water in the estuary, as shown in Figure 3.11. The narrow bar-built mouths of South African estuaries results in flood tide velocities that often exceed the internal gravity wave speed in the mouth, precluding the outflow of low density estuarine water during flood tide. As the inflowing water diverges and slows down in the vicinity of the flood tide

Figure 3.11. **Photograph of the characteristic V-shaped foam line associated with the tidal intrusion front in the Palmiet estuary (from Largier & Taljaard 1991).**

delta, its velocity decreases to a critical point where it plunges beneath the estuarine water. This happens in the deeper channels.

Shear fronts in estuaries are probably best represented by the axial convergence that has been observed in a number of South African estuaries, but only documented in the Sundays River estuary (MacKay & Schumann 1990). These fronts occur in partially or well mixed estuaries, and are a result of the denser seawater on the flood tide intruding preferentially in the centre of the channel, and thus causing a lateral density gradient. This forces the denser mid-channel water to move out toward the sides of the estuary, while the less dense water that was on the channel sides moves at the surface towards the centre of the channel, over the intruding dense water (Simpson & Turrell 1986). A pair of lateral circulation cells are set up with a convergence at the surface down the middle of the estuary channel, as shown in Figure 3.12.

Mixing fronts have not been well documented in South African estuaries, but they undoubtedly do occur in many cases. In deeper water, the bottom stress due to tidal currents is weaker than in shallow water. Furthermore, in deeper water bottom-generated turbulence is further

removed from a mid-depth or near-surface pycnocline. The result is that the deeper parts of an estuary can be stratified while the shallower parts are mixed. This spatial variation in stratification can be explained by the Estuarine Richardson Number (equation 6), since for a given river inflow Q and river–ocean density difference $\Delta\rho$, a transition can be expected from mixed to stratified conditions at particular values of $W u_t^3$. Values of u_t can be derived from tidal models, and by contouring $W u_t^3$, likely locations for tidal mixing fronts can be identified. At these boundaries between two-layered regions and single mixed layers there are sudden changes in water type and density at both the surface and the bottom. As in all types of estuarine fronts, the large-scale structure of these mixing fronts is modified by small-scale frontogenesis. The steep horizontal density gradients have a natural tendency to sharpen up to form horizontal fronts.

As South African estuaries are small and shallow, fronts cannot achieve geostrophic equilibrium, i.e. they are not in an inertial state and they will not persist without continual forcing. Fronts in South African estuaries are transient, and are associated with tidal flows or pulses of river flow and only last as long as the forcing flow

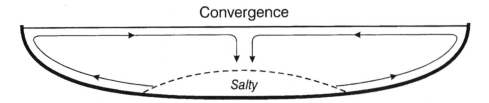

Figure 3.12. **The cross-section of an estuary is depicted at the top showing schematic two-celled transverse current structures caused by the intrusion of more saline water along the centre of the channel, producing convergent surface flow at the centre. The photograph,** **taken in November 1987, shows such a convergence zone indicated by the line of foam in the Sundays River estuary; the inflow comes from the sea on the left (from MacKay & Schumann 1990, with the photograph taken by W.K. Illenberger).**

lasts: several hours for a tidal front and perhaps a day or two in the case of a river flood. These transient fronts are often very important for the transport and ultimate fate of water-borne material such as planktonic larvae, suspended sediment or pollutant loadings, but it is not accurate to think of these fronts as barriers between different water types (Largier 1993).

Stratification in closed estuary basins

The foregoing discussion relates to estuaries that have an open connection to the sea. Of particular interest in this section is the type of estuary that experiences temporary closure, as well as coastal lakes.

Some estuaries are open to the sea seasonally, closing later in the dry season. In these estuaries, there is a potential for persistent stratification and extremely long resi-

dence times in the bottom waters. For instance, at times the salinity and temperature structures in closed KwaZulu-Natal systems during the dry season indicate that the slow inflow of freshwater acts to intensify stratification (Begg 1978). A similar situation resulted in bottom residence times of three months being recorded in the tiny and shallow Palmiet estuary in the Western Cape (Slinger & Largier 1991). At times even anoxic conditions have been recorded in this system.

There are indications of similar conditions in many other seasonally closed South African estuaries. Recent experiments in the management of water quality in the Great Brak estuary (Slinger et al. 1994) have resulted in reservoir release procedures that are partially effective in flushing old anoxic water out of the deeper pools of the middle and upper estuary. Complete renewal of the water in these reaches is dependent on the flushing by seawater which occurs once the mouth is open.

In estuarine basins that are only weakly connected to the sea, any saline water that intrudes into them may remain resident at depth indefinitely. As an example, saline water that has intruded into the 20 m deep Swartvlei has formed a permanent saline layer and persistent anoxia occurs at the base of the water column (Allanson & Howard-Williams 1984). In the absence of strong tidal currents, vertical mixing in such basins is due only to wind stress on the surface, and concepts such as the Wedderburn Number (equation 7) are useful in estimating the likelihood that wind can bring about upwelling and turnover; in Swartvlei, it is only the strongest winds that may be effective in producing mixing across the deep halocline (Allanson & Howard-Williams, 1984).

Another example is Lake Sibaya, one of several lakes situated on the coastal plain of northern KwaZulu-Natal with a tenuous connection to the sea, but with evidence of a cyclical input of saline water to its deeper reaches. Allanson & van Wyk (1969) found that a weak pattern of temperature stratification existed in summer because periods of calm were rare, and the lake was swept by southerly and northerly winds which caused substantial mixing. In the cool season, the lake tended to isothermal conditions.

Seasonal variations, episodic events and longer-term changes

Seasonal variations which have an impact on the physical character of estuaries are associated with rainfall and the subsequent runoff from catchment areas, and insolation and influent water temperatures. Seasonal winds also play an important role in mixing processes, coastal upwelling, and changes in ocean wave characteristics (height, period and direction), while the seasonal variation in the frequency of storm events can be very important in ocean–estuary exchanges.

However, substantial estuarine impacts also result from non-regular, episodic events. Such events may take place over a very short period, for instance those due to an exceptionally severe storm, but their effect may last for a long time. Other changes take place on inter-annual and longer time scales, though it is often difficult to gauge these effects, particularly if suitable long time series of data are not available.

The effect of all these influences varies according to the location of the estuary concerned on the South African coast, and here the biogeographic regional classification helps in assessing their importance. Moreover, the impact

on an estuary depends on its intrinsic character, e.g. whether it is permanently open, temporarily open, a river mouth, an estuarine lake or a marine bay.

Rainfall, freshwater inflow and salinity structures

The principal effect of seasonality in rainfall and runoff is to alter the condition of an estuary mouth, as well as the degree of stratification. Different types of estuaries respond in various ways to the seasonal changes in freshwater input, with marine bays exhibiting the seasonal changes of the ocean, and river mouths those of the parent rivers. The responses of permanently open estuaries, temporarily open estuaries and estuarine lakes will differ depending on their geographical location, size and configuration.

Permanently open estuaries

Extreme seasonality is exhibited by the large, permanently open estuaries of the cool temperate biogeographic zone. For instance, both the Olifants and Great Berg estuaries exhibit tidal variations at distances of 32 km and 65 km, respectively, from their mouths in summer when the influence of freshwater flow is reduced; in fact the Olifants would be tidal further upstream were it not for a causeway constructed at that point. Salinities of more than 5 have been measured further than 37 km upstream in the Great Berg in late summer (Slinger & Taljaard 1994). Under normal winter conditions these saline waters are pushed downstream by the influx of freshwater, and can be expelled from the estuary by the sustained high flows. Both the Great Berg and Olifants then run fresh to the sea on the ebb tides; Figure 3.13 demonstrates these variations for the Great Berg estuary.

Large systems of the southern portion of the warm temperate zone also exhibit strong seasonality. In the Breede River estuary salinities of 5 occur 17 km upstream of the mouth in summer, whereas the estuary can run fresh to the mouth in winter (Flemming & Martin 1994).

In contrast, permanently open estuaries located in the year-round rainfall region (south Eastern Cape coast) and further north in the summer rainfall areas of the east coast, exhibit variations in tidal influences and salinity distributions with rainfall events, or dry and wet periods, rather than with seasonal changes. This means that heavy rainfall, extreme tides or sustained dry periods will often exercise a greater influence than the corresponding seasonal variations.

Temporarily open estuaries

The condition of the mouth of a temporarily open estuary is governed by the competition between the closing forces of wave- and flood-tide-driven sediment suspension and deposition in the mouth on the one side, and the scouring forces of ebb-tidal currents on the other. The latter are

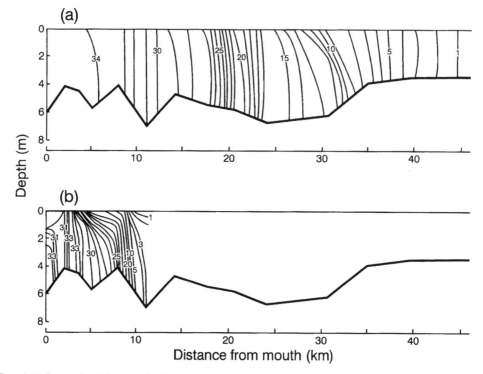

Figure 3.13. **Seasonal variation in the longitudinal salinity sections at flood tide in the Great Berg River estuary in (a) February 1990** (summer) and (b) September 1989 (winter) (from Slinger & Taljaard 1994).

enhanced with higher river flows, while the former appear to increase under higher wave conditions.

Both river flows and wave conditions are markedly seasonal in the warm temperate zone of the South African coastline, and in summer temporarily open estuaries are usually open at times of increased river flow and low wave conditions (or when the intensity and frequency of storms is lower). Higher wave conditions and a lower average rainfall and runoff occur in winter, and the mouths tend to be closed during this time, or at least to be closed more frequently. This leads to the situation where seasonality is exhibited primarily in the depth and state of the mouth and consequently is reflected in the water levels, salinities and water column structure.

The numerous small estuaries of the KwaZulu-Natal coast generally also fall into this category. Their mouths tend to close in winter, and then burst open with the inflow of river water in early summer, the wet season. In many instances the runoff has been reduced by human activities, leading to longer closure periods (Begg 1978).

Estuarine lakes

In contrast to the permanently and temporarily open estuaries, the seasonal behaviour of estuarine lakes is determined primarily by their configuration. If they are deep they are likely to exhibit persistent stratification. A wide,

shallow system, however, could react to other seasonal changes such as winds, as well as altered depth of the mouth associated with enhanced or reduced marine influence (tidal mixing). Their behaviour is thus largely site-specific; South African examples are the Bot River lagoon and Swartvlei (Heydorn & Tinley 1980).

Oceanic inputs to estuaries

The seasonal variations in winds and insolation cause marked changes in coastal oceanic conditions, which can then also have a marked impact on adjacent estuaries. In particular, the south Western Cape is recognised as a major upwelling region (Shannon 1985), with the process generated by the strong southeasterly winds in summer. These winds cause an offshore Ekman transport in the surface waters, allowing deeper waters to be brought to the surface close to the coast. High concentrations of nutrients within these deeper waters can then enter the estuaries; this is discussed in Chapter 4.

The deeper waters are also colder, and it means that coastal sea temperatures on the South African west coast can actually be lower in summer than in winter. Consequently the temperature difference between river and seawater is greater in summer than in winter, and this is enhanced further by the increased freshwater flow from the rivers. The result is that stratification in west coast

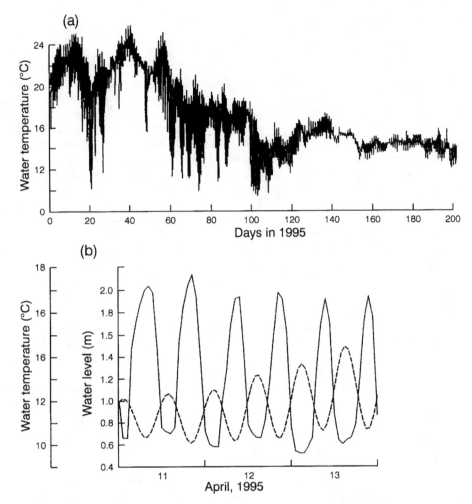

Figure 3.14. **(a) Water temperature just inside the mouth of the Knysna River estuary over the first 200 days of 1995, and (b) a short** section in April showing variations in water temperature (solid line) and water level (broken line).

estuaries such as the Great Berg and Olifants systems is more pronounced in winter than in summer (Slinger & Taljaard 1994).

Intense thermoclines are found over the Agulhas Bank in summer, and easterly winds also generate upwelling at selected sites (Schumann *et al.* 1988). As an example of the effects of the resulting cold water, Figure 3.14*a* shows temperature time series measured near the mouth of the Knysna estuary. In summer the impact of the cold water can be dramatic, with temperatures dropping by more than 14 °C over short periods. In winter there is a much smaller temperature difference between inflowing oceanic water and resident estuary water. This is because upwelling in the ocean is not as frequent, and since there is no marked thermocline the bottom waters are actually warmer than in summer. Moreover, resident estuarine waters are also colder and therefore closer to the temperature of the oceanic water.

Figure 3.14*b* shows details of the temperature fluctuations with the tide over a short period. Thus the colder water brought in by the flood tide caused abrupt changes of more than 10 °C, while on the ebb tide this cold oceanic water was again expelled, to be replaced by warm water found in the estuary in summer. Similar temperature variability to that shown in Figure 3.14 has been measured in the Kowie estuary on the southeast coast, with the added complication that upwelling also occurs as a result of dynamic processes in the bottom boundary layer of the Agulhas Current (Schumann 1987). Schumann & de Meillon (1993) were able to follow the movement of such an input of cold seawater into a marina located on the Kromme River estuary.

Mouth closure

The configuration of the mouth of an estuary is determined by the scouring/erosive action of outflowing water (both riverine and tidal), and the action of the flood-tidal currents and waves in transporting and depositing marine sand in the inlet region. The resulting sill and flood-tidal deltas then form or are eroded, depending on the relative magnitudes of these competing processes.

An extreme consequence of the accretion of sediment in the mouth region is eventual closure, and this has been discussed in Chapter 2. The actual dynamic processes involved in closure are complex, are not yet fully understood, and will not be discussed in detail. Recent investigations of the closure of selected South African estuaries have indicated that the wave-driven on–off-shore movement of sediment may be a primary mechanism causing mouth closure, rather than the longshore movement of sediment (Huizinga 1994). Confirmation of these results is being sought for more general conditions, though it is likely that the details will vary from estuary to estuary.

As examples of the differences that do occur in the balance between erosion and accretion, the mouth of the Great Brak River River can remain open under river flows of the order of 0.5 m^3 s^{-1}, while the mouth of the Mgeni River estuary has been known to close when river flows are of the order of 7 m^3 s^{-1} (Huizinga 1994). Factors that play a part in this balance are the wave climate, sediment sizes, beach slopes and morphological beach compartments, i.e. the presence of rocky headlands or outcrops.

In most South African estuaries, net erosion only occurs during greater than normal river flow, i.e. during flood conditions. Because of the seasonality of the rainfall many estuaries are not open to the sea on a permanent basis, and mouth closure is common in the dry season; this is particularly the case on the coast of KwaZulu-Natal because of the limited flow of the rivers in winter. On the other hand, larger systems may undergo mouth closure over longer periods in drought conditions (Begg 1978). The increased frequency of high waves in winter may also contribute to estuarine mouth closure along the south and east coasts. In the winter rainfall region this effect is less marked because of increased scouring from the greater freshwater inflow, and it is the change in wave direction combined with reduced runoff which can cause mouth closure in late summer.

Episodic events and interannual variability

Episodic events are those that occur in a non-periodic manner, and generally refer to extreme conditions. For estuaries such episodic events refer particularly to unusually heavy rainfall leading to flooding, and exceptionally strong winds and/or very high wave conditions which can cause coastal surges in sea level. Events which take a longer time to develop, but nonetheless can have severe impacts on estuaries are also discussed here; in particular, drought and all its consequences come under the heading of inter-annual variability.

Floods

The importance of floods for the long-term functioning of many estuaries has been discussed in Chapter 2. Essentially, the flood waters serve to increase the erosive effect of the ebb tide, and can therefore scour out flood-tidal deltas established over many previous tidal cycles. They serve to re-set the mouth condition in terms of depth of the sill as well as the position of the bar-built estuary mouth. In many cases floods also bring enough freshwater to flush seawater out of an estuary completely, thereby temporarily changing its character into that of a river mouth. Reddering and Esterhuysen (1987) report on a flood in the Nahoon River estuary with a peak discharge of 1400 m^3 s^{-1}, which served to scour out the flood-tidal delta with freshwater flowing to the sea; however, within a week a seawater wedge 1–4 m thick had already penetrated 4 km upstream from the mouth. This can be contrasted with the seasonal variations in the Great Berg estuary shown in Figure 3.13. With such a flood, stratification occurs seaward of the mouth and sediment brought down by the floodwaters is deposited offshore. Such offshore discharge plumes may be large enough to be affected appreciably by the Earth's rotation and be deflected to the left, forming a coastal buoyancy current; such effects were noted during the 1988 Orange River floods (Shillington et al. 1990).

Coastal surges

Coastal surges result from weather-related events such as CTWs, strong winds and waves, and can cause dramatic rises in sea level, and consequent inundation of estuaries by marine waters. One such case was recorded by Zhang et al. (1995), when conditions on the Cape south coast combined to produce an estimated sea level increase of about 1.15 m above the expected spring high tide. This served to cause massive flooding of the Gamtoos River estuary by inflowing marine waters, in places resulting in substantial erosion. Direct measurements of conditions at the Gamtoos are not available, but offshore waves were over 7 m high and winds measured at Port Elizabeth attained speeds of over 25 m s^{-1} (gusts of over 36 m s^{-1}) at the time of the surge.

Another such incident was recorded in the closed Great Brak estuary in May 1992 (Taljaard & Slinger 1993). Exceptionally high wave conditions resulted in overtopping of the sand bar at the mouth of the estuary, and the consequent influx of seawater caused the water levels in the estuary to rise by more than 0.6 m. Water column surveys indicated that renewal, or partial renewal, of the bottom water of the estuary had occurred more effectively than under previous freshwater releases from the Wolwedans Dam in the lower catchment.

Droughts

The drought scenario is one where there is a gradual reduction in freshwater inflow into an estuary, whether due to natural or anthropogenic causes. In estuarine terms, the concept of drought is a low-inflow condition in which the estuary is net evaporative, i.e. there is a net loss of freshwater, dependent on both local rainfall and that over the whole catchment. The consequences depend on the characteristics of the estuary, but in systems where the mouth remains open it can be expected that there will be a gradual movement upstream of the main stratification zone in the estuary (Schumann & Pearce 1997). A continued decrease in freshwater inflow will eventually result in a well mixed saltwater estuary with a suppressed or absent estuarine circulation.

A further consequence is that the residence times in the estuary will increase, since no flushing will take place from freshwater throughflow, and tidal flushing will only penetrate a limited distance upstream. Following equation (6), vertical mixing can be expected with an absence of a density-driven estuarine circulation. Longitudinal dispersion will be due only to tidal motions, which will be weak in the inner estuary, and zero at the head of the estuary. Although the Langebaan Lagoon is not an estuary, conditions approximate an estuary with no freshwater inflow, and residence times of over a month have been observed in the inner basin (Largier *et al.* 1997*b*).

Hypersaline conditions will occur further upstream if the net loss of freshwater through evaporation persists, resulting in inverse longitudinal density gradients in an **inner hypersaline estuary** (Largier *et al.* 1997*a*). While limited evidence exists for inverse density-driven circulation, the movement of higher salinity water seawards along the bottom of such an estuary is important in preventing excessive salinities in shorter estuary basins that are in free connection with the ocean. Salinity values of over 45 are unlikely, unless water is trapped by sandbars or vegetation.

In the case of the St Lucia Estuary the axis of the estuary is very long with limited exchanges possible with the sea, and consequently salinities well over 100 have been recorded during severe droughts (Begg 1978). The natural salinity variability of the system is shown in Figure 3.15, and dramatically portrays the substantial interannual changes that can occur, particularly in the upper reaches.

In a number of South African estuaries, low-inflow periods result in complete closure of the mouth, thereby disconnecting the estuary from the sea. Extreme salinities can then occur, though it is important to consider each system individually, since even under drought conditions some small systems freshen when closed.

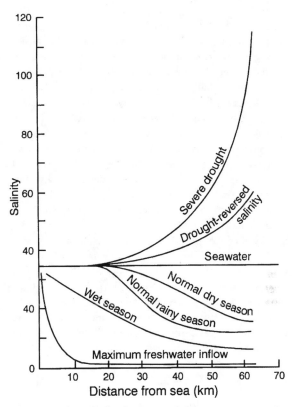

Figure 3.15. **Interannual salinity fluctuations in the St Lucia Estuary (from Begg 1978).**

Sea level rise

It is now generally accepted that excessive inputs of anthropogenically generated gases into the atmosphere are resulting in a gradual warming of the Earth's atmosphere (Leggett 1990). Such warming is causing a rise in sea level due to expansion of the ocean's waters, though additional factors such as postglacial rebound and possible melting of the polar ice sheets are also involved.

The present rate of mean sea level rise based on a number of long tide gauge records is about 1.8 ± 0.3 mm per year (Douglas 1991). Hughes *et al.* (1991*a*) have analysed sea level rise in the South African context, and though long-term sea level records are limited, three sites (Port Nolloth, Luderitz and Simon's Town) show increases of up to 1.23 mm per year. Considering that models of global isostasy indicate that, as a result of long-term crustal adjustments, sea level around South Africa should be falling (Hughes *et al.* 1991*a*), these results tend to confirm that South African sea level trends are in good agreement with other southern hemisphere and global estimates.

Future sea level scenarios indicate that rises of between 0.8 m and 1.5 m can be expected in the next 100 years

(Hughes *et al.* 1991a). Such a rise will result in an effective erosion seaward of coastal land, and will change the mouth morphology of a large number of present-day estuaries, and large areas of low-lying floodplain will be inundated. To determine the impacts, each estuary needs to be assessed individually; as an example, Hughes and Brundrit (1991) investigated the vulnerability of the False Bay coastline, including some estuaries, to a projected rise in sea level.

However, it is probably not the direct rise in sea level which will have the greatest influence on low-lying coastal regions, in particular areas such as estuarine flood plains. An increase in the background sea level causes the return period of storm events to decrease, with the result that a particular extreme storm will have a much greater frequency of occurrence (Hughes *et al.* 1991b). Thus a rise in sea level of 20 cm would mean that the impact of a 1-in-10 year storm event would be equivalent to the impact of what was previously a 1-in-200 year storm. At this stage the consequences can be deemed to be speculative, but it is nonetheless clear that, when considering scenarios longer than 20 or 30 years, the rise in sea level deserves serious attention.

Estuarine modelling and anthropogenic effects

Modelling provides the means of formally describing an estuarine system. However, because of the complexities involved, it is not possible to have a model encompassing all the processes, and various models have been developed dealing with specific aspects. Even so, approximations and assumptions have to be made, and the outputs of such models have to be carefully evaluated and calibrated for each estuarine system; such calibration involves the input of basic variables measured on site.

Once a model has been formulated and calibrated, and is deemed to be giving reliable results, it can be used in a predictive capacity. Frequently a number of models may be coupled or linked through the exchange of appropriately 'averaged' data to provide more holistic predictions, also on longer time scales. Nonetheless, it is the modelling of the physical environment which provides the basis for all such linkages.

Modelling approaches
Without going into details, modelling the physical processes in an estuary involves the solution of a series of mathematical equations depicting various known dynamic relationships under different boundary conditions. As indicated above, specific and limited objectives have to be set, though even here parameterisation of

complex processes is often necessary. The resulting equations are often too complex to solve analytically, and therefore numerical techniques have to be used. This generally involves setting up a suitable grid of points in space and time, and approximating the equations in terms of the grid structures.

One-, two- and three-dimensional models
One-dimensional models commonly consider variations along the length of an estuary, but may focus on variations with depth at one position. Two-dimensional models generally address the longitudinal and lateral changes important in wider estuaries, or the longitudinal and vertical changes important in deep estuaries, whilst three-dimensional models include longitudinal, lateral and vertical variations. The degree of complexity increases markedly with two- and three-dimensional models, and though these are available, one-dimensional models are well suited to the characteristically fairly small and narrow South African systems (e.g. the Great Berg River estuary) compared to the broad and deep estuaries of Europe, North America and Asia. Nonetheless, the limitations of these results must be recognised.

The first South African estuarine modelling study was undertaken by Hutchison (1976), and Hutchison and Midgeley (1978) for Lake St Lucia. This study investigated the viability or otherwise of supplementing inflows from the estuary to the lake for the alleviation of hypersaline conditions by constructing a channel from the nearby Mfolozi estuary. The next study was undertaken on the Knysna estuary, where the influence of embankments on the circulation and salinities was ascertained (Huizinga and Haw 1986). Subsequently, many studies involving marina circulation problems, prediction of salinity distributions and/or water levels under different inflow conditions have been successfully undertaken (CSIR 1987, 1990a, b, 1993).

One such one-dimensional estuarine modelling system is the *Mike-11* hydrodynamic, transport-dispersion and water quality model (DHI 1992). The St Venant equations (Leendertse 1967) and biogeochemical process equations are solved and time series outputs of water volume, water levels, volume fluxes, salinities, temperatures, dissolved oxygen (DO) and certain dissolved nutrient concentrations can be produced. Less confidence is placed in the simulations of DO and dissolved nutrients at present, owing to only limited application testing having been undertaken (Slinger 1996a). Recent implementation sites for these modelling tools include the Great Berg, Kromme, Keurbooms, Swartkops, and the Great Fish and Sundays River estuaries.

Two- and three-dimensional models are being applied to coastal waters around South Africa, but to date no attempt has been made to model the complicated stratified hydrodynamics of estuaries in South Africa; such an exercise would require extensive and detailed measurements for validation.

The Estuarine Systems Model

A limitation of standard multi-dimensional hydrodynamic models is that they cannot easily accommodate rapid alterations in mouth conditions. Although the effects of different mouth configurations on the tidal flux through the mouth can be investigated, sharp alterations in mouth state cannot be accommodated during a simulation.

Realising the limitations of the hydrodynamic models, and appreciating the importance of medium to long-term predictions required for management decisions such as freshwater needs for maintenance of estuarine character and functioning, led Slinger (1996b) to formulate the estuarine systems model. This is a semi-empirical model which treats the estuary as a basin connected to the sea by a channel of variable height. Time histories of water levels, tidal fluxes, mean salinities, stratification, circulation state, freshwater and tidal flushing and the sill height at the mouth are simulated. The provision of concurrent information on mouth condition and the associated effects on salinities and water levels is a significant improvement in prediction for South African estuaries, as is the capability to simulate the process of mouth closure and breaching.

The model has been implemented for the Great Brak River estuary, and the role of episodic wave events in causing closure of the mouth has been highlighted. The model demonstrates that reduced freshwater flows lead to an increase in sensitivity of the system to marine forcing, and sounds a warning note for the future of many freshwater-deprived systems on the South African coast. A further application was undertaken on the Kromme River estuary, in order to determine the optimum timing for the release of a flood of 2×10^6 m^3 from the Mpofu dam.

Anthropogenic effects

The impact of human activities on the normal functioning of estuaries can be considerable, and there are many cases where the characteristics of such estuaries have been changed completely by developments throughout the river catchment and the estuary itself. Discussion here will be limited to those aspects which affect the hydrology only, since aspects influencing the biota and water quality are treated in other chapters; effects on water circulation are due to changes in the shape of the estuary basin and/or changes in inflow. Particular consideration is given to the impact of reducing freshwater inflow because of the abstraction of water

in the catchments, and the consequences of the building of roads and bridges.

Reduction in freshwater inflow

Reduction in freshwater inflow due, for example, to abstraction of water in upstream catchments, may result in conditions similar to that already described for droughts. However, the difference now is that a *permanent* change in the characteristics of the estuary will occur. In particular, if tidal exchanges continue, it is likely that many estuarine features will be lost, and the estuary could become an arm of the sea undiluted by freshwater inflow.

An example of such a situation is that of the Kromme River estuary, where semi-permanent hypersaline conditions occur in the upper reaches, and the marine macrophyte *Zostera* is now prevalent over much of the tidal reach (Adams & Bate 1994). The problem has been recognised, and ameliorating measures are being sought in the calculated release of freshwater from storage dams to mimic a flood and flush out the long-resident salt water; as explained already, modelling has been used to assess the impact of such a release from the Mpofu dam.

If tidal exchange does not continue and the mouth of the estuary closes more frequently and for longer periods, the hydrodynamic character of the estuary may alter progressively to that of a coastal lagoon. This is considered to be a major problem facing many smaller South African estuaries.

Agricultural inputs

The upper reaches of estuarine floodplains, where salt intrusions are minimal, are recognised as rich agricultural lands. As such, the natural vegetation has generally been cleared, and where freshwater is available intensive cultivation takes place with the use of fertilisers and pesticides.

In such cases dissolved nutrient and pesticide loads can enter the groundwater and penetrate into estuaries as diffuse sources. Recent investigations in the Gamtoos estuary have indicated that such inputs can be appreciable, though the natural flushing mechanisms in that case were capable of removing any unacceptable buildups of the pollutants (Pearce & Schumann 1997). However, a greater reduction in freshwater inflow, or complete closure of the mouth, will also mean that the flushing of the estuary will decrease, possibly with unacceptable consequences such as eutrophication (see Chapter 4).

Roads and bridges

Problems arise here when the construction of roads or bridges results in restrictions to flows within an estuary. In some cases it is the normal tidal flows which are affected, resulting in reduced exchange between different areas of an estuary or even substantial reductions in ocean–estuary exchanges. Another instance in which constructions can have a major impact is during flood flows when back-flooding can occur, resulting in bridge failure or excessive scour around bridge piers.

However, the effects of floods on bridges and roads can be predicted reasonably well through mathematical and/or physical modelling techniques, allowing improved designs to minimise any such potentially catastrophic consequences. The problems in South African estuaries lie with the relics of the past, where bridges and roads were often located in the mouth regions of estuaries.

A final comment

This chapter has, as the name implies, concentrated on those physical aspects of estuaries involved in water exchanges through the tidal inlets, freshwater inflows, and the consequent interactions of these different water bodies. The descriptions have perforce been brief, but it is hoped that the complexity of estuarine structures has been revealed and, moreover, that the variability on a wide range of time and space scales has been elucidated.

There have been substantial advances in the understanding of estuarine hydrodynamics over the last decade or two, even though there have been few extensive physical investigations of South African estuaries. An appreciation has been built up of the unity of the land, estuary and sea system, and that any management plans need to take cognisance of all the interactions. The models that have been developed have been aimed at specific processes or problems, but have nonetheless sought to improve the overall understanding of estuaries. Still, it is patently clear that each estuary needs to be considered on its own merits, and that generalisations must

be applied with care. It is also apparent that physical investigations have concentrated on a few estuaries, and that there are a large number, particularly on the east coast, where very little work has been done and the overall understanding of estuarine processes is limited. Developments in instrumentation and data processing have enabled new insights into estuarine processes, and indeed substantially new results have been forthcoming. These have been particularly concerned with frontal phenomena, though it is important to note that the smaller scale processes, and their influences on the larger scale effects, still need elucidation. Relevant results from investigations in other countries and regions can be applied, though it is important to realise the uniqueness of regions throughout the world.

Of particular concern are the irreversible alterations that are occurring to South African estuaries because of ill-informed management and developments in the catchment regions of parent rivers. As has been indicated, these developments are leading to limitations in freshwater input, changes in the stratification patterns and constrictions of the mouth in terms of both position and ocean–estuary exchanges. Such physical changes will have profound effects on the biota, and indeed on whether such systems can continue to function as estuaries; terrestrial–marine exchanges will also be constrained.

A thorough understanding of these physical processes underpins prediction and quantification of the associated biological effects. The pursuit of such an integrated understanding, particularly the incorporation of the different time scales of variability of the estuarine circulation and water column structure, and the effects on biota, will lead to an enhanced ability to determine the longer term effects of catchment developments on individual estuarine systems. However, the challenge of determining the effects of anthropogenic influences on the estuarine resources as a whole remains.

References

Adams, J.B. & Bate, G.C. (1994). *The freshwater requirements of estuarine plants incorporating the development of an estuarine decision support system*. WRC Report, Project K5/292/0/1. Port Elizabeth: Water Research Commission.

Allanson, B.R. & Howard-Williams, C. (1984). A contribution to the chemical limnology of Swartvlei. *Archiv für Hydrobiologie*, **99**, 133–59.

Allanson, B.R, & van Wyk, J.D. (1969). An introduction to the physics and

chemistry of some lakes in northern Zululand. *Transactions of the Royal Society of South Africa*, **38**, 217–40.

Aubrey, D.G. & Friedrichs, C.T. (1988). Seasonal climatology of tidal non-linearities in a shallow estuary. In *Hydrodynamics and sediment dynamics of tidal inlets*, ed. D.G. Aubrey & L. Weishar, pp. 103–24. Springer-Verlag Lecture Notes on Coastal and Estuarine Studies. New York: Springer-Verlag.

Begg, G. (1978). *The estuaries of Natal*. Natal

Town and Regional Planning Report Vol 41. Pietermaritzburg: Natal Town and Regional Planning Commission.

Bickerton, I.B. (1981). Report No. 3: Groen (CW7). In *Estuaries of the Cape*, Part II. *Synopsis of available information on individual systems*, ed. A.E.F. Heydorn & J.R. Grindley. CSIR Research Report 402. Stellenbosch: CSIR.

CSIR (1987). *A mathematical model of the Swartkops River for flood studies calibration and simulation results*. CSIR Report C/SEA 8768. Stellenbosch: CSIR.

CSIR (1990a). *Great Brak River environmental study with reference to a management plan for the Wolwedans dam.* CSIR Report EMA-C 9036. Stellenbosch: CSIR

CSIR (1990b). *One-dimensional hydrodynamic and water quality model for St Lucia estuary.* CSIR Research Report 688. Stellenbosch: CSIR.

CSIR (1993). *Salinity distributions and aspects of flooding in the Berg River estuary obtained using the Mike 11 modelling system: model application and preliminary results.* CSIR Report EMAS-C/SEA 93037. Stellenbosch: CSIR.

DiLorenzo, J.L. (1988). The overtide and filtering response of small inlet/bay systems. In *Hydrodynamics and sediment dynamics of tidal inlets,* ed. D.G. Aubrey & L. Weishar, pp. 24–53. Springer-Verlag Lecture Notes on Coastal and Estuarine Studies. New York: Springer-Verlag.

Danish Hydraulics Institute (1992). Mike 11. A microcomputer based modelling system for rivers and channels. In *Reference Manual* (version 3.01). Holstrom, Denmark: DHI.

Douglas, B.C. (1991). Global sea level rise. *Journal of Geophysical Research,* **96,** 6981–92.

Fischer, H.B., List, E.J., Koh, R.C.Y., Imberger, J. & Brooks, N.H. (1979). *Mixing in inland and coastal waters.* New York: Academic Press.

Flemming, B.W. & Martin, A.K. (1994). Hydrology, sedimentology and Holocene evolution of the microtidal Breede River estuary. In: *ECSA 24 Symposium: Northern and southern estuaries and coastal areas,* pp. 5–6. Aveiro, Portugal.

Heydorn, A.E.F & Tinley, K. (1980). *Estuaries of the Cape. Part 1. Synopsis of the Cape Coast – Natural features, dynamics and utilization.* CSIR Research Report 380. Stellenbosch: CSIR.

Hughes, P. & Brundrit, G.B. (1991). The vulnerability of the False Bay coastline to the projected rise in sea level. *Transactions of the Royal Society of South Africa,* **47,** 519–34.

Hughes, P., Brundrit, G.B. & Shillington, F.A. (1991a). South African sea-level measurements in the global context of sea-level rise. *South African Journal of Science,* **87,** 447–53.

Hughes, P., Searson, S. & MacArthur, C.I. (1991b). Stormy waters ahead: the missing link between storminess and sea level rise. *Conserva,* **6,** 16–22.

Huizinga, P. (1994). *Recent advances in the understanding of estuary mouth dynamics.* Conference of Aquatic Ecosystems: Ecology, Conservation and Management. Port Elizabeth.

Huizinga, P. & Haw, P.M. (1986). A mathematical transport–dispersion model of the Knysna estuary. *The Civil Engineer in South Africa,* **28,** 267–70.

Hume, T.M. & Bell, R.G. (1993). *Methods for determining tidal flows and material fluxes in estuarine cross-sections.* Water Quality Centre Publication No. 22. Hamilton, New Zealand: Water Quality Centre.

Hutchison, I.P.G. (1976). *St Lucia Lake research report* (Vol 5). *Mathematical modelling and evaluation of ameliorative measures.* Johannesburg: Hydrological Research Unit, University of the Witwatersrand.

Hutchison, I.P.G. & Midgley, D.C. (1978). Modelling the water and salt balance in a shallow lake. *Ecological Modelling,* **4,** 211–35.

Imberger, J. & Patterson, J.C. (1990). Physical Limnology. *Advances in Applied Mechanics,* **27,** 303–475.

Largier, J.L (1986). Structure and mixing in the Palmiet estuary, South Africa. *South African Journal of Marine Science,* **4,** 139–52.

Largier, J.L. (1992). Tidal intrusion fronts. *Estuaries,* **15,** 26–39.

Largier, J.L. (1993). Estuarine fronts: how important are they? *Estuaries,* **16,** 1–11.

Largier, J.L., Hearn, C.J. & Chadwick, D.B. (1997a). Density structures in 'low inflow estuaries'. In *Buoyancy effects on coastal and estuarine dynamics,* ed. D.G. Aubrey & C.T. Friederichs. Washington, DC: American Geophysical Union (in press).

Largier, J.L. & Slinger, J.H. (1991). Circulation in highly stratified Southern African estuaries. *Southern African Journal of Aquatic Sciences,* **17,** 103–15.

Largier, J.L., Slinger, J.H. & Taljaard, S. (1992). The stratified hydrodynamics of the Palmiet – a prototypical bar-built estuary. In *Dynamics and exchanges in estuaries and the coastal zone,* ed. Prandle, pp. 135–53. Washington, DC: American Geophysical Union.

Largier, J.L., Smith, S.V. & Hollibaugh, J.T. (1997b). Seasonally hypersaline estuaries in mediterranean-climate regions. *Estuarine, Coastal and Shelf Science* (in press).

Largier, J.L. & Taljaard, S. (1991). The dynamics of tidal intrusion, retention and removal of seawater in a bar-built estuary. *Estuarine, Coastal and Shelf Science,* **33,** 325–38.

Leendertze, J.J. (1967). *Aspects of a computational model for long period water wave propagation.* Memorandum RM-5294-PR, Rand Corporation. Santa Monica, California: Rand Corporation.

Leggett, J, (ed.) (1990). *Global warming: the Greenpeace report.* New York: Oxford University Press.

MacKay, H.M. & Schumann, E.H. (1990). Mixing and circulation in the Sundays river estuary, South Africa. *Estuarine, Coastal and Shelf Science,* **31,** 203–16.

MacKay, H.M. & Schumann, E.H. (1991). Tidal and long-period water level variations in the Sundays river estuary, South Africa. *South African Journal of Science,* **87,** 597–600.

Milliman, J.D. & Meade, R.H. (1983). World-wide delivery of river sediment to the oceans. *Journal of Geology,* **91,** 1–21.

Nunes, R.A. & Lennon, G.W. (1987). Episodic stratification and gravity currents in a marine environment of modulated turbulence. *Journal of Geophysical Research,* **92,** 5465–80.

O'Donnell, J. (1993). Surface fronts in estuaries: a review. *Estuaries,* **16,** 12–39.

Pearce, M.W. & Schumann, E.H. (1997). *The effect of land use on Gamtoos estuary water quality.* Water Research Commission Report No. 503/1/97.

Pond, S. & Pickard, G.L. (1986). *Introductory dynamical oceanography.* Oxford: Pergamon Press.

Preston-Whyte, R.A. & Tyson, P.D. (1988). *The atmosphere and weather of Southern Africa.* Cape Town: Oxford University Press.

Reddering, J.S.V. & Esterhuysen, K. (1987). The effects of river floods on sediment dispersal in small estuaries: a case study from East London. *South African Journal of Geology,* **90,** 458–70.

Rossouw, J. (1984). *Review of existing wave data, wave climate and design waves for South African and South West African (Namibian) coastal waters.* CSIR report T/SEA 8401. Stellenbosch: CSIR.

SAN Hydrographer, (1997). *South African tide tables.* Tokai: The Hydrographer, South African Navy.

Schulze, B.R. (1986). *Climate of South Africa, Part 8. General survey.* Pretoria: Weather Bureau.

Schumann, E.H. (1987). The coastal ocean off the east coast of South Africa.

52 / E.H. Schumann, J.L. Largier & J.H. Slinger

Transactions of the Royal Society of South Africa, **46**, 215–29.

Schumann, E.H. & Brink, K.H. (1990). Coastal-trapped Waves off the coast of South Africa: generation, propagation and current structures. *Journal of Physical Oceanography*, **20**, 1206–18.

Schumann, E.H. & de Meillon, L. (1993). Hydrology of the St Francis Bay marina, South Africa. *Transactions of the Royal Society of South Africa*, **48**, 323–37.

Schumann, E.H. & Martin, J.A. (1991). Climatological aspects of the coastal wind field at Cape Town, Port Elizabeth and Durban. *South African Geographical Journal*, **73**, 48–51.

Schumann, E.H. & Pearce, M.W. (1997). Freshwater inflow and estuarine variability in the Gamtoos estuary, South Africa. *Estuaries*, **20**, 124–33.

Schumann, E.H., Ross, G.J.B. & Goschen, W.S. (1988). Cold water events in Algoa Bay and Cape south coast, South Africa, in March/April, 1987. *South African Journal of Science*, **84**, 579–84.

Shannon, L.V. (1985). The Benguela ecosystem. Part I. Evolution of the Benguela, physical features and processes. *Oceanography and Marine Biology Annual Review*, **23**, 105–82.

Shillington, F.A. (1984). Long period edge waves off southern Africa. *Continental Shelf Research*, **3**, 343–57.

Shillington, F.A., Brundrit, G.B., Lutjeharms, J.R.E., Boyd, A.J., Agenbag, J.J. & Shannon, L.V. (1990). The coastal current circulation during the Orange River flood 1988. *Transactions of the Royal Society of South Africa*, **47**, 307–30.

Simpson, J.H. & Nunes, R.A. (1981). The tidal intrusion front: an estuarine convergence zone. *Estuarine, Coastal and Shelf Science*, **13**, 257–66.

Simpson, J.H. & Turrell, W.R. (1986). Convergent fronts in the circulation of tidal estuaries. In *Estuarine variability*, ed. D.A. Wolfe, pp. 139–52. San Diego: Academic Press.

Slinger, J.H. (1996a). *Final report of the predictive capability sub-project. A coordinated research programme on decision support for the conservation and management of estuaries*. WRC Report S77/3/96. Pretoria: Water Research Commission.

Slinger, J.H. (1996b). *Modelling of the physical dynamics of estuaries for management purposes*. PhD thesis, University of Natal, Pietermaritzburg.

Slinger, J.H. & Largier, J.L. (1991). The evolution of thermohaline structure in a closed estuary. *South African Journal of Aquatic Sciences*, **16**, 60–77.

Slinger, J.H. & Taljaard, S. (1994). Preliminary investigation of seasonality in the Great Brak estuary. *Water SA*, **20**, 279–88.

Slinger, J.H., Taljaard, S. & Largier, J.L. (1994). Changes in estuarine water quality in response to a freshwater flow event. In *Changes in fluxes in estuaries: implications from science to management*, ed. K.R. Dyer & R.J. Orth, pp. 51–6. Fredensberg, Denmark: Olsen and Olsen.

Swart, D.H., Crowley, J.B., Moller, J.P. & de Wet, A. (1990). Nature and behaviour of the flood at the river mouth. *Transactions of the Royal Society of South Africa*, **47**, 217–45.

Taljaard, S. & Slinger, J.H. (1993). *Investigation into the flushing efficiency of 1. A freshwater release, and 2. Seawater overwash in the Great Brak estuary*. CSIR Research Report 713. Stellenbosch: CSIR.

Tyson, P.D. (1986). *Climatic change and variability in Southern Africa*. Cape Town: Oxford University Press.

Whitfield, A.K. (1992). A characterization of South African estuarine systems. *Southern African Journal of Aquatic Sciences*, **18**, 89–103.

Whitfield, A.K. & Wooldridge, T.H. (1994). Changes in freshwater supplies to southern African estuaries: some theoretical and practical considerations. In *Changes in fluxes in estuaries: implications from science to management*, ed. K.R. Dyer & R.J. Orth, pp. 41–50. Fredensborg: Olsen & Olsen.

Wolfe, D.A. & Kjerve, B. (1986). Estuarine variability: an overview. In *Estuarine variability*, ed. D.A. Wolfe, pp. 3–17. San Diego: Academic Press.

Zhang, P., Schumann, E.H. & Shone, R.W. (1995). Note on the effect of wind-induced storm set-up in the Gamtoos estuary, May 1992. *South African Journal of Science*, **91**, 57–67.

4 Chemistry

The primary chemical structure
Brian Allanson

Biogeochemical processes
Deo Winter

The influence of Man
Deo Winter and Brian Allanson

Breede River estuary

Introduction

The chemistry of estuaries depends not only upon tidal pulses and river flow but also upon the hydrodynamic and autochthonous biological processes within them. And while there has been considerable advance in our understanding of these processes, work in South Africa's estuaries has not always kept pace with advances in other coastal regions.

This review of recent South African work attempts to expose the nature and magnitude of the correlations between climate, geomorphology, tidal and fluvial patterns, both natural and as a result of human interference, and biological complexity in the determination of the chemical properties of bar-built estuaries in South Africa. The chapter is divided into three interdependent sections.

1   The primary chemical structure
2   Biogeochemical processes
3   The influence of Man

We have been guided to this handling of an abundance of often quite disparate material by three independent reviewers; their help is warmly acknowledged.

THE PRIMARY CHEMICAL STRUCTURE

In common with other coastal systems, much of the chemistry of estuaries is modulated initially by features of climate and geomorphology. The latter allows South African estuaries to be divided into five easily recognisable types defined initially by Whitfield (1992). Each has a reasonable distinct mixing and salinity regimes, summarised in Table 4.1. Further modifications have been provided in recent years by changes in river flow through Interbasin

Transfer, of which the Great Fish River estuary (Grange & Allanson 1995) is a singular example.

The geographic positions of many of the estuaries used by Whitfield (1992) in his synthesis and referred to in this Chapter are given in Figure 4.1. They have been superimposed upon the Coastal Regions map drawn by De Villiers and Hodgson (Chapter 8, this volume)

An important consequence of the climatological and geomorphological features of the South African coastal zone is that the estuaries exhibit an array of patterns of stratification. From the large axial variations in salinity and tidal variation of the Great Berg River, to the fjord-like stratification of the Palmiet River estuary in the Western Cape (Largier & Taljaard 1991), through the marine-dominated, partially mixed systems in which the region of stratification is usually only a small linear and volumetric part of the estuary, e.g. the Knysna River estuary of the Western Province (Korringa 1956; Haw 1984) and the Kariega estuary in the Eastern Cape Province (Allanson & Read 1995; Grange & Allanson 1995), to those in which, during periods of high and sustained rainfall in the catchment, all traces of stratification are lost and the estuary becomes in effect a river mouth in the sense of Begg (1978) and Cooper *et al.* (1993), e.g. the Tugela and Mgeni River estuaries in KwaZulu-Natal. The typical estuarine stratification in these cases is transported seaward, and in large rivers, e.g. the Orange, is often found in the coastal seas.

Few of these microtidal systems maintain a permanent pattern of stratification so that switching from one or other state is not uncommon (Read 1983; Largier & Slinger 1991) and depends upon the frequency and intensity of rainfall and consequent river flow. South African workers have tended to investigate systems which are clearly in one

Table 4.1. Some generalised physical characteristics of the estuaries referred to in Figure 4.1 (Whitfield 1992)

	Tidal prism	Mixing areas	Average salinity
Estuarine bay	large	tidal	20–35
Permanently opened estuary	moderate	tidal/riverine	10–>53
River mouth	small	riverine	<10
Estuarine lake	negligible	wind	1–35
Temporarily closed estuary	absent	wind	1–35

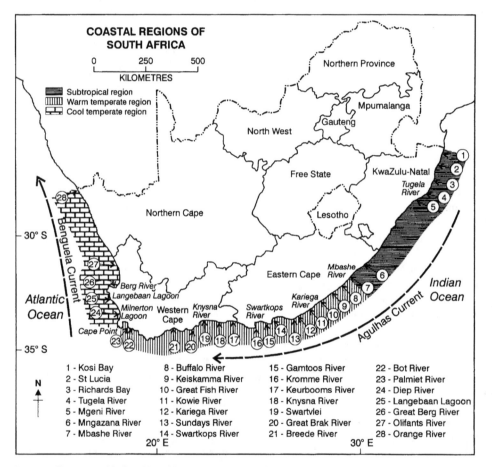

Figure 4.1. **The geographical position of important estuaries in South Africa. The rainfall pattern is given in italics. (Modified from Whitfield 1992.)**

or other of these states or have studied their properties during periods of change. Table 4.2 records a number of important characteristics of those estuaries which are most frequently referred to in the text.

Partially mixed systems

Where saltwater mixes with river water, and the tidal force is sufficiently strong to prevent the river flow from domi-

nating the circulation (Biggs & Cronin 1981), mixing of the major dissolved constituents (Na, K, Ca, Mg, Cl and SO_4) is conservative and in all respects described by the theoretical dilution line. In the partially mixed estuaries of South Africa, in the sense of Dyer (1973) similar features are evident. Figure 4.2 illustrates an array of mixing plots for these constituents in the Bot River estuary (34° 21′ S; 19° 06′ E) in the Western Cape Province, and considered typical of South African estuaries in general.

Because of the largely 'neutral' position of these conser-

Table 4.2. Some properties of a number of estuaries referred to in the text

Estuary	Catchment (km²)	MAR (10⁶ m³)	Length (km)	Area (km²)	Tidal Prism (10⁶ m³)	Depth (m)	Mean retention time (h)
Tugela	28 702	5071	–	5.5	1.0[a]	c. 1	
Mgeni	4871	707	2.5	–		2	
Msikaba	1629	251	3.3	0.2		3–35[b]	
Mngazana	630	110	6	–		0.1–0.3	
Keiskamma	2745	170	29	3.65		>2	
Sundays	22 063	29	20	2.68	2.2	3–4	
Great Fish	29 937	479	11	1.99	1.6	>1	17
Kariega	688	15	17	1.98	2.1	>1	21
Swartkops	1438	84	16	11	2.9	3	25
Kromme	1125	105	14	2.75	3.6	2.5	
Keurbooms/Bitou	1085	160	7	3.73	1.8	>1	
Knysna	335	110	19	18	19.0	1.5–6	
Palmiet	500	310	1.7	0.18	0.2	>1	

[a] The estuarine phase of the Tugela is very brief; most of the year the river is fresh to the sea (Day 1981).
[b] Deepest estuary in South Africa.

vative elements in the chemistry of the water column, by far the greatest effort has been put into describing the varying nutrient status of our estuaries in order to obtain a clear picture of their response to enrichment, and more recently to decreasing inflows with its corollary of how much freshwater is required to sustain these microtidal systems.

The primary source of the essential nutrients of those estuaries whose catchments drain the ancient Table Mountain Sandstones (TMS) of the Cape Fold Mountains remains the sea, with varying quantities of nitrate and phosphate coming from seasonal river flows largely derived from agricultural lands to complement this source. Bally and McQuaid (1985), in a report on the influence of periodic input from the seas by either natural or artificial means on the Bot River estuary, point to the importance of river flushing in diluting ions originally of marine origin. This raises the important concept of flushing or retention time which is defined (Officer 1983) as

the time necessary to replace any given conservative quantity in a given volume of an estuary at the rate at which the quantity is being injected into the volume.

Both Officer (1983) and Kennish (1986) record flushing times varying from 4 days for the Mersey River estuary to 76 days for the Bay of Fundy! Obviously under these conditions the potential for chemical change to occur within the estuary will be governed by the flushing rate. Short rates will tend to restrict chemical change so that nutrients such as nitrate and even phosphate may appear to act conservatively. Balls (1994), working in the estuaries of the North East coast of Scotland, has also shown that conservative mixing is a function of estuarine flushing time which

controls the extent to which internal processes, both biological and abiological, modify nutrients (see also p. 70). Thus the Tweed, Don and Ythan estuaries are flushed rapidly so their nutrients behave conservatively. The microtidal estuaries of South Africa, whilst they might be expected to have short retention times, experience such marked variation in characteristic features as tidal intrusion fronts, bottom density currents and long-residence deep water (Largier & Slinger 1991) that any attempt at determining residence times may be confounded.

It is appreciated that a linear response of a nutrient along a salinity gradient does not imply that the compounds are not participating in biogeochemical processes within the estuary, but rather that the compounds in the estuary are in a 'steady state' (Biggs & Cronin 1981): the assumption of the 'steady state' condition for South African estuaries is not, however, well founded.

In the Kariega and Great Fish River estuaries there is evidence of conservative mixing of river-derived nitrate (Figures 4.3a, e) when river flow is sustained. It would seem that in this respect nitrate is behaving in a manner similar to that observed in North America and British estuaries during periods of winter flood (Park et al. 1975, cited by Morris 1985; Head 1985). During periods of minimum flow or when the flow of the river stops, as in the Kariega River estuary, the sea becomes the only source of new nitrate.

The mixing plots of phosphate do not show such a marked degree of conservatism as nitrate in the Kariega River estuary. Provided that river flow is sustained phosphate departs markedly from the theoretical dilution line (TDL) as shown in Figure 4.3b. A year earlier, in December 1984 (Figure 4.3c), the drought had become so severe that a reverse salinity gradient was established in the Kariega River estuary and phosphate concentrations in the estuary increased linearly with increase in salinity.

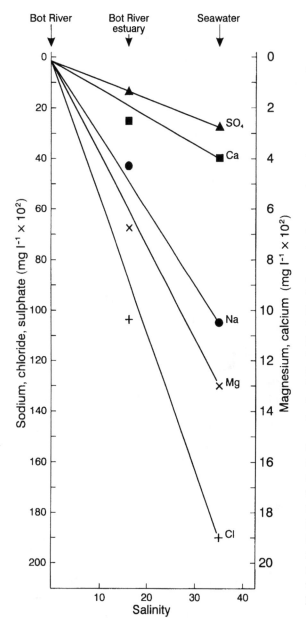

Figure 4.2. **Mixing plots for conservative constituents in the Bot River estuary, Western Cape Province.**

In the Great Fish River estuary mixing plots for phosphate exhibited some degree of conservatism even in the drought referred to above. This was due largely to the augmented river flow. Once the drought broke and river flows increased as on 31 January 1985, phosphate concentrations rose in the estuary, with a possible suggestion of a sink in the middle reaches of the estuary (Figure 4.3f) due to phytoplankton growth.

Silicate mixing plots in the Kariega River estuary over the full range of salinity showed a mark source in the

middle salinity range (Figure 4.3d), dispelling an often held view that silica acts conservatively. A similar, though less striking, plot for the Great Fish River estuary on 21 June 1986 is shown in Figure 4.3g.

Other mixing plot studies for the Orange, Olifants, Berg and Breede River estuaries reported by Eagle and Bartlett (1984) found that, in general, nitrate concentrations varied randomly in the Orange River mouth; in the Olifants River estuary they were affected by upwelling as well as agricultural activity. In the Berg River estuary, although nitrate decreased with salinity the relationship was not linear, and winter flow concentrations were significantly higher than summer values, due to land runoff. The Breede River estuary gave results for nitrate which suggested that some addition was taking place in the estuary.

These authors' study of phosphate mixing behaviour showed that in the Berg and Breede River estuaries there was a linear increase in phosphate concentration with increasing salinity so that phosphate behaved conservatively in these two estuaries at the time of sampling. They noted that the winter and summer levels of phosphate in the Berg River estuary did not vary greatly and in this respect were consistent with the findings of Butler and Tibbits (1972) for the Tamar River estuary in Great Britain who have suggested some form of buffering mechanism in estuaries.

The description of mixing has served to introduce some features of partially mixed estuaries in South Africa, but such estuaries are usefully grouped for purposes of emphasising particular features into (i) those with marked axial variation in salinity, e.g. the Great Berg River estuary; (ii) those affected by Interbasin Transfer, e.g. the Great Fish River estuary; and (iii) those in which the marine influence is overwhelming, e.g. the estuary of the Knysna embayment and the Kariega River. In KwaZulu-Natal, so many of the estuaries have been modified by decades of siltation that they have been converted into brackish water coastal lagoons which only occasionally experience tidal action.

Those with marked seasonal axial gradients
The Great Berg River estuary is a unique example. It is 45 km long and exhibits marked fluvial seasonality. During winter the estuary is dominated by river flow: a maximum average daily flow rate of 398 m^3 s^{-1} in September is reported by Slinger and Taljaard (1994). Under these conditions the tidal intrusion is limited to about 10 km of the mouth. During summer, river inflow even at its lowest (0.5 m^3 s^{-1}) is critical in limiting the upstream migration of seawater.

As the influence of winter river flow in the Great Berg River estuary decreases, with its high levels of nitrate and near oxygen saturation, the effects of tidally-induced dis-

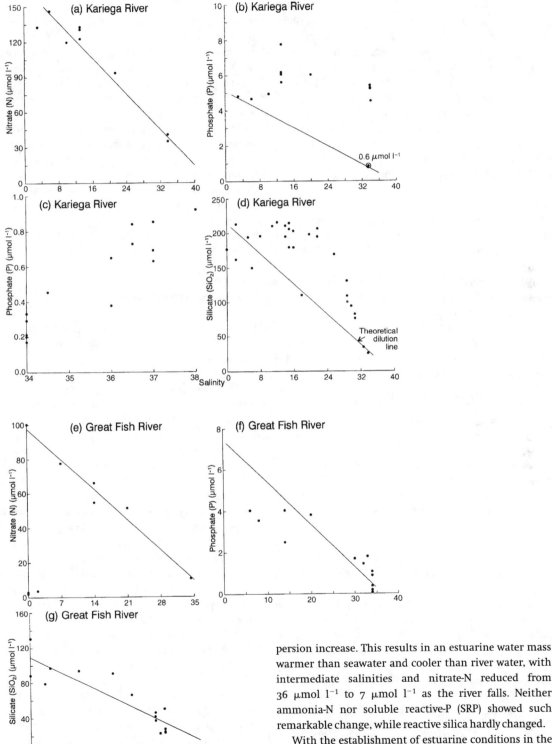

Figure 4.3. **Mixing plots for nitrate, phosphate and silicate for the Kariega River estuary (a,b,c,d) and Great Fish River estuary (e,f,g) during periods of river inflow and drought.**

person increase. This results in an estuarine water mass warmer than seawater and cooler than river water, with intermediate salinities and nitrate-N reduced from 36 μmol l^{-1} to 7 μmol l^{-1} as the river falls. Neither ammonia-N nor soluble reactive-P (SRP) showed such remarkable change, while reactive silica hardly changed.

With the establishment of estuarine conditions in the middle reaches, the increase in SRP in the middle section is due to regeneration, which confirms that the water is resident during the summer months and that flood-tide intrusion only results in limited renewal of the water in the estuary. The consequent low N:P ratio of 1.7:1 implies that during summer the estuary is nitrogen limited, which appears to be a common feature of many estuaries in South Africa not seriously impacted by enriching effluents.

Slinger and Taljaard (1994) have pointed out that the position of the mouth of the Berg River was fixed in 1966. Prior to this perturbation the mouth of the estuary opened to the south and was shallow during low flow conditions, owing to the deposition of marine sands. This shoaling in the entrance would have possibly constrained the extent of the axial tidal influence, so that the present hydrodynamic features may well be as a result of human interference!

Those under the influence of Interbasin Transfer (IBT)
In those estuaries such as the Great Fish River which experience sustained inflows due to interbasin transfer, in this case from the distant Orange River, the high nutrient loads associated with the freshwater inflow of 5–15×10^6 m^3 per month (Grange & Allanson 1995) appear to prevent nitrate limitation (N:P = 20.6:1) and even though they are turbid, algal productivity is high.

While none of South Africa's rivers has the magnitude of the Mississippi or the Po, the change in nitrogen and phosphorus ratios in relation to silica in these very large rivers has been shown by Justic *et al.* (1995) to impose a real potential for coastal eutrophication. Thus interbasin modification affecting such mainstream river systems as the Berg, Breede, Great Fish and Tugela will change the nutrient ratios in ways whose impact we have little understanding of as yet (but see p. 323). In the context of South Africa, however, such IBTs are more likely to reduce river flow to estuaries with concomitant effects of increased marine dominance. And the estuaries become mirrors of these changes if we are prepared to read their reflections!

Those with dominant marine influence
This invariably means that, in the absence of anthropogenic inputs, the chemical properties of the water column are defined almost entirely by the chemical condition of the adjacent seas and mesoscale oceanographic events, such as upwelling and the occurrence and frequency of an eastward translocation of a coastal trapped wave originating along the near or far coast. Mitchell-Innes (1988) reports that nitrate concentrations may increase by factors of two or three during upwelling, and Taylor (1987) has shown that these local upwelling events along the southeast coast transport nitrate-rich water into the estuaries. Current studies in the Knysna River Embayment illustrate the dominance of the marine influence on the volumetrically small estuary of the Knysna River. As the river inflow falls (<5 m^3 s^{-1}), the isohalines become less steep, until a narrow wedge of low salinity water (<5) occurs above the halocline for a short distance downstream of the ebb and flow at the Charlesford Weir. In effect the partially mixed estuary has been replaced by the well-mixed water column of the marine embayment!

While these remarks are valid for partially mixed estuaries which characterise the west, south and southeastern seaboard of South Africa, the chemical structures of the other types of estuaries which are interspersed among this array are equally dependent upon their characteristic hydrodynamic features and have also been subject to investigation because of their environmental importance and the warning they provide of impending trouble! These range from highly stratified estuaries, e.g. those of the Western Cape Province, to those possessing deep basins such as the estuarine lakes of the Wilderness Embayment on the Western Cape South coast and the upper lakes of the Kosi system on the northeast border of KwaZulu-Natal.

Strongly stratified estuaries

A well worked example of such a system is that of the Palmiet River estuary (Largier 1986; Taljaard 1987; Largier & Slinger 1991; Largier & Taljaard 1991) in the vicinity of Hermanus, Western Cape. The marked seasonality in rainfall and therefore inflow, which is highest in winter and greatly lowered in summer, creates a marked fjord-like salinity stratification in the estuary (see Chapter 3). The dynamical description of Largier and Slinger (1991) and Largier and Taljaard (1991) requires that the intrusion of salinity is controlled by internal hydraulics at the mouth. This salinity is spread as a non-mixing density current, and vertical salinity transport is controlled by stratified shear flow and salinity removal in the fast flowing surface layer. The persistence of long-residence deep water is a particular feature of this estuary type.

It is to be expected that the chemical properties of the water column will to some degree reflect this dynamical state. In Table 4.3*a* and *b* the values for salinity, pH, dissolved oxygen and nutrients as well as dissolved organic nitrogen (DON) and carbon (DOC) are reported for a mid-estuarine station during winter when river flow is high and summer when river flow is low. This latter was masked in February 1984 by rain in the catchment prior to sampling.

During a winter spring cycle (Table 4.3*a*) between 19 and 24 August 1986, river flow varied from 15 to 9 m^3 s^{-1} and the flood tide volume from 20 to >60 m^3. The differences in pH, phosphate-P, nitrate-N, nitrite-N, DON and DOC between surface and bottom suggest a stratification during spring tide. During the neap tidal cycle of 25–29 August, the estuary was flushed by high river flows in excess of 30 m^3 s^{-1}, and the tidal range was depressed on 28 August to 0.3 m.

Table 4.3. (*a*) **Mean values for salinity, temperature, pH, dissolved oxygen and nutrients during August 1986 (winter) in the Palmiet River estuary, near the centre of the estuary's length during a spring/neap cycle**

Tide	Temp (°C)	Salinity	pH	DO (mg ℓ^{-1})	PO_4-P (μmol ℓ^{-1})	NO_3-N (μmol ℓ^{-1})	NO_2-N (μmol ℓ^{-1})	NH_4-N (μmol ℓ^{-1})	DON (μmol ℓ^{-1})	DOC (μmol ℓ^{-1})
Spring										
Surface	14.7	1.1	6.8	10.5	0.5	73	2.8	1.4	23	630
Bottom	16.2	25.8	7.5	10.1	0.7	68	22.7	1.4	14	340
Neap										
Surface	13.0	0.0	6.5	10.5	1.2	51	0.6	1.6	38	880
Bottom	13.1	0.0	6.4	10.5	1.2	58	0.6	1.4	34	870

(*b*) **Chemical quality of the Palmiet estuary at a mid-estuary station during spring and neap tide, February 1984 (summer)**

Tide	Temp (°C)	salinity	pH	DO (mg ℓ^{-1})	PO_4-P (μmol ℓ^{-1})	NO_3-N (μmol ℓ^{-1})	NO_2-N (μmol ℓ^{-1})	NH_4-N (μmol ℓ^{-1})	DON (μmol ℓ^{-1})	DOC (μmol ℓ^{-1})
Spring										
Surface	27.6	27.7	8.3	7.7	0.4	2.2	0.4	1.4	7.8	210
Bottom	17.4	34.5	8.0	6.8	0.7	3.7	0.7	0.5	7.3	50
Neap										
Surface[a]	25.2	4.0	7.2	8.1	0.2	11.0	0.3	0.5	10.9	770
Bottom[b]	15.0	32.5	7.7	8.2	0.8	6.8	0.5	—	2.2	180

[a] Rain in catchment prior to sampling.
[b] Upwelling prior to sampling.
Source: (*a*) and (*b*) after Taljaard & Largier (1989).

There was no evidence of a flood tide volume and the measured levels of all parameters were similar throughout the tidal cycle. Thus the river is the nutrient source of the surface water layer during winter, while the sea provides the major source for the bottom layer.

The incorporation of DOC and DON in Table 4.3*a* and *b* and the relatively high concentrations in the surface waters underlines the importance of humic substances in the calcium-poor waters arising from the TMS of the Cape Fold Mountains. These substances in the Palmiet estuary (Taljaard & Largier 1989) during winter seem to be of little importance because of the strong river flow seawards. In summer during lower flows, Taljaard (1987) noted a sharp drop in DON concentration at the head of the estuary once salinity increased to 1. She argues in favour of the formation of insoluble complexes within the DON pool when salinity increases. Allanson and Read (1995) have reported a similar effect of increasing salinity on particulate material in the partially mixed Keiskamma estuary on the southeast coast. Both are indicative of complexing or flocculation under increased ionic activity. While these observations have been recorded, their role in the chemical cycles, particularly in the water column of these estuaries has not been resolved.

The overall water circulation which Largier and Taljaard (1991) describe is similar to that observed in Alaskan fjord systems, e.g. Glacial Bay (Taljaard 1987) except that in the Palmiet estuary the time scale is based upon the spring–neap tide cycle, while in Glacial Bay the bottom water is renewed annually.

Some idea of the further complexity within the water column of stratified estuaries which awaits resolution is given by Atkinson *et al.* (1987). This short study, in the Blackwall Reach of the eutrophic Swan River estuary, Western Australia was based on the premise that physical and biological factors contribute to the spatial variability of organisms and chemicals in the water column. By using a fine scale CTD profiler equipped with a very fast response oxygen sensor of high resolution (5 mmol O_2 m^{-3}, the stratification structure and distribution of oxygen at ebb tide (29 July 1985) was determined (Figure 4.4*a–d*). Simultaneous chlorophyll *a* and phosphate levels were also measured during early and late ebb but not with the same frequency.

The upper surface layer of the Blackwall Reach was approximately 2 m deep (Figure 4.4*a*) and with an average difference between minimum and maximum of 125 mmol O_2 m^{-3} (Figure 4.4*b*) and an average vertical diffusivity (K_z) of 8.2×10^{-4} m^2 s^{-1}, the rate of oxygen production was 92 mmol O_2 m^{-3} h^{-1}. Extrapolating this over 2 m (the thickness of the surface layer) gives a production of 184 mmol O_2 m^{-3} h^{-1}. This matched the high chlorophyll levels (15–25 μg l^{-1}) measured during the experiment at the level of the pycnocline.

In the deeper layers the variability in dissolved oxygen (Figure 4.4*b, c*) during early and late ebb reflects the injection of oxygen from the surface layer or *in vivo* metabolism.

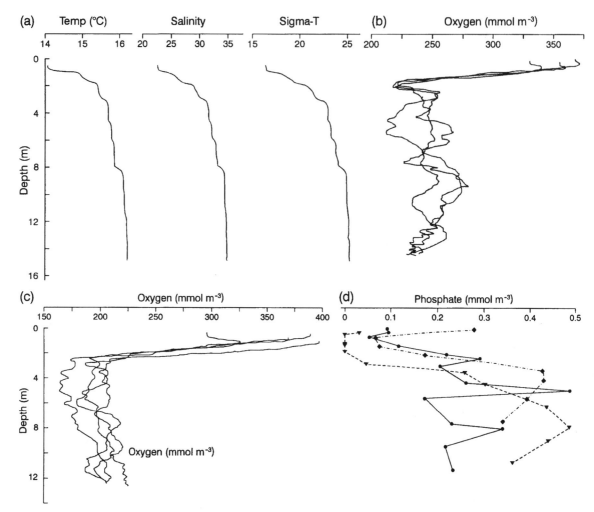

Figure 4.4. (a) **Temperature, salinity and density during ebb tide (29.7.85) in the Blackwall Reach, Swan River estuary, Perth, W. Australia. (b) Variation in dissolved oxygen with depth between 1054 and 1113 h. (c) Similarly at the last stage of ebb tide, 1458–1546 h.**

(d) **The vertical distribution of soluble reactive phosphate during early (●) and late (▼, ◆) ebb tide. (Redrawn from Atkinson et al. 1987.)**

And because oxygen and temperature were not correlated in the deep water, the observed variability in oxygen could also be due to metabolism proceeding at a sufficiently fast rate to create discrete patches of oxygen. These authors estimated that with an average difference between maximum and minimum of 25 mmol O_2 m^{-3} and a vertical diffusivity of 1.7×10^{-4} m^2 s^{-1}, the net oxygen uptake rate was 3.8 mmol O_2 m^{-3} h^{-1}.

In parallel with the observations on dissolved oxygen variation, although at a longer time scale, phosphate exhibited variability (Figure 4.4d) within each profile. The apparent maxima and minima found in any single profile were very probably created by patches of phosphate in very much the same way that similar variations in oxygen were expressions of vertical height in oxygen patches.

The increase in phosphate from early to late ebb in the deep water of the Reach indicates this to be a source of phosphate and that the net decrease in oxygen (40 mmol O_2 m^{-3}) in the deep water in the afternoon (compare Figures 4.4b, c) was consistent with remineralisation of a mixture of benthic and planktonic organic material.

Such an experimental approach, which requires considerable sophistication in measuring equipment, demonstrates that the observed variations cannot be attributed to sampling errors alone. The model which this study suggests is of patches of biogeochemicals growing and coalescing with other patches to create three-dimensional environments of varying size depending upon the nutrient or metabolite involved. To argue, therefore, that the water column is uniformly mixed below a sharp pycnocline is obviously misleading when interactions between such chemicals and their progenitors are required.

(a)

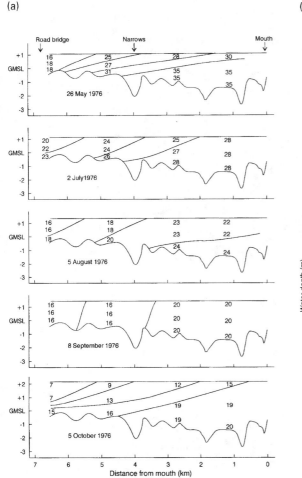

(b)

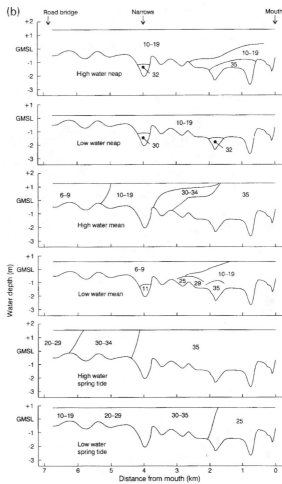

Figure 4.5. **Salinity profiles at neap, mean and spring tides in the Swartvlei estuary during (a) lagoon and (b) tidal phases.**

In estuaries which receive frequent pulses of freshwater inflow, Grange and Allanson (1995) and Allanson and Read (1995) have suggested that the repeated generation of nutrient-rich interfaces, in association with the pycnoclines which are formed and reformed during tidal cycles, is likely to create similar patchy environments conducive to the rapid growth of phytoplankton.

Water column under varying mouth open conditions

There is a spectrum of estuaries, from those which close seasonally and are alleviated by flooding or assisted by Man when water levels in the lagoon threaten homes, to those in which tidal access is limited to short periods separated by long periods of closure, usually 2–5 years.

They are obviously stages in the final development of barrier lagoons or lakes of various types.

Among this array are estuaries ranging from those such as the Klein River at Hermanus (Scott *et al.* 1952), in which freshwater inflow is balanced by evaporation so that normal salinity gradients are maintained, to those such as the Milnerton estuary (Millard & Scott 1954; Taljaard *et al.* 1992) in which evaporation exceeds inflow so that hypersalinity develops. The chemical processes which occur in such systems are in general poorly understood, as the emphasis has been more upon the impact of hypersalinity on their biology. Fortunately, the work of Liptrot (1978) on the Swartvlei estuary on the Cape South coast, which is subject to periodic closure during autumn and winter, provides details of how the estuary responded to these open and closed conditions, using first the salinity signature as illustrated in Figure 4.5*a* and *b*.

During the closed phase the sharply tilted isohalines

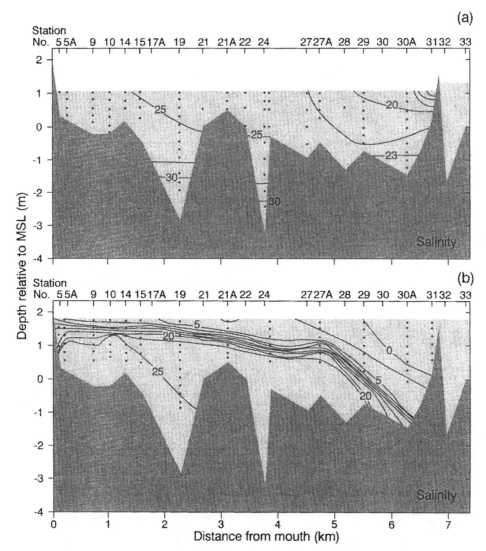

Figure 4.6. **Salinity profiles in the Great Brak River estuary prior to and following a planned flood event which breached the bar. (a) Prior to flooding, (b) prior to breaching the bar, (c) late ebb profiles after breaching of the mouth, (d) profiles during flood tide intrusion.**

(Figure 4.5a) are gradually eroded by wind mixing until by September salinities of <20 are found in the upper estuary. The onset of spring rains in September and October raises the level of the estuary and establishes a wedge of low salinity (7–15) water which, if the inflow from the upper lake is sustained, will result eventually in breaching of the bar. Unfortunately natural breaching is forestalled, as at +2 m MSL low lying property is threatened and the bar is mechanically breached and the tidal influence is re-established.

In the tidal phase (Figure 4.5b) seawater penetration increased from neap to spring tide and frequently this horizontal movement is assisted by southwesterly onshore gales and low barometric pressure. During neaps a saltwedge is established some 3 km upstream of the mouth, followed by horizontal displacement upstream during mean tides, until HWST fills the estuary with seawater for some 4 km, followed by partial mixing with water from Swartvlei, an estuarine lake at the head of the estuary. Under these conditions seawater exchanged with dense water from the scour holes in the estuarine floor very much in the way that Taljaard *et al.* (1986) and Slinger *et al.* (1995) have reported.

In the lagoon phase of 1977 during which stratification was maintained, dissolved oxygen fell to zero in the bottom water and the inorganic carbon rose to >40 g m^{-3}, as a result of decomposition of *Zostera* and *Enteromorpha*, in the scour pools (Liptrot 1978). Recent studies of the nitrate

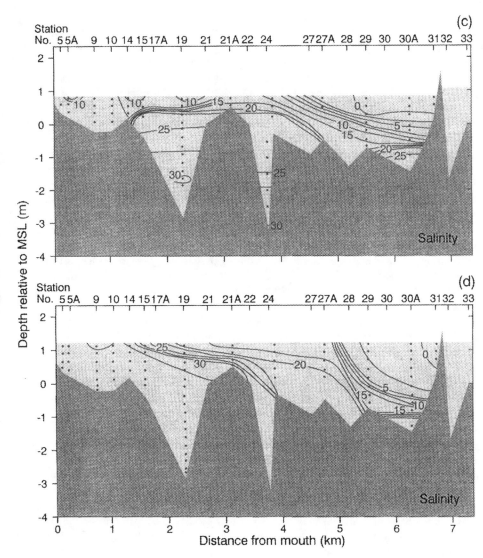

Figure 4.6 (cont.)

and phosphate concentrations under tidal and non-tidal conditions have been reported by Allanson (1989) and have shown a significant decrease in nitrate during the lagoon phase.

The recent whole estuary experimental study of Slinger et al. (1995) on the Great Brak River estuary, illustrated in part in Figure 4.6a–d, has further elucidated the effect or impact of freshwater inflows upon the chemistry of stratified systems. Prior to a predetermined freshwater release from an upstream dam and breaching of the mouth (Figure 4.6a), salinity exceeded 20 and the water column was stratified. Temperature stratification was also evident and lowered dissolved oxygen occurred in the deeper layers (1.5 mg l^{-1}), while anoxic conditions were measured

at the bottom of the deepest station (-3 m MSL). This was associated with elevated ammonium (568 µg l^{-1}), soluble reactive phosphorus (636 µg l^{-1}) and reactive silicate (1146 µg l^{-1}). Otherwise the levels of dissolved nutrients lay within the range of marine and freshwater sources. Following release of freshwater from the dam and immediately prior to breaching of the mouth the freshwater formed a wedge of low salinity water ($<$5) along the surface of the estuary (Figure 4.6b).

When the mouth was breached (Figure 4.6c), Slinger et al. (1995) reported that much of the low salinity water was selectively withdrawn while the estuary drained down to $+$0.785 m MSL. The older saline water remained trapped in the estuary, and neither its temperature nor dissolved oxygen were markedly changed following the flooding. Only in the upper estuary, where the freshwater inflow had sufficient inertia to expel low oxygen, saline water throughout the water column, did effective flushing take place.

Following the establishment of flood tides, new seawater intruded along the bottom and was contained beneath the halocline while the new freshwater layer occurred above it. Their further observations (Figure 4.6d) showed that one week after the flood event and the associated breaching of the mouth the old estuarine water was replaced in the deeper pockets or basins.

The interpretation which Slinger et al. (1995) give is that the flood event did not materially influence the deep saline water and that the transient elevation of faecal coliforms observed in the surface water was due to inundation of the intertidal marshlands. Likewise any elevation in nutrients would be transitory. The replacement of deeper, poorer quality water required the intrusion of cool, denser seawater which lifted the anoxic layers into the subsequent ebb flow causing its removal seaward. The parallel with Taljaard's work on the Palmiet River estuary is immediately obvious. Thus flooding events, comparable with tidal prism volume, enhance the scour of the mouth. And if this is timed to coincide with spring tide, maximum current velocities remove sediment from the mouth. The resultant tidal intrusion becomes the principal mechanism for flushing of the water column and the maintenance of high chemical quality in the estuary. Long periods of mouth closure typical of many South African estuaries are inimical to the maintenance of acceptable water column quality. The addition of man-made wastes under such conditions would raise an already subcritical to a critical condition until either the natural flood or human intervention breached the bar and re-established tidal flow, although this is not always successful.

But perhaps the most critical estuarine climate is found in KwaZulu-Natal. Earlier chapters have shown the impact of the geological history upon its present structure and drainage. The elevation of the Natal Monocline (King 1972) has resulted in the Province possessing a larger number of estuaries per 100 km coastline than any other part of the South African coastline.

With the exception of the largest river systems, e.g. Tugela and Mgeni, the majority close off from the sea during the year. Many, because of the build-up of terrigenous sediments on the estuarine floor, become river mouths in the sense of Begg (1978) and Whitfield (1992) when they open by flood scour. It follows that the chemistry of such systems is controlled by the quality of river flow to a greater degree here than possibly on the southeastern and south coasts. As agriculture and poor pastoral practice have increased the release of topsoil, the rivers are rich in suspensoids (silts), which if not passed to the

coastal waters are deposited on the floor of the estuary. A measure of the change in suspensoids is given by the standardised average yield in tonnes per square kilometre per annum from three regions with coastal boundaries (Rooseboom et al. 1992): thus the Western Cape rivers along the south coast yield 35, those of the Eastern Cape 185 and of KwaZulu-Natal 155 tonnes.

The rivers of KwaZulu-Natal are well watered through summer rain, but with the increasing need for storage of surface runoff and interbasin transfer engineering, even the largest rivers are showing the effects of these manipulations. Certainly the largest of the estuarine-lake systems in South Africa, St Lucia, has experienced the influence of drought exacerbated by storage dams in the catchments of its major rivers, and the impact of Cyclone Demoina on normal river flows. As a consequence the system during the past twenty years has swung from being hypersaline (82) to hyposaline, i.e. filled with freshwater. While a good deal is known of the biology of the system, its chemistry is poorly documented except for the impact of phases of hypersalinity upon gross chemical quality of the open water. The most northerly estuarine-coastal lake system, the Kosi system, is strongly influenced by the sea. Flooding is infrequent as the system is in reality a series of segmented linear barrier lagoons with a connection to the sea, which are supplied less by surface river flow than by subterranean drainage.

Fortunately the pioneering work of Begg (1978) has given some insight into the modern structure of the more conventional systems of the Province and reported on aspects of their chemical quality, particularly as a result of pollution. Development of his work has seen the publication of data on chemical quality for a number of estuaries measured over short time periods (Cooper et al. 1993). The purpose of this investigation was to provide acceptable comparison of chemical quality without undue interference from variable climatological conditions. A summary of these results is given in Table 4.4 and described in more detail on page 77.

These data are an extract from the most complete record we have of the chemical quality of the KwaZulu-Natal estuaries (Cooper et al. 1993) and although many of them exhibit serious pollution, e.g. the Sipingo River estuary, a few have survived which provide some indication of the quality of a comparatively pristine temporary closed estuary, e.g. the Mbizana system.

The meromictic condition in estuarine lakes

A number of estuaries in KwaZulu-Natal and the Western Cape Province are connected either primarily or secondarily to coastal lakes of Pleistocene origin. All these lakes are

Table 4.4. Summary of the chemical quality of the surface water of 13 Kwazulu-Natal estuarine-lagoon (temporarily closed) systems with WQI> 5 compared with the Sipingo estuary which has the lowest WQI of all the KwaZulu-Natal streams

Estuary	WQI	Salinity		DO (mg ℓ^{-1})		Chl-*a* (mg ℓ^{-1})	NH$_4$-N (µg ℓ^{-1})	NO$_3$-N (µg ℓ^{-1})	SRP (µg ℓ^{-1})	E. coli (100 mℓ^{-1})	OA (mg ℓ^{-1})
		Surface	Bottom	Surface	Bottom						
Mbizana	8.8	8	8	7.3	7.3	1.8	22	49	8	4	
Mhlali	8.3	7	7	7.9	7.8	5.9	64	48	10	9	
Sandlundlu	8.1	17	17	7.5	7.5	5.8	17	137	0	19	2.8
Mzumbe	7.8	0	0	7.5	7.5	6.8	102	50	0	10	6.6
Damba	7.8	6	7	7.5	4.3	0.0	14	24	4	10	6.7
Mzinto	7.6	8	7	7.0	7.2	5.3	15	0	14	20	5.8
Zinkhasi	6.9	16	16	6.1	4.9	11.0	172	37	57	10	7.2
Mahlongwa	6.8	9	9	6.0	6.0	3.3	31	28	6	20	3.0
Mpambanyoni	6.6	22	23	5.3	5.2	2.4	0	11	4	30	3.8
Koshwana	6.9	4	4	5.5	2.4	15.0	120	30	5	5	6.0
Mahlongwana	5.8	9	9	?	4.5	2.4	70	111	95	10	4.2
Intshambili	5.7	6	6	3.9	3.0	1.8	14	51	35	15	8.4
Boboyi	5.7	18	18	3.6	3.3	13.1	77	38	47	19	5.8
Sipingo	1.4	4	4	2.4	2.4	9.1	1704	954	594	21760	10

WQI, water quality index : *see* page 77.
DO, dissolved oxygen; SRP, Soluble reactive phosphorus; OA,oxygen absorbed from alkaline KMnO$_4$ in 4 h at room temperature.
Source: After Cooper *et al.* (1993).

pristine in nature, the chemical properties being defined by their inherent limnological features.

Those lakes along the Western Cape coast between Sedgefield and Wilderness receive water via streams which arise in the sandstones of the Outeniqua Mountains. At least one is an endorheic pan, the water source of which is seepage from groundwater and direct precipitation. The surface flows are acidic and deeply stained with humates (phenolic aldehydes) and tannins and their nutrient levels are naturally very low.

As they arose by erosion of Pleistocene river beds and subsequent reworking and transport of marine sands through their estuaries following the last glaciation, a number have retained their deep basins which have become filled with seawater over which less dense surface flows pass. This classical ectogenic meromixis has been described by Allanson and van Wyk (1969), and by Ramm (1992) for the Kosi system and by Robarts and Allanson (1977) and Allanson and Howard-Williams (1984) for the Swartvlei system.

The centre of Ramm's (1992) interest in the Kosi system is a small meromictic lake, Mpungwini (Figure 4.7), which links the upper Kosi lakes with the marine environment. He established that the high levels of hydrogen sulphide below the halocline were due to the dissimilatory reduction of sulphate mediated by bacteria.

$$\text{organic carbon} + SO_4^{2+} \longrightarrow CO_2 + \text{sulphides}$$

Increased concentrations of chloride and sulphate below the halocline reflect the intrusion of more dense seawater.

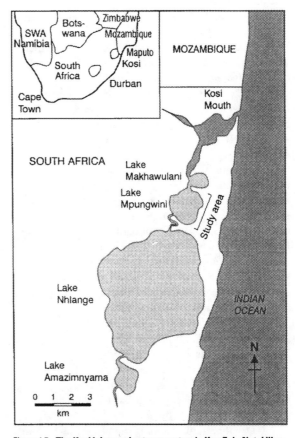

Figure 4.7. **The Kosi lakes and estuary system in KwaZulu-Natal illustrating the position of Lake Mpungwini between the larger Lake Nhlange and the Kosi estuary.**

And when the molar ratios of sulphate-sulphur to chloride were computed (Ramm 1992), sulphate decreased with respect to chloride below the halocline, indicating the reduction of sulphate. If sulphide-S is added to sulphate-S, the ratio of both forms of sulphur and chloride remains relatively constant with depth. Furthermore, Ramm (1992) makes the point that accumulation of free sulphide in the anoxic waters of the lake is attributable to low concentrations of iron and manganese in the system.

An important consequence of Ramm's (1992) work was the finding through the use of ^{35}S-labelled sodium sulphate that sulphate enrichment equivalent to a 5 g kg^{-1} increase in lake water salinity resulted in a 2–4 fold increase in the ambient rate of sulphate reduction. In control experiments enrichment with sodium chloride did not cause an increase in sulphate reduction. From these studies a one-dimensional simulation model was developed which was used to evaluate the potential impact of increased rates of sulphide production on the lake. Vertical mixing coefficients were estimated using the rate of change in chloride mass with respect to time and depth. The simulation established that the potential existed for sulphide buildup to accumulate as the supply of available sulphate rose with increasing marine influence.

As these surface waters provide the only access either to the sea or from the sea to the upper lakes, contamination by free sulphide through upwelling of bottom water (discounting at this stage the effect of increased sulphate through elevation in sea level) would become an effective barrier in the channel, preventing upstream and downstream migrations, and cause mortality of resident species.

An opportunity to test the predictive capability of the model arose in September 1989 during equinoctial spring tides. These tides were followed immediately by a rapid shift in wind direction. High winds from the northeast (4 m s^{-1}) backed to the southeast when wind speeds increased to 6 m s^{-1}. On both occasions high concentrations of sulphides (>10 mg l^{-1}) upwelled to the surface, and the largest fish kill ever reported in a KwaZulu-Natal estuary took place. All the major fish species of the Kosi system were represented in the mortality.

The studies by Robarts and Allanson (1977) and by Allanson and Howard-Williams (1984) of the meromictic lake, Swartvlei, showed that an extensive deoxygenated saline monimolimnion occurred below a fairly sharp halocline at 6 m (Figure 4.8). As with Lake Mpungwini, the primary origin of the stratification was the flow of seawater over a shallow sill into a deep basin. Contrary, however, to the sharp upwelling signature which Ramm (1992) observed, wind from either the southwest or southeast, the latter predominantly in summer, only gradually

erodes the stability of the meromixis so allowing the transfer of oxygen-rich waters into the monimolimnion until, as Figure 4.8 illustrates, full circulation occurs and oxygen is distributed throughout the depth of the lake.

As these conditions in Swartvlei are invariably associated with non-tidal lagoon conditions in the lower estuary, reinforcement of the dense saline layer is prevented, so that wind-induced turbulence becomes effective in mixing the water column. Once the bar is breached and tidal conditions are set up, the saline monimolimnion is rebuilt.

Liptrot (1978), Taljaard and Largier (1989) and Slinger et al. (1995) have reported that somewhat similar meromictic conditions develop in estuarine channels where scour holes occur in the bed of the estuary. The consequent deoxygenation of the deep saline water leading to the production of free sulphide must follow the same chemical pathways described by Ramm (1992).

The sediments

The sedimentological character of South African estuaries has been presented in Chapter 2. The contribution of sediments to the chemistry of these microtidal estuaries is, however, less well known. In those estuaries which tend to close, there is an obvious decrease in the percentage of sand in the deeper basins of the incipient or actual lagoon. Willis (1985) has shown in the Bot River estuary that all sediments above mean sea level are composed of >70% sand; the sediments in the deeper basin −1.5 m MSL are composed of >90% mud. Where there is sustained tidal flow as in permanently open estuaries this distribution is reversed, with muddy sediments occurring in the intertidal zone and the coarser sediments flooring the tidal channel.

The subsequent mineralogical analysis (Willis 1985) by X-ray diffraction showed that the major minerals present are quartz, clay minerals, feldspar, carbonates and pyrite. Illite is the most abundant of the clay minerals and pyrite is found in some of the organic-rich muds. Further numerical analysis making use of multivariate principal component techniques of the chemical and particle size data has shown that the sediments of the estuary can be considered as simple mixtures of quartz only, quartz + carbonate or quartz + clay.

In view of the common origin of these materials along the Western Cape south coast, it would be surprising if the other estuaries even as far as the Kariega River, east of Port Elizabeth, would differ significantly from this pattern. A further feature of the sand fraction present is the paucity of very fine sands (<0.125 mm) which, Willis (1985) notes,

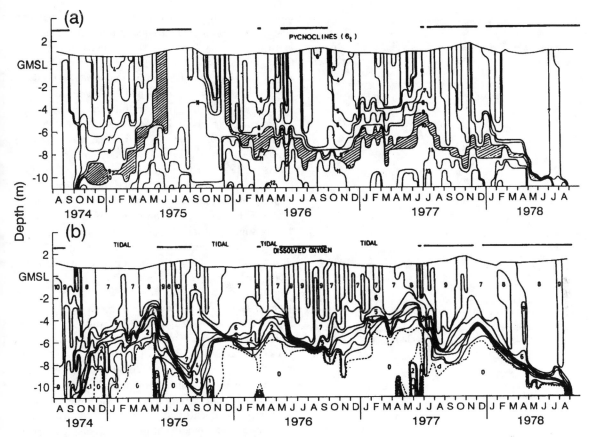

Figure 4.8. **The pattern of pycnoclines and dissolved oxygen iso-pleths in Swartvlei between 1974 and 1978. The horizontal bars represent periods when the mouth of the estuary was closed.**

have a critical erosion velocity below that generated by wave action in the estuary. In view of this, very fine sands would be eroded from the estuary during the tidal phase. The major element composition of the Bot River estuary sediments is given in Table 4.5.

The predominance of quartz sands above +0.5 m, i.e. the margins of the estuary, and the correspondingly low trace element concentrations are likely to be found in the sediments of similar lagoonal estuaries. The increase in aluminium oxide indicates that clay minerals and mud fractions, with their elevated pyrite and phosphorus, would form the basis of a phosphorus cycle, either geochemically or via biochemical routes, between the water column and sediments when pH and redox conditions permit.

While this mineralogical analysis serves to establish in a generic sense the inorganic chemical structure of the sediments of many bar-built estuaries along the southern and southeastern seaboard, the presence of muds accreting high trace element concentrations has important consequences. They sequester those trace elements which

indicate human impact either directly in the estuary or at remote locations within the river catchment.

Willis (1985) points out that the highest concentrations for trace elements, except for strontium, occur in the mud because of their association with clay minerals or mud fractions. It is in this regard that the bench mark studies of Watling and Watling (1980, 1982a–e, 1983a, b) have been so valuable. They have provided a detailed data base on trace elements in sediments of the Western Cape south coast and Eastern Cape estuaries, and not only at the surface but through cores up to 1 m in length. In Table 4.6 data sets for the Knysna and Keurbooms River estuaries are reported from Watling's array because the estuaries are relatively unpolluted, particularly by industrial effluents. In both Knysna and Keurbooms isolated sites appear to have anomalous element concentrations but in the case of Knysna they relate to specific nearby sources of urban input. Zinc, for example, is highest at sites associated with urban or industrial activity. Anomalies in the Bitou River estuary, which joins that of the Keurbooms (Table 4.6) just above the Plettenberg Bay lagoon, are considered by Watling and

Table 4.5. Average major element composition (weight percentage) of two sediment types from the Bot River estuary

Sediment type	Marginal +0.5 m MSL	Central muds <−1.5 m MSL
SiO_2	97.67	55.14
TiO_2	0.09	0.79
Al_2O_3	0.76	19.16
$Fe_2O_3T^a$	0.44	6.93
MnO	0.02	0.49
CaO	0.04	0.49
Na_2O	0.09	0.30
K_2O	0.13	2.96
P_2O?	0.03	0.17
S	0.07	0.52
CO_2	0.20	0.00
LOI^b	0.59	10.36
H_2O	0.57	1.98

^{a} Total iron expressed as Fe_2O_3.
^{b} Loss at 1000 °C excluding CO_2.
Source: From Willis (1985).

Table 4.6. Concentrations (μg g^{-1}) of trace elements from the surface sediments in the Knysna and Keurbooms estuaries

Element	Knysna estuary (Stn 11)^a	Keurbooms estuary (Stn 7)
Cu	0.41	1.2
Pb	5.2	1.4
Zn	2.4	10.6
Fe	1170	3800
Mn	21	60
Co	0.3	<0.1^b
Ni	0.4	1.8
Cd	0.005	<0.1^b
Na	1490	2400
K	334	560
Ca	172	64000
Mg	439	256
Sr	8	4600
Al	1230	2800
Cr	1.6	1.2

^{a} Referred to by Watling & Watling (1980) as 'background levels'.
^{b} Due to computer program.
Source: From Watling & Watling (1980, 1982d).

Watling (1982b) to be largely of geochemical origin. At the extreme southwestern end of the lagoon elevated metals (Cu = 5.6 μg g^{-1}, Pb 7.0 μg g^{-1} and Zn 33.2 μg g^{-1}) are likely as a result of urban contamination (see also p. 83).

Watling and Watling (1980) have shown that predominant inter-element relationships are based on the clay facies elements, potassium, magnesium, aluminium and chromium, with carbonate developed as a sub-facies. These authors comment that iron and nickel appear to be strongly attached to the clay web so that co-precipitation with clay minerals may be presumed, as for the Knysna River estuary.

On the other hand, in the Keurbooms River estuary, while the clay and carbonate facies are well developed (Watling & Watling 1982b), the cobalt, nickel and zinc

matrix web is not related to clay facies. Consequently adsorption of these elements onto clay minerals is not the main distributary factor for these metals in the lower Keurbooms section of the estuary. The catchments' rocks are rich in iron, so that when iron dissolved in the river water comes in contact with the higher salinities of the estuary, iron precipitates carrying other metals with it.

The composition of the interstitial waters of sediments has been reported on extensively by Taljaard (1987) in the bar-built Palmiet estuary. The surface 100 mm of the sediments (Table 4.7) exhibited high concentrations of phosphate, ammonium, nitrite, nitrate and organic matter.

Taljaard (1987) points out that although these nutrients were higher in the sediments than in the seawater during summer, there was no indication of transfer of nutrients from the sediments to the overlying water, either as a result of bioturbation or as a result of primary production requirements establishing the necessary concentration gradients. Certainly, and in view of the generally high redox of the systems, phosphate release will depend almost entirely upon molecular diffusion.

In general there was a decrease in nutrient concentrations upstream from the mouth. This gradient reflects the effects of import of organic material from the sea (kelp) and from the river, mainly palmiet (*Prionum serratum*) which Branch and Day (1984) record contribute 143 000 kg per year! This allochthonous input is very much greater than that supplied by autochthonous algae, e.g. *Cladophora* with a production of 30 360 kg y^{-1}. But, notwithstanding these inputs, the organic matter remains low due to the high winter river flows which increase the coarseness of the sediments and prevent the accumulation of organic material. The biological activity of the sand prawn *Callianassa krausii*, in the more stable lagoon-like section, concentrates fine particles and organic material in the lining of its tubes (Branch & Day 1984).

The overall impression gained from this work is that the sediments, certainly during summer, act as a sink for nutrients, providing the requirements of the relatively abundant but ephemeral stock of *Cladophora* and, through this alga, the needs of the deposit feeders.

Thus far we have reviewed the significant work on estuaries whose catchments largely consist of the TMS series with some coastal variation mostly in the form of Enon sedimentary series, the sands of which are predominantly of marine origin and which through time have become mixed or infiltrated by organic-rich material. Along the eastern seaboard, the estuaries in particular of KwaZulu-Natal are subject to quite different sedimentation regime as described in Chapter 2. Unfortunately the chemistry of these largely terrigenous suspensoids, when they precipitate under the influence of increased ionic activity (seawater), has not been sufficiently well described.

Table 4.7. Surficial (100 mm) levels of nutrients and total organic matter (TOM) in the interstitial waters of sediments in the Palmiet River estuary (μmol l^{-1}). Concentrations of nutrients in the water above the sediments at HWST are shown in parentheses

Stn No.	PO$_4$-P	NO$_3$-N	NO$_2$-N	NH$_4$-N	TOM
1 Sand flats near mouth	67.81 (0.82)	166.39 (3.14)	18.26 (0.15)	555.36 (0.40)	7.86 (198)
2 Sand flats upstream of 1	124.07 (0.57)	161.76 (2.64)	23.18 (0.54)	462.35 (0.62)	3.56 (110)
3 Adjacent to scour hole, north bank	84.61 (0.71)	82.15 (3.69)	15.26 (0.69)	289.70 (0.51)	2.88 (57)
4 At 3 m depth, sand bank opposite boat house	32.49 (0.32)	55.58 (0.97)	8.54 (0.10)	359.16 (0.57)	3.3 (355)

Table 4.8. Sediment analysis of the Kosi estuary system

Site	Depth (m)	% particle <63 μm	OA[a] (mg ℓ^{-1})	Kjeldahl nitrogen (mg g^{-1})	Total phosphorus (μg g^{-1})	Sulphide (mg g^{-1})	Humic acids (mg g^{-1})	Iron (mg g^{-1})
Tidal basin	1.5	0	0.1	0.1	24	0.0	0.7	2.0
Ukhalwe inlet	1.8	organic debris	24.6	3.7	192	0.6	12.8	2.9
L. Mpungwini	6.5	32.4	9.9	0.3	216	0.1	0.7	2.0
L. Sifungwe	16.5	organic debris	37.8	7.2	377	0.8	14.9	6.3
L. Nhlange	24.0	4.2	8.7	1.0	43	0.0	2.1	22.0
	17.0	organic debris	63.7	22.1	47	0.4	16.6	13.5
L. Amanzimnyama	2.0	organic debris	121.8	9.4	202	0.6	97.9	95.3
Malangeni	2.5	organic debris	200.8	10.7	1020	0.0	232.0	51.3

[a] Oxygen absorbed from alkaline KMnO$_4$ in 4 h at room temperature.

In the Keiskamma River estuary studies by Allanson and Read (1987) demonstrated that fluvial suspensoids decreased from 6150 mg l^{-1} in river water to 280 mg l^{-1} in the estuary where the salinity increased from 0.5 to 8. This was linked to an equally striking change in the relative contribution of POC and DOC in the total organic carbon (TOC) pool (Allanson & Read 1995). Before flocculation of particulate organic matter (POC) was 81% of the available pool but fell to 28% in the estuary. The fluvial contribution to the organic resources of the sediments is obviously substantial.

Of particular interest and importance in this respect are the blackwater rivers of the Western Cape and other surface drainage such as the podzolic soils of the Maputaland peneplain. They provide a rich source of humic aggregates when they mix with seawater. Some indication of the magnitude of this component comes from the work of Liptrot (1978) during the closed phase of the Swartvlei estuary. Using simple experimental procedures in which humic-rich river water was subject to increasing salinity and pH, he established that flocculation of the humates occurred over a range of salinities 7.5–17.5 and at pH values ranging from 8 to 10. Given the concentration of humic material and an estimated flow of rivers into the Swartvlei system, some 3040 tonnes of particulate humic materials annually could be deposited in the sediments of the lake and estuary or washed out to sea (see also p. 71).

A similar allochthonous input of humic material has been reported by Hemens et al. (1971), who specifically measured the humic acids and other chemical features in the sediments of the estuarine lakes of the Kosi system. A summary of their unpublished results is given in Table 4.8.

Clearly the deep basins of the lakes which were originally formed by segmentation of deep river channels during the Holocene are repositories or sinks for substantial quantities of allochthonous organic material, more derived from surface inflows than generated within the estuarine lakes. The Kosi system is in this respect very similar to the estuarine lakes of the Wilderness embayment, which removes it somewhat from the general facies of sedimentation in the other estuaries of KwaZulu-Natal.

Cooper et al. (1988), in their assessment of sediment conditions in the estuaries of this Province, point to a distinct difference in the processes operating in large and small catchments. Large estuaries experience considerable scour of both bed and bank material during severe floods. The scoured bed is rapidly filled by sediments, in a well-defined sequence, which in sedimentological terms indicates an equilibrium between sedimentation and erosion, and which will maintain bedloads at their present levels provided that the relative levels of land and sea remain constant. In the smaller estuaries of the province, the sedimentological effects of flooding are relatively minor, involving the removal of small amounts of fine sediment. Little scour is expected. Thus, while the sedimentological facies are well understood in the strongly fluviatile estuaries of KwaZulu-Natal, their chemistry is not adequately interpreted, except in the simplest of pollution parameters.

BIOGEOCHEMICAL PROCESSES

Estuarine sediments are the site of substantial biological and chemical activity. Most of the inorganic substances fixed by plants and converted to organic form are deposited in the sediments. Up to 60% of the net carbon fixed by *Spartina alterniflora* is deposited in the soil through growth of roots and rhizomes (Wiegert *et al.* 1981), while Howard-Williams (1977) found that by far the largest standing stocks of nitrogen and phosphorus in Swartvlei are in the sediments.

Microorganisms are involved in all the phases of nutrient cycling. Aerobic microbial degradation takes place in plant material which becomes exposed on the banks of estuaries when the tide recedes, but anaerobic degradation occurs in the material that becomes buried in the sediment. Aerobic degradation involves widespread and extremely numerous heterotrophic organisms (primarily bacteria) which first occur as early colonisers on growing plants. *Vibrio, Pseudomonas, Flavobacterium* and, rarely, *Proteus* bacterial genera have been isolated from fresh *Ruppia* in a Bot River study (Roberts *et al.* 1983). The same study reported anaerobic degradation involving acidic, black, peat-like sediments, which were highly reducing due to the presence of hydrogen sulphide and iron sulphides. The microbial population consisted almost exclusively of dissimilatory sulphate-reducing bacteria (*Disulphovibrio* spp.) which required organic carbon sources (e.g. organic acids) and generated acetic acid (Roberts *et al.* 1983). Seasonal die-back of macrophyte beds creates an increase of detrital material for bacterial use, which is typically low in nitrogen. A Langebaan Lagoon study showed that as the detrital material becomes nitrogen-enriched by bacterial colonisation, the nitrogen requirements of bacterial growth on plant materials can be subsidised by inorganic nutrients taken from the environment or by bacterial dinitrogen fixation. Aerobic diazotrophs (e.g. *Azotobacter* spp.) possess physiological mechanisms which shield their nitrogenase (which is extremely oxygen-labile) from contact with oxygen (Tibbles *et al.* 1994).

While the position along the axis of an estuary often defines the substratum type, the other major parameter which changes longitudinally in estuaries is salinity. Salinity plays an important role in the biogeochemistry of estuaries, as it modifies the chemical environment. Gardner *et al.* (1991) observed that the percentage of mineralised nitrogen compared with that released as ammonia

was affected by the presence of salts. Ammonia release was enhanced in the presence of seawater or brack (30% seawater) water. Salinity and substrata often combine to dictate the fringing vegetation, and Harvey and Odum (1990) suggest that the transport of solutes is strongly influenced by shoreline type. Long soil contact times in marshes provide greater opportunity for immobilisation of nutrients by plants, microbes and adsorption on sediment. The nature of vertical accretion (which can affect the substrate type) in the fringing zone is also variable, ranging from the influx of foreign matter and episodic events, to the situation reported for Louisiana marshes where most vertical accretion was via deposition of *in situ* organic matter rather than the influx of foreign matter (Childers & Day 1990). Steinke and Ward (1990) estimated litter production in Transkei mangrove estuaries to be between 3.1 and 4.5 t ha^{-1}y^{-1}, and apart from forming particulate material degraded by the microbial system, some fraction probably goes towards vertical accretion. Schleyer and Roberts (1987), in a study of the Siyaya Lagoon in KwaZulu-Natal, calculated that 90% of the detritus input accumulates in the lagoon. Siltation and trapping of material by vegetation also act to make estuaries more shallow where current velocities are low.

Carbon

There are no published studies specifically on microbial mediated carbon cycle from South African estuaries. The following synopsis of some of the significant literature is an attempt to draw attention to this gap. DOC is the primary form in which carbon enters the metabolic part of the microbial transformation system, and is supplied by POC from dead and decomposing organisms, exuded by living organisms and leached from living plant tissue. While sedimentary POC stocks are more stable, there are seasonal variations in DOC. Childers *et al.* (1993) found increased DOC release from a saltmarsh during winter and spring. They suggested that forest stream inputs of DOC were not as important as previously thought. In a southern Texas estuary Koepfler *et al.* (1993) found pore water DOC concentrations five times higher than water column DOC. The pore water DOC concentrations were inversely related to bacterial production and directly related to macrofauna biomass, sediment POC, sediment C:N ratios and O_2 metabolism. Their conclusion was that variability of DOC was controlled by organism activity and detrital quality. Howes and Goehringer (1994), in a New England *Spartina alterniflora* marsh study reported high DOC losses (>1 mol C m^{-2}y^{-1}) through the creek banks as pore water seeped out on the receding tide.

Two decades ago a South African report (Robarts 1976) stated that photosynthetic bacteria contribute significantly to total pelagic primary production in Swartvlei, and after the development of the 'microbial loop' concept (Azam *et al.* 1983) several studies have been conducted to characterise the factors which control bacterial dynamics in aquatic environments. Cole *et al.* (1988) showed that bacterial biomass and production are positively correlated with phytoplankton production in most aquatic ecosystems. While quality and quantity of suspended particles could determine the proportion of attached bacteria (Pedrós-Alió & Brock 1983), temperature and salinity variation and dissolved inorganic nutrient availability have been highlighted as important factors influencing bacterial growth (e.g. Palumbo *et al.* 1984; Iriberri *et al.* 1987; Krambeck 1988). Studies carried out on the microbial food web of estuaries, however, do not allow a generalisation of the parameters controlling bacterial dynamics, partly due to the high environmental variability of these ecosystems. In some estuaries, for instance, the dominance of attached bacteria was related to high seston loads (Bell & Albricht 1981; Bent & Goulder 1981; Healey *et al.* 1988), while in other highly turbid systems attached bacteria were found to play a minor role (Palumbo *et al.* 1984; Painchard & Therriault 1989).

Nitrogen

The primary processes involving nitrogen are ammonification (including mineralisation and denitrification) and nitrification, and these processes are occurring continuously at rates determined by various biotic and abiotic factors. High rates of remineralisation of organic nitrogen within macrophyte beds, relative to unvegetated sediments, can be sufficient to supply the nitrogen required for growth of the plants (*Zostera marina* and *Potamogeton perfoliatus*) (Caffrey & Kemp 1990). In their 1992 study, Caffrey and Kemp found denitrification rates to be greater in vegetated (*Potamogeton perfoliatus*) sediments than bare sediments, while denitrification in the root zone comprised about 16% of total denitrification. In vegetated sediments, about 15% of nitrogen lost from sediments was denitrified, and 25% was taken up by the plants. *Potamogeton* had a significant influence on sediment nitrogen cycling by direct uptake of ammonia and nitrate, and by indirect mechanisms leading to enhanced nitrification and denitrification.

Concurrence comes from Tibbles *et al.* (1994), who found higher nitrogenase activity in *Zostera capensis* bed sediments in a South African study. Thus the microbial communities responsible for key nitrogen transformations in the sediments are enhanced by plants through inputs of organic nitrogen and release of oxygen by plant roots. Howard-Williams (1977) reported that the deep water organic sediments of Swartvlei, South Africa, have particularly high nitrogen concentrations and that the metabolic activities of the littoral were not only a major source of organic nitrogen synthesis but could significantly influence the flux of nitrogen from the sediments to the water. Benner *et al.* (1991) reported that the below ground biomass of *Spartina alterniflora* lost 55% of its organic matter during 18 months of decomposition in salt-marsh sediments. Two phases in the nitrogen dynamics were evident: an initial net loss of nitrogen, thereafter nitrogen immobilisation coupled with rates of microbial degradation. Most of the tissue nitrogen was lost in early stages (4 months) of decomposition, followed by slow accumulation of nitrogen so that by the end of the study more nitrogen had accumulated in the dead tissues than was present at the start.

In a 7-year field study White and Howes (1994) studied nitrogen retention in a New England marsh. They reported a 25% loss of added ammonia through nitrification–denitrification with the remainder incorporated into plant tissue. Nitrogen was increasingly sequestered in the dead organic nitrogen pool. Export accounted for 26–44% and denitrification for 54–77% of total nitrogen loss. Recycling of nitrogen through translocation from above-ground to below-ground biomass and remineralisation of dead below-ground biomass was the major pathway in the sediment nitrogen pathway equivalent to 67–79% of annual plant nitrogen demand. Nitrogen losses were balanced by inputs, primarily nitrogen fixation. Long-term nitrogen retention was controlled primarily by competition for dissolved inorganic nitrogen (DIN) between plants and bacterial nitrifiers–denitrifiers and secondarily by the relative incorporation of nitrogen into above-ground versus below-ground biomass.

Phosphorus

Phosphorus is recognised as one of the limiting nutrients to primary production. N:P ratios are used in some definitions of eutrophic systems although there is still some controversy (see Smith 1990 and response to comment).

The redox potential of the environment is one of the most important factors influencing phosphorus exchange, as changes in redox potential alter the quantity of phosphorus held in association with charged particles, primarily Fe, Al and clay and floc particles. A phosphate buffer system is thought to keep dissolved inorganic phosphorus (DIP) in equilibrium with phosphorus sorbed onto inorganic surfaces (Froelich 1988). Deoxygenation results in the release of phosphorus from metal sesquioxide complexes at E_h values below $+200$ mV, while at lower redox potentials H_2S will react with the complexes and liberate phosphorus.

Phosphorus release rates of 2.5 mg P m^{-2} per day when the sediment was anaerobic, and lower (1.6 mg P m^{-2} d^{-1}) rates under aerobic conditions were reported from Swartvlei sediment core data by Silberbauer (1982). More recently, Maher and DeVries (1994) reported phosphorus adsorption in sediments to be lower in deoxygenated estuarine sediments and Chambers et al. (1995) found that under oxidising conditions, DIP release from particulate matter was low, with some uptake by particles with total P of less than 50 μmoles per gram. Release was higher from particles with a total P content greater than 50 μmol g^{-1}.

Foliar release by macrophytes is well known in phosphorus exchange. The role of *Zostera marina* as an active agent in the uptake and excretion of phosphorus was established by McRoy et al. (1972) and confirmed for the *Z. capensis* beds of Swartvlei estuary by Liptrot (1978), but the inclusion of an associated floating algal mat, e.g. *Enteromorpha* spp. which is particularly abundant during the cool season in many closed and open South African estuaries, is another piece of the jigsaw puzzle of how phosphorus is mobilised and transferred between a number of compartments during the lagoon phase.

Specifically, Liptrot observed that the quantity of phosphorus (43 kg) drawn from the water during the lagoon phase was within the limits of that present in *Enteromorpha* spp. (45 kg) at the end of the lagoon phase. It is tempting to suggest that the growth of *Enteromorpha* during closure is one of the compartments with which phosphorus excreted by *Zostera* is transferred, having regard to the equilibrium maintained by sediment–water exchange at 22–26 μg l^{-1} ; strikingly similar to the levels obtained in Doboy Sound, Georgia (Pomeroy et al. 1965) and the Tamar estuary (Butler & Tibbits 1972). The model which Liptrot (1978) constructed is given in Figure 4.9.

Clearly the source of phosphorus is the sediments which, because of the net import of total phosphorus (which includes particulate phosphorus) during the tidal phase, become progressively richer in phosphorus. During the lagoon phase this 'imbalance' is redressed by phosphorus being lost from the sediment in three processes.

1 Direct sediment–water exchange, which is essentially a physical process.

2 Uptake by *Zostera* and excretion, which stimulates the growth of *Enteromorpha* spp. as a mat above the *Zostera*.

3 The loss of this accumulated phosphorus through flooding at the end of winter when current speeds

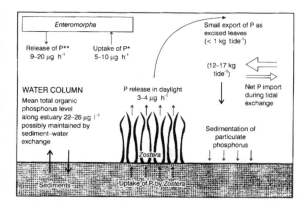

Figure 4.9. **Dynamics of phosphorus flux and cycling in the Swartvlei estuary. *, Uptake of phosphorus from dawn to mid-day; ** release of phosphorus from mid-day to the following dawn (from Liptrot 1978).**

are strong enough >0.5 m s^{-1} to flush out the *Enteromorpha* mats.

More recently Pakulski (1992) reported phosphorus release from *Spartina alterniflora* in a Georgia saltmarsh, amounting to 1.7 g P m^{-2} y^{-1}. There was little diel variation, but the highest rates occurred in midsummer. Foliar release of soluble reactive phosphate from live plants equalled 60% of the annual production of particulate phosphorus from *Spartina* detritus.

Moving from the estuary to the brackish estuarine lake, Swartvlei at the head of the estuary, dense stands of *Potamogeton pectinatus* in the lake provided an opportunity of establishing the dynamics of phosphorus in the littoral of the lake, as it was recognised that phosphorus in solution in the water column above the halocline was close to the limit of detection in this oligotrophic lake. These studies were reported by Howard-Williams and Allanson (1981).

The littoral zone in which *P. pectinatus* is abundant (>1000 shoots m^{-2}) occupies one third of the area of the lake to a depth of 3 m. The use of ^{32}P allowed the determination of (i) those compartments in a multi-compartment system which receive phosphorus, and (ii) how fast? Three types of experiment were undertaken: (a) to measure loss of P from water to intact portions of the *Potamogeton* canopy, (b) the uptake by separate components of the canopy, and (c) the transfer of P from the water to the littoral sediments. The experiments demonstrated that if 100 μg of P is added to an area of *Potamogeton* and sediment and assuming homogenous distribution through the water column, the canopy accumulated 2.6 times as much P as did the sediment by the time equilibrium had been reached.

The highest uptake rate ($R = $ μg P m^{-2} min^{-1}) was found in old canopy in which a complex adnate periphyton had developed on the leaves of *Potamogeton*. Values of $R = 50.4$

and 32.5 for old and mature canopy, respectively, are reported by Howard-Williams and Allanson (1981). This is in contrast to the uptake rate of $R = 9.3$–25.6 for the surface detritus on the sediments and 1.5–0.6 for the first 50 mm of sediment. Such uptakes indicate that soluble fractions of phosphorus in the water column are taken up rapidly in the canopy and that the sediments, even including the thin detrital layer, are relatively much slower, although they possess large nutrient reserves, so that the ability of the littoral zone to sequester essential nutrients points to the vital role it plays in 'old' nutrient pathways.

Sulphur

Sulphate reduction plays an important role in remineralisation and diagenesis of organic substances when the environment is anoxic. Where sulphate concentrations are high, sulphate reduction can dictate the nature and rate of organic carbon remineralisation. In some saltmarsh situations 60–80% of organic remineralisation is accomplished via bacterial sulphate reduction (Howes *et al.* 1984). In addition, the formation and accumulation of free sulphides in water (particularly as H_2S) can cause a severe water quality problem due to its high toxicity, as Ramm (1992) has shown in the Kosi estuarine-lake system (see also pp. 65 and 66).

Biogenic transformations

Macrofauna are an often neglected group in studies on nutrient transformations, and the role they play may be underestimated. Predation, filter- and particle-feeding and scavenging, as well as microbial assimilation, cause new organic compounds and CO_2 to be released into the water. The filter-feeding organisms in the marsh, including zooplankton, bivalves and worms, are responsible for incorporating some POC and DOC into the biotic pool, while some of them mediate the exchange of water between the sediment and the water column (e.g. *Mytilus*: Gontier *et al.* 1992). Particle feeders, including crustaceans and fish, consume relatively large organisms, both living and dead. A number of them, including crabs, shrimp and mullet, probably also assimilate significant quantities of algae and the microbial biomass associated with detrital POC. Scavengers and herbivores accelerate the transformation of material by breaking it into smaller physical fragments, as well as by digestive transformations.

Not all organic substrates in estuarine soils are transformed in place: large amounts of material are brought to the surface and deposited by burrowing macrofauna. At the same time, the burrowing activities are modifying the environment both physically and chemically, thus altering the biogeochemical milieu. Clavero *et al.* (1991) found that the sediment–water phosphate gradient was inversely related to *Nereis diversicolor* abundance, and that phosphate flux increased with abundance. Koepfler *et al.* (1993) reported that pore water DOC was inversely related to bacterial biomass, and directly related to macrofaunal biomass. They isolated detrital quality and organism activity as the most important controlling factors in pore water DOC variability. Rysgaard *et al.* (1995) found that denitrification was correlated to bioturbation while Caffrey (1995) reported that the variation in ammonia concentration with depth she observed may be due to deep dwelling macrofauna.

Law *et al.* (1991) reported a strong covariance between rates of denitrification and the degree of bioturbation by *N. diversicolor*, and ascribed this to increased transport and supply of nitrate via the burrows. In a South African study, Tibbles *et al.* (1994) found that oxygen stimulation of nitrogenase activity was less marked in aerobic sediments around the rim of prawn burrow openings where subsurface sediments had been displaced to the surface. They reported that bioturbation of sediments by benthic infauna, such as the thalassinids *Callianassa kraussi* and *Upogebia africana*, may increase aeration of subsurface sediments and that bioturbation may limit the extent of oxygen-stimulation of nitrogenase activity in surface sediments if microaerophilic or aerobic diazotrophs are displaced from the surface around the burrow opening.

Recent work on the saltmarsh littoral of the permanently open Kariega River estuary by Taylor (1992) has determined the role of this littoral in modifying concentrations of 'new' nutrients, in the sense of Valiela (1991), derived from the sea, particularly during periods of upwelling. The approach used examined the nature of short-term variability (month) in fluxes onto and from the marshes. And because of the extreme variability of the intertidal areas, a controlled experimental approach was adopted using mesocosms to contain the variability within acceptable limits.

The tidal marsh chosen could be conveniently divided into saltmarsh and tidal creek. The designs depended upon measuring the net flux of nitrate and phosphate in mesocosms prepared from these two areas during periods of simulated non-upwelled and upwelled tides. Control mesocosms were inundated with non-upwelled water for all the six tidal periods studied. A summary of the data is given in Table 4.9.

These data show that both tidal creek and saltmarsh when inundated with upwelled water had a marked effect on the net flux of nitrate, but no change of any significance could be detected for the other nutrients. By contrast,

Table 4.9. A comparison of mean net fluxes (μmol N or P m⁻² per tide) of nutrients in mesocosms of a tidal creek and saltmarsh taken from the Kariega estuary during simulated periods of inundation with non-upwelled and upwelled water. Positive values denote net release by the marsh and negative values denote net uptake

Nutrient	Tidal creek		Saltmarsh	
	Non-upwelled	Upwelled	Non-upwelled	Upwelled
NO_3^-	−85 ± 60 (4)	−514 ± 101 (6)	113 ± 80 (9)	−226 ± 164 (6)
NO_2^-	−12 ± 4 (9)	−6 ± 6 (6)	−7 ± 4 (9)	−1 ± 4 (6)
NH_4^+	1 ± 2 (9)	−1 ± 2 (6)	1 ± 2 (9)	−2 ± 2 (6)
SRP[a]	52 ± 19 (9)	64 ± 32 (6)	21 ± 26 (9)	17 ± 14 (6)

[a] Soluble reactive phosphorus.
Source: From Taylor (1992).

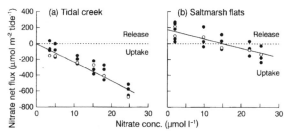

Figure 4.10. **Tidal net fluxes of nitrate vs. initial nitrate concentrations in inundating water for mesocosms of (a) the tidal creek and (b) the saltmarsh at 25 °C (●) and 16 °C (○), in the Kariega River estuary. Regression coefficient, $r^2 = 0.885$ for the tidal creek and 0.897 for the marsh. (After Taylor 1992.)**

during non-upwelling the nitrate flux was small in both regions, with the tidal creek exhibiting a net uptake of nitrate and the saltmarsh a slight net loss. These fluxes are further elaborated in Figure 4.10.

The marked difference in nitrate flux between the two sites is held by Taylor (1992) to be due to the moister, more reduced sediments (E_h <200 mV) and the likely existence of denitrifying microbial assemblages in the tidal creek mesocosms. The lower redox and higher SRP (soluble reactive phosphorus) from the tidal creek compared with the saltmarsh support this view.

The comparability of such experimental systems with natural events has been shown acceptable by Taylor (1992), who determined that over five field tides the nitrate concentrations varied from 5.8 μmol l⁻¹ to 15 μmol l⁻¹, and that the tides with the lowest and highest concentrations of nitrate had the smallest (−7 μmol per tide) and greatest (−153 μmol per tide) uptake, respectively. Taylor (1992) notes that the range of fluxes in the field was narrower than in the mesocosm, which he attributes to drainage features that were not simulated in the mesocosms. For example, as the natural marsh drains its nutrient load is recycled in various ways which would, overall, influence the final value of the flux. What is not at issue is the relationship between the concentration of nutrient, in this case nitrate, in the inundating water and the degree to which it is taken up in the saltmarsh and drainage creeks.

Taylor (1987, 1992) has shown that the positive flux of nitrate in the estuary, assuming a marsh area of 570×10^3 m² (24% of the estuary area) is <1% of the gross input of 150000 g per tide of nitrate to the estuary, and that because of the concentration-dependent nature of such flux, that of upwelled oceanic water entering the estuary plays a major role in the nutrient dynamics of the estuary and its marshes. Of particular significance is the repeated observation (Thompson *et al.* 1995) that nitrogen has been found to limit primary production of saltmarsh vegetation and that in this context Langis *et al.* (1991, cited by Thompson *et al.* 1995) have found that mineralisation of nitrogen in old natural marsh was twice that measured in recently constructed marsh.

A comparable trend is evident in the carbon sources within estuaries, for example, the Palmiet (Branch & Day 1984), Swartvlei (Whitfield 1988), Kariega (Taylor & Allanson 1993) and Kowie (Kokkinn & Allanson 1985) which are characterised by the accretion of allochthonous and autochthonous carbon sources. This carbon is often redistributed through tidal action (Whitfield 1988) and provides carbon subsidies in estuarine localities in which detritus production is minimal, e.g. sandy tidal flats. An indication of the magnitudes of such fluxes is given in Table 4.10.

More recent studies, notably those of Taylor and Allanson (1993, 1995), have determined quantitatively not only the fluxes of organic carbon moieties DOC, POC, TOC between the marsh and the estuary, but also the very significant modulation imposed by the dominant epibenthic crab fauna, principally *Sesarma catenata* and *Cleistosoma (Paratylodiplax) edwardsii*. Potentially, the effects of crabs on the carbon budget of saltmarshes can be partitioned into the effects of the crabs themselves and the effects of their burrows. As the mesocosm experiments reported by Taylor and Allanson (1993) contained pre-existing burrows, the experiments quantified the effect of

Table 4.10. Carbon sources in the Palmiet River estuary and Port Alfred saltmarsh

Palmiet River estuary (kg y^{-1})		Port Alfred saltmarsh (kg y^{-1})	
Cladophora	30 360	Spartina and Ruppia	3270
Phytoplankton	16 000	Phytoplankton	60
Benthic diatoms	400	Epipsammic algae	810
Net imported POC (river & marine)	143 000	Net total imported POC (marine)	3300

Source: From Branch & Day (1984); Kokkinn & Allanson (1985).

Table 4.11. Impacts of crabs Sesarma catenata and Cleistostoma (Paratylodiplax) edwardsii on the daily carbon budget (mg C m^{-2} d^{-1}) at the marsh surface. Symbols not in parentheses represent direction of net flux: + values denote net loss to water column; − values net uptake by marsh surfaces. Effects of crabs are denoted by parenthesized symbols: (+) indicates crabs enhanced the process; (−) indicates crabs reduced the rate. nd: not determined

	Marsh flats (S. catenata)			Tidal creek (C. edwardsii)		
	With crabs	Without crabs	Effect of crabs[a]	With crabs	Without crabs	Effect of crabs[a]
TOC net flux[b]	+376	+304	(+)72	+1180	+723	(+)457
Epibenthic NP[c]	+660	−396	(−)264	−192	−168	(−)24
Sarcocornia NP[d]	−1068	−1068	0	nd	nd	nd
Community NP[e]	−332	−1464	(−)1132	−120	−168	(−)48
Balance d^{-1}[f]	+44	−1160	1204	+1060	+555	505

[a] Mean rates with crabs minus mean rates without crabs.
[b] Calculated from hourly rates, assuming 2.1 and 3.7 h inundation in the marsh flats and tidal creek regions, and both regions inundated twice daily. Includes excretory losses.
[c] Calculated from data in Table 4 of Taylor & Allanson (1993) assuming a 12 h light:12 h dark regime, and subtracting dark respiration from net production.
[d] From Christie (1981), and assumed to be the same in the presence and absence of crabs.
[e] Epibenthic NP plus Sarcocornia NP minus crab respiration.
[f] Community NP minus TOC net flux.
Source: From Taylor & Allanson (1993).

the crabs only and the magnitude of this effect is given in Table 4.11.

The conclusion from these data and their supporting evidence given in more detail by Taylor and Allanson (1993) is that the two crab species enhanced the net loss of carbon from the marsh, although they did so in quite different ways: C. (P.) edwardsii enhances the net loss by increasing TOC flux, while S. catenata does so by reducing epibenthic net primary production.

The presence of macrofauna and interaction with microorganisms can serve to modify substrata indirectly as well, as Gerdol and Hughes (1994) found that the removal of the amphipod Corophium volutator lead to a more stable substratum. They ascribed the increase in stability to the increased quantities of leucopolysaccharide secreted by more abundant microflora, which acted as a binding agent in the sediment. In the presence of C. volutator the biomass of bacteria and diatoms was reduced through predation and resulted in a less stable substratum. In a laboratory study C. volutator has been found to double the oxygen uptake, triple denitrification of nitrate derived from nitrification within the sediment, and cause a 5-fold increase in denitrification of nitrate derived from

the overlying water (Pelegri et al. 1994). Harding (1994) implicated the polychaete worms (Ficopomatus enigmaticus) in Zandvlei, near Cape Town, of contributing towards the maintenance of water quality, allowing the estuary to be used for recreational activities in spite of ecological near-collapses resulting from anthropogenic inputs and modifications.

Meiofauna have not been investigated much in the context of biogeochemistry, but might prove to be very important. Kennedy (1994) reported that nematode community production was exceeded by carbon consumption, and that carbon cycling within the meiofauna must thus account for a significant proportion of benthic production.

An important consequence of these studies is to question the applicability of the now classical 'Outwelling Hypothesis' of E.P. Odum. The studies of Branch and Day (1984), Kokkinn and Allanson (1985), Baird and Winter (1992) and Taylor and Allanson (1995) have provided evidence that, by and large, microtidal estuaries are not exporters, but importers of organic carbon. Furthermore, Taylor and Allanson (1993) have shown that there are frequently large areas of intertidal marsh considered as

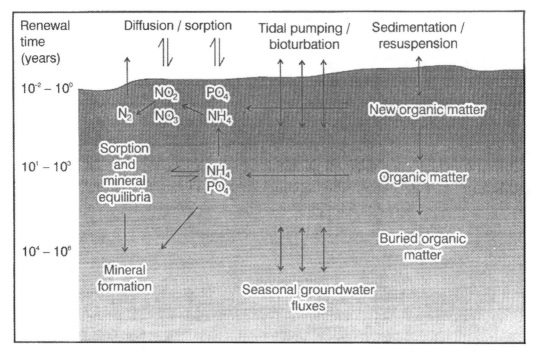

Figure 4.11. **Summary of processes responsible for nutrient exchange at the interface between sediment and the water column. (Redrawn from Fisher *et al.* 1982.)**

'high' marsh which trap detached plant material, so facilitating decomposition within the marsh rather than in the adjoining estuary, a consequence of considerable import in view of the quantity of autochthonous organic debris which accumulates in these microtidal estuaries (cf. Whitfield 1988). A measure of the complexity of the transport within an estuary of, for example, particulate organic carbon (POC) has been elegantly analysed by Imberger *et al.* (1983) who showed that the time scales for this transport can be divided between fast and slow components, largely determined by the intensity of turbulence. The effect of this is for POC to be recycled in the estuary with only a small fraction exported.

A summary of the processes responsible for nutrient exchange in sediments which are sufficiently stable to allow the accumulation of organic material is given by Fisher *et al.* (1982), and from the available evidence applies to the microtidal estuaries of South Africa (Figure 4.11).

THE INFLUENCE OF MAN

Enrichment

Increasing urbanisation of the littoral of South African estuaries and of their near catchments has brought in its wake the not unexpected conditions of nutrient enrichment, either seasonally or persistently. Of course estuaries, by their very nature, are sinks for the natural transport of soluble and particulate materials, both from river catchments and from the sea. In microtidal estuaries of South Africa this chapter has shown so far that there exists an array of metabolic compartments within, for example, a shallow estuary with extensive intertidal marshes or mangroves (see Steinke, Chapter 6) which provide the means whereby these inputs are metabolised. More recently, the investigations of Taljaard *et al.* (1992) on the Diep River estuary (Milnerton Lagoon) near Cape Town have shown the effectiveness of biological agents such as reedbeds in ameliorating many aspects of diffuse pollution for urban runoff.

We are concerned, therefore, less with concentration and more with the impact of enrichment upon the estuarine ecosystem. Impact in this context is determined by external loading and retention time. Estuaries with short retention times (hours) can receive comparatively high external loadings without deleterious effects. Likewise those estuaries with very large tidal prisms (19×10^6 m^3), e.g. the Knysna embayment at spring tide, allow considerable dilution of mineralised sewage effluent so that volumetric loading and its retention becomes small and of short duration.

Few estuaries in South Africa are sufficiently well

known hydrologically to provide sets of well worked out loading data. The Swartkops estuary, Port Elizabeth, is an exception and through the investigations of Emmerson (1985), Baird (1988) and Baird *et al.* (1986, 1988) and the recent detailed investigation of the effect of urban runoff on the water quality of the estuary by MacKay (1993), an assessment of the impact of nutrient inflows for an estuary with a relatively large mean spring tidal prism (3.06×10^6 m^3: MacKay 1993) is now available.

These studies have stressed the importance of understanding the tidal transport within an estuary if any sensible and definitive statements are to be made about the fate and impact of polluting inputs (MacKay 1993). The impact of man-made structures upon circulation, such as road and rail bridges, is always noticeable. MacKay (1993) has shown that while the Wylde Bridge across the Swartkops estuary is not sufficient to restrict flood flows, the levees used to approach the bridge do restrict exchange of new marine water with the upper estuary. A consequence of this is that the lower estuary acts as a marine embayment with strong currents and good flushing. The bar created by the levees is sufficiently dominant to act like a second mouth. Upstream of this point the Swartkops system shows all the hydrodynamic characteristics of a bar-built estuary! And because of this, exchange of water between the upper and lower sections is limited at neap tide.

This effective translocation of an estuary mouth upstream as a result of man-made structures has important consequences as regards the retention of pollutants in the area of the estuary above this secondary mouth. MacKay (1993) has established for the Swartkops estuary that if the volume of the tidal prism is in excess of high water volume below the Wylde Bridge, the upper estuary will experience some exchange with the new marine water. At neap tides, however, the lower estuary is not entirely flushed so that new seawater is unlikely to gain access to the upper estuary. As discharges of three polluted sources occur above the Wylde Bridge and its constricting levees, they will affect the quality of the estuary for extended periods, at least 10–14 days. After flooding Hilmer (cited by MacKay 1993) showed that up to one month was required for salinity to return to pre-flood levels.

The estuaries of the geographical region previously known as the Transkei, now within the boundaries of the Eastern Cape Province, remain comparatively pristine, with the Mngazana River estuary a prime example. This acceptable ecological scenario gives way in KwaZulu-Natal (KZN) to some of the grossest forms of estuarine destruction possible. The pioneer work of Begg (1978) has illustrated how inextricably siltation of these estuaries, as a consequence of the steep topography of the coastline and high rainfall, is linked to the effects of insensitive encroachment of sugarcane plantings and paper pulp manufacture. This degradation has been added to by the increased accessibility of the coast via transport corridors requiring both road and rail traverses of estuaries close to their sensitive mouths.

These descriptive studies were developed by Cooper *et al.* (1993) and Ramm *et al.* (1994) using an Estuarine Health Index (EHI) which involves the collection and presentation of information on geomorphology, biology, water quality and the aesthetic status of estuaries. The component we are particularly interested in here is water quality. To establish existing water quality, comparable locations in the chosen systems were made within a narrow time window. Fifty-six KZN estuaries were surveyed between 19 October and 12 November 1992. Table 4.12 reports the data for the least disturbed and most enriched of three permanently tidal estuaries.

From the array of parameters measured seven were chosen (Table 4.13) reflecting the water's suitability for aquatic life, human contact and eutrophication potential: dissolved oxygen, oxygen absorbed, ammonia, nitrate, phosphate, chlorophyll *a* and *E. coli*. Each variable is given a water quality rating (*q*) which relates the observed concentration to a corresponding rating value between 0 and 10. In this way, all seven variables are assigned rating values based on comparable scales such that the index

$$\frac{1}{10}\left\{\sum_{i=1}^{n} q_i . w_i\right\}^2$$

takes on the range 0–10 and w_i = the weighting of the ith variable as given in Table 4.13.

Using these variable weightings and the analytical description of the water quality of KZN estuaries or river mouths, examples of which are given in Table 4.4 (which records the grossly polluted Sipingo River estuary) and Table 4.12, Cooper *et al.* (1993) and Ramm *et al.* (1994) developed a water quality index based primarily upon the studies of House (1989) and recently in South Africa by Moore (1990), shown in Table 4.14.

When the original table (Cooper *et al.* 1993) is examined, only nine of the 56 estuaries reported on gave an aquatic life index of >3, and 21 gave an index of <1.5. These values indicate the parlous state of a considerable number (>50%) of KZN's estuaries.

Comparable indexes have just become available for Eastern Cape estuaries, north and south of the estuarine port, East London, which illustrate the value of this methodology in assessing the health of the estuaries. And because this part of the coast (Figure 4.1) has not experienced the ribbon development of the KZN coast, the majority of the estuaries examined were found to be in excellent health!

None of the estuaries which remain open to the sea or

Table 4.12. The chemical water quality at three sampling sites in three tidal estuaries in Natal taken from Table 2 (Results of October–November 1992 surveys) of Cooper *et al.* (1993) to illustrate the magnitude of change from the relatively pristine Mzimkulu River estuary to the enriched Mgeni River estuary

Site	Depth (cm)	Secchi (cm)	S-temp (°C)	B-temp (°C)	S-sal	B-sal	S-DO (mg ℓ⁻¹)	B-DO (mg ℓ⁻¹)	S-CHLA (mg ℓ⁻¹)	B-CHLA (mg ℓ⁻¹)	S-NH₃ (µg ℓ⁻¹)	B-NH₃ (µg ℓ⁻¹)	S-NO₃ (µg ℓ⁻¹)	B-NO₃ (µg ℓ⁻¹)	S-PO₄ (µg ℓ⁻¹)	B-PO₄ (µg ℓ⁻¹)	E. coli (100mℓ⁻¹)	OA (mg ℓ⁻¹)
Mzimkulu-1	230	110	24	21	16	33	7.9	6.8	1.2	0.0	0	25	22	21	22	17	9	0.2
Mzimkulu-2	200	60	24	20	6	33	7.6	7.3	0.9	1.8	0	21	21	15	11	12	20	0.1
Mzimkulu-3	30	–	24	24	5	5	7.8	7.8	0.0	–	27		20		10		60	0.5
Tugela-1	210	70	22	21	20	30	6.8	6.0	9.5	12.4	28	177	0	18	8	10	4	4.4
Tugela-2	100	50	22	22	14	27	7.6	6.9	6.8	23.1	193	53	0	11	0	5	192	6.0
Tugela-3	25		23	23	7	7	8.5	8.5	5.3	0.0	0		0		9		0	5.4
Mgeni-1	80	–	23	23	35	35	6.6	6.6	2.4	5.3	157	43	20	18	46	69	0	4.0
Mgeni-2	140	130	24	24	33	33	6.4	6.4	3.0	1.2	419	46	746	47	780	59	660	3.2
Mgeni-3	140	40	27	27	2	23	5.3	4.2	9.2	10.7	1162	423	1724	606	1804	623	1700	4.2

S-, surface sample; B-, bottom sample; †Temp, temperature; Sal, salinity; DO, dissolved oxygen; CHLA, chlorophyll *a*; NH₃, ammonia nitrogen; NO₃, nitrate nitrogen; PO₄, orthophosphate; OA, oxygen absorbed: 4 h oxidation in alkaline KMnO₄.

Table 4.13. Relative weights provisionally assigned to variables of concern

Category	Variables	Basis for inclusion	Weight
Suitability for aquatic life	Dissolved oxygen	Essential to aquatic fauna	0.20
	Oxygen absorbed	Measure of organic feeding	0.05
	Ammonia nitrogen	Toxicity to aquatic fauna	0.10
			0.35
Suitability for human contact	E. coli	Presumptive evidence for human pathogens	**0.30**
Trophic status	Nitrate nitrogen	Aquatic plant growth stimulant	0.10
	Orthophosphate	Aquatic plant growth stimulant	0.15
	Chlorophyll-a	Indicator of algal growth	0.10
			0.35

Table 4.14. Water quality index for three Natal estuaries

Estuary name		Aquatic life	Human contact	Trophic status
Reference	(10.0)	3.5	3.0	3.5
Tugela	(8.4)	3.0	2.2	3.2
Mgeni	(3.8)	1.5	0.3	1.9
Sipingo	(1.4)	0.2	0.0	1.2
Mzimkulu	(8.1)	3.2	1.5	3.4

Source: From Cooper *et al.* (1993).

are subject to short periods of closure has been found to exhibit persistent eutrophication of the magnitude of the Peel-Harvey inlet, Western Australia as reported by McComb *et al.* (1981), although evidence is accumulating that in Swartvlei estuary during the winter and early spring internal cycling of phosphorus and possibly ammonium are responsible for the buildup of *Enteromorpha* spp. in the shallows. This has similar facies to the findings of McComb *et al.* (1981) and of Lavery and McComb (1991), who report that the extensive growths of the macroalgae *Cladophora*, *Enteromorpha* and *Chaetomorpha* are due to nutrients, following river input, being trapped in the sediment from where they are transferred to the overlying algal mats and recycled within the mats, assisted by reducing conditions at the sediment surface. Under high light intensity and increase in temperature the bloom becomes well established.

It would appear that in general the retention times of the estuaries in South Africa are short, although where secondary marine embayments have been formed (in the sense of MacKay 1993), retention of inflows rich in nitrogen and phosphorus can occur for very much longer periods. But even here the volume of contaminated inflow compared with the very much larger volume of resident estuarine water coupled with turbulent mixing within the water column militates against visible eutrophication in

the form of excessive algal growth. An exception is seen in the estuary of the Hartenbos River, south Western Cape which, because of a dam built 12 km upstream of the mouth, suffers from severe restriction in flow. The inflow from the sewage works ameliorates this effect somewhat, but contributes to the loading of N and P, which because of the infrequent tidal flow results in large and unsightly growths of *Enteromorpha* spp. (Bickerton 1982) and *Nannochloris* sp. (personal observations), a sure sign of eutrophication of what remains of the estuary.

Where intertidal marshes occur, there is ample evidence (see Chapter 5) that they act as major sinks for particulate and dissolved organic matter, PIN, DIN, DIP, PIP, etc. These alter the N:P ratio, providing for an overall ratio usually very much less than the Redfield ratio. And although it is recognised that the complexity of nitrogen and phosphorus metabolism in these marshes and their sediments makes adherence to this ratio (Nixon 1981) as a guide to potential eutrophication unwise, South African estuaries along with estuaries of other continents exhibit nitrogen limitation. This occurs notwithstanding the introduction of new nitrogen and phosphorus supplies via either the river or tides and in some urbanised systems via groundwater flow. An explanation, at least in part, is to be found in the metabolism of the intertidal sediments in which denitrification is an important component.

The immediately important implication is, as Balls (1994) reports for Scottish estuaries, and Eyre (1994) has shown for the tropical Moresby River estuary, Queensland, estuaries are systems which act to modify nutrient input and their transport mechanisms on the journey from the catchment to the sea. It is, therefore, critical that the complete estuary–catchment system be included in any design to evaluate cause and effect levels between land use practices and their impact upon the final recipient region, i.e. the coastal waters, reefs and sediments.

Sedimentation and mouth closure

In South Africa, if comparison has to be made, cutting off or greatly reducing river flow is far more long lasting in its effect than eutrophication. This latter can be relieved by sensible waste management, but the loss of river flow, coupled with closure of the bar which is often an eventual consequence, spells the permanent demise of the estuarine system. The increase in the number of lowland rivers which have been dammed during the past 20 years has resulted in a substantial increase in the number of altered estuaries. The estuaries have without exception become predominantly marine as long as the mouth remains open.

Closure usually results in either marked stratification or mixing principally under the stress of wind. Only recently (Slinger *et al.* 1995) has the feasibility of hydrodynamic management been explored, and this work has given us valuable insights into the relative roles of river and marine inflows in defining the chemistry of the affected system. Thus where damming of a lowland river is unavoidable, the design and operation of the dam must permit propitious water releases to relieve incipient stressful conditions in the estuary.

The additional stress of sedimentation and its associated increase in turbidity is of increasing consequence in South African estuaries. Where sedimentation has exceeded erosion, the estuarine floor is raised consequent upon mouth closure by long-shore drift. These elevated estuary floors are typical of many small estuaries of the KwaZulu-Natal coast. They are associated with brackish lagoons of salinities of 15 or less. This change is responsible for the effective loss of the estuarine environment until floods breach the bar and tidal conditions are re-established. Begg (1978) points out that, in addition, siltation has materially altered the storage capacity of many small estuaries, so that even when the bar is breached, the lagoon drains dry and only the highest tides reach the lagoon channel.

In contrast to this infilling with terrigenous silts are those estuaries, particularly of the south and southwest coasts, whose rivers rise in the Cape Fold mountains. They receive relatively little terrigenous material, and infilling of the tidal channel or basin is largely by marine sands deposited as flood tide deltas. The detail of this is described in Chapter 2. And while this is a natural process proceeding without the intervention of Man, human impact is largely felt where closed estuarine mouths are opened when the head of lagoon water and river flow to sustain it is unable to establish a sufficiently deep tidal channel, so that in a matter of days or weeks the bar is re-formed and the cycle is repeated, the estuary becoming shallower and shallower. Even in the estuary of the Knysna River, which has the largest tidal prism ($19 \times 10^6 \, \text{m}^3$ per tide) of all South African estuaries, the influence of anthropogenic disturbance in many small littoral drainage areas is becoming increasingly obvious through the elevation in suspended solid levels (Russell 1996).

There are no reported data on the impact this substantial sedimentation has upon the chemistry of the water column or sediments. Few if any turbidity maxima have been recorded in detail in these microtidal systems, although Grange and Allanson (1995) have described in some detail the positive influence of freshwater inflows upon the distribution of seston in the Great Fish River estuary. Highest values were recorded towards the upper reaches, while in the Kariega River estuary, in which freshwater inflow is greatly attenuated, suspended particles were more uniformly distributed along the length of the estuary.

Contaminants

The biological, physical and chemical environment influences the severity of organic and metal contaminant pollution. For example, the removal of arsenic from contaminated sediment by three benthic invertebrate species was reported by Riedel *et al.* (1989). In a study of a range of organic micropollutants in the Scheldt estuary, van Zoest and van Eck (1991) found a generalised decrease in pollutant concentration with increase in salinity, while an inverse relationship between grain size and concentrations of lead and mercury has been demonstrated by Coakley *et al.* (1993). Microbial activity has also been implicated in the cycling of estuarine sediment contaminants (Elskens *et al.* 1991). Bio-reduction reactions may have different effects on the cycling of trace metals, as trace metal speciation is a function of redox potential and redox potential is related to heterotrophic bacterial activity.

The most commonly reported organic pollutants are polycyclic aromatic hydrocarbons (PAH), polychlorinated biphenyls (PCB) and dichloro-diphenyltrichloroethane (DDT). The common processes involved in the removal of organic pollutants are volatilisation, degradation and photolysis, although there is considerable variation between different substances. Metal contamination is commonly removed by chemical action, particulate sorption, or biological uptake and immobilisation.

Organic contaminants

Batterton *et al.* (1978) showed that the water-soluble fraction (WSF) of oil is of importance to phytoplankton as this facilitates transport into the cells. South African research using phytoplankton cultures has shown that the WSF of outboard motor exhaust emissions (Hilmer & Bate 1983) and used lubricating oil (Bate & Crafford 1985) are toxic to phytoplankton, inhibiting both carbon assimilation and oxygen evolution. Hilmer and Bate (1987) measured above-background hydrocarbon levels in five out of six stations in the Swartkops estuary. Levels were higher after rainfall than pre-rainfall, and were most likely of terrestrial origin as opposed to sea-borne origin from Algoa Bay. There was no direct incrimination of common oil products such as car and diesel sump oil, outboard engine oil or crude oil as the sources, and runoff containing coal dust may have

Table 4.15. Mean PCB and total DDT concentrations for Eastern Cape bird tissues and eggs. Data are from de Kock & Randall (1984) and de Kock (1985) and are based on wet mass

	Tissues			Eggs		
	PCB (μg g^{-1})	t-DDT (ng g^{-1})	n	PCB (ng g^{-1})	t-DDT (ng g^{-1})	n
Kelp Gull	21.71	5370	8	850	300	15
White-breasted Cormorant	0.03	990	3	890	170	9
Jackass Penguin	0.24	270	7	250	110	31
Roseate Tern				710	400	8
Cape Cormorant				210	500	7
African black Oystercatcher				630	300	5
Grey-headed Gull				600	450	4
Cape Gannet				50	400	4
Darter	0.67	800	4			
Reed Cormorant	0.18	620	7			

Concentrations based on single specimens are not reported here. These are for the following species : Whimbrel, Grey Plover, Great Crested Grebe, Black-necked Grebe, Sacred Ibis, Grass Owl, Common Tern and Greenshank.

been responsible in one sample. Their conclusion was that road runoff *per se* may not be important in the overall input of coal dust-related hydrocarbons into the estuary but that bridges and roads may be important in concentrating this form of pollution, resulting in short input peaks after rains.

Kayal and Connell (1990) issued a caveat regarding the interpretation of laboratory-based data relating to the sediment–water partitioning PAHs in particular – their findings were that field conditions have distinctive differences from laboratory experiments that do not allow direct translation of laboratory-based relationships to the natural environment.

Pesticide-based organic contaminants have been studied intensively in South Africa. Fish and sediment samples from Kosi Bay in 1976 revealed concentrations of DDT and its degradation products, DDE and TDE, higher than those found elsewhere along the KwaZulu-Natal coast. The proportion of DDT relative to its degradation products in most of the tissue specimens, and the measurable presence of these compounds in bottom sediments, indicated that contamination was recent, and the apparent source was the anti-malarial spraying programme being conducted in the region. A second set of samples obtained from the region five years later showed a marked decrease in the levels of all three chlorinated hydrocarbons (Butler *et al.* 1983). This apparent reduction of DDT suggested to the authors that the measures introduced by the malaria-control body succeeded in reducing the environmental impact of their activities. In a study of the degradation of various organophosphorus pesticides (e.g. Malathion) in estuarine water samples, Lacorte *et al.* (1995) reported half-lives of less than one week to about 10 days, and 7–12 days for some degradation products. This is in

dramatic contrast to organochlorines which have half-lives measured in years. In the mid-1980s the levels of chlorinated hydrocarbon pesticides and PCBs in various groups of animals, including estuarine fish and birds were reported on by de Kock (1985), and summarised in Tables 4.15 and 4.16.

Caution should be used in interpreting the mean values reported in Table 4.15, as the sample sizes are low and variability high. The condition of the animal is an important factor in this high variability, since chlorinated hydrocarbon residues tend to accumulate in fatty tissue. Nevertheless, based on the egg data and the feeding habits of the birds involved, de Kock and Randall (1984) concluded that the Eastern Cape estuaries were more polluted than the marine, but less so than the freshwater environments. Fortunately the concentrations of chlorinated hydrocarbons were well below those considered critical for reproductive impairment, i.e. DDE levels greater than about 3–10 μg g^{-1} fresh weight in the eggs, as were the PCB levels.

The estuarine fish component of this study was conducted on an estuary-by-estuary basis. Whereas the recommended limits for safety are generally accepted as 1 μg l^{-1} for t-DDT and 0.5 μg l^{-1} for PCBs, individual specimens from the Nahoon, Swartkops and Wilderness Lakes were found which exceeded these limits. PCBs exceeding the recommended levels were recorded from individuals in the Nahoon and Swartkops estuaries.

Of interest are the high concentrations of PCB residues in fish from the Swartkops estuary. This system is close to the industrialised area of Port Elizabeth, but the Nahoon and Buffalo systems which are straddled by East London's industrial areas exhibited lower concentrations. Similarly the Sundays River and Gamtoos River estuaries, which

Table 4.16. Geometric means of t-DDT and PCB concentrations from Eastern Cape estuarine fish

Species	A DDT n	B DDT n	C DDT n	D DDT n	E DDT n	F DDT n	G DDT n	H DDT n	I DDT n	J DDT n
Myliobatus aquila							194.0 2		ND 1	
Elops machnata		46.4 12			0.7 1		57.7 4			
Pomadasys commersonnii			4.5 3	2.6 7	3.9 8	4.6 3	114.2 5	9.8 17	2.0 1	
Gilchristella aestuaria		3.5 2					7.0 1	66.7 1		
Pomatomus saltatrix		4.6 2				2.9 2	3.0 2		2.2 3	
Monodactylus falciformis		1.3 1	4.9 1	16.9 3	13.6 1	3.3 3	17.7 1		2.6 4	
Lithognathus lithognathus		2.3 1	6 1			1.0 4	11.7 3	7.2 3		
Liza tricuspidens			6.1 1	0.7 1		2.2 3	2.9 5	23.5 12	1.0 1	
Mugil cephalus		21.7 1		5.7 4	7.4 6	2.8 7	9.0 1	19.8 8		18.7 25
Galeichthys feliceps	15.9 5	15.9 9	23.2 6	4.9 5	5.9 13	4.4 5	14.3 5	9.0 7	8.1 9	
Rhabdosargus holubi		2.1 2				ND 1	3.5 12		1.0 1	50.1 2
Liza dumerilii							3.2 9	3.2 1		72.1 4
Solea spp.							9.9 4			
Lichia amia	25.2 5	40.8 6	11.3 3	1.0 1	2.1 5	2.4 2	3.7 1	4.6 3	1.6 8	13.2 1
Liza richardsonii		4.2 7	6.9 3	1.0. 3		1.1 5	3.6 3	13.4 8	0.7 2	17.2 9
Argyrosomus hololepidotus		30.6 3	14.0 3	1.4 7	0.9 8	1.1 10	2.8 2	8.5 11	3.9 4	8.5 1
Cyprinus carpio	55.0 3			5.2 7		2.9 2		82.7 2		
Myxus capensis	8.3 4			1.0 1				9.3 7		
Hepsetia breviceps										16.9 3
Oreochromis mossambicus										4.2 10

Species	A PCB n	B PCB n	C PCB n	D PCB n	E PCB n	F PCB n	G PCB n	H PCB n	I PCB n
Myliobatus aquila							4105.0 2		534.0 1
Elops machnata		101.9 12			ND 1		506.7 4		
Pomadasys commersonnii			21.6 3	5.2 7	13.9 8	8.2 3	283.9 5	ND 17	ND 1
Gilchristella aestuarius		11.3 2					178.0 1	ND 1	
Pomatomus saltator		17.0 2				42.0 2	72.4 2		11.0 3
Monodactylus falciformis		3.8 1	3.2 1	37.1 3	4.3 1	8.0 3	63.7 1		ND 4
Lithognathus lithognathus		4.6 1	54.0 1			8.3 4	60.1 3	ND 3	
Liza tricuspidens		14.2 1		ND 1		3.1 3	59.5 5	ND 12	ND 1
Mugil cephalus		6.5 1		121.0 4	ND 6	33.3 7	58.0 1	7.4 8	
Galeichthys feliceps	13.7 5	19.4 9	19.9 6	13.5 5	11.5 13	21.4 5	42.3 5	11.0 7	16.1 9
Rhabdosargus holubi		17.4 2				33.0 1	28.3 12		ND 1
Liza dumerilii							26.0 9	10.0 1	
Solea spp.							24.5 4		
Lichia amia	71.8 5	47.5 6	29.6 3	ND 1	2.5 5	13.3 2	21.8 1	6.5 3	5.2 8
Liza richardsonii		10.6 7	16.6 3	ND 3		10.2 5	14.1 3	ND 8	ND 2
Argyrosomus hololepidotus		42.8 3	30.4 3	34.6 7	2.6 8	15.9 10	12.4 2	ND 11	14.0 4
Cyprinus carpio	57.9 3			5.0 7		10.5 2		ND 2	
Myxus capensis	26.4 4			ND 1				12.7 7	

Estuaries: A, Gqunube; B,: Nahoon; C, Buffalo; D, Great Fish; E, Bushmans; F, Sundays; G, Swartkops; H, Gamtoos; I, Kromme; J, Wilderness Lakes.
Units are ng.g^{-1} wet mass. ND = not detected.
Data from de Kock 1985.

have extensive agricultural areas in their catchments, had low levels of contamination.

Although organochlorines and PCBs have received the most attention and publicity to date, industrialised societies produce a large quantity and variety of organic contaminants, the effects of which are at present unknown. This future generation of research potential has already started manifesting itself – the presence of anthropogenic brominated and nitrated phenols has been reported in estuarine sediments (Tolosa et al. 1991).

Metals

Metal total concentrations in the sediment are of limited interpretative value in that they do not necessarily reflect the biologically or chemically reactive fraction. It is the kinetic lability and bioavailability which will determine their toxicity or other impact on the biota. Specific metal concentrations can vary in the space of metres, and with grain size. Coakley et al. (1993), for example, found an inverse relationship between grain size and lead and mercury concentrations in the St Lawrence River estuary. Thus varying levels are not necessarily indicative of

Table 4.17. Metal concentrations in surface water samples from southern and eastern Cape estuaries

Estuary	Cu (µg ℓ$^{-1}$)	Pb (µg ℓ$^{-1}$)	Zn (µg ℓ$^{-1}$)	Fe (µg ℓ$^{-1}$)	Mn (µg ℓ$^{-1}$)	Co (ng ℓ$^{-1}$)	Ni (µg ℓ$^{-1}$)	Cd (ng ℓ$^{-1}$)	Cr (µg ℓ$^{-1}$)	Hg (µg ℓ$^{-1}$)	Na (µg ℓ$^{-1}$)	K (µg ℓ$^{-1}$)	Mg (µg ℓ$^{-1}$)	Ca (µg ℓ$^{-1}$)	Sr (ng ℓ$^{-1}$)	Reference
Knysna	0.2	0.6	0.3	81	5	130	0.1	100	0.1	199						a
Bietou	0.1	2.6	2	68	14	100	7	2900		29						b
Keurbooms	0.1	2.5	1.9	61	6.7	300	1	2100		21						b
Kromme	0.6	0.2	0.5	100	5.4	50	0.1	130		10	10767		300	1420	422	c
Gamtoos	0.6	0.6	1.1	372	41	190	0.2	100		11	3878	95	539	192		c
Swartkops	3.9	1.5	3.6	275	41	100	1.7	300		5	9060	282	1184	341		d
Sundays	3.2	0.7	2.4	334	18	100	1	40	1.3	33	4	0.1				e
Bushmans	1.8	0.2	0.5	302	10.2	0.2	0.23	100	0.4	6			0.2	0.03	9	f
Kariega	1.5	0.22	0.51	170	9.5	0.1	0.13	70	0.3	3						f
Kowie	1.7	0.33	0.63	254	14.6	0.18	1.07	50	0.2	15						f
Great Fish	2.4	1.1	2.1	133	52	100	2.1	60	2.1	75	13	0.3	0.1	0.4	13	f
Buffalo	3.1	78	4.7	154	15	500	35	30	0.1	1530	11	0.3	1	0.4	11	g
Blind	3	17	1.2	38	3	300	2.2	260	1.4	70	4	0.2	2	0.2	5	h
Ihlanza	6.5	20	34	36	8	500	3.9	120	9.1	10	7	0.2	1	0.3	12	h
Nahoon	0.5	64	2.6	110	19	400	4.8	90	0.5	2200	12	0.4	1	0.5	12	h
Quinera	1.5	12	5.8	202	36	100	3.4	20	2.7	10	10	0.3	1	0.4	9	h

References: a. Watling & Watling 1982a; b. Watling & Watling 1982b; c. Watling & Watling 1982c; d. Watling & Watling 1982c; e. Watling & Watling 1982e; f. Watling & Watling 1982e; g. Watling et al. 1985; h. Talbot et al. 1985.

Table 4.18. Metal concentrations in surface sediment samples from southern and eastern Cape estuaries

Estuary	Cu (µg g^{-1})	Pb (µg g^{-1})	Zn (µg g^{-1})	Fe (µg g^{-1})	Mn (µg g^{-1})	Co (µg g^{-1})	Ni (µg g^{-1})	Cd (ng g^{-1})	Cr (µg g^{-1})	Hg (ng g^{-1})	Na (mg g^{-1})	K (mg g^{-1})	Mg (mg g^{-1})	Ca (mg g^{-1})	Sr (mg g^{-1})	Al (mg g^{-1})	Reference
Knysna	5	14	17	10120	40	3	7	600	21		7	3	0.4	17	1	1	a
Bietou	3	2	27		54	1	5	200	2		6	1	0.3	13	0.3	11	b
Keurbooms	2	1	17	6600	73	1	7	100	1		3	0.4	1	109	0.3	7	b
Kromme	1	3	4	2653	20	4	2	31	7		3	1	1	1	0.01	2	c
Gamtoos	5	7	16	9180	79	1	8	307	15		2	9	2	70	3	9	d
Swartkops	18	31	55	20800	298	10	13	200	5		12	4	7	11	23	243	e
Sundays	16	18	57	72	360	2	19	59	38		5	2	3	27	0.3	26	f
Bushmans	3	5	13	7330	49	5	5	60	22		3	3	8	44	0.4	15	f
Kariega	8	10.7	27	16048	131	18	24	60	27		9	6	5	1	0.01	39	f
Kowie	17	29.8	96	32300	530	5	11	222	64	30	7	3	3	5	0.1	13	f
Great Fish	7	10	34	10760	140	6	14	830	13		3	3	3	4	26	24	g
Buffalo	24	4	37	31	143	3	7	230	39		5	2	3	51	0.4	9	h
Blind	29	14	124	9	346	0.5	1	80	29		3	0.2	2	124	1	1	h
Ihlanza	29	24	5	2	35		17	260	7		1	4	6	34	0.4	12	h
Nahoon	15	17	47	19	280	4		50	34		3	2	2	4		16	h
Quinera	9	14	19	11	108	601	10		44		4				31		h

References:
a. Watling & Watling 1982a; b. Watling & Watling 1982b; c. Watling & Watling 1982c; d. Watling & Watling 1982c; e. Watling & Watling 1982e; f. Watling & Watling 1982e; g. Watling et al. 1985; h. Talbot et al. 1985.

pollution, and conclusions should be drawn on the basis of overall trends and metal interrelationships rather than on absolute concentrations (Watling & Watling 1983b).

Trace metals do, however, have an effect on microorganisms active in recycling organic material, particularly the nitrifying bacteria, and in general on those involved with nitrogen cycling more than those active in cycling carbon. Metals which have been implicated include Cd, Cr, Ni, Cu, Zn and Pb. Gilbert (1990), investigating nitrogen, phosphorus, heavy metals (Zn, Cu, Pb and Hg) and plant productivity found assimilation of Cu and Hg, with 11.2 times more Hg in plants than in sediment. Cu was more actively assimilated in the roots during fast growth, while Pb and Zn were passively absorbed. In a Louisiana study, Gaston and Young (1992) found macrofauna numbers inversely correlated to Pb, Hg, Cd and to a lesser extent also Cu, Cr and Zn.

South African research has shown the effect of Cu, Pb and Zn on nitrogen regeneration (Talbot 1988). Zinc at concentrations below 500 μg l^{-1} led to increased ammonia in the experimental sediments and decreased nitrate production. This was ascribed to an inhibitory effect of the metal on the microorganisms involved. Inhibition of nitrification was reported as 50–60% at Zn concentrations of 500 μg l^{-1}, 70–80% at 1 mg l^{-1} and 100% at 10 mg l^{-1}. Copper at 10 and 100 μg l^{-1} had no inhibitory effect, but at very low (1 μg l^{-1}) levels there was a slight (15%) inhibition of nitrification for 24 hours. There was evidence that both ammonification and nitrification were inhibited during the first 24 h at 1 μg l^{-1}, but after 48 h full recovery had

taken place. Neither inorganic nor organic forms of Pb showed any effect, although at low concentrations there was an effect similar to Cu. In terms of effects on nitrogen cycling, Zn had the greatest impact, followed by Cu and then Pb. This order of toxicity (Zn > Cu > Pb) has also been documented for freshwater systems (Forstner & Wittmann 1981).

Metals may also affect estuaries at higher levels of organisation: multivariate analyses indicated a strong relationship between nematode community structure and metal concentrations in the Fal system of southwest England (Somerfield et al. 1994). The metal surveys of estuaries in the southern and eastern Cape by John Watling and co-workers from the mid-1970s to the mid-1980s provided a set of baseline data for surface water and surface sediment, and are summarised for the middle reaches in Tables 4.17 and 4.18.

The overall conclusion from these studies (Watling & Watling 1975, 1976, 1977, 1980, 1982a–e, 1983a, b; Talbot et al. 1985) is that the southern and eastern Cape estuaries could not be considered polluted.

In a summary of contaminant investigations using cores of estuarine and coastal sediments, Valette-Silver (1993) shows that concentrations of many contaminants are decreasing in European estuaries. Heavy metals and PAH levels showed a gradual increase from the mid-1800s but most were already showing a decreasing trend from around 1970. PCBs started dropping in the late 1970s after a two-decade long plateau reached from a meteoric climb starting in the mid-1920s, and only Ni and Co still seem to be increasing.

References

Allanson, B.R. (1989). *Bacterial contamination of the Swartvlei estuary. Final Report and recommendations to the Sedgefield Municipality.* 18pp & 2 Appendices.

Allanson, B.R. & Howard-Williams, C. (1984). A contribution to the chemical limnology of Swartvlei. *Archiv für Hydrobiologie,* **99**, 133–59.

Allanson, B.R. & Read, G.H.L. (1987). *The response of estuaries along the south eastern coast of southern Africa to marked variation in freshwater inflow.* Institute for Freshwater Studies, Rhodes University, Grahamstown, Special Report No. 2/87: 40 pp.

Allanson, B.R. & Read, G.H.L. (1995). Further comment on the response of south east coast estuaries to variable freshwater flows. *Southern African Journal of Aquatic Sciences,* **21**, 56–70.

Allanson, B.R. & van Wyk, J.D. (1969). An introduction to the physics and chemistry of some lakes in northern Zululand. *Transactions of the Royal Society of South Africa,* **38**, 217–40.

Atkinson, M.J., Berman, T., Allanson, B.R. & Imberger, J. (1987). Fine-scale oxygen variability in a stratified estuary: patchiness in aquatic environments. *Marine Ecology Progress Series,* **36**, 1–10.

Azam, F., Fenchel, T., Field, J.G., Gray, J.S., Meyer-Reil, L.A. & Thingstad, F. (1983). The ecological role of water-column microbes in the sea. *Marine Ecology Progress Series,* **10**, 257–63.

Baird, D. (1988). Synthesis of ecological research in the Swartkops estuary. In *The Swartkops estuary,* ed. D. Baird, J.F.K. Marais & A.P. Martin, pp. 41–56. South African National Scientific Programmes Report No. 156. Pretoria: CSIR.

Baird, D., Hanekom, N.M. & Grindley, J.R. (1986). Report no 23: Swartkops (CSE 3). In *Estuaries of the Cape, Part 2. Synopsis of available information on individual systems,* ed. A.E.F. Heydorn & J.R. Grindley. CSIR Research report No. 422, 82pp.

Baird, D., Marais, J.F.K. & Martin, A.P. (eds.) (1988). *The Swartkops estuary.* Proceedings of a symposium held on 14 and 15 September 1987 at the University of Port Elizabeth. South African National Scientific Programmes Report No. 156. Pretoria: CSIR.

Baird, D. & Winter, P.E.D. (1992). Flux of inorganic nutrients and particulate carbon between a *Spartina maritima* salt marsh and the Swartkops estuary, eastern Cape. *Southern African Journal of Aquatic Sciences,* **18**, 64–73.

Balls, P.W. (1994). Nutrient inputs to estuaries from nine Scottish East coast rivers. Influence of estuarine processes on inputs to the North Sea. *Estuarine, Coastal and Shelf Science,* **39**,

Bally, R. & McQuaid, C.D. (1985). Physical and chemical characteristics of the waters of the Bot River estuary, South Africa. *Transactions of the Royal Society of South Africa,* **45**, 317–32.

Bate, G.C. & Crafford, S.D. (1985). Inhibition of phytoplankton photosynthesis by the WSF of used lubricating oil. *Marine Pollution Bulletin,* **16**, 401–04.

Batterton, J.C., Winters, K. & van Baalen, C. (1978). Sensitivity of three micro-algae to crude oils and fuel oils. *Marine and Environmental Research,* **1**, 31–41.

Begg, G.W. (1978). *The estuaries of Natal.* Natal Town and Regional Planning Report 41, Pietermaritzburg.

Bell, C.R. & Albricht, L.J. (1981). Attached and free-floating bacteria in the Fraser river estuary, British Columbia, Canada. *Marine Ecology Progress Series,* **6**, 317–27.

Benner, R., Fogel, M.L. & Sprague, E.K. (1991). Diagenesis of belowground biomass of *Spartina alterniflora* in salt marsh sediments. *Limnology and Oceanography,* **36**, 1358–74.

Bent, E.J. & Goulder, R. (1981). Planktonic bacteria in the Humber estuary; seasonal variation in population, density and heterotrophic activity. *Marine Biology,* **62**, 35–45.

Bickerton, I.B. (1982). Report No. 11: Hartenbos (CMS 1). In *Estuaries of the Cape. Part 2. Synopses of available information on individual systems,* ed. A.E.F. Heydorn & J.R. Grindley. CSIR Research Report No. 410. Stellenbosch: CSIR.

Biggs, R.B. & Cronin, L.E. (1981). Special characteristics of estuaries. In *Estuaries and nutrients,* ed. B.J. Neilson & L.E. Cronin, pp. 3–23. Clifton, NJ: Humana Press.

Branch, G.M. & Day, J.A. (1984). Ecology of southern African estuaries: Part 13. The Palmiet river estuary in the south-western Cape. *South African Journal of Zoology,* **19**, 63–77.

Butler, A.C., Sibbald, R.R. & Gardner, B.D. (1983). Gas chromatographic analysis indicates decrease in chlorinated hydrocarbon levels in Northern Zululand. *South African Journal of Science,* **79**, 162–3.

Butler, E.J. & Tibbits, S. (1972). Chemical survey of the Tamar estuary. I. Properties of the waters. *Journal of the Marine Biological Association UK,* **52**, 681–99.

Caffrey, J.M. (1995). Spatial and seasonal patterns in sediment nitrogen remineralisation and ammonium concentrations in San Francisco Bay, California. *Estuaries,* **18**, 219–33.

Caffrey, J.M. & Kemp, W.M. (1990). Nitrogen cycling in sediments with estuarine populations of *Potamogeton perfoliatus* and *Zostera marina. Marine Ecology Progress Series,* **66**, 147–60.

Chambers, R.M., Fourqurean, J.W., Hollibaugh, J.T. & Vink, S.M. (1995). Importance of terrestrially derived, particulate phosphorus to phosphorus dynamics in a west coast estuary. *Estuaries,* **18**, 518–26.

Childers, D.L. & Day, J.W. (1990). Marsh water column interactions in two Louisiana estuaries. I. Sediment dynamics. *Estuaries,* **13**, 393–403.

Childers, D.L., McKellar, H.N., Dame, R.F., Sklar, F.H. & Blood, E.R. (1993). A dynamic nutrient budget of subsystem interactions in a salt marsh estuary. *Estuarine, Coastal and Shelf Science,* **36**, 105–31.

Christie, N.D. (1981). Primary production in Langebaan Lagoon. In *Estuarine ecology with particular reference to Southern Africa,* ed. J.H. Day, pp. 101–16. Cape Town: A.A. Balkema.

Clavero, V., Niell, F.X. & Fernandez, J.A. (1991). Effects of *Nereis diversicolor* O.F. Muller abundance on the dissolved phosphate exchange between sediment and overlying water in Palmones River estuary (southern Spain). *Estuarine, Coastal and Shelf Science,* **33**, 193–202.

Coakley, J.P., Nagy, E. & Serodes, J.B. (1993). Spatial and vertical trends in sediment phase contaminants in the upper estuary of the St Lawrence River. *Estuaries,* **16**, 653–69.

Cole, J.J., Findlay, S. & Pace, M.L. (1988).

Bacterial production in fresh and saltwater ecosystems: a cross-system overview. *Marine Ecology Progress Series*, **43**, 1–10.

Cooper, J.A.G., Harrison, T.D., Ramm, A.E.L. & Singh, R.A. (1993). *Refinement, enhancement and application of the Estuarine Health Index to Natal's estuaries, Tugela–Mtamvuna*. Technical Report, Department of Environment Affairs. Pretoria: CSIR.

Cooper, J.A.G., Mason, T.R., Grobbler, N.G. (1988). Flood effects on Natal estuaries: a sedimentological perspective. In *Interim report on estuarine and coastal processes programmes, Geoscience Project Activities*, pp 1–11. Pretoria: Water Technology CSIR.

De Kock, A.C. (1985). *Polychlorinated biphenyls and organochlorine compounds in marine and estuarine systems*. MSc thesis, University of Port Elizabeth.

De Kock, A.C. & Randall, R.M. (1984). Organochlorine insecticide and polychlorinated biphenyl residues in eggs of coastal birds from the Eastern Cape. *Environmental Pollution*, **35**, 193–201.

Dyer, K.R. (1973). *Estuaries: a physical introduction*. London: John Wiley.

Eagle, G.A. & Bartlett, P.D. (1984). *Preliminary chemical studies in four Cape estuaries*. CSIR Report T/SEA 8307: 46 pp.

Elskens, M., Leermakers, M., Panutrakul, S., Monteny, F. & Baeyens, W. (1991). Microbial activity in sandy and muddy estuarine sediments. *Geo-Marine Letters*, **11**, 194–98.

Emmerson, W.D. (1985). The nutrient status of the Swartkops River estuary, eastern Cape. *Water SA*, **11**, 189–98.

Eyre, B. (1994). Nutrient biochemistry in the tropical Moresby River system, North Queensland, Australia. *Estuarine, Coastal and Shelf Science*, **39**, 15–31.

Fisher, T.R., Carlson, P.R. & Barber, R.T. (1982). Sediment nutrient regeneration in three North Carolina estuaries. *Estuarine, Coastal and Shelf Science*, **14**, 101–6.

Forstner, U., & Wittmann, G.T.W. (1981). *Metal pollution in the aquatic environment*. Berlin: Springer-Verlag.

Froelich, P.N. (1988). Kinetic control of dissolved phosphate in natural rivers and estuaries: a primer on the phosphate buffer mechanism. *Limnology and Oceanography*, **33**, 649–68.

Gardner, W.S., Seitzinger, S.P. & Malczyk, J.M. (1991). The effects of sea salts on the forms of nitrogen released from estuarine and fresh water sediments. Does ion-pairing affect ammonium flux? *Estuaries*, **14**, 157–66.

Gaston, G.R. & Young, J.C. (1992). Effects of contaminants of macrobenthic communities in the Upper Calcasieu Estuary, Louisiana. *Bulletin of Environmental Contamination and Toxicology*, **49**, 922–8.

Gerdol, V. & Hughes, R.G. (1994). Effect of *Corophium volutator* on the abundance of benthic diatoms, bacteria and sediment stability in 2 estuaries in southeastern England. *Marine Ecology Progress Series*, **114**, 109–15.

Gilbert, H. (1990). Nutrient elements (N & P), heavy metals (Zn, Cu, Pb & Hg) and plant productivity in an intertidal freshwater marsh in Quebec City. *Canadian Journal of Botany*, **68**, 857–63 (Abstract only).

Gontier, G., Grenz, C., Calmet, D. & Sacher, M. (1992). The contribution of *Mytilus* sp. in radionuclide transfer between water column and sediments in the estuarine and delta systems of the Rhône River. *Estuarine, Coastal and Shelf Science*, **34**, 593–601.

Grange, N.R. & Allanson, B.R. (1995). The influence of freshwater inflow on the nature, amount and distribution of seston on estuaries of the Eastern Cape, South Africa. *Estuarine, Coastal and Shelf Science*, **40**, 402–20.

Harding, W.R. (1994). Water quality trends and the influence of salinity in a highly regulated estuary near Cape Town, South Africa. *South African Journal of Science*, **90**, 240–46.

Harvey, J.W. & Odum, W.E. (1990). The influence of tidal marshes on upland groundwater discharge to estuaries. *Biogeochemistry*, **10**, 217–36.

Haw, P.M. (1984). *Freshwater requirements of the Knysna estuary*. MSc thesis, University of Cape Town, Rondebosch. Research report No.5.

Head P.C. (1985). Data presentation and interpretation. In *Practical estuarine chemistry: a handbook*, ed. P.C. Head, pp. 278–337. Cambridge: Cambridge University Press.

Healey, M.J., Moll, R.A. & Diallo, C.O. (1988). Abundance and distribution of bacterioplankton in the Gambia River, West Africa. *Microbial Ecology*, **16**, 291–310.

Hemens, J., *et al.* (1971). *Natal coast estuaries:*

environmental surveys, 3: the Kosi Bay estuarine lakes. CSIR/NIWR Project Report 6201/9728.

Hilmer, T. & Bate, G.C. (1983). Observations on the effect of outboard motor fuel oil on phytoplankton cultures. *Environmental Pollution* (Ser. A), **32**, 307–16.

Hilmer, T. & Bate, G.C. (1987). Hydrocarbon levels in the Swartkops estuary: a preliminary study. *Water SA*, **13**, 181–4.

House, M.A. (1989). A water quality index for river management. *Journal IWEM*, **3**, 336–44.

Howard-Williams, C. (1977). The distribution of nutrients in Swartvlei, a southern Cape coastal lake. *Water SA*, **3**, 213–17.

Howard-Williams, C. & Allanson, B.R. (1981). Phosphorous cycling in a dense *Potamogeton pectinatus* L. bed. *Oecologia*, **49**, 56–66.

Howes, B.J., Dacey, J.W.H. & King, G.M. (1984). Carbon flow through oxygen and sulfate reduction pathways in saltmarsh sediments. *Limnology and Oceanography*, **29**, 1037–51.

Howes, B.L. & Goehringer, D.D. (1994). Porewater drainage and dissolved organic carbon and nutrient losses through the intertidal creekbanks of a New England salt marsh. *Marine Ecology Progress Series*, **114**, 289–301.

Imberger, J., Berman, T., Christian, R.R., Sherr, E.B., Whitney, D.E., Pomeroy, L.R., Wiegert, R.G. & Wiebe, W.J. (1983). The influence of water motion on the distribution and transport of materials in a salt marsh estuary. *Limnology and Oceanography*, **28**, 201–14.

Iriberri, J., Unanue, M., Barcina, I. & Egea, L. (1987). Seasonal variation in population density and heterotrophic activity of attached and free-living bacteria in coastal waters. *Applied and Environmental Microbiology*, **53**, 2308–14.

Justic, D., Rabelais, N.N., Turner, R.E. & Dortch, Q. (1995). Changes in nutrient structure of river-dominated coastal waters: stoichiometric nutrient balance and its convergence. *Estuarine, Coastal and Shelf Science*, **40**, 339–56.

Kayal, S.I. & Connell, D.W. (1990). Partitioning of unsubstituted polycyclic aromatic hydrocarbons between surface sediments and the water column in the Brisbane River estuary. *Australian Journal of Marine and Freshwater Research*, **41**, 443–56.

Kennedy, A.D. (1994). Carbon partitioning within meiobenthic nematode communities in the Exe Estuary, UK. *Marine Ecology Progress Series*, **105**, 71–8.

Kennish, M.J. (1986). *Ecology of estuaries*. Vol.1. Physical and chemical aspects i–iv. Boca Raton, Florida: CRC Press.

King, L.C. (1972). *The Natal monocline explaining the origin and scenery of Natal, South Africa*. Durban: University of Natal.

Koepfler, E.T., Benner, R. & Montagna, P.A. (1993). Variability of dissolved organic carbon in sediments of a seagrass bed and an unvegetated area within an estuary in southern Texas. *Estuaries*, **16**, 391–404.

Kokkinn, M.J. & Allanson, B.R. (1985). On the flux of organic carbon in a tidal salt marsh, Kowie River estuary, Port Alfred, South Africa. *South African Journal of Science*, **81**, 613–17.

Korringa, P. (1956). *Oyster culture in South Africa. Hydrological, biological and ostreological observations in the Knysna lagoon, with notes on conditions in other South African waters*. Investigational Report of the Division of Sea Fisheries of South Africa No. 20, 1–85.

Krambeck, C. (1988). Control of bacterioplankton structures by grazing and nutrient supply during the decline of an algal bloom. *Verhandlungen der Internationale Vereinigung für theoretische und angewandte Limnologie*, **23**, 496–502.

Lacorte, S., Lartiges, S.B., Garrigues, P. & Barcelo, D. (1995). Degradation of organophosphorus pesticides and their transformation products in estuarine waters. *Environmental Science and Technology*, **29**, 431–8.

Largier, J.L. (1986). Structure and mixing in the Palmiet estuary. *South African Journal of Marine Science*, **4**, 139–52.

Largier, J.L & Slinger J.H. (1991). Circulation in highly stratified southern African estuaries. *Southern African Journal of Aquatic Sciences*, **17**, 103–15.

Largier, J.L. & Taljaard, S. (1991). The dynamics of tidal intrusion, retention, and removal of seawater in a bar-built estuary. *Estuarine, Coastal and Shelf Science*, **33**, 325–38.

Lavery, P.S. & McComb, A.S. (1991). Macroalgal–sediment nutrient interactions and their importance to macroalgal nutrition in a eutrophic estuary. *Estuarine, Coastal and Shelf Science*, **32**, 281–95.

Law, C.S., Rees, A.P. & Owens, N.J.P. (1991). Temporal variability of denitrification in estuarine sediments. *Estuarine, Coastal and Shelf Science*, **33**, 37–56.

Liptrot, M.R.N. (1978). *Community metabolism and phosphorus dynamics in a seasonally closed South African estuary*. MSc thesis, Rhodes University, Grahamstown.

McComb, A.J., Atkins, R.P., Birch, P.B., Gordon, D.M. & Lukatelich, R.J. (1981). Eutrophication in the Peel-Harvey estuary system. In *Estuaries and nutrients*, ed. B.J. Neilson & L.E. Cronin, pp. 323–43. Clifton, NJ: Humana Press.

MacKay, H.M. (1993). *The impact of urban runoff on the water quality of the Swartkops estuary: implications for water quality management*. WRC Report KV 45/93. Pretoria: Water Research Commission.

McRoy, C.P., Barsdale, R.J. & Nebert, M. (1972). Phosphorus cycling in an eelgrass (*Zostera marina* L.) ecosystem. *Limnology and Oceanography*, **17**, 58–67.

Maher, W.A. & DeVries, M. (1994). The release of phosphorus from oxygenated estuarine sediments. *Chemical Geology*, **112**, 91–104.

Millard, N.A.H. & Scott, K.M.F. (1954). The ecology of South African estuaries. Part 6. Milnerton estuary and the Diep river, Cape. *Transactions of the Royal Society of South Africa*, **34**, 279–324.

Mitchell-Innes, B.A. (1988). Changes in phytoplankton populations after an incursion of cold water along the coast at Tsitsikamma National Park. *South African Journal of Marine Science*, **6**, 217–26.

Moore, C.A. (1990). *Classification systems and indices for reporting and determining the effect of management on water quality in South African waterbodies*. Summary report, CSIR Division of Water Technology. Pretoria: CSIR.

Morris, A.W. (1985). Estuarine chemistry and general survey strategy. In *Practical estuarine chemistry: a handbook*, ed. P.C. Head, pp. 1–60. Cambridge: Cambridge University Press.

Naidoo, G. & Raiman, F. (1982). Some physical and chemical properties of mangrove soils at Sipingo and Mgeni, Natal. *South African Journal of Botany*, **1**, 85–90.

Nixon, S.W. (1981). Remineralization and nutrient cycling in coastal marine ecosystems. In *Estuaries and nutrients*, ed. B.J. Neilson & L.E. Cronin, pp. 111–38. Clifton, NJ: Humana Press.

Officer, C.B. (1983). Physics of estuarine gradation. In *Ecosystems of the world: Estuaries and enclosed seas*, ed. B.H. Ketchum, pp. 15–41. Amsterdam: Elsevier.

Painchard, J. & Therriault, J. (1989). Relationship between bacteria, phytoplankton and particulate organic carbon in the upper St Lawrence Estuary. *Marine Ecology Progress Series*, **56**, 301–11.

Pakulski, J.D. (1992). Foliar release of soluble reactive phosphorus from *Spartina alterniflora* in a Georgia (USA) salt marsh. *Marine Ecology Progress Series*, **90**, 53–60.

Palumbo, A.V., Ferguson, R.L. & Rublee, P.A. (1984). Size of suspended bacterial cells and association of the heterotrophic activity with size fractions of particles in estuarine and coastal waters. *Applied and Environmental Microbiology*, **48**, 157–64.

Pedrós-Alió, C. & Brock, T.D. (1983). The importance of attachment to particles for planktonic bacteria. *Archiv für Hydrobiologie*, **98**, 354–79.

Pelegri, S.P., Nielsen, L.P. & Blackburn, T.H. (1994). Denitrification in estuarine sediment stimulated by the irrigation activity of the amphipod *Corophium volutator*. *Marine Ecology Progress Series*, **105**, 285–90.

Pomeroy, L.R., Smith, E.E. & Grant, C.M. (1965). The exchange of phosphate between estuarine water and sediments. *Limnology and Oceanography*, **1**, 167–72.

Ramm, A.E.L. (1992). Aspects of the biogeochemistry of sulphur in Lake Mpungwini, Southern Africa. *Estuarine, Coastal and Shelf Science*, **34**, 253–61.

Ramm, A.E.L., Cooper, J.A.G., Harrison, T.D. & Singh, R.A. (1994). *The Estuarine Health Index: a new approach to scientific information transfer*. In Classification of Rivers and Environmental Health Indicators, ed. M. Uys. Proceedings of a Joint South African Australian Workshop, pp. 271–80. Cape Town: Water Research Commission Report No.TT 63/94.

Read, G.H.L. (1983). The effect of a dry and wet summer on the thermal and salinity structure of the middle and upper reaches of the Keiskamma estuary, Ciskei. *Transactions of the Royal Society of South Africa*, **45**, 45–62.

Riedel, G.F., Sanders, J.G. & Osman, R.W. (1989). The role of 3 species of benthic invertebrates in the transport of arsenic from contaminated estuarine sediment. *Journal of Experimental Marine Biology and Ecology*, **134**, 143–55.

Robarts, R.D. (1976). Primary production of the upper reaches of a South African estuary (Swartvlei). *Journal of Experimental Marine Biology and Ecology*, **24**, 93–102.

Robarts, R.D. & Allanson, B.R. (1977). Meromixis in the lake-like upper reaches of a South African estuary. *Archiv für Hydrobiologie*, **80**, 531–40.

Roberts, C.H., Robb, F.T. & Branch, G. (1983). Microbial ecology of the Bot River Estuary. *South African Journal of Science*, **79**, 155–55.

Rooseboom, A., Verster, E., Zietsman, H.L. & Lotriet, H.H. (1992). *The development of the new sediment yield map of South Africa*. WRC Report No.297/2/92. Pretoria: Water Research Commission.

Russell, I.A. (1996). Water quality in the Knysna estuary. *Koedoe*, **39**, 1–8.

Rysgaard, S., Christensen, P.B. & Nielsen, L.P. (1995). Seasonal variation in nitrification and denitrification in estuarine sediment colonized by benthic microalgae and bioturbating infauna. *Marine Ecology Progress Series*, **126**, 111–21.

Rysgaard, S., Risgaard-Petersen, N., Nielsen, L.P. & Revsbech, N.P. (1993). Nitrification and denitrification in lake and estuarine sediments measured by the ^{15}N dilution technique and isotope pairing. *Applied and Environmental Microbiology*, **59**, 2093–98.

Schleyer, M.H. & Roberts, G.A. (1987). Detritus cycling in a shallow coastal lagoon in Natal, South Africa. *Journal of Experimental Marine Biology and Ecology*, **110**, 27–40.

Scott, K.M.F., Harrison, A.D. & Macnae, W. (1952). The ecology of South African estuaries. Part 2: the Klein River estuary, Hermanus, Cape. *Transactions of the Royal Society of South Africa*, **33**, 282–331.

Silberbauer, M.J. (1982). Phosphorus dynamics in the monimolimnion of Swartvlei. *Journal of the Limnological Society of southern Africa*, **8**, 54–60.

Slinger, J.H. & Taljaard, S. (1994). Preliminary investigation of seasonality in the Great Berg estuary. *Water SA*, **20**, 279–88.

Slinger, J.H., Taljaard, S. & Largier, J.

(1995). Changes in estuarine water quality in response to a freshwater flow event. In *Changes in fluxes in estuaries*, ed. K.R. Dyer & R.J. Orth, pp. 51–6. Denmark: Olsen & Olsen.

Smith, V.H. (1990). Nitrogen, phosphorus, and nitrogen fixation in lacustrine and estuarine ecosystems. *Limnology and Oceanography*, **35**, 1852–9.

Somerfield, P.J., Gee, J.M. & Warwick, R.M. (1994). Soft sediment meiofaunal community structure in relation to a long term heavy metal gradient in the Fal Estuary system. *Marine Ecology Progress Series*, **105**, 79–88.

Steinke, T.D. & Ward, C.J. (1990). Litter production by mangroves. III. Wavecrest (Transkei) with predictions for other Transkei estuaries. *South African Journal of Botany*, **56**, 514–16.

Talbot, M.M.J. (1988). *Trace metals in the marine coastal environment and their effect on nitrogen recycling*. PhD thesis, University of Port Elizabeth.

Talbot, M.M.J.F., Branch, E., Marsland, S., & Watling, R. (1985). Metals in South African estuaries. X. Blind, Ihlanza, Nahoon and Quinera Rivers. *Water SA*, **11**, 65–8.

Taljaard, S. (1987). *Nutrient circulation in the Palmiet River estuary: a summer study*. MSc thesis, University of Port Elizabeth.

Taljaard,S., de Villiers, S., Fricke, A.H. & Kloppers, W.S. (1992). *Water quality status of the Rietvlei/Milnerton lagoon system (Diep River Estuary)*. Data Report EMAS-D 92007. Stellenbosch: CSIR.

Taljaard, S., Eagle, G.A. & Hennig, H.F.-K.O. (1986). The Palmiet estuary: a model for water circulation using salinity and temperature measurements over a tidal cycle. *Water SA*, **12**, 119–26.

Taljaard, S., & Largier, J.L. (1989). *Water circulation and nutrient distribution patterns in the Palmiet River estuary: a winter study*. CSIR (Stellenbosch) Research Report 680. Stellenbosch: CSIR.

Taylor, D.I. (1987). *Tidal exchange of carbon, nitrogen and phosphorus between a Sarcocornia salt-marsh and the Kariega estuary, and the role of salt-marsh brachyura in this transfer*. PhD thesis, Rhodes University, Grahamstown.

Taylor, D.I. (1992). The influence of upwelling and short-term changes in concentrations of nutrients in the water column on fluxes across the

surface of a salt marsh. *Estuaries*, **15**, 68–74.

Taylor, D.I. & Allanson, B.R. (1993). Impacts of dense crab populations on carbon exchanges across the surface of a salt marsh. *Marine Ecology Progress Series*, **101**, 119–29.

Taylor, D.H. & Allanson, B.R. (1995). Organic carbon fluxes between a high marsh and estuary, and the applicability of the Outwelling Hypothesis. *Marine Ecology Progress Series*, **126**, 263–70.

Thompson, S.P., Paeri, H.W. & Go, M.C. (1995). Seasonal patterns of nutrification and eutrophication in a natural and restored saltmarsh. *Estuaries*, **18**, 399–408.

Tibbles, B.J., Lucas, M.I., Coyne, V.E. & Newton, S.T. (1994). Nitrogenase activity in marine sediments from a temperate salt marsh lagoon. Modulation by complex polysaccharides, ammonium and oxygen. *Journal of Experimental Marine Biology and Ecology*, **184**, 1–20.

Tolosa, I., Bayona, J.M. & Albaiges, J. (1991). Identification and occurrence of brominated and nitrated phenols in estuarine sediments. *Marine Pollution Bulletin*, **22**, 603–7.

Valette-Silver, N.J. (1993). The use of sediment cores to reconstruct historical trends in contamination of estuarine and coastal sediments. *Estuaries*, **16**, 577–88.

Valiela, I. (1991). Ecology of coastal ecosystems. In *Fundamentals of aquatic ecology*, ed. R.S.K. Barnes & K. Mann, pp. 57–76. Oxford: Blackwell Scientific Publishers.

van Zoest, R. & van Eck, G.T.M. (1991). Occurrence and behaviour of several groups of organic micropollutants in the Scheldt estuary. *The Science of the Total Environment*, **103**, 57–71.

Watling, H.R. & Watling, R.J. (1976). Trace metals in oysters from Knysna estuary. *Marine Pollution Bulletin*, **7**, 45–8.

Watling, R.J., Talbot, M.M.J., Branch, E. & Marsland, S. (1985). Metals in South African estuaries. IX. Buffalo River. *Water SA*, **11**, 61–4.

Watling, R.J. & Watling, H.R. (1975). Trace-metal studies in Knysna estuary. *Environment RSA*, **2**, 5–7.

Watling, R.J., & Watling, H.R. (1977). Metal concentrations in surface sediments from Knysna estuary. CSIR Special Report FIS 122: 38 pp.

Watling, R.J. & Watling, H.R. (1980). *Metal*

surveys in South African estuaries. II. Knysna estuary. CSIR Special Report FIS 203: 122 pp.

Watling, R.J., & Watling, H.R. (1982a). Metal surveys in South African estuaries. II. Knysna River. Water SA, **8**, 36–44.

Watling, R.J., & Watling, H.R. (1982b). Metal surveys in South African estuaries. IV. Keurbooms and Bietou Rivers (Plettenberg Bay). Water SA, **8**, 114–19.

Watling, R.J., & Watling, H.R. (1982c). Metal surveys in South African estuaries. V. Kromme and Gamtoos Rivers (St Francis Bay). Water SA, **8**, 187–91.

Watling, R.J., & Watling, H.R. (1982d). Metal surveys in South African estuaries. I. Swartkops River. Water SA, **8**, 26–35.

Watling, R.J. & Watling, H.R. (1982e). Metal surveys in South African estuaries. VI. Sundays River. Water SA, **8**, 192–95.

Watling, R.J., & Watling, H.R. (1983a). Trace metal surveys in Mossel Bay, St Francis Bay and Algoa Bay, South Africa. Water SA, **9**, 57–65.

Watling, R.J. & Watling, H.R. (1983b). Metal surveys in South African estuaries. VII. Bushmans, Kariega, Kowie and Great Fish Rivers. Water SA, **9**, 66–70.

White, D.S. & Howes, B.L. (1994). Long term ^{15}N-nitrogen retention in the vegetated sediments of a New England salt marsh. Limnology and Oceanography, **39**, 1878–92.

Whitfield, A.K. (1988). The role of tides in redistributing macrodetrital aggregates within the Swartvlei estuary. Estuaries, **11**, 152–9.

Whitfield, A.K. (1992). A characterization of southern African estuarine systems. Southern African Journal of Aquatic Sciences, **18**, 89–103.

Wiegert, R.G., Pomeroy, L.R., & Wiebe, W.J. (1981). Ecology of salt marshes: an introduction. In The ecology of a salt marsh, ed. L.R. Pomeroy & R.G. Wiegert, pp. 3–19. New York: Springer-Verlag.

Willis, J.P. (1985). The bathymetry, environmental parameters and sediments of the Bot River estuary, S.W. Cape Province. Transactions of the Royal Society of South Africa, **45**, 253–84.

5 Primary producers

Estuarine microalgae
Janine Adams and Guy Bate

Estuarine macrophytes
Janine Adams, Guy Bate and
Michael O'Callaghan

Great Berg River estuary

ESTUARINE MICROALGAE

Microalgae are small unicellular organisms, too small individually to be seen without the aid of a microscope.
Microalgae can occur suspended in the water column
(phytoplankton) or attached to sand grains (episammic
flora), bottom sediments (epipelic flora), rock (epilithic
flora) and other plants below the surface of the water (epiphytic flora). Research in South Africa has focused on phytoplankton, and benthic microalgae attached to sand grains
and bottom sediments. Phytoplankton are dominant in
large channel-like estuaries which have large catchments, a
high mean annual runoff, and where freshwater introduces nutrients and creates stable stratified conditions. In
estuaries with large intertidal regions, benthic microalgae
are important contributors to primary production. This
chapter describes a comparative study on Cape estuaries
where the distribution and biomass of phytoplankton and
benthic microalgae were investigated in relation to fresh-
water-controlled physical factors, i.e. nutrients, water
retention and turbidity. A suitable method for the measurement of benthic chlorophyll-a is also described.

Phytoplankton

Phytoplankton are those microalgae which are suspended
in the water column. Thus, although some species have
limited motility, their spatial distribution in horizontal
and vertical planes is controlled primarily by water motion
(as either mass flow or turbulence). Species of phytoplankton may settle out under calm conditions and be mistakenly
included among the benthic microflora and, likewise,
benthic flora may be disturbed from their normal attached

habitats by various means such as water turbulence and
temporarily assume a planktonic habit. Ecological studies
have to include these aspects and correct for them.

Importance of freshwater input: a comparative
South African study

Both local (Hilmer & Bate 1990, 1991; Allanson & Read 1995;
Grange & Allanson 1995) and overseas studies (Flint 1985;
Drinkwater 1986; Paerl et al. 1990; Cloern 1991a) have shown
that phytoplankton biomass in estuaries is positively correlated with freshwater input. Drinkwater (1986) stated

> Freshwater induces important circulation patterns,
> affects vertical stability, modifies mixing and
> exchange processes, and influences nutrients and
> primary production. Further, light reduction from
> heavy silt loading, prevention of nutrient addition
> to the surface layer by strong vertical stability, or
> rapid advection such that the flushing time exceeds
> the time scale of phytoplankton growth, will act to
> reduce production. On the other hand, in river
> plumes where freshwater and saltwater are actively
> mixed, primary production is usually high.

and

> In regions where there are sufficient surface nutrients,
> vertical stability imposed by runoff helps to
> maintain phytoplankton in the euphotic zone
> thereby enhancing production.

Clearly, the effect of freshwater on the species composition and production of phytoplankton is via a complex interaction of abiotic factors. Smetacek (1986) reviewed the literature on the impact of freshwater on phytoplankton production and wrote:

> Freshwater discharge has a twofold effect: (a) it introduces allochthonous dissolved and particulate materials and (b) it drastically modifies the structure and dynamics of the physical environment. Because of the wide range of variability associated with freshwater discharge (depending on its magnitude and seasonality, the drainage basin and the hydrography/topography of the receiving environment) its impact on the biological processes varies accordingly. Further, such environments are inherently heterogeneous, both spatially and temporally because of the environmental gradient maintained by the admixture of fresh and marine waters.

In South Africa, studies on the relationship between freshwater input and phytoplankton response include the work of Allanson and Read (1995) in the Kariega (33° 41' S; 26° 44' E), Keiskamma (33° 17' S; 27° 29' E) and Great Fish (33° 32' S; 27° 03' E) estuaries, Grange and Allanson's (1995) work in the Great Fish estuary, and Hilmer and Bate's work in the Swartkops (33° 57' S, 25° 38' E) and Sundays (33° 48' S, 25° 42' E) estuaries (Hilmer 1984, 1990; Hilmer & Bate 1990, 1991). The general conclusions from these studies were that in estuaries with large catchments, adequate freshwater inflow supported phytoplankton communities and a pelagic food chain. Submerged macrophyte communities were dominant in smaller estuaries with reduced freshwater input as there is a decrease in turbidity and water velocities and a more stable sediment and salinity environment. Freshwater input supported phytoplankton communities as it brought in nutrients and maintained stable stratified conditions. Strong vertical and horizontal salinity gradients were maintained in open mouth estuaries with adequate tidal exchange. In the Sundays estuary dinoflagellates formed dense blooms during extended neap tides when salinity stratification was most pronounced. Blooms were dispersed on the following spring tide as a result of vertical mixing (Hilmer & Bate 1991). Furthermore, flooding usually produced peaks in phytoplankton biomass and production. Studies were extended to test these responses in other South African estuaries (Adams & Bate 1994a). A description of this work follows.

The study area
A comparative study on the importance of freshwater for estuarine microalgae included the following estuaries;

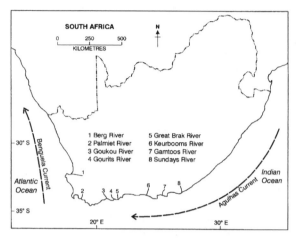

Figure 5.1. **Location of estuaries sampled in August and November 1992 as part of a comparative study on phytoplankton in Cape estuaries.**

the Berg, Palmiet, Goukou, Gourits, Great Brak, Keurbooms, Gamtoos and Sundays (Figure 5.1). These estuaries were chosen because of their accessibility and differences in freshwater input. Physical characteristics of the estuaries are summarised in Table 5.1. The Berg is a permanently open estuary on the west coast of South Africa. During summer months a strong salinity stratification is evident. However, during winter the estuary functions as a river mouth because of the high winter rainfall in the catchment.

The Palmiet is a small, permanently open blackwater estuary (length c. 1.7 km). The river enters the estuary as a fast flowing mountain stream (Taljaard & Largier 1989). During the summer dry season the estuary functions as a highly stratified and strongly marine system with nutrients being imported from the sea. In winter it is mostly a riverine system with a strong freshwater character (cf. Chapter 4). The Goukou is a permanently open estuary but has a constricted tidal inlet. The major part of the catchment consists of privately owned farms where mixed farming is practised (Carter & Brownlie 1990). Although the Gourits has a large catchment area, the catchment is situated in an area of low rainfall (Heydorn 1989). The catchment area drains large areas of the Great Karoo and the whole of the Little Karoo, which is an important area for sheep and goat farming. The estuary extends about 8 km upstream of the mouth. During periods of low flow it is a marine-dominated system.

The Great Brak is a temporarily open blackwater estuary. This estuary has become the focus of interest because of the recent construction of the Wolwedans dam. The Keurbooms is a permanently open estuary, which at present has no dams in its catchment. State forest covers part of the catchment. Management of the highly dynamic

Table 5.1. Physical characteristics of estuaries sampled for phytoplankton study

Estuary	Catchment area (km²)	River length (km)	Mean annual Run-off (10⁶ m³)
1 Berg	7715 (km)	294.3	1035.47
2 Palmiet	535	72.5	200.92
3 Goukou	1550	67.4	106.42
4 Gourits	45715	416.0	539.05
5 Great Brak	190	31.5	38.79
6 Keurbooms	1080	85.3	176.86
7 Gamtoos	34635	644.8	502.51
8 Sundays	20990	481.0	202.26

mouth of the Keurbooms estuary is a subject of controversy (Duvenage & Morant 1984). The Gamtoos is a permanently open turbid estuary with a meandering channel deeply eroded into the wide river floodplain. Extensive agriculture takes place in the catchment area (Heinecken 1981). Land use in the permanently open Sundays estuary is solely agricultural, being mostly sheep farming with some intensive citrus cultivation. The estuary is tidal for about 24 km.

Sampling methods

Sampling was conducted in August and November 1992 in the specified Cape estuaries to establish the relationship between nutrient concentrations and phytoplankton biomass. Sampling was carried out at four stations in each estuary. At each station, water samples were collected using a weighted narrow-necked 1 litre bottle sampler. Water for chlorophyll-a and nutrients was collected at 0.5 m intervals from the surface to 1 m depth, and at 1 m intervals thereafter. All other measurements were made at 0.5 m intervals from the surface to the bottom of the estuary. Salinity, temperature and light attenuation were measured.

Water column chlorophyll-a was measured by filtering 500 ml of water onto Whatman GF/C filters. Chlorophyll was extracted with 10 ml of 95% ethanol for at least 2 hours. After extraction, the samples were cleared by filtration. Absorbances were measured at 665 nm before and after acidification with 0.1 M HCl in a Milton Roy Spectronic Mini 20. Chlorophyll-a concentrations were calculated according to Nusch (1980). Water samples were stored in a cooler box and nitrate-N measured within 24 h using a method adapted from Bate and Heelas (1975).

Water samples were collected for phytoplankton identification and cell number counts. Water samples were fixed with buffered formalin, stained with Rose Bengal and settled in settling chambers. Counts were made using a Zeiss inverted microscope and the species were identified.

Results

Hilmer & Bate (1991) worked on the Sundays estuary and showed that vertical and horizontal salinity gradients resulted in phytoplankton dominance, with macrophytes contributing little to primary production. Hilmer showed that if the salinity gradient calculated as the difference between the mean salinity at the mouth and at the head of the estuary was greater than 10, a horizontal salinity gradient existed. He further described salinity stratification numerically as the difference between the mean top (<0.5 m) and mean bottom (>0.5 m) salinity. If the difference was greater than 5, a vertical salinity gradient existed. From a knowledge of the salinity gradients one could then predict whether phytoplankton would be dominant in an estuary. This work was important as it was the first in estuaries to describe gradients numerically and associate them with a biological response. However, when this model was tested along the south coast, which has a higher rainfall and less catchment agriculture, phytoplankton biomass was insignificant despite well-developed salinity gradients.

On the basis of Hilmer's definition all estuaries displayed a horizontal salinity gradient in both August (Figure 5.2a) and November (Figure 5.2b). Vertical salinity gradients were present in all estuaries in August but only in the Gourits estuary in November. Although the estuaries sampled had both vertical and horizontal salinity gradients in August, Figure 5.3a indicates that high chlorophyll-a concentrations (>20 µg l⁻¹) are better related to water column nitrate concentration than to salinity gradients. A salinity gradient serves as an index of the freshwater input into an estuary, but phytoplankton require high nutrient inputs in order to bloom. Mean chlorophyll-a in the Sundays estuary was 29 µg l⁻¹ in August 1992. This estuary also had the highest mean nitrate concentration, i.e. 7.6 µM. The difference in nitrate concentration between the head and the mouth was 14 µM with the highest at the head indicating a freshwater source of nutrient input. In the Berg estuary the highest chlorophyll-a was found in the mouth (Figure 5.3b). Phytoplankton occurring here were probably the result of a flood tide bringing phytoplankton into the estuary from the nutrient-rich water of the west coast Benguela system. Species identification supported the contention that these were marine coastal species.

Low chlorophyll-a (<1µg l⁻¹) was found in the Palmiet and Gourits estuaries (Figure 5.2b). The Palmiet is a well-flushed system, water residence time is short (Taljaard & Largier 1989) and phytoplankton biomass is low. Although the mean chlorophyll-a levels in the Gamtoos estuary in August (Figures 5.2a and 5.3a) were low (c. 5 µg l⁻¹), previous sampling in 1990 established that phytoplankton blooms do occur but mainly in the summer months. That bloom

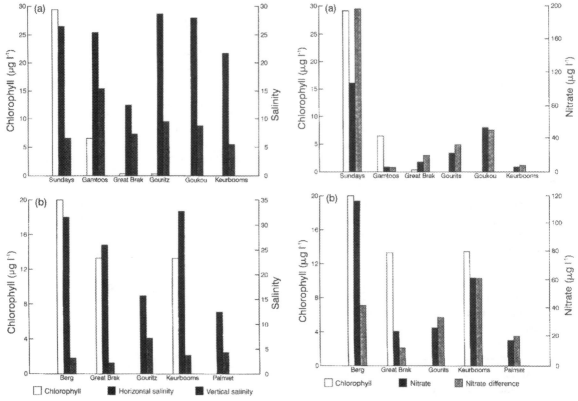

Figure 5.2. **The relationship between mean chlorophyll-*a*, horizontal salinity gradient and vertical salinity stratification for estuaries sampled in August (a) and November (b).**

Figure 5.3. **The relationship between mean chlorophyll-*a*, nitrate concentration and the difference between head and mouth nitrate concentration for estuaries sampled in August (a) and November (b).**

was located in the stratified water column of the middle reaches of the estuary and persisted for one month.

Rainfall and freshwater runoff increased between August and November 1992 along the Cape south coast. A comparison between August and November data for the Keurbooms and Great Brak estuaries (Table 5.2) showed that an increase in freshwater input caused a decrease in mean salinity, an increase in the horizontal salinity gradient, and an increase in nitrate and chlorophyll-*a* concentrations. In the Gourits estuary the mean salinity in November was higher than in August and there was only a slight increase in nitrate concentration from August to November (Table 5.7, below). Chlorophyll-*a* concentrations were low (<1 μg l^{-1}) for both months.

Cells in the water samples taken from the different estuaries were enumerated into the following groups: diatoms, flagellates, dinoflagellates, euglenoids, green and blue-green algae. In general, flagellates were the dominant group in all estuaries sampled. The Great Brak estuary also had a number of Cyanophyta (blue-green algae). There were 195 diatom taxa present in the eight estuaries studied. Certain diatom species were found to occur over a wide salinity range from freshwater (salinity

0) to marine conditions (salinity 35). These species included *Amphora coffeaeformis* Agardh, *Aulacoseira* sp., *Camplylodiscus* sp., *Cocconeis* sp., *Coscinodiscus* sp., *Cylindrotheca closterium* (Ehrenberg) W. Smith, *Entomeis* sp., *Extubocellulus spinifera* (Hargraves & Guillard) Hasle, van Stosch & Syvertsen, *Haemiaulus* sp., *Leptocylindrus* sp., *Melosira* sp., *Navicula tripunctata* (O.F. Müller) Bory, *Nitzschia* sp., *Tabularia* sp. and *Thalassiosira decipiens* (Grunow) Jorgensen. Of the taxa listed, more (29) were found in only saline (salinity 35) water than freshwater (salinity 0). Sixteen taxa were associated with brackish water (salinity 5–20) and not marine or freshwater conditions. Dominant diatom species, i.e. occurring in seven of the eight estuaries sampled, included *Extubocellulus spinifera*, *Nitzschia seriata* Cleve and *Cylindrotheca closterium*. Common genera included *Amphora*, *Cocconeis*, *Navicula* and *Nitzschia*.

Discussion

Nutrient effects

Nutrient availability appears to be the major freshwater related factor controlling phytoplankton biomass, increased nutrient availability leading to increased bio-

Table 5.2. The difference between August and November chlorophyll, salinity and nitrate concentrations in the Keurbooms, Great Brak and Gourits estuaries

	Great Brak		Gourits		Keurbooms	
	Aug	Nov	Aug	Nov	Aug	Nov
Mean chlorophyll-a (μg l^{-1})	0.2	13.3	0.1	0	0	13.3
Mean salinity	18.1	3.3	21.3	26.8	14.1	11.5
Horizontal salinity gradient	12.2	25.8	28.7	15.7	21.2	32.8
Vertical salinity stratification	7.1	2.2	9.3	7.0	5.2	3.7
Mean nitrate (μM)	0.8	1.7	1.5	1.8	0.3	4.4

mass. Increased nutrient availability in estuarine waters may result from a number of processes, namely increased nutrient loading in the freshwater input, increases in the quantity of freshwater input, or increases in fresh/seawater flow rates such that benthic regenerated nutrients are mixed into the euphotic zone (Flint 1985; Malone et al. 1988).

Malone et al. (1988) studied the mesohaline reaches of Chesapeake Bay (salinity 8–14) and found that maximum chlorophyll-a concentrations occurred in spring when salinity was low and nitrate inputs were high due to peaks in river flow. The level of summer productivity was sustained by recycling the nitrogen that was delivered the previous spring. In South African estuaries rainfall is erratic and thus definite seasonal patterns of estuarine phytoplankton production, i.e. regular spring blooms, are absent. Responses in our estuaries are more opportunistic and related to episodic freshwater discharges (Allanson & Read 1987, 1995; Hilmer 1990). In the Great Brak and Keurbooms estuaries, nutrients brought in by a freshwater pulse between August and November 1992 boosted phytoplankton biomass in the short term. Chlorophyll-a concentrations increased from <1 μg l^{-1} to 13 μg l^{-1} (Adams & Bate 1994a). These freshwater pulses may be important in maintaining the pelagic food chain.

Consistent high nutrient loading from freshwater input occurs in the Sundays River estuary as a result of agricultural fertiliser flow through (Emmerson 1989). In most other estuaries, increased nutrient availability results from an increase in the quantity of freshwater input. In the Sundays estuary high chlorophyll-a concentrations (defined as >20 μg l^{-1}) are maintained if the nitrate concentration is greater than 14 μM. A similar situation where freshwater input results in consistently high nutrient concentrations and, therefore, large phytoplankton biomass occurs in the Great Fish estuary. Mean chlorophyll-a in the Great Fish estuary has been reported as 20.5 ± 31.5 μg l^{-1} (Allanson & Read 1995).

Water motion effects

Estuarine circulation is important as it controls water residence and phytoplankton turnover time. In some estuar-

ies, rates of physical removal may dominate over rates of in situ production. For example, if water residence time is short (on the order of hours or days), development of phytoplankton concentrations by in situ processes may be restricted because maximum or full blooms may take longer (days or weeks) to develop. Water column stability can promote blooms by inhibiting vertical mixing (Cloern 1991b). The typical response of phytoplankton biomass to increasing river flow is a reduction in their concentration by increasing dilution rate. Water residence times are also important in terms of nutrient utilisation. If retention times are short (days), then there is insufficient time for the phytoplankton to trap the nutrients, and the latter will be washed out to sea.

Hilmer and Bate (1991) found that phytoplankton blooms formed in the upper reaches of the Sundays estuary when there was a water residence time of 6–7 tidal cycles (Mackay & Schumann 1990). Stable tidal conditions allowed the phytoplankton (especially dinoflagellates) to bloom as the water was present for longer than the doubling time requirements to produce the bloom. Water residence times will depend on freshwater flow rates per se as well as tidal flow and the condition of the mouth. Residence time is a difficult parameter to measure. This poses problems as the realistic simulation of phytoplankton populations will always be dependent upon accurate descriptions of mixing processes in estuaries (Cloern 1991b).

Turbidity effects

River water inflow affects estuarine turbidity either by bringing in suspended particulates or dissolved humics, or by modifying turbulence and thus altering the extent to which particulates are maintained in suspension. Turbulence, or mixing depth, also determines the duration to which phytoplankton cells are exposed to high near-surface irradiances or mixed below the euphotic zone. Turbulence may be induced by rapid flow on the one hand, or reduced by salinity induced stratification on the other (cf. Chapter 3).

Depth of light penetration depends on the concentration of suspended particulate matter, both living and non-living. Generally the concentrations of suspended particles in estuarine water are high and light is rapidly attenuated. Peterson and Festa (1984) found that concentrations of suspended sediments of 10–100 mg l^{-1} strongly suppressed phytoplankton biomass, and these values are typical for estuaries.

Although turbidity is thought to be a major factor controlling phytoplankton production and chlorophyll-a distribution in estuaries, these conclusions are based on data from large overseas estuaries where horizontal and vertical dimensions are usually an order of magnitude larger than those of South African estuaries. South African estuaries are shallow and, therefore, high turbidity due to sediments brought in by river water flow does not necessarily mean light limitation for phytoplankton. Indeed, in the St Lucia estuary (28° 18′ S, 32° 26′ E) Fielding et al. (1991) found that even under conditions of extreme turbidity, chlorophyll-a was present in the water. Average depth of St Lucia is 0.96 m and the water column is well mixed by wind-driven wave action.

Agricultural return flow of Orange River water into the Great Fish estuary has increased the input of nutrients to the system. There has also been an increase in turbidity as a result of the transport of suspended sediments from the Orange River system. Excessive silt loads have been identified as being lethal to fishes in these systems (Whitfield & Paterson 1995). This increase in turbidity does not appear to affect phytoplankton biomass because light is not a limiting factor for phytoplankton growth, owing to the shallowness of the estuary. Turbulence resulting from tidal movements causes the cyclical resuspension of sedimented phytoplankton into the euphotic zone (Grange & Allanson 1995). Lucas (1986) has proposed that it is by this mechanism that phytoplankton populations in the Great Fish estuary are able to remain viable and productive despite the high turbidity.

Species composition

A comparative study of Cape estuaries (Adams & Bate 1994a) showed that diatoms were the dominant benthic microalgal component whereas higher numbers of flagellates occurred in the phytoplankton. Although the microflora comprises diatoms, dinoflagellates and other microalgal groups, diatoms are probably the most important because they respond more quickly to changes in water quality. There have been a number of diatom taxonomic studies in the past (e.g. Archibald 1981), but none related diatom species composition to physical characteristics of the estuaries.

According to Margalef (1978), diatoms are favoured in well-mixed, nutrient-rich waters during spring tides. Flagellates would increase in abundance when stratification sets in and nutrients became depleted. Red tide species of dinoflagellates are favoured when a stable, stratified water column is present, but when nutrient concentrations are still high. South African studies support this: when strong estuarine phytoplankton blooms occur during stable stratified conditions they generally consist of dinoflagellates. A common bloom dinoflagellate species in the Sundays estuary is Katodinium rotundatum (Lohmann) Loeblich and in the Gamtoos estuary Prorocentrum sp. Hilmer and Bate (1991) reported on the horizontal displacement and vertical migration of a bloom dominated by K. rotundatum and the prasinophyte Micromonas pusilla (Butcher) Parke et Manton in the Sundays estuary. This bloom formed above the halocline in the upper reaches of the estuary where the retention time of water was greatest (Mackay & Schumann 1990).

The Great Fish estuary also has high phytoplankton biomass but freshwater diatoms are dominant. According to Lucas (1986), algal blooms of riverine origin enter and accumulate within the estuary in the same manner as sediments. A major bloom (210 µg chlorophyll-a l^{-1}) on 20 November 1995 consisted of the diatom Cyclotella meneghiniana Kutzing. This diatom indicates fresh to slightly brackish waters and enriched conditions. Dinoflagellate blooms have not been reported in the Great Fish estuary, possibly as a result of the low water retention times. There is a continuous flow of water into the Great Fish estuary as the result of the Orange River transfer system. Flows below 1 m^3 s^{-1} rarely occur, whereas before the implementation of the transfer system low flow conditions were common (Huizinga 1996).

Both Katodinium rotundatum and Prorocentrum sp. are categorised as red tide species. Red tide blooms of the raphidophyte Heterosigma akashiwo Hada, which frequently causes serious damage to the aquaculture fisheries of Japan, have also been noted in the Sundays River estuary (Hilmer & Bate 1991). The occurrence of red tide species in South African estuaries requires further investigation as they may impact activities such as angling and emergent shellfish industries along our coast.

Zooplankton grazing can affect species composition, as large species such as dinoflagellates are relatively immune to grazing (Moss et al. 1989). In the Sundays estuary consistently high chlorophyll-a concentrations result in high zooplankton numbers. However, the concentration of the phytoplankton bloom does not appear to be significantly reduced by zooplankton grazing in that system (Jerling & Wooldridge 1995).

Comparative productivity

The literature indicates that estuaries are rich and productive systems and in situ production usually exceeds the

Table 5.3. Range of water column chlorophyll-a values (phytoplankton biomass) published for South African estuaries

| Estuary | chlorophyll-a (μg ℓ^{-1}) | | Reference |
	Min	Max	
Palmiet	2	8	Branch & Day 1984
Bot	0	6	Bally et al. 1985
Sundays	12	23	Hilmer & Bate 1991
Sundays	>100 (bloom)		Hilmer & Bate 1990
Kariega	1	8	Allanson & Read 1995
Great Fish	0	52	Allanson & Read 1995
Great Fish	106	210 (bloom)	Lucas 1986
Keiskamma	0	19	Allanson & Read 1995
Nahoon	1	6	Campbell et al. 1991
Gqunube	5	15	Campbell et al. 1991
Kwelera	0	10	Campbell et al. 1991
St Lucia	0	16	Fielding et al. 1991

rates of import of organic material. Published estuarine phytoplankton rates generally fall between 13 and 346 g C m^{-2} per year, while macrophyte productivities typically range from 200 to 3000 g C m^{-2} y^{-1} (Kennish 1986). Rates of production are extremely variable. So also are the units that are used and the relationships between biomass, cell numbers and chlorophyll concentrations. The most common biological variable measured alongside primary production is chlorophyll, which is used as an estimate of biomass. Water column chlorophyll-a values determined for a number of South African estuaries are given in Table 5.3 and show that phytoplankton biomass is low for most estuaries.

In St Lucia estuary, estimates of phytoplankton production ranged between 218 and 252 mg C m^{-2} per day (Fielding et al. 1991). For the Sundays River estuary an overall mean value of 900 mg C m^{-2} d^{-1} was computed, which is equivalent to 328 g C m^{-2} per year (Hilmer 1990). This places the Sundays estuary into the range of more productive estuaries. Values of 90 g C m^{-2} y^{-1} for the Wassaw Sound estuary (Turner et al. 1979), and between 6.8 and 164.9 g C m^{-2} y^{-1} for the Bristol channel (Joint & Pomeroy 1981), have been recorded for turbid estuaries. These are low values compared with 307 g C m^{-2} y^{-1} for the Delaware estuary (Pennock & Sharp 1986), 400 g C m^{-2} y^{-1} in Tomales Bay and 450 g C m^{-2} y^{-1} for the Great South Bay estuary (Lively et al. 1983).

Rates of phytoplankton primary production in the Gqunube (32° 56' S, 28° 02' E), Kwelera (32° 55' S, 25° 05' E) and Nahoon (32° 59' S, 27° 57' E) estuaries were 147, 32 and 116 mg C m^{-2} h^{-1} respectively. The estuary with the greatest standing stock (Gqunube) was that with both the greatest freshwater inflow and a large tidal influx (Campbell et al. 1991). Nutrient input from agricultural activities in the catchment of the Gqunube contributed to the high production rates.

Contributions by macrophytes and phytoplankton to total system production are variable owing to the diversity in morphology, site meteorology, chemistry and biology of estuaries. On the whole, phytoplankton productivity per unit area is an order of magnitude less than that of macrophytes. However, the area of an estuary or brackish lake which is occupied by phytoplankton is often so large that phytoplankton contribute more to total system primary production than do macrophytes. Furthermore, unlike macrophytes, phytoplankton may contribute substantially more to grazing food chains.

In Swartvlei (34° 00' S, 22° 48' E), a southern African brackish estuarine lake, Howard-Williams and Allanson (1981) found that, despite covering 72% of the area, phytoplankton contributed only 6% to total production. Similarly, in the nearby, mesotrophic Bot estuary (34° 21' S, 19° 04' E) which had been closed to the sea for 4 years, Bally et al. (1985) found that phytoplankton contributed only 7%. However, in the Swartkops estuary Hilmer (1984) estimated that phytoplankton contributed 53% of the total phytomass production of 5.6×10^8 g C per year. The saltmarsh grass *Spartina maritima* contributed 44% and the seagrass *Zostera capensis* 3%. Annual phytoplankton production was estimated to be 93 g C m^{-2} y^{-1} and 3×10^8 g C y^{-1} for the total estuary (Hilmer 1984). These Swartkops values occupy an intermediate position with respect to other South African systems where phytoplankton primary productivity has been measured, and indicate the important contribution phytoplankton make to the primary productivity of what would seem to be a saltmarsh-dominated system.

Benthic microalgae

In shallow aquatic ecosystems, or those with large intertidal regions, benthic microalgae are often important contributors to primary production. Benthic microalgae have certain features in common with emergent and submerged macrophytes by virtue of their habitat. Exposure (desiccation), light availability and salinity are factors that affect distribution and biomass. Within an estuary specific freshwater, brackish and marine benthic microalgal communities can be identified. Some studies have shown that benthic microalgal biomass increases with an increase in riverine nutrient input but can eventually be shaded out by phytoplankton blooms. Water currents and sedimentary disturbance are important factors controlling biomass accumulation. Water speeds also determine benthic sediment texture and the accumulation of nutrient-rich organics – higher flow rates leading to coarser, nutrient-poorer sediments which will support smaller microalgal populations. Benthic microalgal production in mud can be twice that of sand populations.

Grant *et al.* (1986) studied the effect of weather and water movement on diatom films and noticed that these were present on the sediment surface after calm weather and absent after storms. In the Dutch Wadden Sea hard wind with rain and heavy sea ruptured the coherent sediment layer usually stabilised by a film of diatoms and resulted in the resuspension of benthic diatoms (Colijn & Dijkema 1981). Diatom cells attach to sand grains with mucus strands and these films are affected by erosion, burial and exposure. Benthic microalgae may be important in binding sediments at the lower end of the littoral zone (Kennish 1986).

Benthic microalgae can be an important food source in estuaries because they are available year-round. The suspension of large numbers of cells in the water column makes them accessible to filter feeders as well as benthic grazers. Studies in both the Swartkops and Swartvlei estuaries have shown that benthic diatoms are important food items of mullet (Masson & Marais 1975; Whitfield 1988).

A comparative South African study

Studies on benthic microalgae were initiated to test the hypothesis that benthic microalgal biomass is high in estuaries with reduced freshwater input. This study investigated the measurement, distribution and abundance of benthic microalgal communities in Cape estuaries. Hilmer (1990) showed in the phytoplanktonic system of the Sundays River estuary that accurate water column chlorophyll-*a* concentrations could be estimated using routine spectrophotometric techniques. For estuarine sediments, however, where the degradation products of chlorophyll-*a* as well as other pigments are present, the estimation of chlorophyll-*a* is not a simple measurement. Appropriate methods to measure microbenthic chlorophyll-*a* were investigated in detail in the Swartkops estuary (Rodriguez 1993). After establishing a reliable method, a number of estuaries were sampled during August 1992 (i.e. Berg, Goukou, Gourits, Great Brak, Keurbooms, Gamtoos and Sundays).

Determination of optimum conditions for the extraction of benthic chlorophyll-*a*

High performance/pressure liquid chromatography (HPLC) allows qualitative and quantitative determinations of various pigments in samples with great accuracy and speed. HPLC determination was undertaken on sediments from the Swartkops estuary using a Spherisorb reverse phase C_{18} HPLC column attached to a Waters Lambda-Max 481 LC spectrophotometer and Waters LM-45 solvent delivery system. A 30% methanol:70% acetone mixture was used as carrier. The system was calibrated using pure

chlorophyll-*a* (Sigma chemicals). Concentrations were calculated with the aid of a Waters 740 data module.

When chlorophyll-*a* concentrations from the same sample were determined by both HPLC and spectrophotometrically, there was a linear relationship between the readings obtained ($r^2 = 0.63$, $p < 0.001$). The spectrophotometer overestimated chlorophyll-*a* by a mean value of 20% for the top 10 mm of all intertidal sediments. Chlorophyll and the degradation products of chlorophyll absorb or fluoresce in the same range of the visible spectrum. Therefore, in systems with high concentrations of degradation products, traditional spectro-fluorometric techniques will be unreliable and the use of chromatographic techniques is essential. These data illustrate that the techniques used to determine chlorophyll-*a* should be chosen wisely and tested carefully with known standards in relation to the system being measured.

Extraction efficiency of different volumes of ethanol added to sand, silt and mud samples was tested. The relationship between chlorophyll-*a* concentration and solvent volume could be presented as a significant third degree polynomial regression ($p < 0.01$). Extraction efficiency was shown to decrease if ethanol volumes were too low or too high. In extensive tests in the Swartkops estuary chlorophyll-*a* was efficiently extracted from 20 mm internal diameter cores, 10 mm deep, using 30 ml of ethanol (Merck 4111). This ratio held for sand, silt and mud samples. Greater quantities of ethanol might be more efficient in extracting chlorophyll-*a* from very wet sediment samples but greater quantities also yield more dilute samples which are then not efficiently measured.

The Swartkops estuary was sampled to investigate the distribution of microalgal biomass in relation to sediment type. At each of 16 sampling sites, eight replicate samples were taken at two subtidal and two intertidal sites for benthic chlorophyll-*a* determination. Sediment samples were taken with 20 mm corers and immediately extracted in 30 ml of 95% ethanol for 2 hours (Rodriguez 1993). The samples were centrifuged at 3000 r.p.m. for 10 min and the supernatant was filtered using glass-fibre filters (Schleicher & Schull No. 6). The absorbances were measured at 665 nm in a Milton Roy Spectronic Mini 20 corrected to a Phillips PU 8700 UVVis spectrophotometer reading. The data from the field spectrophotometer was also correlated to values produced by HPLC.

Results showed that the distribution of chlorophyll-*a* in estuarine sediments is very uneven and before reliable estimates can be measured in the horizontal plane, the variance of the area should be tested and the required number of replicate samples determined (Rodriguez 1993). In the case of sand samples there was no significant difference in chlorophyll-*a* concentrations for surface sediments to a depth of 40–50 mm (Figure 5.4). In silt and mud

Table 5.4. *Microbenthic algal chlorophyll-a values (n = 4) obtained in various Cape estuaries. Max = maximum, Min = minimum ± SD. Estuaries are ranked by geographical position from west to east*

	Chlorophyll-a (mg m^{-2})					
	Intertidal			Subtidal		
Estuary	Max	Min	Mean (± SD)	Max	Min	Mean (± SD)
Berg	95	0	56 (20)	42	0	26 (10)
Goukou	159	95	115 (15)	275	148	205 (35)
Gourits	183	106	138 (27)	257	64	170 (44)
Great Brak	297	85	163 (76)	297	106	167 (44)
Keurbooms	191	106	138 (27)	257	64	170 (44)
Gamtoos	201	60	106 (51)	275	64	170 (44)
Sundays	256	160	197 (31)	233	105	135 (21)

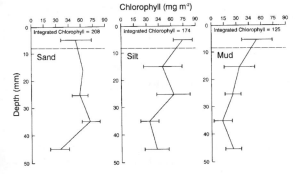

Figure 5.4. **The content of chlorophyll-*a* in the Swartkops estuary at different sediment depths determined spectophotometrically (n = 5, error bars = 2 × SE). Broken lines indicate the depth of the photic zone according to Fenchel & Straarup (1971).**

samples there was a progressive decrease in chlorophyll-*a* concentration with depth. Highest concentrations were obtained for the top 10 mm in both silt and mud samples. This was not the case for sand where a maximum was observed at a depth of 30–40 mm (Figure 5.4). These data illustrate the importance of determining the characteristics of each system to ensure that adequate samples are taken. Silty sediments had the highest chlorophyll-*a* concentrations on the surface (*c.* 76 mg m^{-2}) compared with the sandy and muddy substrata (*c.* 55 mg m^{-2}) (Figure 5.4). The density of microbenthic algal cells at different depths was also investigated. The greatest number of cells was found in the top 0–10 mm layer (34 × 10^6 cells m^{-2}) with a pronounced decrease by 10–20 mm depth (4 × 10^6 cells m^{-2}). Diatoms made up over 75% of the benthic microalgal population while flagellates accounted for 23% and coccolithophorids for 2%. Common genera included *Amphora*, *Navicula* and *Nitzschia*.

Overall results indicated that the surface layer down to 10 mm depth included most of the chlorophyll-*a*. This agrees with other studies (De Jonge 1980; Shaffer & Onuff 1983; Lukatelich & McComb 1986). In Langebaan Lagoon

Fielding *et al.* (1988) found that the highest chlorophyll-*a* concentrations occurred in the top 1 mm of sediment, declining to 35% of surface concentration 5 cm below the surface. In the same system the highest biomass occurred in the sheltered areas and reached 18.9 g m^{-2}.

Chlorophyll-*a* concentrations were higher in the middle reaches of the Swartkops estuary compared with the mouth and upper reaches. In this area, sediment deposition is possibly greater than resuspension, thus allowing microalgal colonisation. Low chlorophyll-*a* values recorded in the mouth area were possibly a result of tidal resuspension of cells or due to wave action.

Distribution of microbenthic chlorophyll-*a* in selected Cape estuaries

Subtidal and intertidal sediment cores were collected at four sites along the length of each estuary. Chlorophyll-*a* was measured in the field using a portable spectrophotometer (Milton Roy Spectronic Mini 20). Table 5.4 shows microbenthic algal chlorophyll-*a* concentrations obtained in the various estuaries. Maximum mean chlorophyll-*a* concentration for intertidal sites was obtained in the Sundays estuary (197 mg m^{-2}) and for subtidal sites in the Goukou estuary (205 mg m^{-2}). Data from this study are comparable to other studies done in South Africa. Mean benthic chlorophyll-*a* for all the estuaries studied was estimated to be 119.1 mg m^{-2}. In the Swartvlei estuary benthic chlorophyll-*a* concentrations were less than 80 mg m^{-2} (Whitfield 1989), and less than 100 mg m^{-2} in Langebaan Lagoon (Fielding *et al.* 1988). Microbenthic chlorophyll-*a* concentrations obtained in this study were in the same range as those obtained by Cadee and Hegeman (1974) for the western Dutch Wadden Sea. They measured a maximum chlorophyll-*a* concentration of 435 mg m^{-2}, one of the highest values reported in the literature. The highest microbenthic chlorophyll-*a* value reported in this study (424 mg m^{-2}) was found in the Great Brak estuary at station 4, 5.3 km from the sea.

Table 5.5. Phytoplankton and microbenthic chlorophyll-*a* biomass calculated for the whole estuary in kg per estuary. Estuaries are ranked by geographical position from west to east

Estuary	Phytoplankton (kg chl-*a*/estuary^{-1})	Microbenthic (kg chl-*a*/estuary)
Goukou	0.07	23
Gourits	0.04	16
Great Brak	0.40	62
Keurbooms	0.06	20
Gamtoos	17.00	14
Sundays	86.00	14

Chlorophyll-*a* contents were calculated for both the water column and sediments for whole estuaries (Table 5.5). Chlorophyll-*a* biomass is expressed in terms of kg chl-*a* per estuary based on the area of the estuary. Higher chlorophyll-*a* concentrations were observed in the sediment than in the water column for all systems except for the Sundays and Gamtoos estuaries.

These data illustrate an important finding along a section of coastline where the quality of the freshwater flowing into estuaries differs. In the Keurbooms, Great Brak, Goukou and Gourits estuaries the freshwater has a much lower nutrient content than in the Gamtoos and Sundays rivers, which are both enriched by fertiliser throughflow. The high nutrient status of the Gamtoos and Sundays water expresses itself as a high phytoplankton biomass. Where the river water is relatively low in nutrients, the benthic biomass can be up to three orders of magnitude greater than the whole estuary phytoplankton biomass.

The Sundays estuary had the highest mean microbenthic algal chlorophyll-*a* value. However, microbenthic chlorophyll-*a* biomass for the whole estuary was lower than that calculated for the phytoplankton biomass (Table 5.5) as this system has narrow intertidal areas. Benthic microalgal biomass may be influenced by nutrients brought in by freshwater input. This aspect is currently being studied in the Gamtoos estuary as more data are required to substantiate these indications. Indeed, in a shallow southwestern Australian estuary benthic microalgal biomass increased with the onset of riverine nutrient input and decreased when phytoplankton blooms occurred in the water column. Sediment chlorophyll was significantly and positively correlated with salinity, ammonium, nitrate and organic phosphorus and nitrogen concentration in the water column (Lukatelich & McComb 1986).

Comparative productivity

Primary production of benthic diatoms in the Bot estuary (Bally *et al.* 1985) was calculated as 787 tonnes dry mass per year or 58 g m^{-2} y^{-1}. Annual production was much lower in the nearby Palmiet estuary (1.5 g C m^{-2} y^{-1}: Branch & Day 1984). Epibenthic algal production has also been measured in the Swartkops estuary, on a sandy beach near the mouth and a muddy beach in the upper reaches in spring and autumn using light and dark respirometers (Dye 1977). Estimated means were 53 and 116 g C m^{-2} y^{-1} for the two beaches. This productivity is similar to that of the phytoplankton of this estuary.

Fielding *et al.* (1988) found that benthic microalgal production contributed 22% to the total primary production of the Langebaan Lagoon. However, in the Bot estuary Bally *et al.* (1985) estimated benthic microalgae to contribute only 7% when the mouth was closed. Production values for benthic microalgae generally lie in the same range as for phytoplankton but are lower than for macrophytes.

ESTUARINE MACROPHYTES

Research on estuarine macrophytes undertaken since Day's (1981) summary has included studies on ecological, ecophysiological and physiological aspects of the plants. O'Callaghan's (1994*a*) analysis has shown that saltmarsh structure is determined by environmental gradients of salinity and inundation. Ecophysiological work has concentrated on the plants' responses to exposure, water level fluctuations and salinity. By studying the ecophysiological responses of selected important species such as *Zostera capensis* Setch., *Ruppia cirrhosa* (Petag.) Grande, *Spartina maritima* (Curtis) Fernald, *Sarcocornia perennis* (Mill.) A.J. Scott and *Phragmites australis* (Cav.) Steud, this research has attempted to understand the freshwater requirements of estuarine plants (Adams & Bate 1994*a*, *b*, *c*, *d*, 1995). Physiological research has addressed the salinity and waterlogging tolerance mechanisms of estuarine plants (Naidoo & Rughunanan 1990; Naidoo & Naicker 1992; Naidoo & Naidoo 1992; Naidoo & Mundree 1993; Naidoo 1994). Taylor's studies (Taylor 1988, 1992; Taylor & Allanson 1993, 1995) have provided valuable information on the exchanges of carbon, nitrogen and phosphorus between a *Sarcocornia* saltmarsh and the water in the Kariega estuary.

This review describes the estuarine vegetation and discusses the ecological importance of macrophytes. Freshwater input controls many physical factors to which the plants respond, namely salinity, water level, water flows, sedimentation and reduced nutrient input. The effect of these physical factors on plant distribution and growth is discussed.

Estuarine vegetation and ecological importance

Macroalgae

Macroalgae may be intertidal (intermittently exposed) or subtidal (submerged at all times) and may be attached to hard or soft substrata, or they may be floating. According to Day (1981), macroalgae are not well represented in South African estuaries, and are limited to a small number of widespread genera that can withstand the deposition of silt and the absence of wave action. However, recent surveys of a number of Eastern Cape estuaries indicate a substantial proportion of macroalgae in terms of biomass. This is particularly true in the Kromme (34° 09′ S, 24° 51′ E), Kowie (33° 36′ S, 26° 54′ E) and Kariega estuaries which have a strong marine influence. Species such as *Caulerpa filiformis* (Suhr) Hering, *Codium tenue* Kuetzing, *Gracilaria verrucosa* Hudson, *Ulva rigida* C. Agardh, *Hypnea viridis* Papenfuss and *Gelidium pristoides* (Turner) Kuetzing are found in these estuaries. In the Kromme estuary *C. filiformis* covered an area of 2.3 ha with a dry biomass ranging from 66 to 179 g m^{-2} (Adams *et al.* 1992).

Another group of macroalgae are the opportunistic species that can tolerate fluctuating salinities and are common in closed or temporarily closed estuaries characterised by slow water movement. In the Wilderness Lagoon (34° 00′ S, 22° 35′ E), epipsammic mats *of Entero-morpha intestinalis* (Linnaeus) proliferated during times when the lagoon mouth was closed but were washed out to sea whenever the estuary mouth opened (Weisser *et al.* 1992). Macroalgae such as *Enteromorpha*, *Chaetomorpha* and *Cladophora* often form part of algal mats and generally have a filamentous morphology. These algae all belong to the family Chlorophyta. It has been found that Chlorophyta extend further into estuaries than Rhodophyta or Phaeophyta (Gessner & Schramm 1972). This may be related to their wide range of salinity tolerance. Reed and Russell (1979) found that certain genotypes of *Enteromorpha intestinalis* from intertidal pools and estuaries could tolerate salinities of 0 to 51, while Gordon *et al.* (1980) found that *Cladophora* aff. *albida* in Peel Inlet, Australia, tolerated an annual salinity range of 2 to 50 .

Floating macrophytes

Floating angiosperms have aerial portions which are rarely submersed and roots which are submersed but not anchored in the sediment. These plants are generally restricted to the fresh and oligohaline (<5) sections of estuaries and to zones of quiet water. Typical species include *Eichhornia crassipes* (Mart.) Solms-Laub. (water hyacinth), *Azolla filiculoides* Lam. (water fern) and *Salvinia molesta* D.S. Mitchell (kariba weed), all of which are exotic to South Africa and are regarded as nuisance weeds (Howard-Williams 1980; King *et al.* 1989).

Submerged macrophytes

Submerged macrophytes are plants rooted in both soft subtidal and low intertidal substrata, whose leaves and stems are completely submersed for most states of the tide. Species that occur in South African estuaries and their distribution are listed below.

In more saline waters:

Zostera capensis (Olifants estuary on the west coast to Mozambique/Madagascar)

Halophila ovalis (R.Br.) Hook (Indian Ocean as far south as Knysna)

Thalassodendron ciliatum (Forssk.) Den Hartog (Northern KwaZulu-Natal to the tropical Indian Ocean)

Halodule uninervis (Forssk.) Aschers (Northern KwaZulu-Natal to the tropical Indian Ocean)

In less saline waters:

Potamogeton pectinatus L. (cosmopolitan)

Ruppia cirrhosa (pantemperate to cosmopolitan)

Ruppia maritima L. (cosmopolitan, occurs in less saline habitats than *R. cirrhosa*)

Zannichellia palustris L. (cosmopolitan)

Pseudalthenia aschersoniana (Graebn.) Den Hartog (endemic, Lamberts Bay to Still Bay)

Althenia filiformis Petit (Namibia to Port Elizabeth)

Chara globularis

Some of these species also occur in freshwater ponds, rivers or saline ponds.

The seagrass *Zostera capensis* occupies the intertidal mudbanks of most permanently open Cape estuaries. *Halophila ovalis* (R.Br.) Hook.f. often occurs in association with *Z. capensis* but is seldom as abundant (Day 1981). It acts as an opportunitistic species occupying estuaries in immediate post-flood conditions (Talbot *et al.* 1990) as it can rapidly colonise sandy substrata. *Ruppia cirrhosa* is common in temporarily open estuaries characterised by fluctuating salinities, but can also be found in the calm brackish upper reaches of permanently open estuaries (Adams *et al.* 1992; Adams & Bate 1994*b*). Because of its stronger morphological structure and ability to tolerate daily periods of exposure, *Z. capensis* has a competitive advantage over *R. cirrhosa* in estuaries where there is tidal exchange and stronger water currents (Adams & Bate 1994*c*).

Ecological importance

Submerged macrophytes help to oxygenate the water column, anchor the sediment and increase the depth of the oxidized microzone at the sediment surface (thus reducing phosphate and ammonia release). By trapping suspended sediment and damping wave action, submerged macrophytes can also improve water clarity (Fonseca *et al.* 1982). They also play an important role in nutrient trapping and recycling. When these plants declined in Chesapeake Bay, nutrient recycling within the system was not as efficient as before (Correll *et al.* 1975). In Zandvlei the *Potamogeton pectinatus* beds and associated filter-feeding polychaetes are considered to be the essential biotic component responsible for the maintenance of good water quality (Davies *et al.* 1989; Harding 1994; Quick & Harding 1994). In Swartvlei (34° 00′ S, 22° 48′ E), Howard-Williams and Allanson (1978) found that *P. pectinatus* acts as a nutrient pump, utilising the sediment as a phosphorus source and releasing it into the water after decay. Phosphorus is retained within the littoral zone through the rapid uptake by other components of the system, including surface detrital floc, epiphytic algae and littoral fauna (cf. Chapter 4).

Submerged macrophytes provide food for epifaunal and benthic invertebrate species in the form of detritus, diatoms and filamentous algae that are trapped in the beds (Whitfield 1989). The plants also provide food indirectly through their diverse and abundant invertebrate faunas which are consumed by carnivorous fish species (Whitfield 1984; Connolly 1994). Submerged macrophytes can be grazed directly but food is mostly provided indirectly through consumers feeding on epiphytic algae. Blaber (1974) showed that the sparid fish *Rhabdosargus holubi* could digest the largely diatomaceous periphyton, but not the tough cellulose of *Ruppia* leaves (Day 1981). Epiphytic biomass can be quite high and in the Swartkops estuary accounted for 28% of the total biomass (*Zostera* and epiphyton) in summer (Talbot & Bate 1987).

Submerged macrophyte beds form nursery areas for juvenile fish by providing food, shelter and protection from predators (Branch & Grindley 1979; Beckley 1983; Whitfield 1984). These beds support more diverse and abundant invertebrate and juvenile fish communities than bare soft-bottomed habitats and marshes (Whitfield *et al.* 1989; Connolly 1994). In the Swartvlei estuary Whitfield (1989) found that the biomass and diversity of macroinvertebrates in eelgrass sites were higher than in adjacent sandy sites. Similarly, the abundance of dominant benthic invertebrate species in the Berg River estuary has also been reported to fluctuate in response to seasonal growth of *Zostera* (Kalejta & Hockey 1991). There is evidence that, if *Zostera* is removed from an estuary, the composition of the fish fauna can change. In Richards Bay (28° 49′ S; 32° 05′ E), the loss of *Zostera* resulted in the disappearance of several species (e.g. *Rhabdosargus sarba, R. holubi, Acanthopagrus berda* and *Pelates quadrilineatus*: Whitfield *et al.* 1989).

Bait organisms are mainly found in the *Zostera* zone,

e.g. *Upogebia africana* (mudprawn), *Alpheus crassimanus* (cracker shrimp), *Arenicola loveni* (bloodworm) and *Solen cylindraceus* (pencil bait). *Zostera* beds can also provide a habitat for rare and endangered species, e.g. the Knysna seahorse (*Hippocampus capensis*) in the Knysna estuary.

Dynamics

Submerged macrophyte communities are highly variable on both spatial and temporal scales. Die-back of macrophytes in the Wilderness lakes (34° 00′ S, 22° 35′ E) occurred between 1979 and 1981 (Weisser *et al.* 1992). A number of factors were thought to have contributed to this decline in submerged macrophytes. These included the high water level management policy, phytoplankton blooms, the development of periphyton on submerged plants and uprooting of plants by wave action during strong winds (Weisser & Howard-Williams 1982). The recent collapse of the *Potamogeton pectinatus* standing stock in Zandvlei is thought to be related to increased water depths and decreased salinity (Harding 1994; Quick & Harding 1994). Recovery of submerged macrophytes in the Wilderness lakes was fairly rapid and mean biomass per square metre doubled between winter 1982 and winter 1983 (Weisser *et al.* 1992). However, in the nearby Swartvlei system, after the collapse of the macrophytes in 1979 (Whitfield 1984) the Characeae and *Potamogeton* beds had still not fully recovered some 10 years later (Weisser *et al.* 1992).

In estuaries subject to episodic flooding the related sedimentary disturbances appear to be the most important factor determining the state of seagrasses. Intense temporal variability in the distribution of intertidal and submerged macrophytes in response to flooding was recorded in the Kwelera and Nahoon estuaries by Talbot *et al.* (1990). Heavy floods with an approximate frequency of 15 years result in the complete removal of *Zostera capensis* beds. Moderate floods (2–3 years) and heavy fluvial deposits smother beds, whereas light (1 year) floods lead to temporary smothering of beds, resulting in impaired growth and shortened leaf lengths. A flood in November 1985 removed *Z. capensis* populations from both estuaries. Recovery of *Z. capensis* was sluggish and lagged behind the process of post-flood sedimentation.

Water storage dams result in a reduction in flooding frequency. For example, in the case of the Kromme estuary, the upriver dams reduce the effect of all floods smaller than a 1-in-30 year flood (Bickerton & Pierce 1988). A consequence of this was an increase in the growth and expansion of submerged macrophytes because of increased sediment stability and improved water clarity that was related to a lack of freshwater input. The surface area cover and biomass of *Z. capensis* increased following the construction of a second dam in the catchment (Adams & Talbot 1992).

Saltmarsh vegetation

Distribution

Saltmarshes occur only in certain estuaries and embayments along the coast of South Africa. North of the Kei River the subtropical climate favours the development of mangrove swamps. Saltmarshes cover approximately 17 000 ha of land surface, with more than 75% of this area confined to five systems, i.e. Langebaan Lagoon, Knysna Lagoon (34° 03′ S, 23° 04′ E), the Swartkops River, the Berg River and the Olifants River (31° 42′ S, 18° 11′ E) (O'Callaghan 1994a). Langebaan Lagoon is a marine embayment and not an estuary as it has no river input. However, this embayment contains over 30% of all South Africa's saltmarsh areas, most of which have been protected since 1988 as part of the West Coast Nature Reserve (O'Callaghan 1990, 1994a). The distribution of saltmarsh angiosperm species shows little affinity with geographical position along the coast. The presence or absence of species in a saltmarsh is related to specific environmental habitats within the system which are determined by patterns of tidal inundation and salinity. Each habitat supports a distinctive saltmarsh community. However, unlike most terrestrial environments, each community in a saltmarsh consists of only one or a few species.

Zonation

Saltmarsh plants show distinct zonation patterns along tidal inundation and salinity gradients (Figure 5.5). Zonation tends to be better developed in areas with a large tidal range. Where there is a small tidal range, communities are often arranged in mosaics rather than distinct zonal bands (Adam 1981; Figure 5.7). The marshes at Langebaan Lagoon will be used as the basis for description. The absence of a riverine influence and the dead-end nature of the lagoon results in relatively stable tidal conditions. The broad intertidal area and the uniformly sloping shoreline result in easily distinguishable zonation patterns. Generally, MHWN = MSL + 25 cm and MHWS = MSL + 75 cm (MHWN, Mean High Water Neap Tide; MSL, Mean Sea Level; MHWS, Mean High Water Spring Tide). However, the position of a species in a marsh is related to relative periods of inundation rather than height above standard MSL.

From Mean Sea Level to Mean High Water Neap

The grass *Spartina maritima* occurs from about MSL to below MHWN (Table 5.6) and is found in 18 of the larger, permanently open Cape estuaries. In the Swartkops estuary, *S. maritima* covers 82 of the 363 ha of intertidal marsh (Pierce 1979; Baird *et al.* 1988). This grass can grow to heights of 80 cm in the lower parts of its zone (Figure 5.6). In the upper parts of its zone, and particularly where it

Table 5.6. Saltmarsh zonation pattern common in Langebaan Lagoon. Generally MHWN = MSL + 25 cm and MHWS = MSL + 75 cm. MHWN, Mean High Water Neap Tide; MSL, Mean Sea Level; MHWS is Mean High Water Spring Tide

Species	Position in relation to elevation above MSL				
	MLWS	MLWN	MSL	MHWN	MHWS
Disphyma crassifolium					XX
Sarcocornia pillansii					XXXX
Suaeda inflata					XXX
Chenolea diffusa					XXX
Limonium linifolium					XX
Cotula coronopifolia				XXX	
Triglochin striata				XXXX	
Sarcocornia perennis				XXXXXXXXX	
Spartina maritima			XXXXXXXXXX		
Zostera capensis	XXXXXXXXXXXXXX				

overlaps with the *Sarcocornia–Triglochin* zone, the stature of this plant is usually no more than 10 or 15 cm.

Pierce (1982) suggested that *S. maritima* was introduced into South Africa at Algoa Bay in *c.* 1830. However, the matter has not been fully resolved as the taxonomic history and ecology of the species does not seem to support this postulate. Pierce (1982) warned that the extension of *S. maritima* would enhance silt accumulation and restrict the scouring action of floods. *Spartina* is well known for its ability to accrete large volumes of sediment. Its stout stems and leaves slow down the ebb and flow of tidal water and trap the sediment it carries, mostly in the axils of the leaves where they sheath the stem. As the aerial portions senesce, this sediment is deposited around the base of the plant and is bound together by the extensive rhizome network (Thompson 1991).

From Mean High Water Neap to Mean High Water Spring

This part of the marsh is characterized throughout by *Sarcocornia perennis* but numerous other species are often present. *Triglochin bulbosa* L. is often found to co-dominate with *S. perennis* at the lowest parts of the zone. A related species, *T. striata* Ruiz & Pav. can dominate along creek banks in a saltmarsh. The annual *Salicornia meyeriana* Moss is usually present if the marsh has been disturbed (e.g. by trampling or through the deposition of wrack). *Cotula coronopifolia* L. often forms patches about midway in this zone and can also function as a facultative annual (often with *S. meyeriana*) in those parts of a saltmarsh that are exposed to an annual draw-down of water.

One of the *Limonium* species (sea lavenders) appears three-quarters of the way up this zone. *Limonium depauperatum* (Boiss.) R.A. Dyer is found mostly in west coast estuaries, *L. linifolium* (L.f.) Kuntze along the south coast and *L. scabrum* (Thunb.) Kuntze along the east coast. *Chenolea diffusa* Thunb. is often found as single plants throughout the upper parts of this zone, or as small shrubs near the top of the zone with *Suaeda inflata* Aell.

Above Mean High Water Spring

Occasional flooding by salt water followed by evaporation results in the substratum above MHWS becoming salinised, with little free water available. The area is usually dominated by the short, gnarled shrub *Sarcocornia pillansii* (Moss) A.J. Scott, although numerous annuals, e.g. *Puccinellia angusta* (Ness) C.A. Sm. & C.E. Hubb, and drought-tolerant species of terrestrial origin might also be present. At the top of this zone, particularly in sandy areas, the succulent *Disphyma crassifolium* (L.) L. Bol. and *Plantago crassifolia* Forssk. make an appearance, while higher up terrestrial species start to dominate (Figure 5.8).

The foregoing describes the species generally found in saltmarshes, but a number of other important species occur according to habitat: *Juncus kraussii* can occur near MHWS in parts of estuaries where the salinities rarely exceed 20 (e.g. at the Breede River (34° 24' S, 20° 51' E), some distance from the mouth). *J. kraussii* is also found in shallow parts of estuaries which remain fresh due to mouth closure, e.g. Kleinmond Lagoon (34° 25' S, 19° 18' E), seepage of freshwater into the system, e.g. near the mouth of the Uilkraals River (34° 36' S, 19° 25' E) or strong seasonal inflow of fresh water into the system, e.g. about 8 km from the mouth of the Berg River. Under these conditions, the habitat is often seasonally flooded and this species survives as a short-growing emergent macrophyte. These areas can develop into relatively large sedgefields with numerous associated species such as *Samolus porosus* (L.F.) Thunb. nearer the river channel, *S. valerandi* L. and *Apium graveolens* L. in more open spaces, and *Orphium frutescens* (L.) E. Mey. in the slightly drier areas. Some of the most important associated species, include *Sporobolus virginicus* (L.) Kunth which prefers sandy substrata, and *Sarcocornia natalensis* (Steud.) Dur. & Schinz which requires freshwater for at least some part of the year. These species are able to withstand flooding for various periods and often form a minor component of tidally flooded marshes.

Ecological importance

Saltmarshes have a number of important functions which include sediment stabilisation and bank protection; they function as filters of sediment and pollutants and provide feeding and shelter areas for both marine and estuarine organisms. Saltmarshes serve as zones of nutrient production and retention and are important inorganic and organic nutrient sources for estuarine ecosystems. Plant biomass decays on the marsh surface, and its energy enters the estuarine food chain as detritus. The estuarine water levels and the degree of tidal flushing are important in determining the amount of nutrient release into the

water column (Childers & Day 1990). A supratidal marsh with little tidal flooding will be less important with respect to nutrient exchange than an intertidal marsh or submerged macrophyte bed (cf. Chapter 4). However, supratidal marshes can function as a sink for organic carbon. During high spring tides, detrital plant material originating from low saltmarsh areas is deposited onto high saltmarsh areas. Long periods of exposure allow heterotrophic breakdown and decomposition in this environment where available oxygen is not limiting. Carbon dioxide produced by respiration is then released to the atmosphere (Taylor & Allanson 1995).

Reeds and sedges

Reeds, sedges and rushes form an important component of the vegetation in freshwater and brackish zones of estuaries, usually as emergent plants lining the banks. These plants are rooted in soft intertidal or shallow subtidal substrata. However, they have photosynthetic aerial portions which are partially and/or periodically submersed. Plants occurring in freshwater zones include *Schoenoplectus littoralis* (Schrad.) Palla, *Typha capensis* (Rohrb.) N.E. Br., *Cyperus laevigatus* L. and *Schoenus nigricans* L. *Bolboschoenus maritimus* (L.) Palla and *Juncus kraussii* Hochst. are usually found in the mid-reaches of an estuary on parts of marsh which are occasionally flooded (Figure 5.9).

Schoenoplectus triqueter (L.) Palla and *Phragmites australis* grow in brackish zones (<15). *Schoenoplectus triqueter* can grow in deeper water and at slightly higher salinities than the common reed *P. australis*, a species which indicates points of freshwater seepage into an estuary from surrounding land. This plant survives in estuaries such as the Kromme and Keurbooms where it is tidally inundated by saline water (salinity 35) as the root system is located in fresh or brackish water (Adams & Bate 1994a). Groundwater seepage plays an important role in modifiying the salinity in these reedbeds and creates nodes of biotic diversity at these sites where both brackish and saline species may occur. In the Goukou estuary *P. australis* grows adjacent to the salt-tolerant grass *Spartina maritima*.

Phragmites-dominated marshes provide a habitat for many bird, invertebrate and fish species. Fish spawn in these areas while birds nest and feed here. Reed and sedge communities also provide food for detritivores and a substratum for periphyton and bacteria. In the Mhlanga estuary (29° 42' S, 31° 06' E), the detrital aggregates arising from the fringing *Phragmites* beds provide the most important food resource for 90% of the fish community (Whitfield 1980). Both *P. australis* and *Schoenoplectus triqueter* die back in late summer, releasing particulate matter into the water. Decomposition products fertilise and enrich the water. However, reeds also remove large quantities of nutrients from the water column during the growing season and are so effective that they are used for water purification in wastewater treatment systems (Brix 1993, 1994).

Emergent communities serve the valuable ecological function of protecting banks from erosion. Destruction of *P. australis* stands by boating and swimming activities in Europe has been shown to require costly shore rehabilitation programmes (Weisser & Howard-Williams 1982). Reed and sedge communities have an important utilitarian value, particularly in rural areas of Natal (Begg 1986). The rush *Juncus kraussii* is used for the construction of sleeping mats and numerous craftwork products. Hut-building and thatching material is obtained from *Phragmites*.

Comparative productivity

In situ production usually exceeds the rates of import of organic material from both terrestrial and marine sources. A marked variation in river inflow, influenced by the rainfall regime and the presence of dams, can reduce the import of organic carbon from terrestrial sources (Kokkinn & Allanson 1985). Some estuaries have, therefore, become reliant on autochthonous (locally produced) plant material and marine imports. The primary production of *Spartina maritima* in the Swartkops estuary has been estimated at over 10 000 kJ m^{-2} per year (602 g m^{-2}y^{-1}) (Pierce 1979; Day 1981). Only half of this is estimated to accumulate in the detritus chain and the remainder is either deposited in the estuary or washed out to sea (Day 1981). Kokkinn and Allanson (1985) showed that the import of organic carbon from marine sources is substantially higher than that exported. This suggests that the organic carbon produced by the marshes remains in the estuary, bound in the food chain, and is utilised by the organisms which live there.

Attached macrophytes are usually the dominant primary producers, having very high production rates per unit area (200–3000 g C m^{-2}y^{-1}: Kennish 1986). Maximum standing biomass attained by emergent macrophytes is generally greater than that attained by other forms of estuarine macrophyte. The same is true of productivity per unit area (Westlake 1963). Comparing productivities within a brackish coastal lake system (Wilderness), Howard-Williams (1980) found productivies (g dry mass m^{-2}y^{-1}) of the various macrophytes to be of the order *Typha latifolia* > *Phragmites australis* > *Schoenoplectus littoralis* (all emergents) > *Potamogeton pectinatus* > *Chara globularis* > *Ruppia cirrhosa* (all submerged). However, in the Swartvlei, Howard-Williams and Allanson (1981) found productivites (g C m^{-2}y^{-1}) of the major plant groups to be

of the order *Chara mariscus* > *Potamogeton pectinatus* (submerged) > *Phragmites australis* = *Schoenoplectus littoralis* (emergent) > phytoplankton. In Langebaan Lagoon, Christie (1981) found that the productivities of the more oligohaline emergents (*Phragmites* and *Typha*) and the salt-marsh emergents (*Sarcocornia* and *Spartina*) were all similar. The submerged macrophyte *Zostera* was only about a third as productive.

Typical annual dry mass production for emergent macrophytes, both oligohaline and polyhaline species, is about 2500 g m^{-2}. This is equivalent to 250 g C m^{-2}y^{-1}, a carbon content of 45% being typical of these macrophytes. In general, there is considerable variation between systems (Table 5.7). Most authors ascribe this to differences in environmental conditions. Howard-Williams (1980) found that the productivities of all the macrophyte groups in the Wilderness lakes were low compared with those for the same species in other systems. He speculated that fluctuating and occasional high salinities and possibly nutrient limitation were responsible. On the other hand, some of the variation may be due to differences in the methods used to estimate production.

The percentage contribution to total primary production by the different macrophyte groups is even more variable, being primarily a function of topography. In Swartvlei, Howard-Williams and Allanson (1981) found that, despite covering 72% of the area, phytoplankton only contributed 6% to total production. Because of their high productivities the two submerged macrophytes *Potamogeton* and *Chara* contributed 85% to total production, while the emergents contributed only 9%.

The effect of freshwater controlled physical factors on plant distribution and growth

Estuarine macrophyte species composition, biomass and productivity, like all plant communities, is determined by a complex interaction between numerous abiotic and biotic factors: irradiance, temperature, nutrient availability, grazing and competition. By the nature of their habitat, estuarine macrophytes are also subjected to varying salinity and fluctuating water levels. This section reviews the documented responses of estuarine macrophytes to various abiotic factors with particular emphasis on the relationship to freshwater inflow.

The severe drought of the 1980s highlighted the scarcity of South Africa's precious freshwater supplies. The need to store the maximum amount of water for human requirements focused attention on the research needed to determine how much freshwater should be released from dams to supply estuarine plants and

Table 5.7. Published biomass and annual production (dry mass) of various estuarine macrophytes

Species	Biomass (g m^{-2})	Production (g m^{-2} y^{-1})	Reference
Macroalgae			
Codium tenue	85		Knoop et al. 1986
Hypnea viridis	178		Knoop et al. 1986
Emergent macrophytes			
Typha latifolia		1249	Howard-Williams 1980
Schoenoplectus littoralis		1004	Howard-Williams 1980
Schoenoplectus littoralis		960	Bally et al. 1985
Phragmites australis		930	Howard-Williams 1980
Phragmites australis	1500	1621	Howard-Williams & Allanson 1981
Phragmites australis		1790	Bally et al. 1985
Spartina maritima		680	Pierce 1983
Submerged macrophytes			
Chara globularis		231	Howard-Williams 1980
Chara mariscus	150	263	Howard-Williams & Allanson 1981
Chara sp.		2230	Bally et al. 1985
Potamogeton pectinatus		415	Howard-Williams 1980
Potamogeton pectinatus	1950	2506	Howard-Williams & Allanson 1981
Potamogeton pectinatus		4124	Bally et al. 1985
Ruppia cirrhosa		83	Howard-Williams 1980
Ruppia maritima		710	Bally et al. 1985
Zostera capensis	226	–	Christie 1981
Zostera capensis	185	–	Adams et al. 1992

animals. A study of the freshwater requirements of estuarine plants resulted in the development of a decision support system to predict how different plants would respond to different amounts of freshwater (Adams & Bate 1994a). Use of this expert system together with other models showed that if one can predict the physical responses of an estuary to changes in freshwater inflow, then one can predict much of the vegetation response (Slinger & Breen 1995).

Sediment related effects

Sediment stability of a system influences macrophyte colonisation. In systems where the sediment is constantly modified by dynamic processes, submerged macrophytes are absent. This is the case in the Palmiet River estuary where strong water currents and frequent flooding result in the removal of established rooted plants (Branch & Day 1984). In the estuaries of Natal, Begg (1984a, b) found that submerged macrophytes occurred only in systems where a semi-permanently closed mouth resulted in stable water levels and sediment for approximately one year. Reduction of freshwater inflow into the Kromme estuary over the decade (1981–91) has led to increased sediment stability. This has resulted in an increase in Zostera capensis biomass and area distribution (Adams & Talbot 1992). Prior to this, Hanekom and Baird (1988) found that fluctuations in the biomass of this species in the Kromme were related to flood events. Other studies in South African estuaries have shown that changes in Z. capensis biomass appear to be linked to flood activities rather than seasonal influences (Edgecumbe 1980; Talbot et al. 1990). Zostera capensis beds may disappear completely after a major flood, but gener-

ally recover after a period of 1–3 years (Talbot et al. 1990).

Edaphic factors that influence emergent macrophyte production include soil drainage, sediment oxidation, exchange of interstitial water, nitrogen nutrition and sulphide concentration (Burdick et al. 1989). Grain size affects nutrient movement and retention within the sediment (Anderson & Kalff 1988). However, Hanekom and Baird (1988) found that spatial variation in the distribution and biomass of Z. capensis, in the Kromme estuary, was not significantly related to sediment particle size.

Light related effects

The underwater light environment, determined by water turbidity and depth, is perhaps the most important factor determining the distribution of submerged macrophytes. Light attenuation in the water column is often severe due to high turbidity associated with suspensoids. Macrophytes can also be heavily colonised with epiphytic algae which reduce the light available to macrophytes for photosynthesis. This is particularly a problem in systems which have high nutrient loading (Stevenson 1988). According to Chambers and Kalff (1985), light limits the maximum depth of colonisation and also influences the depth of maximum biomass. Congdon and McComb (1979) found that experimental shading of in situ Ruppia populations, in a shallow (< 1 m deep) Australian estuarine basin, severely reduced productivity. An 80% reduction in irradiance completely precluded Ruppia after 100 days. Warne (1994) indicated that the morphological and photosynthetic response of Z. capensis to light suggests that it is relatively more shade-adapted than Ruppia cirrhosa, which may explain its penetration into deeper water in the field.

When these species co-occur *Zostera* is generally found in deeper water, below the *Ruppia* zone (Howard-Williams & Liptrot 1980).

Water clarity has been found to influence the distribution of submerged macrophytes in a number of South African estuaries. In Swartvlei, the lower depth limit of the macrophytes was at 5–10% of surface irradiance. Historical data showed that the macrophyte beds encroached into deeper water during dry periods, when water clarity improved, and regressed during rainy periods, when input of suspensoids from the Wolwe River increased (Howard-Williams & Allanson 1981). Vertical distribution of submerged macrophytes in the adjacent Wilderness lake system was also limited by light availability in deeper water (Howard-Williams & Liptrot 1980). High turbidity was the reason given by Day (1951) for the absence of *Z. capensis* from subtidal areas in the Berg River estuary.

In Natal estuaries, which are notably turbid, Begg (1984*a*, *b*) reported that submerged macrophytes were never dominant, although they did occur in closed, clear estuaries. *Zostera capensis* was only found in one estuary. *Potamogeton pectinatus* was more common but, in terms of energy input, the filamentous alga *Chaetomorpha*, which grew in winter when water transparency was high, was more important.

Nutrient related effects

The original source of most nitrogen and phosphorus is weathering of rocks and leaching of soils on land (Day 1981). However, at present, anthropogenic activity is a predominant source of nutrient elements in some systems. Increased sewage discharge, agricultural fertilisation and urbanisation have resulted in accelerated inflow of nutrient elements (Kennish 1986). Addition of nutrients to the water column usually leads to increased macrophyte productivity. Blooms of water hyacinth (*Eichhornia crassipes*) can occur in salinities less than 5 and blooms of macroalgae, particularly species of *Ulva*, *Enteromorpha* and *Cladophora*, have long been associated with eutrophic conditions (Josselyn & West 1985). Excessive macroalgal populations may lead to nocturnal water deoxygenation and subsequent ecological damage (e.g. fish deaths). Upon decay they can cause secondary organic pollution and noxious odours (Robinson & Hawkes 1986). This causes a deterioration in the quality of the area, which may be in demand for recreational purposes and aesthetically pleasing residential areas. Decaying mats of filamentous algae have been shown to impact adversely the social acceptability of water in the Great Brak estuary (P. Huizinga, CSIR, personal communication) and Kleinmond estuary (O'Callaghan 1994*b*) and is very often the reason for the manipulation opening of the estuary mouth.

Emergent brackish water macrophytes have a large capacity for nutrient uptake and have been used as artificial filters for sewage water (Brix 1993). However, while emergent reedswamp vegetation may reduce eutrophication by absorbing nitrogen and phosphate from inflowing polluted water, it is sometimes claimed to increase eutrophication following the release of nutrients during decomposition. Simpson *et al.* (1983) concluded that marshes are a sink for nutrients but only during the growing season.

Saltmarshes are generally considered to depend on recycled nutrients rather than those carried in the freshwater (Schelske & Odum 1961). However, most of the nutrients are derived from muddy sediment and silt carried down in the rivers and deposited in saltmarsh areas. Studies on the nutrient requirements of saltmarsh plants show there is no lack of phosphate in the sediment but that ammonia or nitrate nitrogen is the limiting nutrient during periods of rapid growth (Day 1981).

Salinity related effects

A primary effect of reduced freshwater input is an increase in water column salinity owing to evaporation and a reduction in the dilution of seawater by freshwater. A reduction of freshwater flow into the Kromme estuary following dam construction upriver resulted in salt accumulation in the intertidal marshes (Adams *et al.* 1992). Saltmarsh macrophytes predominate at salinities between 10 and 35. Marsh productivity decreases as salinity increases and the shoots of all rooted vegetation die in salinities above 55 (Day 1981). Conservation of intertidal saltmarshes is dependent on freshwater input and tidal exchange. Freshwater dilutes sediment salinities, preventing dry hypersaline conditions that reduce macrophyte germination and growth. Lack of freshwater input reduces macrophyte diversity as there is evidence that the marine submerged macrophyte communities extend into the normally brackish upper reaches and displace the brackish communities (Adams & Talbot 1992; Adams *et al.* 1992).

Despite their general association with euhaline conditions (salinity 30–40), it seems that high salinity is not physiologically optimal for saltmarsh plants. Chapman (1960) showed that few species will germinate in salinities greater than 20 and Ungar (1962) found that germination in succulent halophytes dropped when the particular limits of salt tolerance were reached for each species. Studies on South African species such as *Triglochin bulbosa* and *T. striata* have shown that germination is best in distilled water and at salinities less than 15 (Naidoo & Naicker 1992). Both field and laboratory studies have also shown that the growth rates of many saltmarsh species increase as salinity decreases (Phleger 1971; Smart & Barko 1978; Price *et al.* 1988). This generally seems to apply to species which have a low salinity range of tolerance, e.g. *Cyperus* and *Phragmites* (Millard & Broekhuysen 1970). However, even the more salt-tolerant species such as *Spartina maritima* (Adams & Bate 1995), *Sarcocornia perennis* (Adams & Bate 1994d) and *Sporobolus virginicus* (Naidoo & Mundree 1993) show better growth in lower salinity under laboratory conditions.

Field observations (Hoese 1967) and laboratory growth studies (Chapman 1960; Phleger 1971; Adams & Bate 1994d, 1995) indicate that even the most salt-tolerant emergent macrophytes such as *Spartina* spp. do not grow in a salinity of 45. They can survive these conditions for short periods but not in a persistent salinity in excess of 40. Some authors believe that such hypersaline conditions would lead to barren salt pans, while others have suggested that the removal of the macrophyte canopy and improved water clarity associated with the lack of fresh or marine water inflow could lead to increased phytoplankton and benthic macroalgal productivity (Hoese 1967).

Within the salinity range 0–40, long-term salinity changes in either direction lead ultimately to changes in marsh community composition. Too much freshwater may eliminate the saltmarsh community. O'Callaghan (1990) has shown how poorly developed the halophytic vegetation is in the False Bay estuaries, because of human impacts and reduced saline inputs. Management for the maintenance of halophytic communities must therefore include management of the freshwater inputs, so that the timing and duration of freshwater discharge do not exceed the threshold at which species composition will change (Zedler 1983).

All plants have an optimal/suboptimal salinity range of tolerance. The optimal salinity range is that at which maximum plant production occurs. The plants grow rapidly, flower and germinate within this range. Within the suboptimal salinity range, productivity is lower and plant cover sparse. Freshwater management decisions have to consider the maximum time that plants can be exposed to lethal salinities. For example, controlled laboratory experiments have shown that the seagrass *Zostera capensis* dies after three months at a salinity of 55 and after one month at a salinity of 75. *Z. capensis* survived better under extended freshwater than under extended hypersaline conditions, as after six weeks of incubation the plants in freshwater began to show signs of stress (Adams & Bate 1994b).

In the estuarine environment, the ability to tolerate fluctuating salinity is important. Although *Ruppia cirrhosa* grows best in brackish water, laboratory studies showed that it can survive in salinity ranging from freshwater to 75, can tolerate fluctuating salinity and can recover from high salinity treatments (Adams & Bate 1994b). *Ruppia cirrhosa* is, therefore, well adapted to its habitat in temporarily open South African estuaries. Erratic freshwater input into these estuaries causes salinity to fluctuate between brackish and hypersaline conditions, particularly in dry areas such as the Eastern Cape.

Water column salinities in which various estuarine macrophytes have been observed, are given in Table 5.8. Although many references may be available for a single species, the references are those which report the widest salinity ranges. These salinity ranges do not necessarily reflect physiological limits, or the duration for which extremes can be tolerated. However, they give some indication of salinity ranges in which the species are likely to be found. Species distribution will be determined not only by the salinity tolerances of the mature plants but also by competitive interactions and by the salinity tolerances for flowering and seed germination. Furthermore, nutrient availability, irradiance and temperature may influence salinity tolerances. Salinity tolerances under fluctuating salinity levels may also differ from those under stable conditions.

Salinity tolerance mechanisms

Tolerance mechanisms used by plants to adapt to salinity can be separated into those that allow the growing cells of the plant to avoid and those that permit the cells to cope with high ion concentrations. Salt avoiders or excretors include plants (e.g. the mangrove *Avicennia marina*, and the saltmarsh plants *Limonium* and *Sporobolus*) that possess salt glands or trichomes that are capable of removing excess salts to the exterior. In certain halophytes, there are readily observable salt glands on the leaves, sometimes consisting of only two cells (e.g. *Spartina* and *Sporobolus*), a basal collecting cell and a cap excreting cell. Salt glands are multicellular in dicotyledons, e.g. *Limonium*. Ionic excretion of salt glands consists mainly of sodium and chloride. Salt glands are assumed to transport and excrete ions against a concentration gradient. It is generally accepted that this excretion is an active process. The high content of organelles at sites of excretion, especially mitochondria, supports this hypothesis.

Plants can regulate their salt content by allowing leaves to absorb water so that salt concentrations do not increase. This leads to the development of succulence (a high volume/surface ratio). At the end of the season the succulent portion may dry up and be sloughed off as in *Salicornia* spp. In *S. herbacea*, succulence was induced by growing the plant in media with NaCl. Succulence was due to the development of larger spongy mesophyll cells and a multilayer palisade tissue. For *Triglochin maritima* it was found that Na^+ was actively transported out across the root plasmalemma at external concentrations above 10 mM (Jefferies 1972).

Exclusion mechanisms are effective at low to moderate

Figure 5.5. **Saltmarsh zonation in Langebaan Lagoon. From right to left: *Zostera capensis*, *Spartina maritima*, *Sarcocornia perennis* and *Chenolea diffusa*. (Photo: M.O'Callaghan.)**

Figure 5.6. ***Spartina maritima* (foreground) and *Zostera capensis* in the Swartkops River estuary. The effects of trampling are evident. (Photo: F.M.Du Plessis.)**

Figure 5.7. **Mosaic of salt marsh plants, *Limonium linifolium* (with small flower), *Chenolea diffusa* (grey colour), *Sarcocornia perennis* (bright green) and the annual *Salicornia meyeriana* (red). (Photo: F.M.Du Plessis.)**

Figure 5.8. **Supratidal saltmarsh showing *Sarcocornia pillansii* in the foreground, *Juncus kraussii* and terrestrial species in the background. (Photo: J.B.Adams.)**

Figure 5.9. **Reeds and sedges at a freshwater seepage site in the Gouritz River estuary. (Photo: J.B.Adams.)**

Figure 5.10. **(below)** *Zostera capensis* **growing in the intertidal zone of Knysna River estuary. The effects of trampling and boats are evident. (Photo: J.B.Adams.)**

Table 5.8. Published water column salinity ranges in which various estuarine macrophytes have been reported (* indicates species that are not found in South African estuaries but have been included in the table for comparative purposes)

Species	Salinity	Reference
Floating macrophytes		
Eichhornia crassipes	<2.5	Haller et al. 1974
Myriophyllum spicatum	<13.3	Haller et al. 1974
Salvinia molesta	<4	Howard-Williams 1980
Emergent macrophytes		
Phragmites australis	<25	Starfield et al. 1989
Bolboschoenus maritimus	7–12	Bally et al. 1985
Bolboschoenus maritimus	0.5–5	Lubke & van Wijk 1988
Saltmarsh macrophytes		
Triglochin striata	<23	Naidoo 1994
Triglochin bulbosa	<23	Naidoo 1994
Juncus kraussii	<15	Heinsohn & Cunningham 1991
Salicornia virginica*	<45	Pearcy & Ustin 1984
Sarcocornia perennis	12–42	Adams & Bate 1994d
Spartina maritima	<35	Adams & Bate 1995
Spartina alterniflora*	<45	Linthurst & Seneca 1981
Sporobolus virginicus	<30	Breen et al. 1976
Submerged macrophytes		
Chara sp.	5–15	Howard-Williams & Liptrot 1980
Potamogeton pectinatus	5–15	Howard-Williams & Liptrot 1980
Potamogeton pectinatus	<19	Ward 1976
Ruppia cirrhosa	2–40	Verhoeven 1975
Ruppia cirrhosa	<50	Ward 1976
Ruppia cirrhosa	0–55	Adams & Bate 1994b
Ruppia maritima	2–18	Verhoeven 1975
Zostera capensis	5–40	Day 1981
Zostera capensis	< 45	Ward 1976
Zostera capensis	10–45	Adams & Bate 1994b
Macroalgae		
Cladophora albida	2–50	Gordon et al. 1980
Enteromorpha intestinalis	0–51	Reed & Russell 1979
Caulerpa filiformis	25–45	Adams 1989

levels of salinity. Ion accumulation is the primary mechanism used by halophytes at high salt levels, presumably in conjunction with the ability to isolate ions into vacuoles and in the cell wall. The vacuole comprises 95% of the mature leaf cell volume and it is a commonly held view that ions are accumulated here for osmotic adjustment. Accumulation of ions in the vacuole can occur because of changes in membrane permeability and ion transport properties that facilitate transport against electrochemical gradients and membrane transport selectivity. There is an accumulation of a controlled amount of salt as an osmoticum. In high salinity environments, organic solutes also accumulate and provide part of the osmoticum required for turgor maintenance and the protection of enzymes against ion toxicity (Lerner 1985).

Studies on Sarcocornia natalensis (Naidoo & Rughunanan 1990) showed that ion accumulation accounted for 86% of the osmotic adjustment. The osmotic organic solute glycinebetaine was also produced. The nitrogenous organic compound proline is another common solute that has been found to accumulate in a number of South African estuarine plants. According to Naidoo (1994), osmoregulation and synthesis of the nitrogenous organic compound proline could be responsible for the slow growth of Triglochin striata and T. maritima at high salinity, because of N-limitation. Proline accumulation in R. cirrhosa and Z. capensis roots and leaves was measured in plants grown at different salinities. Proline concentrations increased with an increase in treatment salinity for both plants, but the proline concentrations were higher in R. cirrhosa than in Z. capensis and this may explain why R. cirrhosa can withstand wide salinity fluctuations (Adams & Bate 1994b).

Water movement related effects

Water movement is the underlying factor controlling many other physical factors of importance to estuarine macrophytes. Such factors include water clarity, sediment and mouth dynamics. Water movement also affects macrophytes directly and is, therefore, considered separately. Strong water currents (>1 m s^{-1}) lead to the removal of submerged plants and at 0.5 m s^{-1} growth is significantly reduced owing to mechanical damage. At water velocities below 0.1 m s^{-1} aquatic macrophytes grow and establish themselves well. Seagrasses Zostera marina and Thalassia testudinum showed maximum standing crop densities where the current velocity averaged 0.05 m s^{-1} (Zieman & Wetzel 1980). Similarly, Ruppia was more productive in sheltered lagoons and bays where water currents and wave action were low (Congdon & McComb 1979).

In the Wilderness lake system decreases in submerged macrophyte biomass were related to increases in wave action and turbidity (Weisser & Howard-Williams 1982). Furthermore, Branch and Day (1984) found virtually no attached macrophytes in the Palmiet estuary owing to the relatively high water speeds. During winter, when maximum water speeds exceeded 1.14 m s^{-1}, *Enteromorpha* was the only macrophyte present and that was on rocky substrata near the mouth. Howard-Williams and Liptrot (1980) reported that submerged macrophytes were absent from the main channel of the Swartvlei estuary because of the current speeds of $1.2–1.4 \text{ m s}^{-1}$ which had been measured over spring tides.

Mouth dynamics and water level fluctuations

Most of South Africa's estuaries are periodically closed at the mouth. In some estuaries freshwater abstraction has led to an increase in the frequency of mouth closure as the effect of floods, which flush out accumulated marine sediment, is reduced. Opening and closing of the mouth results not only in changes in nutrient exchange and salinity, but also in fluctuations in water depth over a time scale which may severely impact macrophyte communities.

In high rainfall areas mouth closure can result in increased water levels. Each plant species has a distinct range of tolerance to depth of water and duration of tidal flooding. Inundation during the growing season is generally harmful to most species (Olff *et al.* 1988) and prolonged inundation can cause die-back of marsh plants (Adams & Bate 1994d). Studies have shown that submergence of marsh plants, e.g. *Spartina alterniflora*, may lead to root oxygen deficiencies, elevated soil sulphide concentrations and decreased plant N uptake, which eventually reduces plant productivity (Wilsey *et al.* 1992).

In temporarily closed South African estuaries, e.g. Great Brak estuary, during times of mouth closure, water levels rise and saltmarshes may be inundated for extended periods which results in a die-back of plants. The *Sarcocornia natalensis* community was reduced during 1989–92 (CSIR 1992). Water level data indicate that this community was submerged for 2–3 months at a time during 1989 and 1991. Management decisions have to be made about the length of time intertidal plants can survive under prolonged submerged conditions before the mouth of the estuary needs to be mechanically opened. Laboratory studies on *S. perennis*, an important lower intertidal saltmarsh succulent, have shown that the plants are stressed two weeks after inundation. Completely submerged plants decomposed rapidly in low salinity treatments (Adams & Bate 1994d). If *S. perennis* communities are flooded for prolonged periods, plants will not flower (O'Callaghan 1992) and produce seed. Although propagation usually occurs vegetatively in established saltmarshes, in bare areas seedling establishment from a resident seed bank may be more important and, therefore, prolonged flooded conditions should be avoided if possible (Adams & Bate 1994d). Although die-back of lower marsh plants can occur under prolonged inundated conditions, the plants are fairly resilient. For instance, in the Great Brak estuary *Triglochin* has the ability to colonise suitable habitats rapidly (CSIR 1992) and in the Seekoei estuary ($34°\ 05'$ S, $24°\ 55'$ E) *Sarcocornia* spp. re-establish rapidly from the resident seed bank when water levels drop.

In dry areas, mouth closure can result in a drop in water level. Macrophytes normally submerged are exposed and die back. *Zostera capensis* can be exposed above water without drying for a longer time than can either *Ruppia cirrhosa* or *Potamogeton pectinatus* (Adams & Bate 1994c). *Zostera* has a leaf sheath which protects the basal meristems. Overlapping leaves also provide the plant with a high degree of desiccation tolerance (Gessner 1971). Seagrasses can trap water during low tides (Powell & Schaffner 1991), which prevents desiccation. Talbot and Bate (1987) found that waterlogged conditions of creek sediments in the Swartkops estuary allowed *Z. capensis* to meet the evaporative demand of their large leaves during tidal emergence.

Studies in which two temporarily open estuaries, the Seekoei and Kabeljous ($34°\ 00'$ S, $24°\ 56'$ E), were monitored, have shown that after three months there was complete die-back of *R. cirrhosa* following a drop in water level that exposed the beds (Adams & Bate 1994d). Resistance of the vegetative parts to drying is very low and, after exposure, all plant parts except ripe seeds die within a few days (Verhoeven 1979). However, due to its large production of seeds the plant is adapted to an ephemeral habit. After localised rainfall events, water levels rise and the plants grow rapidly from a persistent seed bank. Van Vierssen *et al.* (1984) found that the viability of *R. maritima* seeds was hardly affected by 2–10 weeks' desiccation.

A drop in water level will also adversely effect the emergent macrophytes, as abundant soil water plays an important role in maintaining high productivity. Water is not only used directly by the plants but also holds the nutrients in a dissolved state in the sediment. Adams and Bate (1995) showed that a dry treatment inhibited growth of *Spartina maritima*, suggesting a requirement by the plant for waterlogged conditions. If a mouth of an estuary closes due to reduced freshwater input there is the possibility that *S. maritima* communities may disappear. *S. maritima* is absent from temporarily closed South African estuaries and this may be attributed to the plant's requirement for tidal flooding and saturated substrata.

An extensive marsh area on the south bank of the Orange River mouth (28° 38′ S, 16° 27′ E) has been lost. One of the main reasons was thought to be a decrease in inundation frequency. When the mouth closed naturally, back flooding of low salinity water would have inundated the marsh. This back flooding no longer occurs as the mouth is now kept permanently open by artificial means. This has caused a drier marsh environment, higher salinity of the substratum and die-back of the marsh plants. High dust levels from the nearby mining works have coated the foliage of the plants and accelerated die-back of the saltmarsh (Burns 1991).

In the Wilderness Lagoon, Weisser and Howard-Williams (1982) found that encroachment by emergent macrophytes was associated with dropping water levels during periods of artificial mouth opening. To reduce reed encroachment it was recommended that the water level be kept as high as possible, as often as possible. Implementation of this was impractical owing to the encroachment of housing developments too close to the water edges. Onrus Lagoon (34° 25′ S, 19° 11′ E) was a deep estuary until the 1940s. When the water upstream was dammed, the annual floods could no longer sweep the estuary clear of silt deposition. The mouth consequently closed, salinity decreased and the silt was colonised by dense growths of *Phragmites* (Branch & Branch 1985).

Phragmites australis has been observed to expand rapidly in systems that are exposed to increased silt loads (Siyaya Lagoon, 28° 58′ S, 31° 45′ E) and/or lowered salinities (e.g. Onrus River). Before catchment restoration took place in the Siyaya lagoon, *P. australis* beds had been expanding at approximately 0.15 ha per year (Weisser & Parsons 1981; Benfield 1984). Reduced freshwater input resulted in the invasion by *P. australis* and caused anaerobic conditions as the system was unable to cope with the detritus load (Schleyer & Roberts 1987).

Anthropogenic impacts

In addition to the impacts caused by a reduction in freshwater inflow, South African estuaries and their plant communities have also been affected by industrial, residential and recreation development impacts. Wetlands have had a long history of destruction and alteration as a result of human activities. More subtle impacts on coastal marshes result from the restriction of normal tidal flushing following the construction of roads, causeways, bridges and impoundments (Clark 1977; Roman et al. 1984).

Seagrasses (e.g. *Zostera capensis*) are especially vulnerable to degradation by human-induced stress as they inhabit the shallow margin areas of estuaries. Activities which affect seagrass beds are bait digging, damage by boats, land reclamation and dredging (Figure 5.10). Once the rhizome systems in existing seagrass beds are damaged by human activities, the bare areas can take years to recover (Day 1981). In some countries bait digging in these beds has been declared illegal. The passage of propeller driven boats through seagrass beds can produce erosion channels; plants are ripped up and sediment disturbed, increasing the turbidity of the water (Clark 1977; Orth 1976; Zieman 1976). Observation of *Thalassia* flats gouged by motor boat propellers showed a recovery period of 2 to 5 years and it is believed that this was due to a change in the redox potential of the disturbed sediments (Zieman 1976). Redox potential affects oxygen uptake by plants and when the redox potential is changed regrowth of the plants may be delayed.

Submerged macrophytes are often removed to make way for recreational activities (e.g. canoeing, sailing and swimming). In Zandvlei, dense *Potamogeton pectinatus* beds occur as a result of nutrient enrichment. These beds are harvested in order to provide open areas for recreation (Stewart & Davies 1986), with the result that the nutrient abstraction function of the beds is removed and the high nutrient laden water flows on to the next ecosystem.

Because of development effects and recreation impacts in estuaries there is a need to evaluate estuarine and coastal resources and to identify sensitive areas where careful planning and management are required (Begg 1984a; Allanson 1992). Important estuaries should be identified and development (including impoundments) in the catchment should be restricted to the extent that the estuarine environment is not degraded. The Consortium for Estuarine Research and Managment (CERM) is working towards incorporating all the botanical, zoological, physical and socio-economic factors that contribute towards the importance of an estuary, into an overall importance rating for South African estuaries (CERM 1994).

A scoring system has been developed to rank estuaries according to their botanical importance (Coetzee et al. 1996). The area cover of each estuarine plant community, its association with the estuary, the condition of the plant community and the community richness within the estuary formed the basis of the scoring system. Thirty-three estuaries were studied and scores allocated to each according to their importance along the whole Cape coast (Coetzee et al. 1997). The condition of the estuarine plant communities included an assessment of the number of impacts affecting the different communities. The most common impact in supratidal saltmarshes was encroachment by invasive plants followed by trampling

Table 5.9. Impacts listed according to their occurrence in supratidal and intertidal salt marshes of 33 Cape estuaries

Impacts	Saltmarsh type	
	Supratidal frequency (%)	Intertidal frequency (%)
Presence of invasive plants	75	50
Trampling and footpaths	69	81
Bridge or weir through habitat	31	25
Grazing	63	38
Retaining walls or canalisation	13	6
Effluent discharge upstream	0	25

Source: From Coetzee *et al.* (1997).

and footpaths, which was the most common impact affecting intertidal marshes (Table 5.9). Water-saturated low marsh sediments and associated plants are highly susceptible to trampling and recover slowly. In a study of the False Bay estuaries, O'Callaghan (1990) showed that alien vegetation encroachment was a major factor causing changes to the vegetation. This was also supported by Coetzee *et al.* (1997). These aliens can range from aquatic plants (e.g. water hyacinth, water fern, parrot's feather) to sesbania, wattle and *Acacia* species.

Acknowledgements

The Water Research Commission is thanked for funding the following research projects on estuarine plants, i.e. The freshwater requirements of estuarine plants (WRC Report No. 292/94) and projects K5/601 & K5/812 on the botanical importance of estuaries. The Commission also kindly provided funds for the publication of colour photographs included in this chapter.

References

Adam, P. (1981). The vegetation of British salt marshes. *New Phytologist*, **88**, 143–96.

Adams, J.B. (1989). *The effect of light and salinity on the photosynthetic rates of lower intertidal and estuarine* Caulerpa filiformis *(Caulerpales, Chlorophyta)*. Honours Project, Department of Botany, University of Port Elizabeth.

Adams, J.B. & Bate, G.C. (1994a). *The freshwater requirements of estuarine plants incorporating the development of an estuarine decision support system*. WRC Report No. 292/1/94. Pretoria: Water Research Commission.

Adams, J.B. & Bate, G.C. (1994b). The ecological implications of tolerance to salinity by *Ruppia cirrhosa* (Petagna) Grande and *Zostera capensis* Setchell. *Botanica Marina*, **37**, 449–56.

Adams, J.B. & Bate, G.C. (1994c). The tolerance to desiccation of the submerged macrophytes *Ruppia cirrhosa* (Petagna) Grande and *Zostera capensis* Setchell. *Journal of Experimental Marine Biology and Ecology*, **183**, 53–62.

Adams, J.B. & Bate, G.C. (1994d). The effect of salinity and inundation on the estuarine macrophyte *Sarcocornia perennis* (Mill.) A.J. Scott. *Aquatic Botany*, **47**, 341–8.

Adams, J.B. & Bate, G.C. (1995). Ecological implications of tolerance of salinity and inundation by *Spartina maritima*.

Aquatic Botany, **52**, 183–91.

Adams, J.B. & Talbot, M.M.B. (1992). The influence of river impoundment on the estuarine seagrass *Zostera capensis* Setchell. *Botanica Marina*, **35**, 69–75.

Adams, J.B., Knoop, W.T. & Bate, G.C. (1992). The distribution of estuarine macrophytes in relation to freshwater. *Botanica Marina*, **35**, 215–26.

Allanson, B.R. (1992). *An Assessment of the SANCOR Estuaries Research Programme from 1980 to 1989*. Committee for Marine Science Occasional Report No. 1. Pretoria: CSIR.

Allanson, B.R. & Read, G.H.L. (1987). *The response of estuaries along the south eastern coast of southern Africa to marked variation in freshwater inflow*. Institute of Freshwater Studies: Final report. Grahamstown: Rhodes University.

Allanson, B.R. & Read, G.H.L. (1995). Further comment on the response of south east coast estuaries to variable freshwater flows. *Southern African Journal of Aquatic Sciences*, **21**, 56–71.

Anderson, M.R. & Kalff, J. (1988). Submerged aquatic macrophyte biomass in relation to sediment characteristics in ten temperate lakes. *Freshwater Biology*, **19**, 115–21.

Archibald, R.E.M. (1981). *An investigation into the taxonomy of the diatoms Bacillariophyta of the Sundays and Great Fish Rivers with ecological observations in*

Sundays. PhD thesis, Rhodes University, Grahamstown.

Baird, D., Marais, J.F.K. & Martin, A.P. (eds.) (1988). *The Swartkops estuary. Proceedings of a symposium held on 14 and 15 September 1987 at the University of Port Elizabeth*. South African National Scientific Programmes Report No. 156. Pretoria: CSIR.

Bally, R., McQuaid, C.M. & Pierce, S.M. (1985). Primary productivity of the Bot River estuary, South Africa. *Transactions of the Royal Society of South Africa*, **45**, 333–45.

Bate, G.C. & Heelas, B.B. (1975). Studies on the nitrate nutrition of two indigenous Rhodesian grasses. *Journal of Applied Ecology*, **12**, 941–52.

Beckley, L.E. (1983). The ichthyofauna associated with *Zostera capensis* Setchell in the Swarkops estuary, South Africa. *South African Journal of Zoology*, **18**, 15–24.

Begg, G.W. (1984a). *The Estuaries of Natal*. Part 2. Natal Town and Regional Planning Report 55.

Begg, G.W. (1984b). *The comparative ecology of Natal's smaller estuaries*. Natal Town and Regional Planning Report 62.

Begg, G.W. (1986). *The wetlands of Natal* (Part 1). An overview of their extent, role and present status. Natal Town and Regional Planning Report 68.

Benfield, M.C. (1984). *Some factors*

influencing the growth of Phragmites australis *(Cav.) Trin ex Steudel*. MSc thesis, University of Natal.

Bickerton, I.B. & Pierce, S.M. (1988). *Estuaries of the Cape*, Part II. *Synopses of available information on individual systems. Report No. 33: Krom (CMS 45), Seekoei (CMS 46) and Kabeljous (CMS 47)*, ed. A.E.F. Heydorn & J.R. Grindley. CSIR Research Report 432. Stellenbosch: CSIR.

Blaber, S.J.M. (1974). Field studies of the diet of *Rhabdosargus holubi* (Pisces: Teleostei: Sparidae). *Journal of Zoology, London*, **173**, 407–17.

Branch, M. & Branch, G. (1985). *The living shores of Southern Africa*. Cape Town: C. Struik.

Branch, G.M. & Day, J.A. (1984). The ecology of South African estuaries. Part XIII. The Palmiet River estuary in the south-western cape. *South African Journal of Zoology*, **19**, 63–77.

Branch, G.M., & Grindley, J.R. (1979). Ecology of southern African estuaries: Part XI. Mngazana: a mangrove estuary in the Transkei. *South African Journal of Zoology*, **14**, 149–70.

Breen, C.M., Everson, E.C. & Rogers, K. (1976). Ecological studies on *Sporobolus virginicus* (L.) Kunth with particular reference to salinity and inundation. *Hydrobiologia*, **54**, 135–40.

Brix, H. (1993). Wastewater treatment in constructed wetlands: system design, removal processes and treatment performances. In *Constructed wetlands for water quality improvement*, ed. G.A. Moshiri, pp. 9–22. Florida: Lewis Publishers.

Brix, H. (1994). Functions of macrophytes in constructed wetlands. *Water Science Technology*, **29**, 71–8.

Burdick, D.M., Mendelssohn, I.A. & McKee, K.L. (1989). Live standing crop and metabolism of the salt marsh *Spartina patens* as related to edaphic factors in a brackish mixed marsh community in Louisiana. *Estuaries*, **12**, 195–204.

Burns, M. (1991). *Environmental rehabilitation: Orange River salt marshes*. CSIR Report CMA-C 91165. Pretoria: CSIR.

Cadee, C.C. & Hegeman, J. (1974). Primary productivity of the benthic microflora living on tidal flats in the Dutch Wadden Sea. *Netherlands Journal of Sea Research*, **8**, 260–91.

Campbell, E.E., Knoop, W.T. & Bate, G.C. (1991). A comparison of phytoplankton biomass and primary production in three east Cape estuaries, South Africa. *South African Journal of Science*, **87**, 259–64.

Carter, R.A. & Brownlie, S. (1990). *Estuaries of the Cape*, Part II. *Synopses of available information on individual systems. Report No. 34: Kafferkuils (CSW 24) and Duiwenhoks (CSW 23)*, ed. A.E.F. Heydorn & P.D. Morant. CSIR Research Report 433. Stellenbosch: CSIR.

CERM (1994). *A co-ordinated research programme on decision support for the conservation and management of estuaries: progress report for importance rating sub-project*. April 1993–March 1994. Consortium for Estuarine Research and Management Progress Report.

Chambers, P.A. & Kalff, J. (1985). Depth distribution and biomass of submersed macrophyte communities in relation to secchi depth. *Canadian Journal of Fisheries and Aquatic Sciences*, **42**, 701–9.

Chapman, V.J. (1960). *Salt marshes and salt deserts of the world*. New York: Interscience Publishers.

Childers, D.L. & Day, J.W. (1990). Marsh–water column interactions in two Louisiana estuaries. II. Nutrient dynamics. *Estuaries*, **13**, 404–17.

Christie, N.D. (1981). Primary production in Langebaan Lagoon. In *Estuarine ecology with particular reference to southern Africa*, ed. J.H. Day, pp. 101–15. Cape Town: A.A. Balkema.

Clark, J.R. (1977). *Coastal ecosystem managment: a technical manual for the conservation of coastal zone resources*. New York: John Wiley.

Cloern, J.E. (1991a). Annual variations in river flow and primary production in the South San Francisco Bay Estuary (USA). In *Estuaries and coasts: spatial and temporal intercomparisons*, ed. M. Elliott & J.-P. Ducrotoy, pp. 91–6. ECSA 19 Symposium.

Cloern, J.E. (1991b). Tidal stirring and phytoplankton bloom dynamics in an estuary. *Journal of Marine Research*, **49**, 203–21.

Coetzee, J.C., Adams, J.B. & Bate, G.C. (1996). A botanical importance rating system for estuaries. *Journal of Coastal Conservation*, **2**, 131–8.

Coetzee, J.C., Adams, J.B. & Bate, G.C. (1997). A botanical importance rating of selected Cape estuaries. *Water SA*, **23**, 81–93.

Colijn, F. & Dijkema, K.S. (1981). Species composition of benthic diatoms and distribution of chlorophyll-*a* on an intertidal flat in the Dutch Wadden Sea. *Marine Ecology Progress Series*, **4**, 9–21.

Congdon, R.A. & McComb, A.J. (1979). Productivity of *Ruppia*: seasonal changes and dependence on light in an Australian estuary. *Aquatic Botany*, **6**, 121–32.

Connolly, R.M. (1994). The role of seagrass as preferred habitat for juvenile *Sillaginades punctata* (Cuv. & Val.) (Sillaginidae, Pisces): habitat selection or feeding. *Journal of Experimental Marine Biology and Ecology*, **180**, 39–47.

Correll, D.L. (1978). Estuarine productivity. *Bioscience*, **28**, 646–50.

Correll, D.L., Faust, M.A. & Severn, D.J. (1975). Phosphorus flux and cycling in estuaries. In *Estuarine Research* **1**, ed. L.E. Cronin, pp. 108–36. New York: Academic Press.

CSIR (1992). *Great Brak Estuary Management Programme*. CSIR Report EMAS-C92083. Pretoria: CSIR.

Davies, B.R., Stuart, V. & de Villiers, M. (1989). The filtration activity of a serpulid polychaete population [*Ficopomatus enigmaticus* (Fauvel)] and its effects on water quality in a coastal marina. *Estuarine, Coastal and Shelf Science*, **29**, 613–20.

Day, J.H. (1951). The ecology of South African estuaries. Part I. A review of estuarine conditions in general. *Transactions of the Royal Society of South Africa*, **33**, 53–91.

Day, J.H. (ed.) (1981). *Estuarine ecology with particular reference to Southern Africa*. Cape Town: A.A. Balkema.

De Jonge, V.N. (1980). Fluctuations in the organic carbon to chlorophyll-*a* ratios for estuarine benthic microalgae. *Marine Ecology Progress Series*, **2**, 345–53.

Drinkwater, K.F. (1986). On the role of freshwater outflow on coastal marine ecosystems; workshop summary. In *The role of freshwater outflow in coastal marine ecosystems*, ed. S. Skreslet, pp. 429–38. NATO ASI Series, Vol. 67. Berlin: Springer-Verlag.

Duvenage, I.R. & Morant, P.D. (1984). *Estuaries of the Cape*, Part II. *Synopses of available information on individual systems. Report No. 31: Keurbooms/Bitou system (CMS 19) and Piesang (CMS 18)*, ed. A.E.F. Heydorn & J.R. Grindley. CSIR Research Report 430. Stellenbosch: CSIR.

Dye, A.H. (1977). Epibenthic algal

production in the Swartkops Estuary. *Zoologica Africana*, **13**, 157–8.

Edgecumbe, D.J. (1980). Some preliminary observations on the submerged aquatic *Zostera capensis* Setchell. *South African Journal of Botany*, **46**, 52–66.

Emmerson, W.D. (1989). The nutrient status of the Sundays River estuary, South Africa. *Water Research*, **23**, 1059–67.

Fenchel, T. & Straarup, B.J. (1971). Vertical distribution of photosynthetic pigments and the penetration of light in marine sediments. *Oikos*, **22**, 172–82.

Fielding, P.J., Damstra, K.St.J. & Branch, G.M. (1988). Benthic diatom biomass, production and sediment chlorophyll in Langebaan Lagoon, South Africa. *Estuarine, Coastal and Shelf Science*, **27**, 413–26.

Fielding, P.J., Forbes, A.T. & Demetriades, N.T. (1991). Chlorophyll concentrations and suspended particulate loads in St Lucia, northern Natal. *South African Journal of Science*, **86**, 252–5.

Flint, R.W. (1985). Long-term estuarine variability and associated biological response. *Estuaries*, **8**, 158–69.

Fonseca, M., Fisher, J.S., Zieman, J.C. & Thayer, G.W. (1982). Influence of the seagrass *Zostera marina* L. on the current flow. *Estuarine, Coastal and Shelf Science*, **15**, 387–64.

Gessner, F. (1971). The water economy of the seagrass *Thalassia testudinum*. *Marine Biology*, **10**, 258–60.

Gessner, F. & Schramm, W. (1972). Salinity and plants. In *Marine ecology*, Vol. 1, Part 2, ed. O. Kinne, pp. 705–820. London: Wiley Interscience.

Gordon, D.M., Birch, P.B. & McComb, A.J. (1980). The effect of light, temperature and salinity on photosynthetic rates of an estuarine *Cladophora*. *Botanica Marina*, **23**, 749–55.

Grange, N. & Allanson, B.R. (1995). The influence of freshwater inflow on the nature, amount and distribution of seston in estuaries of the Eastern Cape, South Africa. *Estuarine, Coastal and Shelf Science*, **40**, 403–20.

Grant, J., Bathmann, U.V. & Mills, E.L. (1986). The interaction between benthic diatom films and sediment transport. *Estuarine, Coastal and Shelf Science*, **23**, 225–38.

Haller, W.T., Sutton, D.L. & Barlowe, W.C. (1974). Effects of salinity on growth of several aquatic macrophytes. *Ecology*, **55**, 891–4.

Hanekom, N. & Baird, D. (1988). Distributions and variations in seasonal biomass of eelgrass *Zostera capensis* in the Kromme estuary, St Francis Bay, South Africa. *South African Journal of Marine Science*, **7**, 51–9.

Harding, W.R. (1994). Water quality trends and the influence of salinity in a highly regulated estuary near Cape Town, South Africa. *South African Journal of Science*, **90**, 240–6.

Heinecken, T.J.E. (1981). *Estuaries of the Cape*, Part II. *Synopses of available information on individual systems. Report No. 7: Gamtoos (CMS 48)*, ed. A.E.F. Heydorn & J.R. Grindley. CSIR Research Report 406. Stellenbosch: CSIR.

Heinsohn, R.D. & Cunningham, A.B. (1991). Utilization and potential cultivation of the saltmarsh rush *Juncus kraussii*. *South African Journal of Botany*, **57**, 1–5.

Heydorn, A.E.F. (1989). *Estuaries of the Cape*, Part II. *Synopses of available information on individual systems. Report No. 38: Gourits (CSW 25)*, ed. A.E.F. Heydorn & P.D. Morant. CSIR Research Report 437. Stellenbosch: CSIR.

Hilmer, T. (1984). *The primary production of different phytoplankton size fractions in the Swartkops estuary*. MSc thesis, University of Port Elizabeth.

Hilmer, T. (1990). *Factors influencing the estimation of primary production in small estuaries*. PhD thesis, University of Port Elizabeth.

Hilmer, T. & Bate, G.C. (1990). Covariance analysis of chlorophyll distribution in the Sundays River estuary, eastern Cape. *Southern Africa Journal of Aquatic Sciences*, **16**, 37–59.

Hilmer, T. & Bate, G.C. (1991). Vertical migration of a flagellate-dominated bloom in a shallow South African estuary. *Botanica Marina*, **34**, 113–21.

Hoese, H.D. (1967). Effects of higher than normal salinities on salt marshes. *Contributions in Marine Science, University of Texas*, **12**, 249–61.

Howard-Williams, C. (1980). Aquatic macrophyte communities of the Wilderness Lakes: community structure and associated environmental conditions. *Journal of the Limnological Society of Southern Africa*, **6**, 85–92.

Howard-Williams, C. & Allanson, B.R. (1978). *Swartvlei Project Report, Part III. Community metabolism and phosphorus dynamics in the Swartvlei estuary.*

Institute for Freshwater Studies Special Report No. 78/4. Grahamstown: Rhodes University.

Howard-Williams, C. & Allanson, B.R. (1981). An integrated study on littoral and pelagic primary production in a southern African coastal lake. *Archiv für Hydrobiologie*, **92**, 507–34.

Howard-Williams, C. & Liptrot, M.R. (1980). Submerged macrophyte communities in a brackish South African estuarine-lake system. *Aquatic Botany*, **9**, 101–16.

Huizinga, P. (1996). *Great Fish and Sundays Rivers: the effects of different run-off scenarios on the salinity distributions in the estuaries*. CSIR Report ENV/S-C 96040. Stellenbosch: CSIR.

Jefferies, R.L. (1972). Aspects of salt marsh ecology with particlar reference to inorganic plant nutrient. In *The estuarine environment*, ed. R.S.K. Barnes & J. Green. London: Applied Science.

Jerling, H.L. & Wooldridge, T.H. (1995). Relatively negative ^{13}C ratios of mesozooplankton in the Sundays River estuary; comments on potential carbon sources. *Southern African Journal of Aquatic Sciences*, **21**, 71–8.

Joint, I.R. & Pomeroy, A.J. (1981). Primary production in a turbid estuary. *Estuarine, Coastal and Shelf Science*, **13**, 303–16.

Josselyn, M.N. & West, J.A. (1985). The distribution and temporal dynamics of the estuarine macroalgal community of San Francisco Bay. *Hydrobiologia*, **129**, 139–52.

Kalejta, B. & Hockey, P.A.R. (1991). Distribution, abundance and productivity of benthic invertebrates at the Berg River estuary, South Africa. *Estuarine, Coastal and Shelf Science*, **33**, 175–91.

Kennish, M.J. (1986). *Ecology of estuaries*, Vol. 1. *Physical and chemical aspects*. Florida: CRC Press.

King, J., De Moor, F., Botha, A. & Coetzer, A. (1989). Water quantity requirements of invertebrates, macrophytes and other mesobiota. In *Ecological flow requirements for South African rivers*, ed. A.A. Ferrar, pp. 57–70. South African National Scientific Programmes Report No.162. Pretoria: CSIR.

Knoop, W.T., Berger, M.G., Talbot, M.M.B. & Bate, G.C. (1986). Standing macrophyte biomass and total estuarine production. In *East London Programme – final report*, ed. T. Wooldridge, pp. 442–61. University of Port Elizabeth

Institute for Coastal Research Report No. 7.

Kokkinn, M.J. & Allanson, B.R. (1985). On the flux of organic carbon in a tidal salt marsh, Kowie River estuary, Port Alfred, South Africa. *South African Journal of Science*, **81**, 613–17.

Lerner, H.R. (1985). Adaptations to salinity at the plant cell level. *Plant and Soil*, **89**, 3–14.

Linthurst, R.A. & Seneca, E.D. (1981). Aeration, nitrogen and salinity as determinants of *Spartina alterniflora* Loisel. growth response. *Estuaries*, **4**, 53–63.

Lively, J.S., Kaufman, Z. & Carpenter, E.J. (1983). Phytoplankton ecology of a barrier island estuary: Great South Bay, New York. *Estuarine, Coastal and Marine Science*, **16**, 51–68.

Lubke, R.A. & Van Wijk, Y. (1988). Estuarine plants. In *A field guide to the Eastern Cape*, ed. R.A. Lubke, F.W. Gess & M.N. Bruton, pp. 133–45. Grahamstown: The Grahamstown Centre of the Wildlife Society of Southern Africa.

Lucas, A.B. (1986). *The distribution of chlorophyll pigments in relation to cyclic, sporadic and episodic events in the Great Fish River estuary*. MSc thesis, Rhodes University, Grahamstown.

Lukatelich, R.J. & McComb, A.J. (1986). Distribution and abundance of benthic microalgae in a shallow southwestern Australian estuarine system. *Marine Ecology Progress Series*, **27**, 287–97.

Mackay, H.M. & Schumann, E.H. (1990). Mixing and circulation in the Sundays River estuary, South Africa. *Estuarine, Coastal and Shelf Science*, **31**, 203–16.

McRoy, C.P. & Helfferich, C. (1980). Applied aspects of seagrasses. In *Handbook of seagrass biology*, ed. R.C. Phillips & C.P. McRoy. New York: Garland STPM Press.

Malone, T.C., Crocker, L.H., Pike, S.E. & Wendlerm, B.W. (1988). Influences of river flow on the dynamics of phytoplankton production in a partially stratified estuary. *Marine Ecology Progress Series*, **48**, 235–49.

Margalef, R. (1978). Life-forms of phytoplankton as survival alternatives in an unstable environment. *Oceanologica Acta*, **1**, 493–509.

Masson, H. & Marais, J.F.K. (1975). Stomach content analysis of mullet from the Swartkops estuary. *Zoologica Africana*, **10**, 193–207.

Millard, N.A.H. & Broekhuysen, G.J. (1970). The ecology of South African estuaries. Part 10: St Lucia: a second report. *Zoologica Africana*, **34**, 157–79.

Moss, B., Booker, I., Balls, H. & Monson, K. (1989). Phytoplankton distribution in a temperate floodplain lake and river systems I. Hydrology, nutrient sources and phytoplankton biomass. *Journal of Plankton Research*, **11**, 813–38.

Naidoo, G. (1994). Growth, water and ion relations in the halophytes *Triglochin bulbosa* and *T. striata*. *Environmental and Experimental Botany*, **34**, 419–26.

Naidoo, G. & Mundree, S.G. (1993). Relationship between morphological and physiological response to waterlogging and salinity in *Sporobolus virginicus* (L.) Kunth. *Oecologia*, **93**, 360–6.

Naidoo, G. & Naicker, K. (1992). Seed germination in the coastal halophytes *Triglochin bulbosa* and *Triglochin striata*. *Aquatic Botany*, **42**, 217–29.

Naidoo, G. & Naidoo, S. (1992). Waterlogging responses of *Sporobolus virginicus* (L.) Kunth. *Oecologia*, **90**, 445–50.

Naidoo, G. & Rughunanan, R. (1990). Salt tolerance in succulent, coastal halophyte, *Sarcocornia natalensis*. *Journal of Experimental Botany*, **41**, 497–502.

Nusch, E.A. (1980). Comparison of different methods for chlorophyll determination. *Arch. Hydrobiol. Beih. Ergebn. Limnol.*, **14**, 14–36.

O'Callaghan, M. (1990). The ecology of the False Bay estuarine environment, Cape, South Africa. 1. The coastal vegetation. *Bothalia*, **20**, 105–12.

O'Callaghan, M. (1992). The ecology and identification of the southern African Salicornieae (Chenopodiaceae). *South African Journal of Botany*, **58**, 430–9.

O'Callaghan, M. (1994a). *Saltmarshes of the Cape (South Africa): vegetation dynamics and interactions*. PhD thesis, University of Stellenbosch.

O'Callaghan, M. (1994b). The marsh vegetation of Kleinmond lagoon. *Bothalia*, **24**, 235–40.

Olff, H., Bakker, J.P. & Fresco, L.F.M. (1988). The effect of fluctuations in tidal inundation frequency on a salt-marsh vegetation. *Vegetatio*, **78**, 13–19.

Orth, R. (1976). The demise and recovery of eelgrass, *Zostera marina* in the Chesapeake Bay, Virginia. *Aquatic Botany*, **2**, 141–59.

Paerl, H.W., Rudek, J. & Mallin, M.A. (1990). Simulation of phytoplankton production in coastal waters by natural rainfall inputs: nutritional and trophic implications. *Marine Biology*, **107**, 247–54.

Pearcy, R.W. & Ustin, S.L. (1984). Effects of salinity on growth and photosynthesis of three California tidal marsh species. *Oecologia*, **62**, 68–73.

Pennock, J.R. & Sharp, J.H. (1986). Phytoplankton production in the Delaware estuary: temporal and spatial variability. *Marine Ecology Progress Series*, **34**, 143–55.

Peterson, D.H. & Festa, J.F. (1984). Numerical simulation of phytoplankton productivity in partially mixed estuaries. *Estuarine, Coastal and Shelf Science*, **19**, 563–89.

Phleger, C.F. (1971). Effects of salinity on growth of a salt marsh grass. *Ecology*, **52**, 908–11.

Pierce, S.M. (1979). *The contribution of Spartina maritima (Curtis) Fernald to the primary production of the Swartkops Estuary*. MSc thesis, Rhodes University, Grahamstown.

Pierce, S.M. (1982). What is *Spartina* doing in our estuaries? *South African Journal of Science*, **78**, 229–30.

Pierce, S.M. (1983). Estimation of the non-seasonal production of *Spartina maritima* (Curtis) Fernald in a South African estuary. *Estuarine, Coastal and Shelf Science*, **16**, 241–54.

Powell, G.V.N. & Schaffner, F.C. (1991). Water trapping by seagrasses occupying bank habitats in Florida Bay. *Estuarine, Coastal and Shelf Science*, **32**, 43–60.

Price, J.S., Ewing, K., Woo, M.K. & Kershaw, K.A. (1988). Vegetation patterns in James Bay coastal marshes. II. Effects of hydrology on salinity and vegetation. *Canadian Journal of Botany*, **66**, 2586–94.

Quick, A.J.R. & Harding, W.R. (1994). Management of a shallow estuarine lake for recreation and as a fish nursery: Zandvlei, Cape Town, South Africa. *Water SA*, **20**, 289–97.

Reed, R.H. & Russell, G. (1979). Adaptation to salinity stress in populations of *Enteromorpha intestinalis* (L.) Link. *Estuarine and Coastal Marine Science*, **8**, 251–8.

Robinson, P.K. & Hawkes, H.A. (1986). Studies on the growth of *Cladophora glomerata* in laboratory continuous-flow culture. *British Phycological Journal*, **21**, 437–44.

Rodriguez, F.D.G. (1993). *The determination*

and distribution of microbenthic chorophyll-a in selected south Cape estuaries. MSc thesis, University of Port Elizabeth.

Roman, C.T., Niering, W.A. & Warren, R.S. (1984). Salt marsh vegetation change in response to tidal restriction. Environmental Management, **8**, 141–50.

Schelske, C.L. & Odum, E.P. (1961). Mechanisms for maintaining high productivity in Georgia estuaries. Proceedings of the Gulf Caribbean Fisheries Institute, **14**, 75–80.

Schleyer, M.H. & Roberts, G.A. (1987). Detritus cycling in a shallow coastal lagoon in Natal, South Africa. Journal of Experimental Marine Biology and Ecology, **110**, 27–40.

Shaffer, G.P. & Onuff, C.P. (1983). An analysis of factors estimating primary production of the benthic microflora in a Southern California Lagoon. Netherlands Journal of Sea Research, **17**, 126–44.

Simpson, R.L., Good, R.E., Leck, M.A. & Whigham, D.F. (1983). The ecology of freshwater tidal wetlands. BioScience, **33**, 255–9.

Slinger, J.H. & Breen, C.M. (1995). Integrated research into estuarine management. Water Science and Technology, **32**, 79–86.

Smart, R.M. & Barko, J.W. (1978). Influence of sediment salinity tolerance and nutrients on the physiological ecology of selected salt marsh species. Ecology, **61**, 630–8.

Smetacek, V.S. (1986). Impact of freshwater discharge on production and transfer of materials in the marine environment. In The role of freshwater outflow in coastal marine ecosystems. NATO ASI Series, Volume 67, ed. S. Skreslet, pp. 85–106. Berlin: Springer-Verlag.

Starfield, A.M., Farm, B.P. & Taylor, R.H. (1989). A rule-based ecological model for the management of an estuarine lake. Ecological Modelling, **46**, 107–19.

Stevenson, J.C. (1988). Comparative ecology of grass beds in freshwater, estuarine and marine environments. Limnology and Oceanography, **33**, 867–93.

Stewart, B.A. & Davies, B.R. (1986). Effects of harvesting on invertebrates associated with Potamogeton pectinatus L. in the Marina da Gama, Zandvlei, Western Cape. Transactions of the Royal Society of South Africa, **46**, 35–50.

Talbot, M.M.B. & Bate, G.C. (1987). The

distribution and biomass of the seagrass Zostera capensis in a warm-temperate estuary. Botanica Marina, **30**, 91–9.

Talbot, M.M.B., Knoop, W.T. & Bate, G.C. (1990). The dynamics of estuarine macrophytes in relation to flood/siltation cycles. Botanica Marina, **33**, 159–64.

Taljaard, S. & Largier, J.L. (1989). Water circulation and nutrient distribution patterns in the Palmiet River estuary: a winter study. CSIR Research Report No. 680. Stellenbosch: CSIR.

Taylor, D.I. (1988). Tidally-mediated carbon, nitrogen and phosphorus exchange between a salt marsh and the Kariega estuary and the role of salt marsh brachyura in the transfer. PhD thesis, Rhodes University, Grahamstown.

Taylor, D.I. (1992) The influence of upwelling and short-term changes in concentrations of nutrients in the water column on fluxes across the surface of a salt marsh. Estuaries, **15**, 68–74.

Taylor, D.I. & Allanson, B.R. (1993) Impacts of dense crab populations on carbon exchanges across the surface of a salt marsh. Marine Ecology Progress Series, **101**, 119–29.

Taylor, D.I. & Allanson, B.R. (1995) Organic carbon fluxes between a high marsh estuary, and the applicability of the outwelling hypothesis. Marine Ecology Progress Series, **126**, 263–70.

Thompson, J.D. (1991). The biology of an invasive plant. Bioscience, **41**, 393–402.

Turner, R.E., Woo, S.W. & Jitts, H.R. (1979). Phytoplanton production in a turbid, temperate salt marsh estuary. Estuarine, Coastal and Marine Science, **2**, 311–22.

Ungar, I.A. (1962). Influence of salinity on seed germination in succulent halophytes. Ecology, **55**, 763–4.

Ungar, I.A. (1974). Inland halophytes of the United States. In Ecology of halophytes, ed. R.J. Reimold & W.H. Queen, pp. 235–305. New York: Academic Press.

Van Vierssen, W., Van Kessel, C.M. & Van der Zee, J.R. (1984). On the germination of Ruppia taxa in western Europe. Aquatic Botany, **19**, 381–93.

Verhoeven, J.T.A. (1975). Ruppia-dominated communities in the Camargue, France. Distribution and structure in relation to salinity and salinity fluctuations. Aquatic Botany, **1**, 217–41.

Verhoeven, J.T.A. (1979). The ecology of Ruppia-dominated communities in

western Europe. I. Distribution of Ruppia representatives in relation to their autecology. Aquatic Botany, **6**, 197–268.

Ward, C.J. (1976). Aspects of the ecology and distribution of submerged macrophytes and shoreline vegetation of Lake St Lucia. In St Lucia Scientific Advisory Council Workshop, Charters Creek, February 1976, ed. A.E.F. Heydorn. Pietermaritzburg: Natal Parks Board.

Warne, R.H. (1994). The effects of light intensity on the growth and photosynthesis of Zostera capensis and Ruppia cirrhosa. MSc thesis, University of Natal, Durban.

Weisser, P.J. & Howard-Williams, C. (1982). The vegetation of the Wilderness Lakes system and the macrophyte encroachment problem. Bontebok, **2**, 19–40.

Weisser, P.J. & Parsons, R.J. (1981). Monitoring Phragmites australis increase from 1937 to 1976 in the Siyai Lagoon (Natal, South Africa) by means of air photo interpretation. Bothalia, **13**, 553–6.

Weisser, P.J., Whitfield, A.K. & C.M. Hall. (1992). The recovery and dynamics of submerged aquatic macrophyte vegetation in the Wilderness lakes, southern Cape. Bothalia, **22**, 283–8.

Westlake, D.F. (1963). Comparisons of plant productivity. Biological Reviews, **38**, 385–425.

Whitfield, A.K. (1980). A quantitative study of the trophic relationships within the fish community of the Mhlanga estuary, South Africa. Estuarine and Coastal Marine Science, **10**, 417–35.

Whitfield, A.K (1984). The effects of prolonged aquatic macrophyte senescence on the biology of the dominant fish species in a southern African coastal lake. Estuarine, Coastal and Shelf Science, **18**, 315–29.

Whitfield, A.K. (1988). The fish community of the Swartvlei estuary and the influence of food availability on resource utilization. Estuaries, **11**, 160–70.

Whitfield, A.K. (1989). The benthic invertebrate community of a Southern Cape estuary: structure and possible food sources. Transactions of the Royal Society of Southern Africa, **47**, 159–79.

Whitfield, A.K., Beckley, L.E., Bennett, B., Branch, G.M., Kok, H.M., Potter, I.C. & van der Elst, R.P. (1989). Composition, species richness and similarity of ichthyofaunas in eelgrass *Zostera capensis* beds of southern Africa. *South African Journal of Marine Science*, **8**, 251–9.

Whitfield, A.K. & Paterson, A.W. (1995). Flood-associated mass mortality of fishes in the Sundays estuary. *Water SA*, **21**, 385–9.

Wilsey, B.J., McKee, K.L. & Mendelssohn, I.A. (1992). Effects of increased elevation and macro- and micronutrient additions on *Spartina alterniflora* transplant success in salt marsh dieback areas in Louisiana. *Environmental Management*, **16**, 505–11.

Zedler, J.B. (1983). Freshwater impacts in normally hypersaline marshes. *Estuaries*, **6**, 346–55.

Zieman, J.C. (1976). The ecological effects of physical damage from motor boats on turtle grass beds in Southern Florida. *Aquatic Botany*, **2**, 127–39.

Zieman, J.C. & Wetzel, R.G. (1980). Productivity in seagrasses: methods and rates. In *Handbook of seagrass biology: an ecosystem perspective,* ed. R.C. Phillips & C.P. McRoy, pp. 87–117. New York: Garland STPM Press.

6 Mangroves in South African estuaries

Trevor Steinke

Mtentu River estuary

Introduction

Mangroves are trees or shrubs which grow in the tidal, saline wetlands on the coasts of the warmer parts of the world. They are widespread throughout the Indo-Pacific region and also occur extensively in Africa, the Caribbean, the Gulf of Mexico, and in South America. In South Africa, mangroves are restricted to bays and estuaries along the southeastern coastline from East London northwards.

Duke (1992) defined a mangrove as a tree, shrub, palm or ground fern, generally exceeding half a metre in height, and which grows above mean sea level in the intertidal zone of marine coastal environments, or estuary margins. Because this definition is fairly wide, it is not always clear if a plant is a mangrove. Species have often been categorised as 'true' mangroves and other taxa, often called 'mangrove associates' (Mepham & Mepham 1985). For the habitat, the terms 'mangrove' or 'mangrove forest/ swamp', or even 'mangal', are used, although the last term is no longer popular.

This chapter describes the characteristics of the different mangrove species found in South Africa, their distribution, aspects of their ecophysiology, the plants and animals associated with them, and their importance in estuarine ecosystems.

The mangrove environment

Mangroves are found throughout tropical regions of the world and extend into subtropical and even temperate areas. Their distribution pattern corresponds to the presence of warm and cold oceanic currents; consequently, latitudinal ranges tend to be broader on eastern continental margins and more restricted on their western sides (Duke

1992). Temperature is regarded as the most important factor governing distribution. Duke (1992) indicated that mangroves generally match the winter 20 °C isotherm, suggesting the importance of water temperature to the mangrove habitat. There are, however, several important exceptions to this pattern in the southern hemisphere: namely, along the coastlines of eastern South Africa and South America, around Australia and across the North Island of New Zealand.

In South Africa Macnae (1963) proposed, probably on the basis of his recorded mangrove distribution, that mangroves occur only in areas where the mean air temperature does not drop below 19 °C. However, transplants of *Avicennia marina* and *Bruguiera gymnorrhiza* to Nahoon have not only survived, but also flowered and fruited, while growth of *A. marina* to maturity takes place under significantly colder conditions in Australia and New Zealand, where this species extends to approximately 38° S latitude (Bridgewater 1982; Crisp *et al.* 1990). This suggests that temperature *per se* is not the most important factor limiting the distribution of mangroves in the southeastern areas of South Africa. Research in this country and in Australia indicates that paucity of propagules towards the southern limits of mangroves, their mortality and restricted dispersal range may be more important reasons for the limited distribution along our southeast coast (Steinke 1972, 1975, 1986; Clarke & Myerscough 1991; Clarke 1993; Clarke & Allaway 1993).

Mangroves occur usually between mean sea level and mean high water spring tide level. At high tides their roots and lower stems may be submersed, while at low tides they may be exposed for several hours. However, the extent of their submersion varies according to the tidal cycle and their position on the shore. During neap high tides usually only the lower reaches of the mangrove swamp are inundated, while at spring high tides the waters may reach the outer fringes on higher ground. The tidal rise and fall creates a continually changing environment. At high tides the roots may be immersed in water of high salinity, but when rivers or heavy rains bring water from catchment areas the mangroves come into contact with water that is almost fresh. The varying salinity of the water is an important change, but water movements can affect also temperature, nutrients and oxygen levels in the soil and water.

Mangrove soils are poorly drained, saline, anoxic, fine-grained and rich in organic matter (Lear & Turner 1977; Naidoo 1980). The high organic content originates largely from plant debris, much of which is produced by the mangroves themselves. In addition to organic matter, the primary constituents of the soils are variable and range from fine sediments (silt and clay), or sand-sized particles, to relatively coarse material such as rock fragments and coral and shell debris (Lear & Turner 1977; Naidoo 1980). Shell and other calcareous debris is important as a source of calcium which, particularly in areas of high salinity, reduces the level of sodium uptake by the mangroves and so prevents damage from an excess of sodium ions (Lear & Turner 1977). These soils form by accretion of river-borne sediments to which is added material brought in from the sea with the rising tide. Soil constituents and other suspended material carried by rivers and tides settle in mangroves, due partly to reduced current velocities and reduced turbulence which result from the dense growth of mangrove aerial roots and other fringing vegetation. The soils are waterlogged at high tide and tend to remain so even when the tide is low because of the poor drainage. Consequently, mangrove sediments are typically anaerobic or anoxic with oxygen present only in the surface layer and around roots. Under anaerobic conditions sulphate-reducing bacteria produce hydrogen sulphide by reduction of organic sulphate from debris and reduction of sulphate in the estuarine waters. This gas gives mangrove soils their pungent odour and also reduces ferric iron compounds to a variety of hydrated ferrous sulphides which give the soils their characteristic dark colour.

Mangrove taxonomy and description

According to Cronquist (1981), there are 19 families with mangrove representatives, of which only two families are exclusively mangrove. Clearly, therefore, mangroves are not a genetic entity, but an ecological one. The Australasian and Indo-Malesian regions, with approximately 48 taxa, are the richest in the world (Duke 1992).

Table 6.1. A classification of mangrove species in South Africa

Family	Genus	Species	Common names
Pteridaceae	Acrostichum	aureum	mangrove fern
Meliaceae	Xylocarpus	granatum	cannonball mangrove
Combretaceae	Lumnitzera	racemosa	Tonga mangrove, Tonga wortelboom, isiKhaha-esibomvu (Z)
Rhizophoraceae	Bruguiera	gymnorrhiza	black mangrove, swartwortelboom, isiKhangazi, isiHlobane (Z), isiKhangathi (X)
	Ceriops	tagal	Indian mangrove, Indiese wortelboom, isiNkaha (Z)
	Rhizophora	mucronata	red mangrove, rooiwortelboom, umHlume (Z), umHluma (X)
Avicenniaceae	Avicennia	marina	white (or grey) mangrove, witseebasboom, isiKhungathi (Z,X)

X, Xhosa; Z, Zulu.

Tropical East Africa has 10 taxa, while South Africa, where mangroves reach their southernmost limits on the African continent, has only seven taxa (Table 6.1).

Avicennia marina (Forssk.) Vierh.

This is the most widespread and common mangrove along our coastline and, in Australia and New Zealand, is a mangrove which occurs at high latitudes (c. 38° S) (Bridgewater 1982; Crisp et al. 1990). It is commonly called the white or grey mangrove because of the greyish-white undersurface to the relatively small, closely packed, opposite leaves and whitish, smooth trunk (Figure 6.1a). Towards the water's edge the trees tend to be large and spreading, whereas in a closed community they are tall and upright (10 m). The root system is well adapted to an intertidal, estuarine habitat. Shallow, horizontal cable roots which radiate from the base of the tree, provide anchorage in the soft substratum. Growing up from these cable roots are large numbers of pencil roots (or pneumatophores) which have lenticels on their surface for gaseous exchange (Figure 6.1b). At low tide, when these pneumatophores are exposed, there is an exchange of oxygen and carbon dioxide with the atmosphere.

The flowers are small, scented and clustered into pedunculate heads. The fruit is a compressed, ovoid capsule with a buff-green to yellow pericarp (Figure 6.1a). Each fruit contains a single seed comprising an embryo with two large cotyledons folded longitudinally to enclose a hypocotyl.

A. marina is generally regarded as the pioneer mangrove and is often seen as a monospecific pioneer community on newly-accreting mudbanks in estuaries (T.D.S., personal observations). It becomes established rapidly, has

Figure 6.1. (a) Leaves and propagules of *Avicennia marina*; (b) *A. marina*, showing well-developed pneumatophores; (c) leaves and propagules of *Bruguiera gymnorrhiza*; (d) buttresses and knee roots of *B. gymnorrhiza* on a receding tide.

a wide environmental tolerance and also establishes well only in open areas, where there is abundant light, rather than shady situations. In addition the small, usually buoyant, propagules facilitate dispersal.

Bruguiera gymnorrhiza (L.) Lam

This is also a very common mangrove. The opposite leaves are large, elliptic, glossy and olive-green (Figure 6.1c). There are prominent interpetiolar stipules. The fissured bark is dark in colour. Although in this country trees of this species do not reach the size of those in more tropical areas (Putz & Chan 1986; Clough & Scott 1989), Macnae (1963) reported trees almost 20 m high at Sodwana Bay. Generally trees reach a height of 10–15 m, although in the southern estuaries the largest trees do not exceed 5 m. The root system is also well adapted to the environment. The tree produces buttress roots at the base of the trunk. Cable roots are also produced, but at intervals these emerge from and re-enter the soil to form characteristic knee roots which are also important for gaseous exchange (Figure 6.1d).

The flowers, which are borne singly in leaf axils, have conspicuous calyces that are frequently bright red. The pollen release mechanism is explosive and is triggered by insects (Davey 1975). The viviparous fruit is a berry 25 mm long which remains within the persistent calyx. It germinates to give rise to a green (later brownish), cigar-shaped propagule which comprises a hypocotyl approximately 120 mm long with a tiny plumule at its proximal end. The propagule falls with the calyx attached. The propagules are buoyant and may be widely dispersed by water.

This species is not generally regarded as a pioneer, although it has colonised some estuaries, usually those where the mouth closes occasionally, e.g. Mgababa, Mzamba and Mtentu Rivers. Although B. gymnorrhiza prefers higher ground where inundation is restricted largely to spring tides, under conditions of periodic closure this species survives better than A. marina, the roots of which are sensitive to submersion under high water levels. In its early growth stages B. gymnorrhiza also seems to grow better in shaded conditions.

Rhizophora mucronata Lam.

Rhizophora mucronata is not as common as A. marina and B. gymnorrhiza in South Africa. The opposite leaves are similar to those of the latter species except that at the tip of the leaf there is a distinctive sharp point, called a mucro (Figure 6.2a). Reddish-brown cork warts are evenly scattered over the undersurface of the leaves. Interpetiolar stipules are also present, usually pinkish, and larger than those of the previous species. The trees produce a straight trunk and are usually not as tall as those of B. gymnorrhiza. The bark is grey to dark grey and heavily fissured. The adaptation of the root system in this species takes the form

of aerial roots which originate on the trunk above the ground, arch away from it and then enter the soil (prop roots). Prop roots also bear lenticels which serve for gaseous exchange. In a well-developed stand these prop roots produce a tangled mass of aerial roots which may make passage difficult (Figure 6.2b).

Flowering and propagule production are similar to those of B. gymnorrhiza. Inflorescences of two or more flowers are borne in leaf axils and each flower has a characteristic four-lobed, persistent calyx. The mature fruit is dull green-brown, 30–50 mm long. The hypocotyl is green, commonly up to 300 mm long, with numerous lenticels on the surface (Figure 6.2a). Whereas mature B. gymnorrhiza propagules fall to the ground with the calyx attached, in the case of R. mucronata the propagule on abscission leaves the calyx and fruit on the tree.

R. mucronata tends to form dense stands marginal to creeks and streams within the mangrove community.

Ceriops tagal Perr. C.B. Robinson

This species, although common in the tropical Indo-Pacific region, reaches the southern limit of its natural distribution in the Kosi system. The opposite leaves are glossy, small, yellowish-green, with a rounded apex (Figure 6.2c). Trees in this country do not reach a height greater than 2–3 m, although they can grow tall (25 m) in tropical regions. The bark is smooth, light grey to reddish-brown. Buttress roots are produced (Figure 6.2d). The inflorescences are often on the terminal nodes of a new shoot and are generally up to 10-flowered. The fruit is ovoid, green to brown, up to 25 mm long, with calyx lobes characteristically reflexed. The hypocotyls are thin, ridged and approximately 200 mm in length.

This species, like B. gymnorrhiza, is not a pioneer and is usually found in the interior of the swamp on higher ground, flooded by high spring tides. Seedlings will develop in the shade of other mangroves.

Lumnitzera racemosa Willd.

This species, which also has a wide distribution in the tropical Indo-Pacific, has restricted distribution in South Africa, occurring naturally only in the Kosi system. The leaves are light green, succulent, with a wavy margin and a rounded, often notched tip (Figure 6.3a). Trees reach a height of 2–3 m and even in the tropics are not large (8 m). The bark is fissured, grey to reddish-brown. Buttress roots are produced. The inflorescence is an axillary raceme with usually up to seven white flowers. The calyx tube is green and narrow, extending upwards to form a short cup. The fruit is hard, ovoid, 10–15 mm long and crowned by persistent calyx lobes and style. The mature fruits (propagules) are brown, up to 15 mm in length, and buoyant.

This species is usually found on high ground where

Figure 6.2. (a) Leaves and propagules of *Rhizophora* sp.; (b) prop roots of *Rhizophora* sp. lining a creek; (c) leaves and propagules of *Ceriops tagal*; (d) tree of *C. tagal* to the rear of chopped saplings of the same species.

Figure 6.3. (a) Leaves and young fruits of *Lumnitzera racemosa*; (b) leaves and flower of *Hibiscus tiliaceus*; (c) epiphytic algae on pneu-matophores of *A. marina*; (d) algae on the mud substratum among pneumatophores of *A. marina*.

tidal inundations are infrequent. It often occurs on the landward margins of a mangrove swamp.

Acrostichum aureum L.

The distribution of this species in South Africa is not known, although it has been recorded as far south as Pondoland. This fern has large, tufted, erect fronds approximately 1.5 m in length. The pinnae are petiolate with margins entire to irregularly undulate. Fertile pinnae are borne towards the apex of the frond. This fern is usually found on the landward margin of mangrove swamps in areas where tidal inundations are infrequent.

Xylocarpus granatum Koen.

Only one specimen has so far been found in the Kosi system, although there have been indications that this species might have occurred there before the mass mortality of mangroves (Breen & Hill 1969). The leaves are compound. The bark is smooth and aerial roots take the form of ribbon-like buttresses. The fruits are large, globose,

100–150 mm in diameter. This tree produces flowers and mature fruits, but damage to the latter by the local population has prevented viability tests on the seeds. Typically this mangrove occurs towards the back of the mangrove swamps where it is flooded by high spring tides.

Distribution of mangroves

Mangroves extend from the Kosi system (latitude 26° S) in the north to the Nahoon River (33° S) in the south (Ward & Steinke 1982) (Figure 6.4). The original stand near the mouth at Nahoon arose from material transplanted from Durban Bay, although it is possible that a stand further upstream did arise from natural distribution of propagules (Steinke 1972). In the Kosi system seven mangroves are present: *A. marina*, *B. gymnorrhiza*, *R. mucronata*, *Acrostichum aureum*, *L. racemosa*, *C. tagal* and *X. granatum*. The last three do not occur naturally further south, although *L. racemosa*

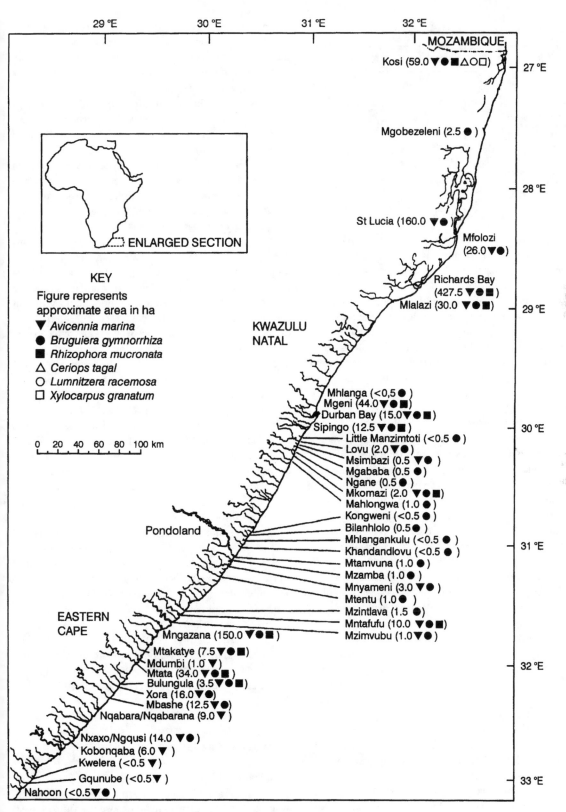

Figure 6.4. **Distribution and approximate areas of mangroves in South Africa.**

Table 6.2. Distribution of mangroves within estuaries

Mangrove	Estuarine location	Intertidal position
Acrostichum aureum	I	H
Lumnitzera racemosa	D,I,U	M,H
Bruguiera gymnorrhiza	D,I,U	M,H
Ceriops tagal	D,I	M,H
Rhizophora mucronata	D,I	L,M
Avicennia marina	D,I,U	L,M,H
Hibiscus tiliaceus[a]	I,U	H

[a]mangrove associate.
D, downstream; I, intermediate; U, upstream estuarine; L, low, M, mid, H, high intertidal.

and *C. tagal* have been transplanted at Beachwood (Mgeni River). *A. marina* is common throughout its range, but *B. gymnorrhiza* is sparse south of the Mbashe River and its natural distribution ends at Wavecrest (Nxaxo-Nqusi Rivers). South of the Kobonqaba River *A. marina* occurs only sporadically and appears to have a rather tenuous existence. Periodic floods and predatory sesarmid crab populations have threatened the existence and spread of some of the most southern stands. *R. mucronata*, which is common as far south as the Mtata River, reaches its southern limit at the Bulungula River. *A. aureum*, the mangrove fern, extends into Pondoland.

In addition to their geographical spread, mangroves are distributed according to at least two other gradients: their location within an estuary, and their position along the intertidal profile (Duke 1992) (Table 6.2). It is clear that certain species, e.g. *A. marina*, can occur over a wide range of conditions, from the mouth to the upper estuarine reaches and from the water's edge to those parts of the swamp inundated only at infrequent very high tides. Other species, e.g. *R. mucronata*, are more restricted in their distribution within an estuary.

Because South Africa has so few mangrove species, distribution patterns are not as complex and clearly defined as those in tropical mangroves (Watson 1928). In some of the southern estuaries *A. marina* is the only mangrove present. Clearly, the fact that there are usually only two common species (*A. marina* and *B. gymnorrhiza*), frequently restricted to a relatively narrow belt of trees along the water's edge in an estuary, is the main reason for this lack of complexity. In these estuaries *A. marina* tends to occur on the seaward as well as the landward edge with *B. gymnorrhiza* more common between these two ranges. At Mngazana and in the Kosi system *R. mucronata* occurs at the edges of channels which infringe on deepish water. In the latter system, where all species of mangroves are present, *C. tagal* usually is present with *B. gymnorrhiza* inland of an *A. marina* fringe along the water's edge, while *L. racemosa* is most common at the landward margin of the

swamps. Mangrove zonation patterns are considered to be the result of physiological adaptations by each species to different portions of a gradient in salinity and frequency of tidal inundation that exists across the intertidal zone (Watson 1928; Macnae 1968) or to tidal sorting of propagules based on differences in their size (Rabinowitz 1978). However, Smith (1987) has suggested that propagule predation by grapsid crabs may also play a significant role in determining zonation patterns.

The area covered by mangroves in each estuary is given in Figure 6.4. This survey was conducted in 1982 (Ward & Steinke 1982) and, with few exceptions, subsequent visits to those estuaries have indicated that there have been few changes since that date. There has been a general tendency for the areas under mangroves to increase over the past twenty years. The most significant increase has been at Richards Bay where *A. marina* and, to a lesser extent, *B. gymnorrhiza* have colonised the alluvial mudflats where the Mhlatuze River has discharged into the Sanctuary area. In the past twenty years mangroves have colonised this area rapidly and now occupy an area in excess of 450 ha. In contrast, in recent years harbour development at Richards Bay and Durban, and flood damage on the Mtata and Mgeni Rivers, have claimed small areas of mangroves. However, it was during the 1960s and 1970s that significant reductions of mangroves took place in South Africa. These reductions occurred, for example, as a result of harbour development at Richards Bay and Durban, in the Kosi system where a mass mortality followed high water levels after closure of the mouth, at Sodwana and Beachwood as a result of high water levels caused by poorly planned bridge construction, and at Sipingo during land development operations (Breen & Hill 1969, Moll *et al.* 1971; Bruton 1980). It is encouraging that in the Kosi system re-establishment of the mangroves is taking place, although there is a need for controlled utilisation of the mangroves and associated biotic resources by the local population (Ward *et al.* 1986).

While there are indications that the estuaries in Eastern Cape are being subjected to increased pressures (Steinke 1972; Wallace & van der Elst 1974; Emmerson 1988), the estuaries are generally still in good condition. To ensure that they remain productive and do not become degraded like many of those in Natal (Begg 1978), they should be protected from further exploitation.

Growth and phenology

Establishment

Following abscission, the fallen fruits of *A. marina* absorb water and the pericarp splits to expose the thick, folded cotyledons. The hypocotyl elongates and, under favour-

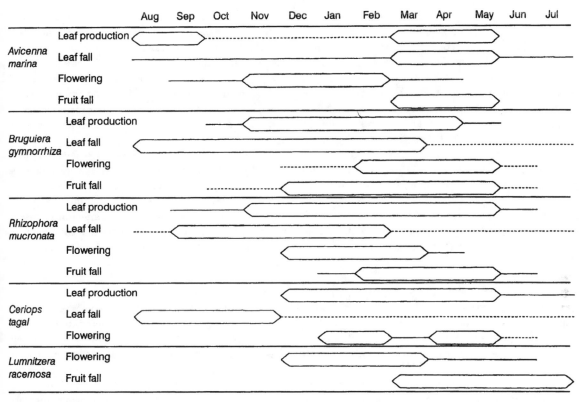

Figure 6.5. **Annual cycles of vegetative and reproductive phenologies in Natal mangroves (boxed areas, solid lines and broken lines indi-** cate high, low but consistent and barely measurable activities, respectively).

able conditions, produces lateral roots at its distal end within 2–3 days. These roots anchor the seedling; the hypocotyl elongates further and then straightens to lift the seedling off the substratum. Elongation of the epicotyl and unfolding of the cotyledons follows. Emergence of the first pair of true leaves can occur within two weeks. Persistence of the cotyledons is variable, but generally they do not appear to be shed before at least the second pair of true leaves has unfolded. During this time the cotyledons serve as a source of reserves, and also for photosynthesis, to provide energy and nutrients used during establishment and early growth (Steinke 1975; Steinke & Charles 1987; Steinke & Naidoo 1991).

On abscission, some propagules of B. gymnorrhiza, R. mucronata and C. tagal may become impaled in the soft, muddy substratum under the parent trees where they become established. Propagules which come to rest horizontally on the substratum produce roots more slowly than the impaled propagules and then only under favourable temperature and moisture conditions. These roots serve to anchor the propagule which eventually becomes raised to the vertical position. Establishment in this way is slow and mortality is high (T.D.S., personal observations). During establishment roots are produced from the distal

end of the propagule and, although early growth even in impaled propagules is not as rapid as in A. marina, a seedling can be firmly established within 2–3 weeks. The plumule then gives rise to an epicotyl bearing two rudimentary true leaves which fall early and are followed by normal leaves. Here too the elongated hypocotyls serve as reserves and also for photosynthesis.

No detailed studies appear to have been conducted on establishment and early growth in L. racemosa. This species does not exhibit vivipary. However, field observations indicate that fruits which come to rest on a moist substratum establish rapidly, with low mortality rates.

Growth

Annual cycles of vegetative and floral phenologies are reflected in Figure 6.5. This diagram represents mean periods for mangroves in KwaZulu-Natal and variations can be expected at more extreme latitudes. A. marina has a bimodal pattern of leaf emergence, whereas the other mangroves show clear trends towards high rates of leaf emergence in the warmer months, with low values in the cooler period of the year (Steinke 1988).

Leaf fall occurs throughout the year in all species, although there is a less marked seasonal influence in

A. marina. In this species a peak in leaf fall coincides with the abscission of fruits. In the other species peak leaf fall occurs in spring and continues into the summer. Groundwater salinities have been suggested as an important factor determining leaf fall (Pool *et al.* 1975) and local research has shown that soil salinities do increase during the dry period, which comes to an end in spring (Naidoo 1989). The difference in leaf fall behaviour between *A. marina* (a 'salt secretor' – see p. 129) on the one hand, and the rhizophoraceous mangroves ('salt excluders') on the other, is an indication of the way in which these two physiological groups cope with their saline environment. This confirms the field observation that salt excluders are not as tolerant of increased groundwater salinities as are salt secretors, e.g. *A. marina*. Accurate data are not available for leaf emergence and leaf fall in *L. racemosa*.

Biomass

Reliable estimates of biomass and growth rates of mangroves are useful for estimating total net primary production in ecological studies, for assessing the yield of commercial products from mangroves, and for the development of sound management practices (Clough & Scott 1989).

In South Africa, although local variations do occur, trees in the northern stands are larger, with a decrease in size towards the southern limits of the mangroves. In the southern estuaries *A. marina* at the upper estuary limits may be reduced to a narrow fringe of low-growing trees (1–2 m high). Size description of mangroves is often relative or restricted to an easily measurable parameter, such as height, because few other quantitative data are available. Most research to determine allometric relationships which can be used to estimate above-ground biomass has been conducted in tropical areas because extensive stands of mangroves are often harvested as a source of timber. In this country too, an accurate method for the assessment of above-ground biomass is essential if management of this resource for a sustainable future is to be successful. Below-ground biomass has received little attention and information is restricted largely to estimates based on core sampling (Putz & Chan 1986; Tamai *et al.* 1986; Komiyama *et al.* 1987; Clough & Scott 1989; Silva *et al.* 1991). Clearly, this is a field which requires further attention.

From work at Beachwood (Mgeni estuary) in a stand dominated by *B. gymnorrhiza* (6 m), mean above-ground living biomass was calculated at 94.5 ± 7.8 t dry matter ha^{-1}, while dead wood contributed a mean mass of 7.6 ± 0.9 t dry matter ha^{-1}. Excavations of roots yielded a below-ground biomass of 9.7 t dry matter ha^{-1} which represented 9.8% of the above-ground value. Further studies to provide a method for the assessment of above-ground biomass

revealed that reliable estimates for individual trees of *B. gymnorrhiza* and *A. marina* of the most abundant local size classes (up to 10 cm diameter of the stem at breast height [DBH]) could be obtained with the allometric relationship $y = ax^b$, where y = above-ground biomass (kg) and x = DBH (cm). Under local conditions height was considered unreliable for size description as there was not a close relationship between height and biomass (Steinke *et al.* 1994).

In the past man has had a severely detrimental impact on mangrove communities in this country and it is unlikely that pressures in the future will be any less. In the opinion of the author, it is advisable that future exploitations should be preceded by environmental impact assessments which would be more meaningful if they included estimates of biomass using the above (or an improved) method.

Flowering and fruiting

Over most of its distribution range, flowering of *A. marina* takes place only in the warm season from September onwards but, in the southern mangrove stands, flowers may be found on trees during most of the year. Fruit drop usually takes place in March/April (Steinke & Charles 1986).

In the case of *B. gymnorrhiza*, peak flowering takes place in late summer. Propagule fall reaches a peak in February/March, although the fall may begin as early as December and, certainly near the limit of its distribution, may end as late as June/July (Steinke & Charles 1986; Steinke & Ward 1988, 1990). In *R. mucronata*, times of flowering and propagule production are similar to those of *B. gymnorrhiza*. In both species a period of approximately 12 months is required for the development of propagules from flowering (Steinke & Ward 1989).

Flowering of *C. tagal* commences in January, approximately one month from the burst of new leaves in the warm season (Steinke & Rajh 1995). In the absence of detailed records, it would appear that a period of at least one year is required for the development of propagules from flowers. In *L. racemosa* flowering occurs mainly in the warmer months and fruits mature within 4–6 months (Steinke & Rajh 1995).

Ecophysiology of mangroves

Because it is not possible in this limited review to deal with all aspects of this wide topic, coverage has been restricted to aspects of the salt balance and photosynthetic physiology which have an influence on the productivity of mangroves.

Salinity

Mangroves are halophytes, plants which naturally complete their life cycles under saline conditions. However, mangrove species exhibit a broad spectrum of growth responses to salinity. Most mangroves will grow in freshwater although growth is stimulated by saline conditions, with the optimal salinities for growth of different species ranging from 5 to 50% seawater. A few Australian species have been shown to be obligate halophytes in that they do not grow to maturity under freshwater conditions. The physiological basis of this apparent requirement of saline conditions for growth is not known (Ball 1988). Regrettably the salinity requirements of South African mangroves have not received much attention.

While they are in general able to tolerate higher internal salt concentrations in their sap than do most land plants, mangroves have several strategies for coping with the high salt of their environment. Mangroves may absorb salt and then excrete it through glands on the leaf surface ('salt secretors'), exclude salt from the roots and leaves ('salt excluders'), and accumulate salt (Scholander *et al.* 1962; Lear & Turner 1977; Ball 1988).

A. marina (a 'salt secretor') absorbs salt with the water taken up by the roots. Salt is later excreted through salt glands in leaves at a rate sufficient to expel most of the salt reaching the leaves via the xylem (Clough *et al.* 1982; Drennan & Pammenter 1982; Clough 1984). However, there may still be an accumulation of salt in the leaves during their lifespan. On the other hand, *B. gymnorrhiza* and *R. mucronata* ('salt excluders') allow less salt to enter their root systems with the water that is taken up. Very little is secreted and excess salt is stored in the leaves, leading to an increase in succulence of older leaves before they fall. This increased succulence apparently maintains leaf ion concentrations at relatively constant levels. There are, however, limits in the extent to which ions can be accumulated in this way and it has been shown in several mangrove species that longevity of leaves declines with increasing salinity (Clough *et al.* 1982; Ball 1988). Evidence points to the fact that the secretion of salt through salt glands and the restriction to the entry of salt within the roots are active processes requiring metabolic energy (Scholander 1968; Lear & Turner 1977; Moon *et al.* 1986; Drennan *et al.* 1992). Consequently, respiratory losses associated with the maintenance of a salt balance could be high, especially during periods of high substratum salinity and high evapotranspiration. It is possible that respiratory losses associated with the maintenance of a salt balance could be higher in the drier than the wet season when substrate salinities are generally lower (Clough & Andrews 1981).

There is, however, a need for mangroves to maintain favourable water relations under saline conditions. A positive water balance can only be maintained if tissue water potentials are lower than the osmotic potentials of the substratum in which the plants are growing. The osmotic potential of seawater is approximately -2.5 MPa, while leaf water potentials in mangroves growing in seawater range from -2.7 to -5.7 MPa (Scholander 1968). These leaf water potentials are achieved through the accumulation of high concentrations of inorganic ions that function in osmoregulation. The vacuole is the major site of ion accumulation, whereas osmotic adjustment in metabolic areas of cells is apparently achieved by the synthesis of compatible organic solutes (Clough 1984; Ball 1988). Thus through intracellular compartmentation of ions, biochemical processes sensitive to high concentrations of NaCl are protected, while water relations favourable for growth are still maintained. However, these favourable water relations are maintained at an increased carbon cost which has a limiting effect upon growth.

Photosynthesis

Rates of net photosynthesis in mangroves are low compared with herbaceous plants, but are comparable with those of most trees. The saline mangrove environment, with its intense light and high temperatures and wind, is physiologically quite dry (Lear & Turner 1977). To overcome this, leaves of most mangroves have anatomical adaptations which restrict CO_2 and water vapour exchange. These include thick cuticles, waxy surface coatings and stomata which are sunken or surrounded by hairs and in general restricted to the lower epidermis. Thus leaf conductance values are also lower than those for herbaceous plants and maximum assimilation rates for mangroves are generally less than 12 μmol CO_2 m^{-2} s^{-1}. However, in South Africa maximal CO_2 exchange rates in *A. marina* and *B. gymnorrhiza* were found to be 9.9 and 5.3 μmol m^{-2} s^{-1}, respectively (Naidoo *et al.* 1997, 1998).

Leaf temperature has been shown to be a critical factor influencing the rate of assimilation. Assimilation rates are maximal at leaf temperatures ranging from 25 to 30 °C, but fall sharply with increase above 35 °C. This reduction in assimilation at high leaf temperatures appears to be due largely to a decrease in leaf conductance, but also to reduced biochemical efficiency (Clough & Andrews 1981).

Mangroves have several adaptations which enable them to maintain relatively low leaf temperatures in an environment where temperatures and solar radiation are usually high: leaf orientation, leaf size, and succulence. Leaf orientation results in leaves at the top of the canopy being nearly vertical, while the lowest leaves in deep shade are held normal to the incoming radiation (Clough & Andrews 1981; Steinke *et al.* 1994). This results in the incident radiation, which often would be more intense than required to saturate photosynthesis if intercepted in the horizontal plane, being shared over a large photosynthetic area, thereby maximising its utility while at the same time reducing the thermal input per unit leaf area and therefore restricting the rise in leaf temperature. Field measurements in Queensland have shown that the vertical leaves at the top of the mangrove canopy may be up to 5 °C cooler than if artificially held horizontally (Clough & Andrews 1981).

Decrease in leaf size enhances boundary layer conductance of heat between a leaf and its environment and results in leaf temperature being closer to ambient air temperature without reducing light interception due to altered orientation. In Australia, leaves of mangroves in hypersaline environments (e.g. *A. marina*, *L. racemosa*, *C. tagal*) are much smaller than those of species (e.g. *B. gymnorrhiza*) which dominate low salinity areas (Ball 1988). Local research has also shown that the more salt-tolerant *A. marina* exhibited higher leaf conductance, CO_2 exchange and transpiration and lower leaf water potentials than *B. gymnorrhiza* (Naidoo *et al.* 1998).

It has been shown that heat capacity per unit area of leaf increases largely with increase in succulence. Although succulence is also involved in the maintenance of favourable internal ion concentrations, there is a tendency for mangrove leaves to be more succulent under conditions in which, owing to intense radiation and/or limitations to evaporative cooling, they are most vulnerable to rapid fluctuations in leaf temperature (Ball 1988).

With decreased leaf size there is a greater proportion of supportive and conductive tissue per unit of exposed leaf area than in large leaves. Similarly, increase in leaf succulence takes place at the expense of expansion in leaf area. Thus these changes reduce the assimilative capacity of the plant and thereby are growth-limiting (Ball 1988).

The photosynthetic characteristics of mangroves are clearly those of plants utilising the C-3 pathway (Andrews *et al.* 1984; Ball 1988). Through the adaptations referred to above, rates of water loss and water-use efficiencies in mangroves are conservative for C-3 species, despite growing in a well-watered environment. This conservative water use has been shown to be at the expense of growth (Ball 1988; Clough & Sim 1989).

In recent years progress has been made towards understanding some of the physiological characteristics that enable mangroves to grow in saline anaerobic soils and in climatic conditions that often are not optimal for the growth of C-3 plants. It is encouraging that research in this country is now contributing information in this field. However, there are still many gaps in our knowledge of the undoubtedly complex mechanisms which mangroves may employ to offset their usually harsh, restrictive and dynamic intertidal environmental conditions.

Associated plant and animal species

Mangroves and adjoining vegetation, seagrass beds and mudflats are able to support a wide diversity of plants and animals. From the upper tidal zone, which is inundated only during high spring tides, to the channels of the subtidal zone is a rich diversity of plant and animal life, including microorganisms, algae, mangroves, fish and birds.

Plants
The so-called freshwater mangrove *Hibiscus tiliaceus* is a frequent associate of mangroves (Figure 6.3b). This species, which usually occupies higher ground than the mangroves, occurs commonly south to the Mbashe River, whereafter it may be found sporadically to the Nahoon River. In addition, in the saltmarshes, which often border the mangroves on inland edges or on higher ground, species such as *Juncus kraussii*, *Sarcocornia spp.*, *Triglochin striata*, *T. bulbosa* and *Sporobolus virginicus* are common. In Eastern Cape estuaries *Cotula coronopifolia* and *Limonium scabrum*, respectively, may also be abundant. These saltmarshes are a harsh environment which only relatively few species have the necessary adaptations to survive. The adaptations of these plants and their role in estuarine ecosystems merit investigation. In the eastern United States saltmarshes are among the most productive natural ecosystems and are also considered to be important in the temperate south and west coast estuaries of South Africa (Branch & Branch 1981). However, in the more northern estuaries, with the exception of areas such as St Lucia, the relatively low extent of most coastal saltmarshes and the nature of the vegetation suggest that they make a relatively small contribution to detritus food chains. On the other hand, associated fringing vegetation, such as *Phragmites australis* and overhanging trees, which usually produce luxuriant growth, may make a significant input of litter into our estuarine ecosystems, although no critical study of this nature has yet been undertaken.

Seagrasses
Frequently associated with mangroves are seagrasses, three of which, i.e. *Zostera capensis*, *Halophila ovalis* and *Halodule uninervis*, are estuarine, where they inhabit intertidal and comparatively shallow subtidal regions. In some estuaries *Ruppia* spp. may also be found, usually under conditions of low salinity. The distribution of *Z. capensis* was outlined by Edgcumbe (1980), although there have been significant changes since that date. Additional records are available for some estuaries (e.g. Nxaxo-Nqusi and Kobonqaba), while recent investigations suggest that this species may have disappeared from others (e.g. Mgbaba, Richards Bay). Clearly, these communities are

not stable. They may suffer periodic reductions as a result of floods and there are also indications of a cyclical pattern of growth (or regrowth), maturity and then dieback in many estuaries (T.D.S., personal observations). Although *H. ovalis* has been recorded in mangrove estuaries in Natal (Kosi system, Durban Bay), Transkei (Mngazana River) and Eastern Cape (Kwelera and Nahoon Rivers), records are incomplete and the species may be more widespread. This species appears to be very sensitive to environmental change and to competition from *Z. capensis* with which it is often found in association (T.D.S., personal observations). *H. uninervis* has been recorded only from the Kosi system, although in Tongaland it is present in pools on rocky intertidal ledges which may also support *Thalassodendron ciliatum* (Ward 1962). In the United States and Europe, seagrass communities are considered highly productive ecosystems. While seagrass beds are used extensively by marine fish as nursery and feeding areas (Wallace & van der Elst 1975), there is unfortunately no estimate of the extent of these transient communities and limited information on their productivity in our northern estuaries (Howard-Williams 1980). However, evidence from Swartvlei of their associated rich invertebrate fauna and periphyton and their high productivity, indicates the importance of seagrasses as estuarine communities (Howard-Williams & Liptrot 1980).

Algae

These plants very often form an important component of the mangrove community. They may be present as epiphytes on the above-ground parts of mangroves (Figure 6.3c) or as mats of blue-green algae (cyanobacteria) on the mud substratum (Figure 6.3d) either in the mangroves or in the adjacent saltmarshes. The epiphytic species are present as a conspicuous algal felt which is termed the bostrychietum (Post 1936). A survey of species epiphytic on the mangroves revealed 12 red algae (Rhodophyceae), 27 blue-green algae (Cyanophyceae), 8 green algae (Chlorophyceae) and 3 brown algae (Phaeophyceae) from the Kosi system to the Nahoon River (Lambert *et al.* 1987, 1989). The greatest numbers were collected from the pneumatophores of *A. marina* which, under local conditions, usually occupies the lower reaches of the swamp where the lowest pneumatophores are constantly submerged, or exposed for only short periods. Most of the other mangroves occupy higher ground which is inundated less frequently and consequently they seldom have good growths of epiphytic algae. The term 'bostrychietum', as coined by Post (1936), included species of only the genera *Bostrychia*, *Caloglossa*, *Catenella* and *Murrayella*, but under local conditions only *Bostrychia* spp. and *Caloglossa leprieurii* have a wide distribution; the remaining red algae tend to be restricted in their spread. *Murrayella periclados*

Table 6.3. The estimated mean annual contribution of litter by mangroves and algal epiphytes in the St Lucia Estuary

Mean litter yield (t ha^{-1} y^{-1})	
A. marina	10.13
B. gymnorrhiza	7.29
Algal epiphytes	2.22
Mean total litter yield from algal epiphytes on pneumatophores along shoreline (t y^{-1})	10.05

has been found only in the Kosi system, while *Catenella nipae* was collected from only two estuaries in Transkei. The blue-green algae present on *A. marina* pneumatophores have a greater species diversity and generally individual species, with a few exceptions, have a wider distribution than most red algae.

A survey of horizontal and vertical zonation of algae epiphytic on pneumatophores of *A. marina* has been carried out at Beachwood. The red algae, chiefly *Bostrychia moritziana*, *B. radicans*, *Caloglossa leprieurii* and *Polysiphonia subtilissima*, occurred predominantly near the mouth of the estuary and decreased upstream, following a horizontal gradient from isosalinity to hyposalinity. The blue-green algae, chiefly *Lyngbya confervoides* and *Microcoleus chthonoplastes*, and green algae, dominated by *Rhizoclonium implexum* and *R. riparium*, were more obvious in the drier areas away from the creek (Phillips *et al.* 1994). Vertical distribution along the pneumatophores was most obvious along the edges of the creek where the highest densities of pneumatophores occurred. Three zones were identified: an upper *Rhizoclonium* zone, a mid *Bostrychia* zone and a lower *Caloglossa* zone. The photophilic blue-green algae were not limited to any particular zone. The vertical distribution corresponded to the height of tidal ebb and flow on the pneumatophores (Phillips *et al.* 1996).

Algae associated with mangroves have a significant role to play in the estuarine ecosystem. Steinke and Naidoo (1990) found at St Lucia that the biomass of algal epiphytes on the pneumatophores of *A. marina* increased in the warmer months, followed by a sharp decrease in February/March which was sustained during the cooler periods (Figure 6.6). It was estimated that with this sharp decrease, which appeared to be due to the sloughing off of the algae as a result of unfavourable environmental conditions, approximately 10.0 t y^{-1} of algal litter was made available (Table 6.3). In addition to the major litter contribution of the mangroves, this provides an important input to the ecosystem.

In a study of nitrogen fixation (acetylene reduction) by heterocystous and non-heterocystous blue-green algae (cyanobacteria) associated with *A. marina* pneumatophores and wet and dry surface sediments at Beachwood, Mann and Steinke (1989) showed that there was a marked seasonal variation in nitrogen fixation. The

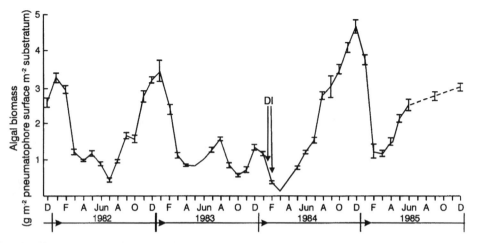

Figure 6.6. **Mean algal biomass on pneumatophores for the period 1981–5 (D and I = cyclones Domoina and Imboa, respectively). Vertical bars represent SE.**

high values obtained in summer were correlated with high algal biomass, high ambient temperatures, high light intensities and long daylengths (Figure 6.7). It was estimated that nitrogen fixation by the blue-green algal communities amounted to as much as 24.3% of the annual nitrogen requirement of the mangroves at Beachwood (Mann & Steinke 1993).

In the past many descriptions of mangrove swamps have either made no reference to the algae or refer to them simply as the 'bostrychietum'. Clearly, in these cases the algal flora has not been seriously studied and only field observations are reported. This treatment is unfortunate as it overlooks the role that the algal component plays in the ecosystem.

Animals

The animals associated with the southern African mangroves and their roles in the estuarine ecosystem have been well documented (Macnae 1968; Day 1974a; Berjak et al. 1977; Branch & Grindley 1979) and will also be dealt with in later chapters of this book. Consequently, only a brief reference needs to be made here to avoid repetition of information.

In the South African mangrove swamps a fauna typical of that found in the tropics is also present. This includes sesarmid crabs, fiddler crabs (Uca spp.), the giant mud crab (Scylla serrata), the mudskipper (Periophthalmus kalolo), and gastropods. However, not all occur throughout the range of distribution of the mangroves.

Of the sesarmid crabs, the most conspicuous and largest is Sesarma meinerti (the red mangrove crab), which has been shown to play an important role in the breakdown of mangrove litter (Day 1974b; Berjak et al. 1977; Emmerson & McGwynne 1992; Steinke et al. 1993b). S. cate-

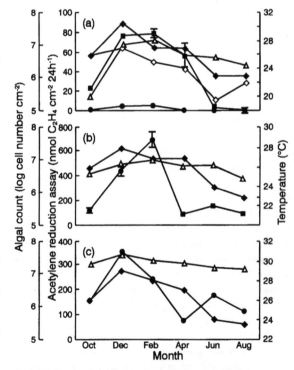

Figure 6.7. **Seasonal variation in nitrogen fixation (acetylene reduction) of blue-green algae at Beachwood. (a) Pneumatophore habitat; (b) wet surface sediments; (c) dry surface sediments; (●) exposed and (■) submerged conditions; (△) total blue-green algal numbers; and temperature under exposed (◆) and submerged (◇) conditions.**

nata is another sesarmid which is present, often in large numbers, and is significant because it feeds on detritus from which organic material and microorganisms are obtained.

The fiddler crabs (Uca spp.) add colour to a mangrove

swamp when they emerge from their burrows to feed and engage in courtship displays during low tides. These crabs are considered to play a significant role in estuarine food webs as they feed on microorganisms on organic detritus and on the surface of the substratum. Dye and Lasiak (1986) have shown that *Uca* spp. in Australia exhibited high assimilation efficiency for bacteria (>98%) as well as strong selection for bacteria over microalgae from the mud surface. Three species, namely *U. annulipes*, *U. chlorophthalmus* and *U. urvillei*, are associated with mangroves to the Nahoon River, although none was apparently present in that estuary at the time of transplanting. In addition, Berjak *et al.* (1977) reported *U. vocans* and *U. inversa* as occurring in estuaries as far south as Beachwood and Inhaca, respectively.

Scylla serrata (the giant mud crab or mangrove crab) is found in mangrove swamps throughout the Indo-Pacific and in South Africa extends to the southern Cape. This large, greenish-brown portunid crab is a scavenger and a predator on molluscs and on other small crabs (Hill 1979). It is widely exploited as a food source in tropical Australasia and is among six species of crab considered as the most important on the world market (Matilda & Hill 1980).

Cerithidia decollata (mangrove snail) is a conspicuous gastropod which occurs on mangroves throughout their range. Characterised by the blunt tip to its spiral, this gastropod alternates between the muddy substratum and the trunks of the mangroves where it is frequently to be seen at high tide. It has been suggested that these movements are related to the animal's feeding habits: this organism obtains food from the extraction of digestible material while feeding on the mud surface at low tide. Their presence on the trunks at high tide is considered to be a mechanism to avoid predation (Cockcroft & Forbes 1981).

Assiminea bifasciata is often present in large numbers on the surface of the substratum near the high water mark (Brown 1971). This widespread, tiny snail thrives on the rich organic detritus characteristic of mangrove muds.

Other common gastropods are *Terebralia palustris* (mangrove whelk) and *Littorina scabra* (periwinkle); the former is present in estuaries in Natal (and further north), whereas the latter occurs commonly on mangroves and saltmarsh vegetation also in the Eastern Cape (Day 1974b). *T. palustris* appears to feed largely on the substratum surface films of algae and other microorganisms. *L. scabra* occurs on the mangrove trees, very often above the highest spring tide levels, where it browses on algae and other material.

Mangroves may also give anchorage to various filter-feeding organisms: *Crassostrea cucullata* (Natal rock oyster), barnacles (mainly *Balanus amphitrite*) and even mussels. The oysters are attached to the lower stem or pneumatophores, while the barnacles are usually found on the pneumatophores in the lower intertidal parts of a swamp. These, in contrast to the molluscs above, filter plankton and nutrients from the estuarine water as it flows past them.

Besides these organisms, mangroves and seagrass beds support numerous species of fish and other marine organisms such as prawns and shrimps. Probably the most common fish are mullet which consume significant quantities of plant material, while other important species, e.g. Cape stumpnose (*Rhabdosargus holubi*), are carnivores which feed on zooplankton and small invertebrates. Larger carnivores, such as the kob (*Argyrosomus hololepidotus*), prey on the smaller fish in estuaries (van der Elst 1981). Mangrove swamps and adjacent mudflats are feeding grounds for a number of coastal birds which include kingfishers, sacred ibis, African fish eagle, etc.

Mangrove food webs

While there is good evidence that some mangrove systems are highly productive, there is equally strong evidence that others are not (Clough 1992). Their productivity may be more than twenty times that of open ocean waters, five times that of rich coastal waters and compare favourably with certain terrestrial ecosystems (Lear & Turner 1977). Unfortunately, translation of the high productivity of mangroves into the energy economy of the ecosystem has received little attention and consequently understanding of the trophic relationships and food webs is incomplete (Redfield 1982).

Mangroves do have a significant role to play in estuarine food webs. These plants provide a source of reduced carbon in the form of leaves, wood and other litter which falls from the trees and contributes to detritus-based food chains in estuaries (Figure 6.8). The significance of mangroves as a source of reduced carbon for marine organisms can be gauged from their litter production. Litter studies conducted in mangroves at St Lucia, Richards Bay, Beachwood (Mgeni estuary) and Wavecrest (Transkei) made possible estimates of production of the various litter components, from which it was possible to extrapolate to other estuaries (Table 6.4) (Ward & Steinke 1982; Steinke & Charles 1986; Steinke & Ward 1988, 1990). These results show that mangroves in South Africa make a significant contribution of litter towards our estuaries and that, with few exceptions, leaf material makes up approximately 60% of this litter. This supports evidence from other parts of the world that mangrove leaf litter makes a major contribution to the nutrient budget of an estuary (Odum *et al.* 1973; Lugo & Snedaker 1974; Teas 1976). Although tidal export of leaves can result in losses of nutrients from

Table 6.4. Estimated litter production of the larger (>5 ha) mangrove communities in South Africa

Estuary	Latitude	Area of mangroves (ha)	Mangrove species present[a]	Total litter (t y^{-1})
Kosi system	26° 55'	59.0	A.m.+B.g.+R.m.+C.t.+L.r.+X.g.	646.5
St Lucia	28° 22'	160.0	A.m.+B.g.	1322.7
Mfolozi	28° 24'	26.0	A.m.+B.g.	214.9
Richards Bay	28° 50'	427.5	A.m.+B.g.+R.m.	4684.1
Mlalazi	28° 57'	30.0	A.m.+B.g.+R.m.	248.0
Mgeni	29° 48'	44.0	A.m.+B.g.+R.m.	334.2
Durban Bay	29° 54'	15.0	A.m.+B.g.+R.m.	164.4
Sipingo	29° 59'	12.5	A.m.+B.g.+R.m.	137.0
Mntafufu	31° 34'	10.0	A.m.+B.g.+R.m.	45.3
Mngazana	31° 42'	150.0	A.m.+B.g.+R.m.	679.5
Mtakatye	31° 51'	7.5	A.m.+B.g.+R.m.	34.0
Mtata	31° 57'	34.0	A.m.+B.g.+R.m.	154.0
Xora	32° 10'	16.0	A.m.+B.g.	72.5
Mbashe	32° 15'	12.5	A.m.+B.g.	56.6
Nqabara-Nqabarana	32° 20'	9.0	A.m.	27.9
Nxaxo-Nqusi (Wavecrest)	32° 35'	14.0	A.m.+B.g.	50.7
Kobonqaba	32° 36'	6.0	A.m.	18.6

[a] A.m., *Avicennia marina*; B.g., *Bruguiera gymnorrhiza*; R.m., *Rhizophora mucronata*; C.t., *Ceriops tagal*; L.r., *Lumnitzera racemosa*; X.g., *Xylocarpus granatum*.

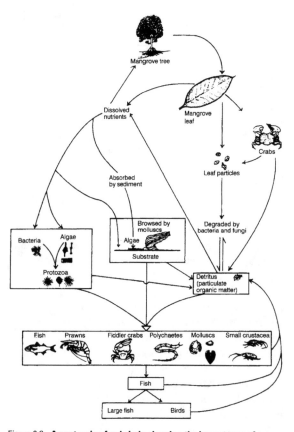

Figure 6.8. **An estuarine food chain showing the importance of a supply of mangrove litter.**

an estuary, observations indicate that, in most South African estuaries, this loss is unlikely to be high (Steinke *et al.* 1993a). However, not only the mangroves but also epiphytes, algae on the soil surface, other macrophytes and phytoplankton in estuarine waters contribute to the production of plant organic matter.

When litter falls, there is an initial rapid loss in mass as a result largely of leaching, followed by a slower, steady decrease for the remainder of the decomposition period (Steinke *et al.* 1993a). Decomposition of intact litter, without the intervention of animals, such as crabs, takes place over a period of 3–6 months, depending upon species, season and position of leaves in relation to tidal inundation (Steinke & Ward 1987). Generally, decomposition was found to be more rapid in leaves of *A. marina* than *B. gymnorrhiza*; in leaves constantly submerged compared with those exposed for long periods; and in the warm season rather than the cool season. During decomposition the concentration of nitrogen increased, although the phosphorus and potassium contents decreased sharply (Steinke & Ward 1987; van der Valk & Attiwill 1984). The higher nitrogen concentrations, due to the colonisation of decomposing leaves by microorganisms and nitrogen fixation by these organisms, and also to the effects of detritus feeders, emphasise the importance of mangrove litter as a source of, and substratum for nutrient release in these ecosystems (Newell 1965; Steinke & Ward 1987; Mann & Steinke 1992). There was also a decrease in tannin content of leaves, which has significant consequences for further decomposition. Loss of tannins from leaves has been shown to coincide with rapid increases in the densities of bacteria on mangrove leaves (Cundell *et al.* 1979; Robertson

1988), while palatability of leaves to crabs increased with decrease in tannin concentrations (Robertson 1988; Steinke *et al.* 1993*b*).

Originally bacteria were considered to be the main decomposers in the marine environment (Fenchel 1972; Ferguson-Wood 1975) and fungi were not considered to have a significant role (Hanson & Wiebe 1977). However, there is growing evidence (Fell & Newell 1981) that marine fungi are important in the decomposition of organic material in estuarine ecosystems. Anaerobic (bacterial) decomposition processes in waterlogged soils are slow and probably contribute little to nutrient cycling, whereas aerobic decomposition of leaves, in which fungi play an important role, has been shown to be rapid (Boto 1982). In South Africa, Steinke and Jones (1993) listed 55 species of marine fungi on mangrove wood, and new records have been added to this list. Most of the lignicolous fungi have tropical/subtropical affinities and are commonly found also in mangrove swamps in tropical Australasia (Hyde 1988; Hyde & Jones 1988; Jones & Kuthubutheen 1989; Kohlmeyer & Volkmann-Kohlmeyer 1991). Jones (1993) has suggested that temperature is the most important factor controlling the distribution of marine fungi. There are some tropical species which have not yet been found in South Africa and it is possible that conditions at this southernmost limit of mangroves are too cool for these fungi. Collections are needed from countries to our north to provide an answer to this. From decomposing leaves approximately 25 species of terrestrial and marine fungi have been isolated (Steinke *et al.* 1990; Singh & Steinke 1992; Steinke & Jones 1993).

Nutrients released through leaching and the actions of microorganisms are available for other estuarine organisms and, through tidal outflow into the marine environment, for phytoplankton which form the basis of marine food chains. Runoff from the land also brings nutrients into an estuary, in the form of plant and animal organic material and dissolved mineral salts released by erosion of the land. Dissolved nutrients (e.g. inorganic salts of nitrogen and phosphorus) in estuarine tidal waters are generally low and, therefore, the decomposition of organic material is an important source of input of these nutrients (Boto 1982). The active and important role mangroves play in this process is indicated by results from Australia that mangroves may, in one year, cycle four times as much nitrogen and half as much phosphorus as there is in the surrounding water (Clough & Attiwill 1975).

Reference has already been made to the important role of crabs, especially *Sesarma meinerti*, and other macroinvertebrates in the degradation of mangrove litter. To these must be added the meiofauna which are also significant in biodegradative processes in estuaries (Dye 1983). Through the actions of these animals and the decomposer organisms, litter is broken down to finely particulate detritus. This detritus continues to support microbial growth which is exploited as a food source by small animals (such as worms) living in the mud, by fiddler crabs, gastropods and young fish. The detrital particles are taken in by these organisms, the microbial growth removed during passage through their digestive tract, and the particles egested, to be colonised again by microorganisms, and so the process is continually repeated.

Carnivorous scavengers and predators form the next stage in the food chain. At the top of the food web is man who utilises many of the estuarine and inshore fish and shellfish nourished by nutrients from the mangrove debris. In South Africa and other parts of the world man has long used mangrove areas as a source of food, fuel, timber and for other purposes. This underlines the need for careful management to ensure the future of these communities.

Uses and management

Mangroves are becoming increasingly recognised as a very significant part of the coastal environment and there is no doubt that they have an extremely important role to play in our estuaries. Mangroves lining waterways form a river/land barrier which protects the shoreline from erosion. Mangrove roots and pneumatophores also trap plant litter and silt and so raise the floor of the estuary to a level where mangrove seedlings can become established and extend the barrier between land and water (Crisp *et al.* 1990; T.D.S., personal observations). It is generally recognized that estuaries are an important breeding- and feeding-ground for marine species. Mangrove detritus provides nutrients for zooplankton, including crustaceans and fish larvae, while the mangrove environment provides food, shelter and breeding sites for numerous species of fish, many of which are commercially important. Whitfield (1993) has shown that, of the fish species found in our subtropical estuaries, which he defines as occurring from the Mbashe River northwards, 70% have a strong estuarine association: i.e. they breed in estuaries, or they breed at sea but their juveniles show dependence upon estuaries. There is a high investment in our coastal resorts and, clearly, sustained yields from fishing are essential for the financial success of these resorts and satellite industries. These sustained yields are dependent upon healthy estuarine ecosystems which are capable of fulfilling their role as nursery areas for a wide variety of marine organisms.

Not only are estuaries a source of fish and other marine life for subsistence and for recreation, but they are also

Figure 6.9. (a) **Fish traps constructed mainly from mangroves in the tidal basin of the Kosi system; (b) cultivation to the edge of a narrow fringe of *A. marina* on the Kobonqaba River; (c) chopped and thinned** *A. marina* **at Wavecrest (Nxaxo-Nqusi Rivers); (d) an example of man's disregard for mangroves and the estuarine environment.**

frequently exploited for other purposes. In parts of Transkei mangroves are an important source of poles for the construction of kraals, while in the Kosi system mangroves are widely used in the extensive fish traps throughout the estuary (Figure 6.9*a*). Estuaries are a source of bait for fishing, in many areas the giant mud crab is harvested by the local population, while in the Kosi system there is repeated mass removal from the mangroves of crabs of all species. This last practice has, unfortunately, resulted also in the destruction of many pioneer trees but, perhaps more importantly, the removal of so many animals has had serious consequences for the mangrove environment (Ward *et al.* 1986). In some areas, associated vegetation, e.g. *Juncus kraussii* (ncema), is harvested by the local people for use in traditional crafts.

It is, therefore, in our interests to ensure that management of estuaries should be based on sound principles to ensure conservation and sustainable use. Unfortunately, mangrove communities have suffered from man's activ-

ities in the past. In addition to being subjected to unfortunate management practices, some of which were referred to earlier, mangroves have been used as a dumping-ground for refuse, while in certain areas extensive chopping of large trees, grazing by domestic livestock and encroachment of cultivation have also led to significant reductions (Figure 6.9) (Moll *et al.* 1971; Steinke 1972; Ward *et al.* 1986). While industrial pollution has as yet had little effect on South African mangroves, unwise agricultural practices, from the estuaries to their catchment areas, have had detrimental effects on many Natal estuaries which, according to Begg (1978), have ceased to function as such.

In the past a problem affecting the management of our coastal resources in general has been the conflict of responsibilities for these resources among government agencies. In addition, because many mangroves occur in small, often isolated, estuarine pockets along the coast, it has proved difficult to control and manage mangrove areas effectively. While certain basic principles can be elaborated at this stage, there is an urgent need for more research on the biology, sustainable utilisation and management of mangroves, including their interaction with other coastal systems. It is only when we have this information that we shall be able to limit, with confidence, the utilisation of mangroves and their associated fauna and flora at levels to ensure sustainable yields in the future. In the interim one option for managing mangrove communities is to protect certain areas by setting them aside in reserves. This is especially important in the southern estuaries (e.g. Mngazana, Mtata, Nxaxo-Nqusi and Kobonqaba), where the mangroves are presently under threat from activities such as wood chopping, and grazing by stock. In these areas chopping should be prohibited and, as far as possible, stock should be kept out of the mangroves. In other areas, e.g. Kosi system and Richards Bay, a rotational system of rest and utilisation should be considered, especially where pressures on the resources are intense. Utilisation of natural resources by local populations should be permitted, as far as possible, on a controlled basis only after sustainable levels for these resources have been established. Frequent monitoring should also be carried out to detect changes in the flora and fauna due to incorrect management practices, pollution, etc. Timeous detection of any deterioration will enable steps to be taken to arrest/reverse changes before irreparable damage is done.

Another option which should be considered for areas where mangroves have been lost, through mismanagement or natural causes, is re-establishment by transplanting or seeding. Research on establishment and early growth of mangroves has shown that this is feasible.

Although re-establishment is possible, not having to do this, by managing mangrove areas correctly, should be the preferred option.

Sound agricultural principles (e.g. veld management, no cultivation of steep areas, use of contour furrows) in catchment areas are necessary to reduce degradation. Expansion of coastal resorts should be restricted and new developments planned so that they cause minimal damage to the environment. Harmful effects of structures such as roads, causeways, and boat-launching ramps should be minimised. Mangroves will survive if these structures allow tidal flow to be maintained. Developments should also be planned to fit in with the natural surroundings and to detract as little as possible from the aesthetic appeal of the coastal environment.

It is almost superfluous to state that the results of research should be speedily and effectively disseminated to policy makers, and other appropriate authorities, and to the public in order that appropriate management strategies can be devised and applied to safeguard the coastal environment.

Acknowledgements

The author wishes to thank Mrs R. Bunsee for typing and A. Rajh for photographic work. To my co-workers, the University of Durban-Westville, Natal Parks Board and Transkei Department of Agriculture and Forestry thanks are also due.

References

Andrews, T.J., Clough, B.F. & Muller, G.J. (1984). Photosynthetic gas exchange properties and carbon isotope ratios of some mangroves in North Queensland. In *Physiology and management of mangroves*, ed. H.J. Teas, pp. 15–23. The Hague: Dr W. Junk.

Ball, M.C. (1988). Ecophysiology of mangroves. *Trees*, **2**, 129–42.

Begg, G.M. (1978). *The estuaries of Natal*. Natal Town & Regional Planning Commission Report No. 41.

Berjak, P., Campbell, G.K., Huckett, B.I. & Pammenter, N.W. (1977). *In the mangroves of Southern Africa*. Durban Wildlife Society of Southern Africa, Natal Branch.

Boto, K.G. (1982). Nutrient and organic fluxes in mangroves. In *Mangrove ecosystems in Australia*, ed. B.F. Clough, pp. 239–57. Canberra: ANU Press.

Branch, G.M. & Branch, M. (1981). *The living shores of Southern Africa*. Cape Town: C. Struik.

Branch, G.M. & Grindley, J.R. (1979). Ecology of southern African estuaries. Part XI. Mngazana: a mangrove estuary in Transkei. *South African Journal of Zoology*, **14**, 149–70.

Breen. C.M. & Hill, B.J. (1969). A mass mortality of mangroves in the Kosi estuary. *Transactions of the Royal Society of South Africa*, **38**, 285–303.

Bridgewater, P.B. (1982). Mangrove vegetation of the southern and western Australian coastline. In *Mangrove ecosystems in Australia*, ed. B.F. Clough, pp. 55–6. Canberra: ANU Press.

Brown, D.S. (1971). Ecology of Gastropoda in a South African mangrove swamp. *Proceedings of the Malacological Society of London*, **39**, 263–79.

Bruton, M.N. (1980). An outline of the ecology of the Mgobezeleni Lake System at Sodwana, with emphasis on the mangrove community. In *Studies on the ecology of Maputaland*, ed. M.N. Bruton & K.H. Cooper, pp. 408–26. Cape Town: Cape & Transvaal Printers.

Clarke, P.J. (1993). Dispersal of grey mangrove (*Avicennia marina*) propagules in southeastern Australia. *Aquatic Botany*, **45**, 195–204.

Clarke, P.J. & Allaway, W.G. (1993). The regeneration niche of the grey mangrove (*Avicennia marina*): effects of salinity, light and sediment factors on establishment, growth and survival in the field. *Oecologia*, **93**, 548–56.

Clarke, P.J. & Myerscough, P.J. (1991). Buoyancy of *Avicennia marina* propagules in south-eastern Australia. *Australian Journal of Botany*, **39**, 77–84.

Clough, B.F. (1984). Growth and salt balance of the mangroves *Avicennia marina* (Forssk.) Vierh. and *Rhizophora stylosa* Griff. in relation to salinity. *Australian Journal of Plant Physiology*, **11**, 419–30.

Clough, B.F. (1992). Primary productivity and growth of mangrove forests. In *Tropical mangrove ecosystems*, ed. A.I. Robertson & D.M. Alongi, pp. 225–49. Washington, DC: American Geophysical Union.

Clough, B.F. & Andrews, T.J. (1981). Some ecophysiological aspects of primary production by mangroves in North Queensland. *Wetlands*, **1**, 6–7.

Clough, B.F., Andrews, T.J. & Cowan, I.R. (1982). Physiological processes in mangroves. In *Mangrove ecosystems in Australia*, ed. B.F. Clough, pp. 193–210. Canberra: ANU Press.

Clough, B.F. & Attiwill, P.M. (1975). Nutrient cycling in a community of *Avicennia marina* in a temperate region of Australia. In *Proceedings of the International Symposium on the Biology and Management of Mangroves, Hawaii*, ed. G. Walsh, S. Snedaker & H. Teas, pp. 137–46. Gainesville: University of Florida Press.

Clough, B.F. & Scott, K. (1989). Allometric relationships for estimating above-ground biomass in six mangrove species. *Forest Ecology and Management*, **27**, 117–27.

Clough, B.F. & Sim, R.G. (1989). Changes in gas exchange characteristics and water use efficiency of mangroves in response to salinity and vapour pressure deficit. *Oecologia*, **79**, 38–44.

Cockcroft, V.G. & Forbes, A.T. (1981). Tidal activity rhythms in the mangrove snail *Cerithidea decollata* (Linn.) (Gastropoda: Prosobranchia: Cerithiidae). *South African Journal of Zoology*, **16**, 5–9.

Crisp, P., Daniel, L. & Tortell, P. (1990). *Mangroves in New Zealand. Trees in the tide*. Auckland: G.P. Books.

Cronquist, A. (1981). *An integrated system of classification of flowering plants*. New York: Columbia University Press.

Cundell, A.M., Brown, M.S., Standford, R. & Mitchell, R. (1979). Microbial degradation of *Rhizophora mangle* leaves immersed in the sea. *Estuarine and Coastal Marine Science*, **9**, 281–6.

Davey, J.E. (1975). Note on the mechanism of pollen release in *Bruguiera gymnorrhiza*. *South African Journal of Botany*, **41**, 269–72.

Day, J.H. (1974a). The ecology of Morrumbene Estuary, Mozambique. *Transactions of the Royal Society of South Africa*, **41**, 43–97.

Day, J.H. (1974b). *A guide to marine life on South African shores*, 2nd edn. Cape Town: A.A. Balkema.

Drennan, P.M., Berjak, P. & Pammenter, N.W. (1992). Ion gradients and adenosine triphosphate localization in the salt glands of *Avicennia marina* (Forsskål) Vierh. *South African Journal of Botany*, **58**, 486–90.

Drennan, P.M. & Pammenter, N.W. (1982). Physiology of salt excretion in the mangrove, *Avicennia marina* (Forssk.) Vierh. *New Phytologist*, **91**, 597–606.

Dye, A.H. (1983). Composition and seasonal fluctuations of meiofauna in a southern African mangrove estuary. *Marine Biology*, **73**, 165–70.

Dye, A.H. & Lasiak, T.A. (1986). Assimilation efficiencies of fiddler crabs and deposit-feeding gastropods from tropical mangrove sediments. *Comparative Biochemistry and Physiology*, **87**A, 341–4.

Duke, N.C. (1992). Mangrove floristics and biogeography. In *Tropical mangrove ecosystems*, ed. A.I. Robertson & D.M. Alongi, pp. 63–100. Washington, DC: Coastal & Estuarine Studies Series, Vol. 41, American Geophysical Union.

Edgcumbe, D.J. (1980). Some preliminary observations on the submerged aquatic *Zostera capensis* Setchell. *Journal of South African Botany*, **46**, 53–66.

Emmerson, W.D. (1988). Estuaries in the Transkei : their past, present and future. In *Research needs in the Transkei and Ciskei coastal zone*, ed. G.M. Branch & L.Y. Shackleton, pp. 19–27. South African National Programmes Report No. 155.

Emmerson, W.D. & McGwynne, L.E. (1992). Feeding and assimilation of mangrove leaves by the crab *Sesarma meinerti* de Man in relation to leaf-litter production in Mgazana, a warm-temperate southern African mangrove swamp. *Journal of Experimental Marine Biology and Ecology*, **157**, 41–53.

Fell, J.W. & Newell, S.Y. (1981). Role of fungi in carbon flow and nitrogen immobilization in coastal marine plant litter systems. In *The fungal community – its organization and role in the ecosystem*, ed. D.T. Wicklow & G.C. Carroll, pp. 665–78. New York: Marcel Dekker.

Fenchel, T. (1972). Aspects of decomposer food chains in marine benthos. *Verhandlungen der Deutschen Zoologischen Gesellschaft*, **65**, 14–23.

Ferguson-Wood, E.J. (1975). *The living ocean – marine microbiology*. London: Croom Helm.

Hanson, R.B. & Wiebe, W.J. (1977). Heterotrophic activity associated with particulate size fractions in a *Spartina alterniflora* salt-marsh estuary, Sapelo Island, Georgia, USA, and the continental shelf waters. *Marine Biology*, **42**, 321–30.

Hill, B.J. (1979). Aspects of the feeding strategy of the predatory crab *Scylla serrata*. *Marine Biology*, **55**, 209–14.

Howard-Williams, C. (1980). Aquatic macrophytes of the coastal wetlands of Maputaland. In *Studies on the ecology of Maputaland*, ed. M.N. Bruton & K.H. Cooper, pp. 42–51. Cape Town: Cape & Transvaal Printers.

Howard-Williams, C. & Liptrot, M.R.M. (1980). Submerged macrophyte communities in a brackish South African estuarine-lake system. *Aquatic Botany*, **9**, 101–16.

Hyde, K.D. (1988). Studies on the tropical marine fungi of Brunei. *Botanical Journal of the Linnean Society*, **98**, 135–51.

Hyde, K.D. & Jones, E.B.G. (1988). Marine mangrove fungi. *Marine Ecology*, **9**, 15–33.

Jones, E.B.G. (1993). Tropical marine fungi. In *Aspects of tropical mycology*, ed. S. Isaacs, J.C. Frankland, R. Watling & A.J.S. Whalley, pp. 73–89. Cambridge: Cambridge University Press.

Jones, E.B.G. & Kuthubutheen, A.J. (1989). Malaysian mangrove fungi. *Sydowia*, **41**, 160–9.

Kohlmeyer, J. & Volkmann-Kohlmeyer, B. (1991). Marine fungi of Queensland, Australia. *Australian Journal of Marine and Freshwater Research*, **42**, 91–9.

Komiyama, A., Ogino, K., Adsornkoe, S. & Sabhasri, S. (1987). Root biomass of a mangrove forest in southern Thailand. I. Estimation by the trench method and the zonal structure of root biomass. *Journal of Tropical Ecology*, **3**, 97–108.

Lambert, G., Steinke, T.D. & Naidoo, Y. (1987). Algae associated with mangroves in southern African estuaries. I. Rhodophyceae. *South African Journal of Botany*, **53**, 349–61.

Lambert, G., Steinke, T.D. & Naidoo, Y. (1989). Algae associated with mangroves in southern African estuaries. Cyanophyceae. *South African Journal of Botany*, **55**, 476–91.

Lear, R. & Turner, T. (1977). *Mangroves of Australia*. St Lucia: University of Queensland Press.

Lugo, A.E. & Snedaker, S.C. (1974). The ecology of mangroves. *Annual Review of Ecology and Systematics*, **5**, 39–64.

Macnae, W. (1963). Mangrove swamps in South Africa. *Journal of Ecology*, **51**, 1–25.

Macnae, W. (1968). A general account of the fauna and flora of mangrove swamps and forests in the Indo-West-Pacific region. *Advances in Marine Biology*, **6**, 73–270.

Mann, F.D. & Steinke, T.D. (1989). Biological nitrogen fixation (acetylene reduction) associated with blue-green algal (cyanobacterial) communities in the Beachwood Mangrove Nature Reserve. I. The effect of environmental factors on acetylene reduction activity. *South African Journal of Botany*, **55**, 438–46.

Mann, F.D. & Steinke, T.D. (1992). Biological nitrogen fixation (acetylene reduction) associated with decomposing *Avicennia marina* leaves in the Beachwood Mangrove Nature Reserve. *South African Journal of Botany*, **58**, 533–6.

Mann, F.D. & Steinke, T.D. (1993). Biological nitrogen fixation (acetylene reduction) associated with blue-green algal (cyanobacterial) communities in the Beachwood Mangrove Nature Reserve. II. Seasonal variation in acetylene reduction activity. *South African Journal of Botany*, **59**, 1–8.

Matilda, C.E. & Hill, B.J. (1980). *Annotated bibliography of the portunid crab Scylla serrata (Forskal)*. Queensland Fisheries Service Technical Report No. 3.

Mepham, R.H. & Mepham, J.S. (1985). The flora of tidal forests – a rationalization of the use of the term 'mangrove'. *South African Journal of Botany*, **51**, 77–99.

Moll, E.J., Ward, C.J., Steinke, T.D. & Cooper, K.H. (1971). Our mangroves threatened. *African Wildlife*, **25**, 103–7.

Moon, G.J., Clough, B.F., Peterson, C.A. &

Allaway, W.G. (1986). Apoplastic and symplastic pathways in *Avicennia marina* (Forssk.) Vierh. roots revealed by fluorescent tracer dyes. *Australian Journal of Plant Physiology*, **13**, 637–48.

Naidoo, G. (1980). Mangrove soils of the Beachwood area, Durban. *Journal of South African Botany*, **46**, 293–304.

Naidoo, G. (1989). Seasonal plant water relations in a South African mangrove swamp. *Aquatic Botany*, **33**, 87–100.

Naidoo, G., Rogalla, H. & von Willert, D.J. (1997). Gas exchange responses of a mangrove species, *Avicennia marina*, to waterlogged and drained conditions. *Hydrobiologia*, **352**, 39–47.

Naidoo, G., Rogalla, H. & von Willert, D.J. (1998). Comparative gas exchange characteristics of *Avicennia marina* (Forssk.) Vierh. and *Bruguiera gymnorrhiza* (L.) Lam. under field conditions during summer. *Mangroves and Salt Marshes* (in press).

Newell, R.C. (1965). The role of detritus in the nutrition of two marine deposit feeders, the prosobranch *Hydrobia ulvae* and the bivalve *Macoma balthica*. *Proceedings of the Zoological Society of London*, **144**, 25–46.

Odum, W.E., Zieman, J.C. & Heald, E.J. (1973). Importance of vascular plant detritus to estuaries. In *Proceedings of the Coastal Marsh and Estuarine Management Symposium*, ed. R.H. Chabreek, pp. 91–114. Baton Rouge: Louisiana State University Press.

Phillips, A., Lambert, G., Granger, J.E. & Steinke, T.D. (1994). Horizontal zonation of epiphytic algae associated with *Avicennia marina* (Forssk.) Vierh. pneumatophores at Beachwood Mangroves Nature Reserve, Durban, South Africa. *Botanica Marina*, **37**, 567–76.

Phillips, A., Lambert, G., Granger, J.E. & Steinke, T.D. (1996). Vertical zonation of epiphytic algae associated with *Avicennia marina* (Forssk.) Vierh. pneumatophores at Beachwood Mangroves Nature Reserve, Durban, South Africa. *Botanica Marina*, **39**, 167–75.

Pool, D.J., Lugo, A.E. & Snedaker, S.C. (1975). Litter production in mangrove forests of southern Florida and Puerto Rico. In *Proceedings of international symposium on biology and management of mangroves, Hawaii*, ed. G.E. Walsh, S.C. Snedaker & H.J. Teas. Gainesville: University of Florida.

Post, E. (1936). Systematische und

pflanzengeographische Notizen zur *Bostrychia-Caloglossa*. Assoziation. *Revue Algologique*, **9**, 1–84.

Putz, F.E. & Chan, H.T. (1986). Tree growth, dynamics and productivity in a mature mangrove forest in Malaysia. *Forest Ecology and Management*, **17**, 211–30.

Rabinowitz, D. (1978). Early growth of mangrove seedlings in Panama, and an hypothesis concerning the relationship of dispersal and zonation. *Journal of Biogeography*, **5**, 113–33.

Redfield, J.A. (1982). Trophic relationships in mangrove communities. In *Mangrove ecosystems in Australia*, ed. B.F. Clough, pp. 259–62. Canberra: ANU Press.

Robertson, A.I. (1988). Decomposition of mangrove leaf litter in tropical Australia. *Journal of Experimental Marine Biology and Ecology*, **116**, 235–47.

Scholander, P.F. (1968). How mangroves desalinate water. *Physiologia Plantarum*, **21**, 251–61.

Scholander, P.F., Hammel, H.T., Hemmingsen, E. & Garey, W. (1962). Salt balance in mangroves. *Plant Physiology*, **37**, 723–9.

Silva, C.A.R., Lacerda, L.D., Silva, L.F.F. & Rezenda, C.E. (1991). Forest structure and biomass distribution in a red mangrove stand in Sepetiba Bay, Rio de Janeiro. *Revista Brasileira de Botanica*, **14**, 21–5.

Singh, N. & Steinke, T.D. (1992). Colonization of decomposing leaves of *Bruguiera gymnorrhiza* (Rhizophoraceae) by fungi, and *in vitro* cellulolytic activity of the isolates. *South African Journal of Botany*, **58**, 525–9.

Smith, T.J. (1987). Seed predation in relation to tree dominance and distribution in mangrove forests. *Ecology*, **68**, 266–73.

Steinke, T.D. (1972). Further observations on the distribution of mangroves in the Eastern Cape Province. *Journal of South African Botany*, **38**, 165–78.

Steinke, T.D. (1975). Some factors affecting dispersal and establishment of propagules of *Avicennia marina* (Forssk.) Vierh. In *Proceedings of the International Symposium on the Biology and Management of Mangroves, Hawaii*, ed. G. Walsh, S. Snedaker & H. Teas, pp. 402–14. Gainesville: University of Florida Press.

Steinke, T.D. (1986). A preliminary study of

buoyancy behaviour in *Avicennia marina* propagules. *South African Journal of Botany*, **52**, 559–65.

Steinke, T.D. (1988). Vegetative and floral phenology of three mangroves in Mgeni Estuary. *South African Journal of Botany*, **54**, 97–102.

Steinke, T.D., Barnabas, A.D. & Somaru, R. (1990). Structural changes and associated microbial activity accompanying decomposition of mangrove leaves in Mgeni Estuary. *South African Journal of Botany*, **56**, 39–48.

Steinke, T.D. & Charles, L.M. (1986). Litter production by mangroves. I. Mgeni Estuary. *South African Journal of Botany*, **52**, 552–8.

Steinke, T.D. & Charles, L.M. (1987). The utilization of storage compounds during early growth of *Avicennia marina* seedlings under laboratory and field conditions. *South African Journal of Botany*, **53**, 271–5.

Steinke, T.D., Holland, A.J. & Singh, Y. (1993a). Leaching losses during decomposition of mangrove leaf litter. *South African Journal of Botany*, **59**, 21–5.

Steinke, T.D. & Jones, E.B.G. (1993). Marine and mangrove fungi from the Indian Ocean coast of South Africa. *South African Journal of Botany*, **59**, 385–90.

Steinke, T.D. & Naidoo, Y. (1990). Biomass of algae epiphytic on pneumatophores of the mangrove, *Avicennia marina*, in the St Lucia Estuary. *South African Journal of Botany*, **56**, 226–32.

Steinke, T.D. & Naidoo, Y. (1991). Respiration and net photosynthesis of cotyledons during establishment and early growth of propagules of the mangrove, *Avicennia marina*, at three temperatures. *South African Journal of Botany*, **57**, 171–4.

Steinke, T.D. & Rajh, A. (1995). Vegetative and floral phenology of the mangrove, *Ceriops tagal*, with observations on the reproductive behaviour of *Lumnitzera racemosa*, in the Mgeni Estuary. *South African Journal of Botany*, **61**, 240–4.

Steinke, T.D., Rajh, A. & Holland, A.J. (1993b). The feeding behaviour of the red mangrove crab *Sesarma meinerti* de Man, 1887 (Crustacea: Decapoda: Grapsidae) and its effect on the degradation of mangrove leaf litter. *South African Journal of Marine Science*, **13**, 151–60.

Steinke, T.D. & Ward, C.J. (1987). Degradation of mangrove leaf litter in the St Lucia Estuary as influenced by season and exposure. *South African Journal of Botany*, **53**, 323–8.

Steinke, T.D. & Ward, C.J. (1988). Litter production by mangroves. II. St Lucia and Richards Bay. *South African Journal of Botany*, **54**, 445–54.

Steinke, T.D. & Ward, C.J. (1989). Some effects of the cyclones Domoina and Imboa on mangrove communities in the St Lucia Estuary. *South African Journal of Botany*, **55**, 340–8.

Steinke, T.D. & Ward, C.J. (1990). Litter production by mangroves. III. Wavecrest (Transkei) with predictions for other Transkei estuaries. *South African Journal of Botany*, **56**, 514–19.

Steinke, T.D., Ward, C.J. & Rajh, A. (1994). Forest structure and biomass of mangroves in the Mgeni Estuary, South Africa. *Hydrobiologia*, **295**, 159–66.

Tamai, S., Nakasuga, T., Tabuchi, R. & Ogino, K. (1986). Standing biomass of mangrove forests in southern Thailand. *Journal of the Japanese Forestry Society*, **68**, 384–8.

Teas, H.J. (1976). *Productivity of Biscayne Bay mangroves*. University of Miami Sea Grant Special Report No. 5, 103–12.

van der Elst, R. (1981). *A guide to the common sea fishes of Southern Africa*. Cape Town: C. Struik.

van der Valk, A.G. & Attiwill, P.M. (1984). Decomposition of leaf and root litter of *Avicennia marina* at Westernport Bay, Victoria, Australia. *Aquatic Botany*, **18**, 205–21.

Wallace, J.H. & van der Elst, R.P. (1974). *Interim report on the juvenile fish sampled in Transkei estuaries, with comments on the need for conservation of the estuarine environment*. Oceanographic Research Institute, Durban.

Wallace, J.H. & van der Elst, R.P. (1975). *The estuarine fishes of the east coast of South Africa. IV. Occurrence of juveniles in estuaries*. Report No. 42, Oceanographic Research Institute, Durban.

Ward, C.J. (1962). *Cymodocea ciliata* (Forsk.) Ehrenb. ex Aschers – a marine angiosperm – in South African waters. *Lammergeyer*, **11**, 21–5.

Ward, C.J. & Steinke, T.D. (1982). A note on the distribution and approximate areas of mangroves in South Africa. *South African Journal of Botany*, **1**, 51–3.

Ward, C.J., Steinke, T.D. & Ward, M.C. (1986). Mangroves of the Kosi System, South Africa: their re-establishment since a mass mortality in 1965/66. *South African Journal of Botany*, **52**, 501–12.

Watson, J.G. (1928). Mangrove forests of the Malay Peninsula. *Malayan Forest Records*, **6**, 1–275.

Whitfield, A.K. (1993). *How dependent are Southern African fishes on estuaries?* Poster paper, 8th Southern African Marine Science Symposium, Langebaan.

7 Estuarine zooplankton community structure and dynamics

Tris Wooldridge

Mdloti River estuary

Introduction

The term **plankton** is a general term that refers to organisms in the water column having limited powers of locomotion relative to water currents. Although the word plankton engenders thoughts of microscopic organisms, size does not decide membership of the plankton. Planktonic organisms exhibit a considerable size range, varying from the ultramicroscopic bacteria to large jellyfish. The primary components of the plankton are **phytoplankton** (free-floating plants, dealt with in Chapter 5), and the **zooplankton** or animal community. Because planktonic organisms are usually captured with the use of nets of various mesh sizes, it is often convenient to classify them on the basis of size (Webber & Thurman 1991). However, organisms smaller than 20 μm are difficult to filter out of the water column using conventional nets and it is more practical to capture them by first collecting water samples before utilising other extraction techniques. These small organisms include the **picoplankton** (0.2–2 μm) and the **nanoplankton** (2–20 μm). The **microplankton** (20–200 μm) and **mesoplankton** (0.2–20 mm) are well represented in the zooplankton, while **macrozooplankters** (2–20 cm) and **megazooplankters** (>20 cm) make up the largest size categories.

Planktonic organisms are also classified according to the portion of the life cycle spent in the plankton community. Those that spend their entire lives in the plankton community are called the **holoplankton** and those that are temporarily planktonic are termed the **meroplankton**. Crustaceans such as copepods, for example, usually dominate the holoplankton while the early life-history stages of many typically benthic organisms such as polychaete larvae, molluscan larvae, prawn and crab larvae are typical of the meroplankton. **Bentho-pelagic** species spend much of the

24-hour day in close association with the substrate and move into the water column temporarily. Included are species of harpacticoid copepods, mysids, amphipods, isopods and cumaceans.

An important feature distinguishing estuaries from adjacent sea and river habitats is the existence of dynamic axial and vertical salinity gradients. Salinity is normally the key abiotic variable regulating spatial structure of the estuarine plankton community (e.g. Cronin et al. 1962). Thus, salinity tolerance of species is a further means of subdividing estuarine zooplankton since it influences the abundance of individual species as well as intraspecific patterns of distribution. In the lower estuary an incursive marine component penetrates the estuary up to a point where the salinity does not fall below about 28. The neritic assemblage is rich in species and includes cnidarians, ctenophores, annelids, crustaceans, chaetognaths and molluscs. They are sometimes referred to as the stenohaline group and occur in lower estuarine regions where suitable salinity conditions prevail. Generally, stenohaline zooplankters are transported back to sea on the ebb tide.

An oligohaline or freshwater associated community may also occur in the upper reaches of estuaries where salinity falls below about 4. However, in terms of biomass, stenohaline and oligohaline communities are not well represented in most South African estuaries. This is in contrast to a true estuarine or euryhaline community that becomes well established (numbers and biomass) above the general volume of mixed estuarine water that reaches the inlet area around low tide. The latter group also includes species not usually found in the open sea or in freshwater.

In this chapter, focus is on the euryhaline mesozooplankton of South African estuaries, with emphasis on

Table 7.1. Geographical ranges of copepods and mysids commonly recorded from South African estuaries and land-locked bays (modified from Grindley 1981). The Orange River and Kosi estuary demarcate the political boundaries between Namibia and Mozambique respectively (Figure 7.1). West coast systems include the Orange up to and including Langebaan; the south coast includes the Breede and Knysna, while the remaining systems provide distributional data from the east coast. Along the east coast, the climatic regime grades from warm temperate into subtropical.

	Orange	Olifants	Berg	Langebaan	Breede	Knysna	Swartkops	Bushmans	Bashee	Mgazana	Msikaba	Mzimkulu	Durban Bay	Umlalazi	Richards Bay	St Lucia	Kosi
Copepods																	
Acartia africana		*	*	*	*	*	*	*									
Acartia longipatella		*		*		*	*	*	*		*						
Acartia natalensis						*	*	*	*	*	*		*	*	*	*	*
Pseudodiaptomus hessei	*	*	*	*	*	*	*	*	*	*	*	*	*	*	*		*
Pseudodiaptomus charteri													*	*	*	*	*
Mysidacea																	
Gastrosaccus brevifissura		*		*	*	*	*	*	*	*	*				*	*	
Gastrosaccus gordonae														*	*	*	
Mesopodopsis africana										*	*			*	*	*	*
Mesopodopsis wooldridgei		*		*	*	*	*	*	*	*	*					*	
Rhopalophthalmus terranatalis		*	*	*	*	*	*	*	*	*	*					*	
Tenagomysis natalensis															*	*	

functional processes. The reader is also referred to an earlier review by Grindley (1981) who provides additional information on the zooplankton of South African estuaries.

Community structure

Species composition

The holozooplankton endemic to southern African estuaries is dominated by a few genera of copepods and mysid shrimps (Table 7.1). Few quantitative data are available on copepod assemblages endemic to west coast estuaries, although *Acartia longipatella*, *A. africana* and *Pseudodiaptomus hessei* are relatively common (Grindley 1981). In warmer east coast estuaries, *A. longipatella* and *P. hessei* are replaced by *A. natalensis* and *P. charteri* (Table 7.1). No data are available on their northward extension into Mozambique estuaries.

The most common copepod species recorded in south and east coast estuaries include *Acartia longipatella*, *A. natalensis*, *Oithona brevicornis*, *Pseudodiaptomus hessei* and *P. charteri* (Grindley 1981). Abundance of *A. natalensis*, *P. hessei* or *O. brevicornis* may on occasions exceed 100 000 per cubic metre of water (e.g. Wooldridge 1977; Wooldridge & Bailey 1982). These genera are all commonly recorded in estuarine waters around the world (e.g. Wellerhaus 1969; Lee & McAlice 1979; Greenwood 1981), whereas pseudodiaptomids are described as circumglobal (Walter 1986).

Although copepods dominate numerically, mysid-shrimps are often more important in terms of biomass in local estuaries. Mysid abundances are easily underestimated in research programmes, largely because of inadequate sampling techniques. In the Sundays River estuary, copepods contributed on average more than 95% to total abundance. Mysids, on the other hand, regularly exceeded 90% of total mesozooplankton dry biomass (Wooldridge & Bailey 1982). Mysids commonly recorded in the estuarine plankton include *Mesopodopsis wooldridgei* (formerly *M. slabberi*), *M. africana*, *Gastrosaccus brevifissura*, *G. gordonae* and *Rhopalophthalmus terranatalis* (Connell 1974; Grindley 1981; Wooldridge & Bailey 1982; Forbes 1989).

Meroplanktonic forms may be seasonally abundant, particularly the larval stages of benthic invertebrates. Species represented in the plankton include the anomuran mudprawn *Upogebia africana* and the crabs *Hymenosoma orbiculare*, *Sesarma* spp. and *Paratylodiplax edwardsii* (Pereyra-Lago 1988; Wooldridge 1994). Pelagic eggs and larvae of fish such as *Gilchristella aestuaria* are also common spring and summer (Wooldridge & Bailey 1982; Harrison & Whitfield 1990; Haigh & Whitfield 1993).

Bentho-pelagic taxa are also well represented in South African estuaries. Species that have strong links with the substratum include peracarid Crustacea such as the isopod *Cirolana fluviatilis*, amphipods of the genus *Grandidierella*, the cumacean *Iphinoe truncata* and the mysid *Gastrosaccus brevifissura* (Read & Whitfield 1989; Schlacher & Wooldridge 1994).

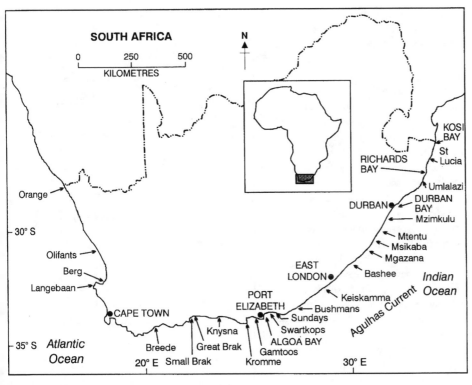

Figure 7.1. **Location of the South African estuaries mentioned in the text.**

The estuarine holoplankton

Cycles of zooplankton dynamics are often described as a function of temperature (e.g. Jeffries 1962; Bayly 1965; Knatz 1978; Lee & McAlice 1979; Greenwood 1981; Ambler *et al.* 1985; Mallin 1991), although river discharge may modify the effect of spring warming in regions having seasonal rainfall cycles. Day (1989) noted that while there are no clear seasonal patterns of zooplankton abundance in estuaries, some regional patterns have been described. In high latitude estuaries, clear seasonal cycles are usually evident. In tropical estuaries there is often a lack of a low seasonal population in winter, while other systems exhibit distinct summer minima for the macrozooplankton. In the temperate zone winter minima usually occur, although summer minima are also described (Day 1989). Miller (1983), in his review on the zooplankton of estuaries, noted that while there were a large number of papers on estuarine zooplankton, few studies focused on distribution where sampling and sampling evaluation adequately demonstrated relationships of species presence and abundance to distribution of physical factors.

The spatial and temporal distribution of the endemic mesozooplankton over time scales ranging from months to years has been investigated in few South African estuaries. Grindley (1981) documented earlier work, including an 8-year qualitative study on the zooplankton of the Swartkops estuary. More recent quantitative investigations (e.g. Wooldridge & Bailey 1982; Allanson & Read 1995) have illustrated the complexity of population response patterns to temporal and spatial variations in environmental gradients.

Estuarine copepod variability

Most of the research on the temporal and spatial dynamics of endemic populations in local estuaries has been undertaken in the Sundays, Gamtoos and Kromme River systems (Figure 7.1). Monthly sampling was always done at night (around new moon) using two slightly modified WP2 nets (57 cm diameter and 200 μm mesh). Surface and midwater-bottom samples were taken at a series of stations from a 4 m flat-bottomed boat. Nets were attached to lateral booms attached to the bow of the boat, with the depth of the lower net regulated by a worker using a vertical pole. Local estuaries seldom exceed 2–4 m depth on average, making it possible to regulate sampling depth. The Gamtoos and Kromme estuaries were sampled on consecutive days over 26 months; thereafter the Kromme study continued for a further 11 months. The Sundays was sampled independently (Wooldridge & Bailey 1982; Jerling & Wooldridge 1991), although the strategies were the same.

The estuaries of the Sundays and Gamtoos Rivers are similar in terms of physical characteristics. Along the coast, tides are microtidal and semi-diurnal. Tidal inlets are permanently open and the estuaries are about 21 and 20 km in length respectively. Flood-tidal deltas are well developed and there is an active dunefield along the western bank in the lower reaches of each estuary. In the Gamtoos estuary, a now inactive inlet channel previously opening to the sea 4 km to the east of the present inlet forms a shallow, lagoon (<1.5 m depth at low water of spring tide) behind a barrier dune and sandy beach. Estuarine tidal flats are poorly developed and their respective intertidal zones average <10 m in width. The estuaries are shallow, having average depths of 2–4 m at low tide. Although river catchment areas are affected by water storage reservoirs, freshwater inflow to the Sundays and Gamtoos estuaries is sufficient to maintain full axial salinity gradients during all seasons. No river gauging stations are located in lower catchment areas, but freshwater inflow to each estuary probably averages up to about 1 m³ s⁻¹ under normal flow conditions (see MacKay & Schumann 1990; Schumann & Pearce 1997).

In comparison to the Sundays and Gamtoos River estuaries, the permanently open Kromme River estuary has patches of intertidal saltmarshes along its 15 km length. The lower estuary is shallow (<2 m depth) and is characterised by extensive sandbanks along the first 5 km. Upstream water depth averages 3–4 m in channel areas. Freshwater inflow to the estuary is significantly attenuated as a result of a large storage reservoir 4 km above the tidal head of the estuary. A second reservoir is also located higher up in the catchment. Together, the combined storage capacity of the two dams (135 × 10⁶ m³) exceeds the Mean Annual Runoff (MAR) (105 × 10⁶ m³: Department of Water Affairs 1986) in the catchment basin. The Kromme River is therefore highly regulated and the estuary becomes marine dominated during low rainfall periods. In the late 1980s and early 1990s the Kromme River catchment area was affected by a severe drought and dam levels were below 30% of capacity (Jury & Levey 1993). Although an annual freshwater allocation of 2 × 10⁶ m³ from the lower dam was instituted to prevent hypersaline conditions developing in the estuary (evaporative demand), the volume of water released monthly during the drought was significantly attenuated before it reached the estuary. During the latter part of the drought, no freshwater was released. Consequently, river flow below the dam became erratic and the estuary received little or no freshwater, except for local runoff after heavy rains. During the 3-year zooplankton study in the Kromme, mean axial salinity range was only 2.9, reflecting marine dominance of the system at that time.

Pseudodiaptomus hessei, Acartia longipatella and

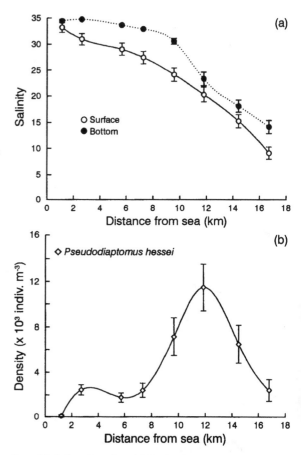

Figure 7.2. **Salinity (a) and spatial distribution of *Pseudodiaptomus hessei* (b) in the Gamtoos estuary. Data are based on monthly sampling over 27 months (see Figure 7.3) at eight stations (surface and near-bottom samples).**

A. natalensis represented the endemic copepod community in all three estuaries investigated. Species composition did not differ between the estuaries, although spatial and temporal structure of the populations showed striking differences. In the Gamtoos River estuary, *P. hessei* (Figure 7.2) attained maximum abundance in the middle and upper reaches ($n = 96$ samples) where surface salinity ranged between 10 and 25. The observed pattern of spatial distribution therefore corresponded to that described from other estuaries in the region having marked axial salinity gradients (Wooldridge & Bailey 1982; Jerling & Wooldridge 1991). Three relatively strong freshwater pulses occurred during the first year of the study in the Gamtoos River estuary, followed by four weaker pulses over the following 15 months. Local rains also reduced salinities in the middle and lower estuary in areas opposite small tributaries. Regression analysis of the Gamtoos data showed a positive relationship between *P. hessei* abundance and distance from the mouth ($p < 0.001$, $r^2 = 0.28$, $n = 200$) and a negative relationship between abundance and salinity (p < 0.05, $r^2 = -0.15$, $n = 200$). There was no relationship between abundance and water temperature. Mean *P. hessei* abundance for all stations was $4574 \pm 7244 \text{ m}^{-3}$ of water in summer (October–March, $n = 112$ samples) and $3929 \pm 6550 \text{ m}^{-3}$ in winter (April–September, $n = 96$).

Time series analysis using periodogram and integrated periodogram procedures on abundance showed significant cycles at the 95% level of 0.125 (8 months) and 0.25 (4 months), corroborating no seasonal cycle during the study period. Peaks of *P. hessei* abundance corresponded instead to the frequency of freshwater pulses flowing into the estuary over the same period (Figure 7.3). Cycles in the abundance of *P. hessei* juveniles followed the same pattern, as did the proportion of adult females in the population that were ovigerous. On occasions, more than 90% of adult females bore eggsacs. These observations on cycles of abundance obviously have no predictive value, but they do link peaks in *P. hessei* abundance with rainfall events rather than with seasonal temperature changes. Nonetheless, rainfall distribution in some years may fall in spring and early summer, leading to an apparent seasonal pattern of *P. hessei* abundance. This occurred in the Sundays River estuary (Jerling & Wooldridge 1991).

A weak axial salinity gradient characterized the Kromme River estuary (Figure 7.4a). Surface water salinity at the extreme limit of tidal influence seldom fell below 32, while hypersaline conditions developed during the first summer (Figure 7.5a). A pulse of freshwater in December 1989 reduced surface salinity to 23, while bottom salinity remained above 32.5. The pulse of freshwater inflow was of short duration, and salinity again exceeded 30 a few weeks later. Euhaline conditions persisted over the following 14 months, although local rains reduced surface water salinity in September 1990 to around 29 in the uppermost reaches of the estuary (Figure 7.5a).

Spatial distribution of *P. hessei* in the Kromme River estuary was relatively homogeneous, with abundance ranging between <100 and >100 000 m^{-3}. Lowest abundance occurred when the estuary was hypersaline (January–September 1992; Figure 7.5a and b). By contrast, peak abundance coincided with the freshwater pulse recorded in December 1989 when surface salinity declined to c. 23–29. The period of reduced salinity was of short duration (Figure 7.5a), as was *P. hessei* response. Abundance peaked at >108 000 m^{-3}, returning to a mean of <5000 individuals per cubic metre of water within a few weeks. Mean summer abundance was $4841 \pm 13 252 \text{ m}^{-3}$ (October–March, $n = 90$ samples) and $2420 \pm 3350 \text{ m}^{-3}$ in winter (April–September, $n = 90$ samples). Abundance showed a negative relationship with surface (p < 0.001, $r^2 = -0.46$, $n = 176$) and bottom (p < 0.001, $r^2 = -0.26$,

$n = 176$) salinity, but no significant relationship with distance from the mouth or water temperature. The lower r^2 recorded for bottom salinity was probably due to vertical stratification and higher salinity nearer the bottom during December 1989. Time series analysis showed no periodicity in the data set, confirming no seasonal trend during the study period.

In summary, distributional patterns of *P. hessei* in the Sundays River estuary (Wooldridge & Bailey 1982; Jerling & Wooldridge 1991) and in the Gamtoos estuary (Figures 7.2 and 7.3) indicate that population abundances are maximal in middle–upper estuarine areas where mesohaline conditions prevail. Under conditions of euhalinity (Kromme estuary, Figures 7.4 and 7.5), abundances remain low and populations are more homogeneously distributed in time and space. *P. hessei* responds positively to freshwater pulsing and it is the first mesozooplankton species to recolonise estuaries flushed by floods (Wooldridge & Bailey 1982). Temporal abundance patterns do not necessarily follow a seasonal cycle, but are instead linked to the frequency of freshwater pulses flowing into an estuary. On occasions, abundances are orders of magnitude higher in comparison to numbers recorded immediately before a freshwater pulse event.

Acartia longipatella and *A. natalensis* occur in different zones along estuarine salinity gradients (Wooldridge & Melville-Smith 1979). In the Swartkops estuary, the two species showed clear cycles of temporal succession. *A. natalensis* appeared in the plankton in spring and reached maximum abundance in mid to late summer in upper estuarine reaches. In autumn, it was replaced by *A. longipatella*. The latter species attained peak population density at the end of winter–early spring downstream of *A. natalensis* (Wooldridge & Melville-Smith 1979). Similar patterns of temporal succession are well described for other copepod taxa (see e.g. Jeffries 1962; Knatz 1978; Lee & McAlice 1979; Greenwood 1981; Ambler *et al.* 1985).

The spatial distribution of the two acartiids in the Gamtoos (Figure 7.6) and Kromme (Figure 7.7) river estuaries followed the same spatial pattern described for the Swartkops estuary. However, their temporal distribution was not well defined in the Gamtoos estuary (Figure 7.8). *Acartia natalensis* was only sporadically present in the Kromme estuary, attaining a maximum population density of only 1425 m^{-3} in March–April 1989. This is considerably lower when compared to their maximum abundance in the Gamtoos estuary (170 000 m^{-3}; Figure 7.8), or in the Swartkops estuary (43 250 m^{-3}: Wooldridge & Melville-Smith 1979). Maximum abundance of *A. longipatella* in the Kromme estuary (Figure 7.9) occurred in December 1989 (36 000 m^{-3}) when a strong freshwater pulse reduced surface salinity. Numbers remained low for the remainder of the study period.

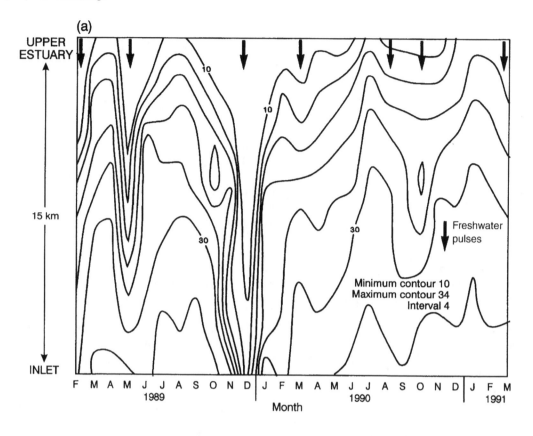

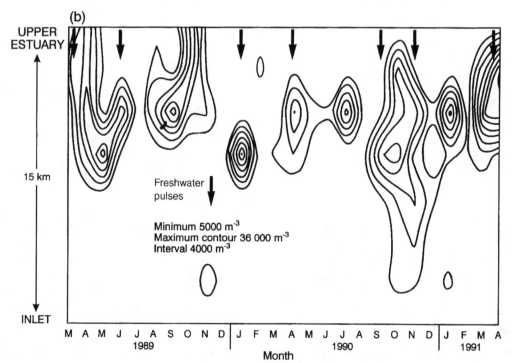

Figure 7.3. **Temporal variability in surface salinity (a) and** ***Pseudodiaptomus hessei*** **abundance (b) in the Gamtoos estuary.** Arrows along the upper margins indicate the intrusion of freshwater pulses into the estuary.

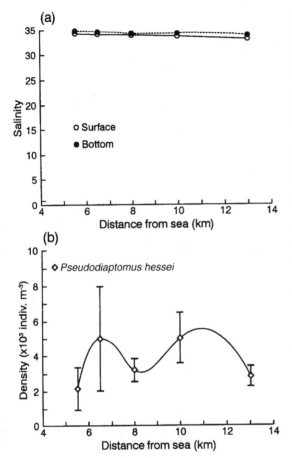

Figure 7.4. **Salinity (a) and spatial distribution of *Pseudodiaptomus hessei* (b) in the Kromme estuary. Data are based on monthly sampling over 36 months (see Figure 7.5) at five stations (surface and near-bottom samples).**

conditions develop during the refractory phase; thereafter eggs can hatch once conditions become favourable (competent phase). Quiescence is a state of delayed development induced by adverse environmental conditions. These eggs resume development immediately after conditions again become suitable. Despite their differences, the two types of resting eggs are functionally similar in that they introduce a lag phase into the life cycle (Marcus 1991).

Production of resting eggs probably occurs in *Acartia natalensis* and in *A. longipatella*. Acartiid eggs are present in muddy sediments of several estuaries in the Eastern Cape where studies have been undertaken, with egg abundance exceeding 14×10^6 m^{-2} in some areas (mean range 0.5–1.0 $\times 10^6$ m^{-2}; $n = 15$). Although large numbers of acartiid nauplii successfully hatch from eggs in controlled laboratory experiments, it has not been possible to rear larvae in order to identify species.

In *A. natalensis*, reappearance in the plankton is best explained by renewed development of dormant eggs in response to spring warming and mesohaline conditions. During prolonged periods of high salinity in the Kromme estuary, *A. natalensis* disappeared from the water column. This is probably due to eggs of this species remaining in a resting state. Consequently, even if water temperatures are suitable for the hatching of *A. natalensis* eggs, and salinity remains above a threshold level, arrested development persists until conditions become favourable. Although the salinity threshold level is not yet known, field data on reappearance in the plankton suggests that it is around 25–30.

In studies on winter acartiids, Ambler *et al.* (1985) concluded that the hatching of *A. clausi* eggs stopped at salinities below 30 and temperatures higher than at least 17.5 °C. Similarly, the production of resting eggs in the dominant winter–spring copepod in Narragansett Bay, *A. hudsonica*, is induced by temperatures greater than 16 °C (Sullivan & McManus 1986).

Although *Acartia longipatella* occurs in areas of relatively high salinity compared to *Pseudodiaptomus hessei* and *A. natalensis*, temporal peaks in abundance correspond to the periodicity of the freshwater pulses. The interactive effects of temperature and salinity as factors influencing temporal and spatial occurrence of *A. longipatella* in the plankton are not clear, although persistence in the plankton appears to become progressively more restricted to the cooler months as salinity decreases. Freshwater supply to local estuaries is therefore also important in regulating the spatial and temporal dynamics of the acartiid species.

Clearly, cycles of copepod abundance in local estuaries need not follow a regular seasonal pattern and it is difficult to make generalisations concerning causal factors. Although not specifically referring to copepods, Day (1989) noted that seasonal abundance of estuarine

The variability in the temporal occurrence and distribution of the two acartiids observed in the three estuaries is primarily determined by the interactive effects of temperature and salinity (Wooldridge & Melville Smith 1979). Other studies have shown that during periods of environmental adversity, eggs of some copepods do not hatch but sink to the bottom where they remain in or on the sediment. Resting eggs are known from at least 24 taxa (e.g. Grice & Marcus 1981; Marcus 1984, 1990, 1991; Uye 1985; Sullivan & McManus 1986; Ianora & Santella 1991), including eight species of *Acartia* (Uye 1985). Resting eggs therefore act as a pool of potential recruits for the planktonic population when favourable environmental conditions again prevail. Grice & Marcus (1981) distinguished two types of resting eggs: diapause eggs and quiescent subitaneous eggs.

Diapause is a genetic adaptive response that ensures long-term viability and synchronises a species' life cycle with the environment. In diapause eggs arrested development is mandatory and cannot resume even if favourable

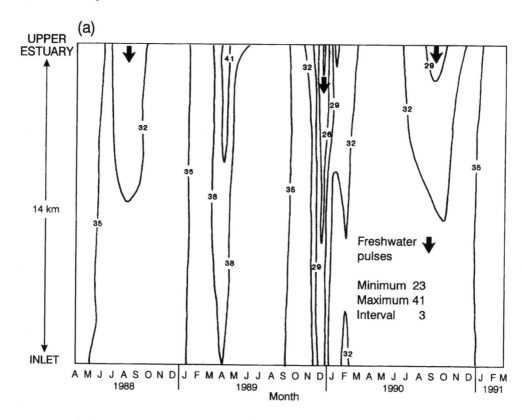

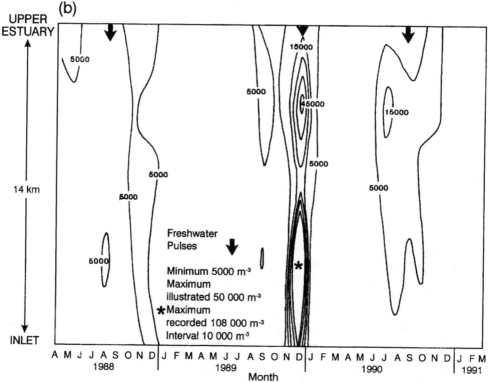

Figure 7.5. **Temporal variability in surface salinity (a) and** *Pseudodiaptomus hessei* abundance (b) in the Kromme estuary.

Arrows along the upper margins indicate the intrusion of freshwater pulses into the estuary.

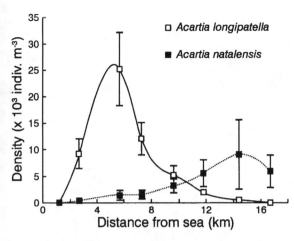

Figure 7.6. **Spatial distribution of *Acartia longipatella* (□) and *A. natalensis* (■) in the Gamtoos estuary. Data are based on monthly sampling over 27 months (see Figure 7.3) at eight stations (surface and near-bottom samples).**

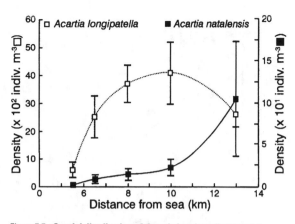

Figure 7.7. **Spatial distribution of *Acartia longipatella* (□) and *A. natalensis* (■) in the Kromme estuary. Data are based on monthly sampling over 36 months (see Figure 7.5) at five stations (surface and near-bottom samples).**

zooplankton is much more variable when compared to marine zooplankton. Whereas seasonal cycles of zooplankton abundance are distinct in high latitude estuaries, temporal variability in the temperate and tropical zones is often less precise owing to a range of factors (Day 1989).

Mysid shrimps

Mysids are abundant during summer in South African estuaries (Connell 1974; Wooldridge 1977; Wooldridge & Bailey 1982; Forbes 1989). Some of the dominant estuarine species also occur in marine nearshore waters. However, abundances of marine populations remain relatively low for species such as *Rhopalophthalmus terranatalis*, *Gastrosaccus brevifissura* and *G. gordonae*. By contrast, *Mesopodopsis wooldridgei* is numerically abundant in the estuary of the Sundays River (Wooldridge & Bailey 1982) and in the nearshore waters of Algoa Bay (Wooldridge 1983). In bay waters, *M. wooldridgei* undergoes diel rhythms of horizontal migration off sandy beaches (Webb & Wooldridge 1990), migrating shorewards to behind the breaker line (*c.* 5 m water depth) at sunset. Here, they exploit phytoplankton accumulations that outwell from the inner surf zone (Talbot & Bate 1987, 1988). At sunrise mysids return to deeper water (>15 m water depth). On occasions, dense swarms (up to 20000 individuals per cubic metre: Wooldridge 1983) pervade estuaries on a nocturnal flood tide, returning on the ebb. Besides these short-term incursions, discrete estuarine populations become well established.

The congeneric *Mesopodopsis slabberi* occurs in abundance in estuarine and coastal marine waters of northwestern Europe, flourishing over a salinity range of 1.3–43 (Tattersall & Tattersall 1951; Wittmann 1992). Whereas

M. wooldridgei undertakes diel shifts in horizontal distribution, Mauchline (1980) notes that populations of *M. slabberi* in the Black and Azov Seas appear to undergo seasonal onshore–offshore migrations, moving to shallow water (<20 m) in summer and down to *c.* 60 m in winter.

The mysid *Neomysis integer* is also recorded in marine, estuarine and brackish water environments in northwestern Europe (Tattersall & Tattersall 1951). The species has a wide salinity tolerance range (e.g. see Bobovich 1976) and is particularly common in estuarine environments. For example, *M. slabberi* and *N. integer* are the two most abundant species in the Tamar (Moffat & Jones 1992), Westerschelde (Mees *et al.* 1993) and Elbe estuaries (Köpche & Kausch 1996). In contrast to the nocturnal onshore migration observed for *M. wooldridgei*, *N. integer* migrates at night from the shallows to deeper offshore regions in the Darß–Zingster Bodden of the Baltic Sea (Debus *et al.* 1992). Mysids then return to the nearshore before sunrise.

Given their distribution in coastal waters in southern Africa, it is not unexpected to find that *Mesopodopsis wooldridgei*, *Rhopalophthalmus terranatalis* and *Gastrosaccus brevifissura* exhibit wide salinity tolerance ranges. The three species are all efficient osmoregulators, maintaining body fluid concentrations at species-specific levels over a general salinity range of 10–35 (Webb *et al.* 1997). In a study of *M. slabberi* in the Tamar estuary, Greenwood *et al.* (1989) and Moffat & Jones (1993) reported that juveniles occurred slightly downstream of adults. Adults comprised most of the population in areas where salinities were less than 10. Salinity tolerance experiments (Greenwood *et al.* 1989) further showed that adults tolerated a range of 3.5–35.0, while successful development of juveniles from eggs required a minimum salinity of 7.0. Thus salinity may

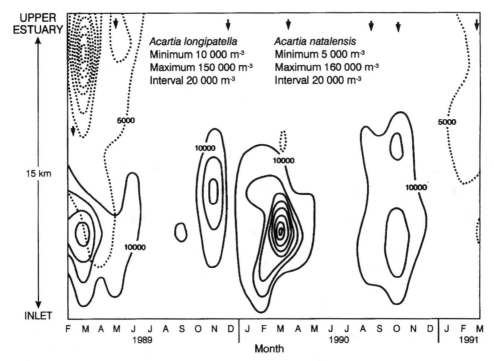

Figure 7.8. **Temporal abundance of *A. longipatella* (solid lines) and *A. natalensis* (hatched lines) in the Gamtoos estuary. Arrows along the upper margin indicate freshwater pulsing into the estuary (refer to** Figure 7.3). Minimum, maximum and interval of abundance (numbers m^{-3} of water sampled) shown in the figure. Sampling monthly, beginning February 1989 and terminating March 1991.

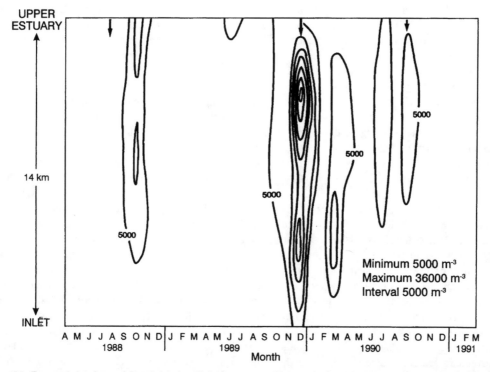

Figure 7.9. **Temporal abundance of *Acartia longipatella* in the Kromme estuary. Arrows along the upper margin indicate freshwater pulsing into the estuary (refer to Figure 7.5). Minimum, maximum** and interval of abundance (numbers m^{-3} of water sampled) shown in the figure. Sampling monthly, beginning April 1988 and terminating March 1991.

impose limits upon the range in which females may successfully rear young (Greenwood *et al.* 1989, Moffat & Jones 1993). The tolerance of *M. slabberi* to low salinity in general was also shown by Mees *et al.* (1995), who recorded this species in water having a salinity of close to zero in the Gironde estuary.

Contrary to their euryhalinity, *Mesopodopsis wooldridgei*, *Rhopalophthalmus terranatalis* and *Gastrosaccus brevifissura* occur in discrete but overlapping zones along estuarine axial gradients. Salinity *per se* does not appear to be the major factor regulating spatial zonation, although low salinity water (<10) may inhibit distribution into upper estuarine areas for some species. Other factors such as predation, substratum type and water depth appear to play an important role in determining mysid distribution. In the Sundays estuary, *R. terranatalis* consistently occurs in the relatively deep middle estuary (3–4 m) where salinity fluctuates between 10 and 28. Unlike juveniles, adults do not migrate to subsurface waters after dark (Wooldridge & Erasmus 1980), remaining below midwater where they are presumably less conspicuous to predators. In the Sundays estuary, population density may exceed 4000 individuals per cubic metre and adults may attain a body length of 2.5 cm.

Rhopalophthalmus terranatalis is omnivorous, with a greater portion of the diet in adults consisting of animal material (Wooldridge & Bailey 1982). Whereas young *R. terranatalis* feed on copepods, larger individuals feed mainly on juvenile *Mesopodopsis wooldridgei* (Wooldridge & Webb 1988). The high density of the predator in the Sundays estuary significantly reduces recruitment into the *M. wooldridgei* population and this is probably the major factor leading to the centre of the *M. wooldridgei* population occurring nearer the tidal inlet compared to *R. terranatalis*.

The bentho-pelagic mysid *Gastrosaccus brevifissura* is substrate specific (Webb *et al.* 1997), often dominating the estuarine mysid fauna in sandy areas (Schlacher & Wooldridge 1994). Consequently, the species occurs near tidal inlets (Wooldridge & Bailey 1982) and in upper estuarine reaches if the substrate is suitable. It is abundant in the upper Gamtoos estuary, but abundance decreases sharply in a seaward direction as mud replaces sand in middle estuarine reaches (Schlacher & Wooldridge 1994). Salinity in the upper Gamtoos regularly falls below 10 (Figure 7.3), corroborating the wide tolerance range to salinity found for this species (Webb *et al.* 1997). Similarly, distribution of the euryhaline *G. spinifer* in European estuaries is correlated with sediment characteristics (Tattersall & Tattersall 1951, Mees *et al.* 1993). *G. spinifer* is a burrowing species (Mauchline 1980) and was the most common mysid in grab samples (Mees *et al.* 1993) collected in the Delta area (SW Netherlands).

Distribution and rhythms of migratory behaviour

Axial displacement in estuaries due to net seaward flow poses a retention problem for the endemic zooplankton community. However, continuous survival of populations that are spatially zoned along axial gradients (see Figures 7.2 and 7.6) provides indirect evidence that organisms can invoke strategies that effectively regulate horizontal position. Among different strategies adopted, rhythmic and directed migration in response to the tidal phase provides a mechanism to promote retention in a particular area. Thus, vertical migration even over a distance of metres may bring about selective tidal transport. This may occur in vertically stratified systems or in well-mixed estuaries where current shear still exists because of bottom friction (Bosch & Taylor 1973; Wooldridge & Erasmus 1980; Orsi 1986; Kimmerer & McKinnon 1987; Hill 1991; Hough & Naylor 1992; Schlacher & Wooldridge 1995). Larger zooplankters such as mysids also migrate laterally to use upstream currents on the flood tide or to reduce downstream transport on the ebb (Wooldridge & Erasmus 1980; Hough & Naylor 1992; Moffat & Jones 1993; Schlacher & Wooldridge 1994). Mysid behavioural adaptations in response to tidal currents may differ between sexes and age classes within species, leading to different degrees of tidal displacement (Mauchline 1970; Siegfried *et al.* 1979; Orsi 1986; Wooldridge & Erasmus 1980; Hough & Naylor 1992; Moffat & Jones 1993; Schlacher & Wooldridge 1994).

In contrast to behavioural adaptations that aid retention of estuarine populations, other planktonic taxa adopt strategies that favour tidal transport out of the estuary (Forward 1987). For example, larval release in the mudprawn *Upogebia africana* coincides with high tide in the early evening (Wooldridge & Loubser 1996). First-stage larvae aggregate near the water surface where ebb current velocities are maximal.

Tidally-phased migratory behaviour is also evident among demersal zooplankton species in the Gamtoos estuary. In the riverine sector of the estuary (where salinity generally remains <10), sediment is mainly fluvial sand. Here, peracarid crustaceans (mainly amphipods, cumaceans and mysids) sometimes become extremely abundant (Schlacher & Wooldridge 1995). Nocturnal movements into the water column show complex inter- and intraspecific differences in spatial and temporal patterns that are tidally related. Some taxa, such as amphipods and cumaceans, restrict nocturnal planktonic activity mainly to the flood tide. This results in significant changes in planktonic community structure and diversity between ebb- and flood-tidal phases.

Besides tidally-phased migration, vertical migration in response to changes in light intensity also occurs widely among the zooplankton (see Cushing 1951; Clutter 1969; Mauchline 1980). Typically, migrants move towards the surface at night and back to deeper waters during the day, although other patterns occur. Numerous hypotheses have been advanced to account for the adaptive advantage of diel vertical migration, with much recent support for the predator avoidance hypothesis (Zaret & Suffern 1976; Alldredge & King 1980; Stich & Lampert 1981; Fancett & Kimmerer 1985; Gliwicz 1986; Bollens & Frost 1989; Bollens et al. 1992; Jerling & Wooldridge 1992).

The demersal copepod Pseudodiaptomus hessei shows strong rhythms of diel vertical migration (Grindley 1972; Hart & Allanson 1976). Distribution is strongly affected by moonlight, with instars showing variations in vertical pattern (Hart & Allanson 1976; Jerling & Wooldridge 1992). In the latter study, undertaken in a tidal estuary, late copepodid stages and adults showed significant differences in vertical distribution around full moon, with more animals in bottom waters. Differences in vertical distribution became progressively more significant with each successive instar and adults nearer the bottom often exceeded surface numbers by an order of magnitude. No difference in vertical distribution was apparent around new moon for the same age classes. The total number of adults present in the water column was also significantly lower on moonlit nights compared with dark nights. The strongest decrease in abundance was shown by ovigerous females. On average, adults were 15 times more abundant in the water column on dark nights (Jerling & Wooldridge 1992). These observations on the temporal appearance and vertical distribution of P. hessei in the plankton of tidal waters are therefore highly complex, regulated at least in part by diel, tidal and lunar cycles that may also act differentially on the various age classes.

Pseudodiaptomus hessei is heavily preyed upon by the estuarine clupeid Gilchristella aestuarius in the Sundays estuary (Wooldridge & Bailey 1982; Talbot & Baird 1985; Harrison & Whitfield 1990). Consequently, predator avoidance may be an important strategy influencing the observed migratory behaviour of P. hessei. Calculated subsurface light intensities on moonlit nights in the Sundays estuary were well above the minimum feeding threshold level found for another visually orientated clupeid (Blaxter 1964). On dark nights, light was below the required minimum reported by Blaxter in his study. In addition to general predator avoidance on moonlit nights, vertical distribution of P. hessei instars is in accordance with the hypothesis that larger size classes are more conspicuous to visual hunters nearer the water surface where light intensity is higher. This leads to a progressive increase in the degree of vertical migration during ontogenetic development.

Although no lunar effect was observed on pseudodiaptomids in a tidal channel in Westernport Bay (Fancett &

Table 7.2. Body dry mass (μg) of male and female *Pseudodiaptomus hessei* at winter (13.6 °C) and summer (25.7 °C) temperatures (from Jerling & Wooldridge 1991). Dry mass of an average egg pack was 5.25 μg, determined by difference between females without eggs and those with eggs. This value did not differ between summer and winter animals and is not included in female mass given below

	Body dry mass (μg)	
Sex class	Winter	Summer
Adult male	13.10	10.65
Adult female	29.60	21.60

Kimmerer 1985), ontogenetic differences in vertical distribution were similar to those described for P. hessei. Vertical migration was also most pronounced in ovigerous females while early copepodids did not migrate. In a laboratory study, predation was heaviest on the ovigerous females (Fancett & Kimmerer 1985), supporting previous conclusions (e.g. Bollens & Frost 1991) that many planktivorous fish exhibit a dietary preference for ovigerous over nonovigerous zooplankters. In Westernport Bay, predation was also greater on pseudodiaptomid adults than on those of the smaller Acartia tranteri. These authors (Fancett & Kimmerer 1985; Bollens & Frost 1991) concluded that the demersal behaviour of Pseudodiaptomus represents a method of mitigating visually directed predation.

Besides diel, tidal and lunar effects, water of reduced salinity also influences the degree of vertical migration in estuarine zooplankton (Lance 1962; Grindley 1964). This has great significance for endemic populations. Low salinity water inhibits migration of many species into surface layers during times of river floods when they would otherwise be flushed to sea (Grindley 1964). Unless completely flushed, a strong discontinuity layer develops. Survival and recovery rates of populations after flooding are enhanced by remaining below the fluvial surface layer.

Zooplankton production

Estimates of production are described for only a few species of zooplankton from South African estuaries. In the copepod Pseudodiaptomus hessei, reproduction is continuous throughout the year (Hart & Allanson 1975; Jerling & Wooldridge 1991). Winter animals attain a larger mass compared with their summer counterparts and females are also significantly heavier than males (Table 7.2).

Total development rates from egg extrusion to the adult stage also differed between sexes, with the shortest time of 14.3 days found for males at 26 °C (Table 7.3). In the Sundays River estuary (Jerling & Wooldridge 1991), production peaked in summer (47.45 mg dry mass m^{-3} d^{-1}) in the

Table 7.3. Total development time (days) for male and female *Pseudodiaptomus hessei* determined at different temperatures from laboratory experiments. Experimentai temperatures span the reported range recorded in the Sundays estuary. A salinity of 20 ± 1 was used and this represented values reported for the middle estuary where animals were collected (Jerling & Wooldridge 1991). Data are not available for the final stage during female development at 26 °C, so that total development time is an estimated value

	Development time (days)			
Sex class	at 16 °C	at 20 °C	at 23 °C	at 26 °C
Male	27.3	21.2	17.6	14.3
Female	27.9	21.8	18.0	(14.9)

upper estuary. Maximum integrated production for all seasons also occurred in the upper reaches (4.34 g dry mass m^{-3}) with an average value of 2.46 g dry mass m^{-3} for all ten stations sampled. Daily production to biomass ratios varied between 0.11 and 0.38, while annual P:B ratios ranged from 78.5 near the mouth to 100.2 in the upper estuary (Jerling & Wooldridge 1991). In the freshwater Lake Sibayi, Hart & Allanson (1975) calculated an annual mean daily P:B ratio of 0.11 (growth increment method) for the same species. A study by Grindley & Wooldridge (1974) on the zooplankton of Richards Bay, South Africa, gave a preliminary daily P:B value of 0.04 for *Pseudodiaptomus charteri*. Daily P:B ratios for pseudodiaptomids are also available for *P. marinus* from the inland sea of Japan, with reported values ranging between 0.032 and 0.242 (Uye *et al.* 1983). These daily P:B values are similar to those quoted for other endemic estuarine or bay-resident species quoted in Heip *et al.* (1995).

Production estimates for mysids recorded in South African estuaries are reported for *Rhopalophthalmus terranatalis*. In the Sundays River estuary three overlapping generations are produced per annum (Wooldridge 1986). Breeding ceases in winter, although in the warmer St Lucia estuary breeding continues throughout the year (Forbes 1989).

Individual females produce up to five broods in the Sundays estuary. Overwintering animals attain a greater body mass and have larger broods compared with their summer counterparts. An overwintering generation has a lifespan of 9–10 months compared to about 6 months for the spring and summer generations. Although summer generation females produce relatively small broods, population densities may exceed 4000 individuals per cubic metre of water compared with fewer than 100 animals per cubic metre in midwinter (Wooldridge 1986). Daily production values of 2.69, 6.93 and 13.97 mg dry mass per m^3 were calculated for the overwintering, spring and summer generations, respectively. Daily P:B coefficients varied between 0.020 and 0.026, while the annual P/B ratio integrated for the Sundays population

was 7.85. Estimates of production are available for few other mysid species, most of which inhabit cold freshwater systems (e.g. Miroshnichenko & Vovk 1973; Hakala 1978; Bremer & Vijverberg 1982; Sell 1982).

Three generations were observed for *Neomysis integer* in a shallow Frisian lake (Bremer & Vijverberg 1982), having maximum daily P:B ratios of 0.015, 0.038 and 0.050 for the overwintering, first summer and second summer generations respectively. The annual P:B ratio estimated by these authors was 4.0. An annual P:B ratio of 3.66 was calculated for *N. americana* from Long Island (Richards & Riley 1967); while for *Mysis relicta*, values varied between 2.2 and 3.8 (Hakala 1978; Sell 1982). Although annual P:B ratios were not determined for *N. mirabilis*, computations over relatively short time periods are higher than those described for *R. terranatalis* or *N. integer*. These varied between 0.13 and 0.17 per day (Shushkina 1973).

Recent estimates of mysid production are provided by Rudstam *et al.* (1986), who calculated the annual production of *Mysis mixta* in the northern Baltic to be 0.6 g C m^{-2} per year. In *Neomysis mercedis*, annual P:B ratios ranged from 9.4 to 13.4 between years in two British Columbia coastal lakes (Cooper *et al.* 1992). The annual ratio averaged over all years and basins was 11.0. In the Westerschelde estuary, Mees *et al.* (1994) found that *N. integer* produced three generations per year, each generation having an average cohort P:B of 3.0. The annual P:B ratio was 6.0. San Vicente and Sorbe (1995) provide information on annual P:B ratios for four species of *Schistomysis* in the Bay of Biscay. Annual P:B values ranged from 6.09 for *S. ornata* and 9.73 for *S. parkeri* (Table 5, San Vicente & Sorbe 1995). Fenton (1996*b*) calculated annual turnover ratios of three co-occurring coastal mysid species from Tasmania, using two different methods. Her data show an annual P:B ratio of 5.36 (size frequency method) and 5.49 (Petrovich method) for *Tenagomysis tasmaniae*, 7.54 and 7.73 for *Anisomysis mixta australis*, and 5.43 and 5.33 for *Paramesopodopsis rufa*, respectively.

Zooplankton trophic relationships

Estuarine copepods

In his review on the diets of calanoid copepods, Kleppel (1993) noted that there was overwhelming evidence to suggest that copepod diets are frequently diverse, composed of a variety of food types. Small phytoplankton and microzooplankton are also increasingly recognized as important components of copepod dietary intake (Kleppel 1993; Heip *et al.* 1995).

Diet diversity is clearly evident in feeding studies undertaken on two of the common euryhaline copepods

(*Pseudodiaptomus hessei* and *Acartia longipatella*) found in South African estuaries (Jerling 1993). The study was undertaken in the Sundays estuary using a direct counting technique that allowed for the determination of the type of food particles available to, and those utilized by the two species. Copepods were sorted into three functional groups: adults, copepodids (C1–C3), and nauplii (N3–N6). Water used in experiments was collected at night from specific estuarine localities and filtered through a 90 μm sieve. Primary production in the Sundays estuary is largely through the phytoplankton (Hilmer & Bate 1990), although the contribution of the microphytobenthos is not clear. Micro- and nanoflagellates, dinoflagellates and euglenoids dominated the phytoplankton, although diatoms were abundant following freshwater flushing of the estuary (Hilmer & Bate 1990; Jerling & Wooldridge 1995a). The microzooplankton consisted mostly of ciliates, zooflagellates and rotifers (Jerling & Wooldridge 1995a).

Results from the Sundays study (Jerling 1993) provided strong evidence for particle selection by adults, and to a lesser extent by copepodids, of *P. hessei*. No clear pattern was evident for other copepod instars. Clearance rates and electivity indices showed that this selection was based on particle size, rather than prey species or abundance. None of the copepod instars studied cleared the more abundant prey at a higher rate than species that were less abundant. There was no evidence for competition for any of the food resources present (Jerling 1993; Jerling & Wooldridge 1995a), and those phytoplankton species consumed regularly attain bloom status in the estuary (Hilmer & Bate 1990). In the Westerschelde estuary, *Acartia tonsa* was also shown to feed selectively on either live phytoplankton or microzooplankton when the organisms were present in blooms (Tackx *et al.* 1995).

At maximum, total consumption of phytoplankton in the Sundays estuary attained about 40% of daily phytoplankton production (H.L. Jerling, unpublished data). This conclusion therefore supports other studies in that herbivorous estuarine zooplankton is not generally considered to be food limited, consuming about 50%, and often much less, of phytoplankton present (Day 1989; Heip *et al.* 1995).

The investigation by Jerling (1993) excluded any possible contribution of carbon ingestion through consumption of detritus, although numerous studies have demonstrated the ability of endemic estuarine species to thrive on detrital particles and associated bacteria (Heip *et al.* 1995). This may have led to an underestimation of total carbon ingestion rates by Jerling (1993), unless the copepods selected strongly against detritus. However, data suggest that daily ingestion rates of phytoplankton and microzooplankton were sufficient to meet the daily energy requirements of adult *Pseudodiaptomus hessei*. Females of the congeneric *P. marinus* needed to ingest 35%

of body carbon per day for general metabolism (Uye & Kasahara 1983). During periods of egg production, females needed to consume 61% of body carbon per day. In Jerling's (1993) study, it was calculated that *P. hessei* consumed on average 46% of body carbon per day in the form of phyto- and microzooplankton. Highly negative values using $\delta^{13}C$ isotope ratios strongly support the conclusion that phytoplankton, rather than detritus derived from macrophytes, acts as the main carbon source for both *P. hessei* and *Acartia longipatella*, including their nauplii stages in the Sundays estuary (Jerling & Wooldridge 1995b).

In estuaries having a relatively low input of freshwater, macrophyte production dominates over phytoplankton with respect to their respective contributions to the organic carbon pool (Hilmer & Bate 1990, Allanson & Read 1995; Grange & Allanson 1995; Heymans & Baird 1995). Reduced freshwater inflow to the Kromme, linked to less disturbance of the sediment, relatively stable salinity and reduced turbidity were considered to have contributed to a four-fold increase over nine years in the biomass of *Zostera capensis* in the estuary (Adams & Talbot 1992). Much of this macrophyte production probably entered the food web as detritus, contributing to copepod diet in the estuary. However, average biomass of the dominant estuarine copepods (*P. hessei*, *A. longipatella* and *A. natalensis*) per unit water volume in the Kromme (Figures 7.5 and 7.7) was considerably lower when compared to that in phytoplankton-dominated systems such as the Sundays and Gamtoos (Figures 7.2 and 7.6). Similar correlations between copepods and chlorophyll-*a* concentrations are documented for other South African estuaries having contrasting inputs of freshwater (Allanson & Read 1995). For example, in the Great Fish estuary (freshwater rich), phytoplankton and zooplankton biomass levels were on average an order of magnitude greater when compared with biomass in the freshwater-starved Kariega estuary (Allanson & Read 1995).

Mysid shrimps

Mysids are generally considered to be omnivorous (see e.g. Tattersall & Tattersall 1951; Mauchline 1980; Fenton 1996a), although herbivory (Bowers & Grossnickle 1978) or carnivory (Cooper & Goldman 1980; Johnston & Lasenby 1982; Fulton 1982) also characterise some species in specific habitats. Thus far, investigations on mysid feeding in South Africa suggest that omnivory is the principal feeding mode, with partitioning of resources within and between species.

In an earlier study, Webb *et al.* (1987) investigated the feeding of *Mesopodopsis wooldridgei* on the surfzone diatom, *Anaulus australis* Drebes *et* Schultz along the Sundays River beach. During the day diatoms form visible accumulations inside the breaker line and adjacent to rip currents (Talbot *et al.* 1990). Cells are continually being added to and eroded from patches because of water currents. Cells eroded from patches may be entrained in major rips and transported behind the breakers where the lack of air bubbles (created by waves inside the breaker line) results in cells sinking and accumulating near the bottom of the water column. This accumulation of *Anaulus* and presumably detrital material is then exploited by *M. wooldridgei* at night (Webb *et al.* 1987). *Mesopodopsis* migrates shorewards from deeper water (15–20 m depth) after dark and concentrations exceeding 15 000 per m^3 of water have been recorded (Wooldridge 1983; Webb & Wooldridge 1990).

Estuarine populations of *M. wooldridgei* were also shown to feed extensively on phytoplankton (Jerling & Wooldridge 1995c). However, when microzooplankton abundance was relatively high, it formed a more important component in the diet compared to phytoplankton. In both the surf zone and the estuary, selectivity patterns are probably related to particle size (Webb *et al.* 1987; Jerling & Wooldridge 1995c), supporting the conclusion of Siegfried & Kopache (1980) for *Neomysis mercedis*. *M. wooldridgei* also fed on copepod instars under experimental conditions, the predation rate increasing linearly with increasing prey concentration (Jerling & Wooldridge 1995c).

Whereas *Mesopodopsis wooldridgei* feeds mainly on phyto- and microzooplankton, *Rhopalophthalmus terranatalis* preys more successfully on larger plankton such as adult copepods and juvenile *M. wooldridgei* (Wooldridge & Webb 1988). Microzooplankton probably constitutes the lower limit of the potential food particle size range consumed by *R. terranatalis* (Jerling & Wooldridge 1995c). A comparative study on the morphology of the feeding appendages of these two mysid species also supported differences in dietary composition and size of particles utilized (Jerling & Wooldridge 1994).

In predatory feeding experiments, newly-emerged *R. terranatalis* (<6 mm body length) fed successfully on the copepod *P. hessei* (Wooldridge & Webb 1988). Larger predators show an increasing preference for juvenile *M. wooldridgei*, even under conditions of high copepod density. The predator swims ventral side uppermost, using the periopods to form a large basket. Prey capture is effected by rapidly darting forward and trapping individuals as the periopods fold against the body. Larger *M. wooldridgei* usually succeed in escaping capture by rapidly darting away.

Other trophic interactions

In the upper Gamtoos estuary, large numbers of sparid juvenile fish occur (mainly *Lithognathus lithognathus* and *Rhabdosargus holubi*) (Schlacher & Wooldridge 1996a). These low-salinity areas probably contribute significantly to the nursery function of estuaries in general and it is important to take cognisance of Whitfield's findings, where densities of larvae and 0+ juveniles were significantly greater

in two estuaries receiving moderate to high riverine inputs, compared to a system where the axial salinity gradient was relatively uniform and closer to seawater (Whitfield 1994). In the Gamtoos estuary, both *L. lithognathus* and *R. holubi* strongly selected for corophioid amphipods, consuming other taxa in low numbers. The steenbras *L. lithognathus* had a narrower prey spectrum, feeding almost exclusively on the tube-dwelling amphipod *Grandidierella lignorum* (Schlacher & Wooldridge 1996a).

Grandidierella lignorum showed prominent behavioural differences between the sexes: males were markedly more active on the sediment surface compared with females, which rarely left their tubes during the day. Males presumably switched readily to an epifaunal mode of existence in search of receptive females, concurrently increasing their exposure to fish predators (Schlacher & Wooldridge 1996a). Consequently, *L. lithognathus* ingested significantly more male than female amphipods, causing a marked bias towards females in the sex ratio and age structure of the amphipod population. Juveniles were less preyed upon, presumably because of lower prey detection or capture efficiency by the predators. These data underline current notions about predation as an important structuring element for benthic communities, and also stress the prominence of size- and sex-selective predation in structuring individual prey populations (Schlacher & Wooldridge 1996a).

Qualitative observations in the upper Gamtoos estuary showed that the reed *Phragmites australis* is the most abundant macrophyte species, forming extensive beds along the fringes of the estuarine channel. Isotope analysis (Schlacher & Wooldridge 1996b) clearly showed that detritus ($\delta^{13}C$ −24.1 ± 0.3°/$_{oo}$) from estuarine sediments largely resemble macrophytes ($\delta^{13}C$ −25.7 ± 0.3°/$_{oo}$) in isotopic composition. This suggests that fringing vegetation is a main source of benthic detrital particles in the upper Gamtoos. However, only the epifaunal crab *Hymenosoma orbiculare* ($\delta^{13}C$ −23.8 ± 1.5°/$_{oo}$) appeared to use benthic detrital particles to any significant extent. Amphipods ($\delta^{13}C$ −28.0 ± 0.6°/$_{oo}$) utilised some benthic detritus, but fed mainly on fine suspended particulate organic matter ($\delta^{13}C$ −31.2 ± 0.5°/$_{oo}$), as did the anomuran *Callianassa kraussi* ($\delta^{13}C$ −32.5 ± 0.3°/$_{oo}$). These highly negative isotope values fall within the same range as those of phytoplankton grazing copepods ($\delta^{13}C$ −30.8 to −32°/$_{oo}$) in the Sundays estuary (Jerling & Wooldridge 1995b). This leads to the conclusion that in the upper Gamtoos estuary, dominant macrobenthic organisms are probably mainly suspension feeders (Schlacher & Wooldridge 1996b). In this respect, published data about the exact trophic status of any of the consumer species studied above are ambiguous. *C. kraussi* is sometimes classified as a deposit feeder (e.g. Day 1981; Branch *et al.* 1994), but it is also regarded as a filter feeder (e.g. Branch & Branch 1981; Day 1981). If deposit feeding were a prevailing mode of foraging in *C. kraussi*, it does not appear to be on benthic detritus below 1 mm particle size (Schlacher & Wooldridge 1996b).

These studies on trophic relationships again suggest the prominence of phytoplankton in driving estuarine primary production in fresh-water rich systems, even in estuaries that harbour large stands of macrophytes. This is also supported by findings in the Swartkops estuary which has large areas of saltmarsh: in this estuary phytoplankton was identified as the single most productive autotroph compartment (46%) (Hilmer *et al.* 1988).

The estuarine meroplankton

At times, the meroplankton constitutes a major component of the estuarine planktonic community. Invertebrate larvae usually dominate this group, although in local estuaries identification is difficult and most larval stages remain undescribed. Attention has recently focused on the larval life-history strategies of benthic crustaceans, including the mudprawn *Upogebia africana* (Wooldridge & Loubser 1996). Mudprawn larvae require a marine phase of development, with first-stage larvae migrating from the estuary on evening ebb tides. Three intermediate stages undergo development at sea before postlarvae return to estuarine mudflats to settle.

Estuarine–marine larval exchange patterns are now well documented for littoral and supralittoral decapods, particularly brachyurans (Sandifer 1975; Christy & Stancyk 1982; Johnson & Gonor 1982; DeCoursey 1983; Epifanio 1988; McConaugha 1988; Dittel *et al.* 1991; Queiroga *et al.* 1994). Larval release patterns are rhythmic, usually in relation to tidal, diel and lunar cycles. Rhythms relating to lunar phase are mostly semi-lunar, with releases mostly linked to the time of spring tides (Forward 1987). If tidal and diel rhythms are also important, then most commonly larvae are released at the new and full moon, in the first few hours after sunset and around the time of high tide (Forward 1987).

The timing of larval reinvasion of estuaries is poorly documented compared with the export phase. Most studies deal with brachyurans. Evidence suggests that movement into and up estuaries is by selective tidal transport, usually at night around spring tides (Epifanio *et al.* 1984; Dittel & Epifanio 1990; Little & Epifanio 1991). In other studies no clear difference in megalopal abundance between spring and neap tide phases was evident (De Vries *et al.* 1994; Queiroga *et al.* 1994).

Upogebia africana is common in muddy substrata of local estuaries where it often dominates the intertidal macrobenthos, particularly at lower tidal levels along channel margins. In the Swartkops estuary, it accounts for >80% of total macrofaunal biomass on the mudflats (Hanekom *et al.* 1988). However, it is absent from estuaries that have been closed off from the sea for 'extended periods' (Day 1981). Peak breeding season occurs around midsummer and first-stage larvae are released into the plankton at night and on the ebb tide (Figure 7.10a).

Maximum release activity and export of larvae to the marine environment follow a semi-lunar cycle (Wooldridge & Loubser 1996) synchronised to the time when high water in the estuary is crepuscular (Figure 7.11a). This occurs after peak spring tidal amplitude. Thus, timing of peak larval release does not maximise for seaward transport on the ebb tide. This is offset, at least in part, by rapid larval release and subsequent larval behaviour. Appearance of Stage 1 in the plankton is abrupt, with peak numbers present at the beginning of the ebb tide (Figure 7.10a). An independent study undertaken in the Swartvlei estuary, using ebb tide water volumes computed hourly (Huizinga 1987) and larval abundance (incorporating data from Figure 7.10a), showed that more than 90% of larvae left the estuary during the initial 3 hours of the protracted (7 h) ebb tide. Export from the estuary is further promoted by near-surface aggregation of larvae where water velocities are maximal. The relative distribution of larvae in the water column is consistent, with significantly more larvae nearer the surface compared to deeper waters (Wilcoxin sign test, $Z = 7.212$, $n = 54$).

Estuarine reinvasion by postlarvae is also nocturnal (Figure 7.10a), and maximum return occurs after neaps when low water at sea is crepuscular (Figure 7.11b). Plankton sampling in water c. 1 m deep indicated that large numbers of postlarvae aggregated in the surf zone at night (Fig 7.11b). Despite greater tidal volumes associated with spring tides, significantly more postlarvae return around neaps (Wooldridge & Loubser 1996). Rhythmic cycles of larval export and postlarval estuarine reinvasion are therefore asynchronous and are best explained by the timing of the change in light intensity. If maximum activity rhythms of Stage 1 and postlarvae are independent of tidal amplitude, then timing of maximum release and reinvasion during the lunar cycle would alter as the time of sunset shifts between solstices. At the beginning of the breeding season in winter, maximum release of larvae would occur nearer spring tides (local sunset occurs around 1730 h at the end of July) compared with summer when local sunset occurs around 1930 h (see Figure 7.11a). Similarly, maximum postlarval reinvasion would shift nearer neaps in winter compared to the December solstice. This would be necessary to maintain synchrony between crepuscular high and low water respectively.

Although not strictly part of the estuarine meroplankton, marine invertebrates also migrate into estuaries

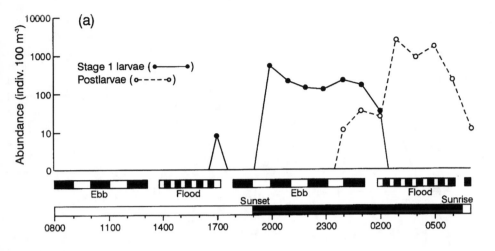

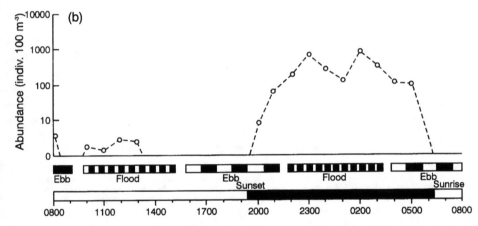

Figure 7.10. **Diel and tidal rhythms of *Upogebia africana* larval export (stage 1) and postlarval return to an estuary (a). Large numbers of postlarvae aggregate in the surfzone at night before eventual** entrainment through the inlet on the flood tide (b). These data collected off the beach close to Swartvlei estuary mouth.

during their early life-history stages. Examples are penaeid prawns and the carid shrimp *Palaemon peringueyi*. In southern Africa, breeding of *P. peringueyi* occurs in marine coastal waters, followed by onshore movement of Stage 6 and postlarvae into tidal pools or estuarine *Zostera capensis* beds. Juveniles spend 9–12 months in the latter habitat, before returning to the marine environment as subadults (Emmerson 1986). In an extensive study of three eastern Cape estuaries, shrimp abundance peaked from November to March (Emmerson 1986). Estuarine recruitment of *P. peringueyi* takes place on both diurnal and nocturnal flood tides, but the highest numbers migrate after dark (Wooldridge 1991). In comparison, emigration of subadults back to the marine environment only takes place at night on the ebb tide.

Abundance of *P. peringueyi* in estuarine *Z. capensis* beds may reach 400 m^{-2}, and it is described as the single most important vagile invertebrate associated with seagrasses in some estuaries (Emmerson 1986). In the Kromme

estuary, *P. peringueyi* is the most important contributor to macroinvertebrate production, followed by *Sesarma catenata* and *Upogebia africana* (Winter & Baird 1988).

Although *P. peringueyi* is not dependent on estuaries to complete its life cycle, large numbers may migrate into these environments where they potentially play an important part in the estuarine food web. For example, more than 2×10^6 late-stage larvae migrated through the Swartvlei estuary mouth over a spring flood tide (Wooldridge 1991).

Estuarine zooplankton response to reducing inputs of freshwater to estuaries

Freshwater is one of South Africa's most limited natural resources. Recent projections suggest that full utilisation of the available supply (without desalination or recycling)

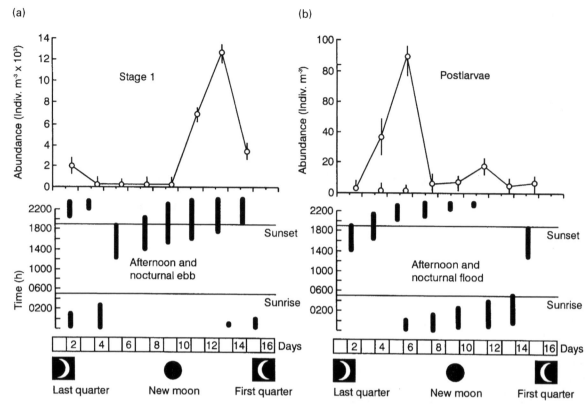

Figure 7.11. **Stage 1 larval export (a) and postlarval import (b) of** *Upogebia africana* **in the Gamtoos estuary. Sampling undertaken on alternate nights beginning at 1700 h and terminating at 0800 h the following morning. Four replicate samples collected every 1.5 h.**

Data points indicate mean abundance (±1 SD) over the ebb (a) or flood (b) phase for each sampling date. These data points comprise between 20 and 28 samples, depending on the duration of the flood or ebb tide. (After Wooldridge & Loubser 1996.)

will occur early in the new century (Davies *et al.* 1993). Steeply increasing demands by a rapidly growing population on this limited commodity have already resulted in a severe reduction of water supplies to natural users such as estuaries. Now, only about 8% of the Mean Annual Runoff (MAR) in the country reaches the coastal zone (Department of Water Affairs 1986). The severity of this is exemplified in the following statement by Davies *et al.* (1993):

> There are few rivers in southern Africa that have not been over-exploited, degraded, polluted, or regulated by impoundments, and we know of many that were once perennial, but which now flow only seasonally or intermittently.

Although freshwater input to estuaries is naturally variable, human activities in river catchment areas represent the main cause for historically recent changes in freshwater quality and temporal delivery patterns to coastal environments. In several catchment basins, the storage capacity of reservoirs exceeds the average runoff of

their rivers. This trend of anthropogenic-induced change in estuaries can only intensify in the future.

Estuarine freshwater dependence is *inter alia* expressed in the following components (from Reddering 1988; Whitfield & Bruton 1989; Schlacher & Wooldridge 1996c).

- Regulation of estuarine sediment characteristics.
- Flooding events that scour accumulated sediments.
- Riverine nutrient inputs necessary for primary production.
- Axial salinity gradients that increase habitat and species diversity.
- Maintenance of open tidal inlets that prevent salinity and temperature extremes.
- Maintenance of open tidal inlets necessary for larval exchange, fish migrations, and tidal flushing of saltmarshes.

In the following discussion, the effects of artificially changing salinity gradients on the estuarine zooplankton and the importance of maintaining open tidal inlets for invertebrate larval exchange are discussed.

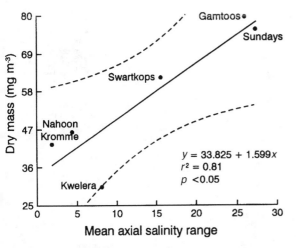

Figure 7.12. **Relationship between copepod biomass and axial salinity range in estuaries along the southeastern Cape coast of South Africa. Each data point represents mean biomass values for those estuaries having at least medium term data sets. Data represent about 1850 samples collected over a total of 20 years from the Kromme, Gamtoos, Swartkops, Sundays, Nahoon and Kwelera estuaries. Note the two estuaries representing extreme values.**

Effects of changing axial salinity gradients

Although a host of abiotic and biotic factors control primary productivity in estuaries, a positive correlation between phytoplankton biomass and the magnitude of freshwater inflow is a recurring one in many systems (e.g. Malone *et al.* 1988; Mallin *et al.* 1993, and references therein). These patterns are also evident in South African estuaries, with freshwater pulses transporting significant amounts of inorganic nutrients to estuarine waters (e.g. Allanson & Read 1995), thereby enhancing phytoplankton growth (Table 5.2, this volume). In estuaries along the south Eastern Cape coast, phytoplankton biomass is positively related to the mean axial salinity gradient regulated through river flow entering a particular system (Figure 4, Schlacher & Wooldridge 1996c). The same pattern holds for pelagic consumers that include copepods (Figure 7.12) endemic to estuaries (refer also to Allanson & Read 1995) and fish (Figure 4, Schlacher & Wooldridge 1996c). Phytoplankton, endemic copepods and fish attained significantly higher biomass in estuaries having pronounced axial salinity gradients compared to estuaries having weak axial salinity gradients.

In individual estuaries, freshwater pulses were associated with rapid and sharp increases in copepod numbers (see e.g. Figures 7.2 and 7.5 for the response of *Pseudodiaptomus hessei* in the Gamtoos and Kromme estuaries). The Kromme and Gamtoos River estuaries contrasted in their freshwater supply patterns, with the former having a major dam located 4 km above the estuary. When freshwater inflow was severely curtailed (Figure 7.4), the normally oligo- to mixo-haline reaches became uniformly

euhaline. During periods of drought, evaporation in the upper reaches exceeded riverine inputs and hypersaline (>35) conditions developed (Figure 7.4). In addition to *P. hessei*, *Acartia natalensis* disappeared from the plankton when eggs entered a state of diapause when salinity exceeded 20–25.

The investigations carried out in the Gamtoos and Kromme River estuaries indicate that copepod distribution and abundance patterns were primarily driven by rainfall events within years, rather than by seasonal changes in water temperature. Freshwater attenuation (reservoir retention) and modification to river flow patterns therefore deprive local estuaries of a key mechanism that regulates temporal heterogeneity of copepod distribution and abundance. These effects will predictably have negative consequences for higher trophic levels dependent on plankton in the water column as a food source.

Axial salinity gradients also influence the magnitude of fish recruitment from marine to estuarine waters. Densities of larvae and 0+ juveniles were significantly greater in two estuaries receiving moderate to high riverine inputs compared to a system where the axial salinity gradient was relatively uniform and closer to seawater. Moreover, diversity of ichthyoplankton was markedly lower when axial salinity gradients were weakly developed and the estuarine water column differed little from adjacent marine waters in salt content and turbidity (Whitfield 1994).

Tidal inlet management and larval exchange

Estuarine–marine exchange processes are directly affected by the increasing attenuation of freshwater in catchment areas. The significance of this statement is better appreciated when considering that only 37 (13%) of 289 river mouths between the Orange River and Kosi Bay maintain permanently open tidal inlets (Reddering & Rust 1990). Most of the remaining 87% of rivers enter the sea via estuaries having tidal inlets that close periodically owing to sandbar development. These unconsolidated marine sediments accumulate naturally in lower estuarine regions and are removed or scoured during river flooding.

A major consequence of river impoundments is the reduced frequency and amplitude of flood events (Reddering & Esterhuysen 1987; Whitfield & Bruton 1989). Estuaries accumulate excessive volumes of marine sediment and inlets become severely choked. Athough opening and closing of tidal inlets is a natural occurrence, many estuaries are closing more frequently and for longer periods owing to less effective scour of inlet channels during floods. As the frequency of effective river flooding decreases, shoaling also takes place over longer periods, requiring a flood of greater erosive capacity to remove the increased volume of sediment (Reddering & Esterhuysen

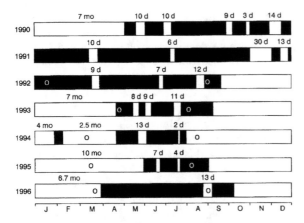

Figure 7.13. **The open (□) and closed (■) phases of the Great Brak tidal inlet, January 1990–August 1996. Data from Huizinga (personal communication). Circles indicate sampling dates.**

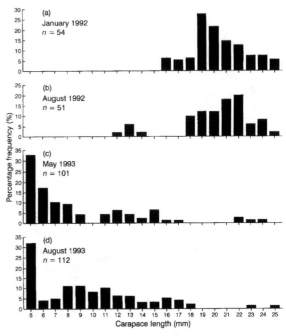

Figure 7.14. **Size class distribution of *Upogebia africana* in the Great Brak estuary on the first four sampling occasions (indicated by circles shown in Figure 7.13). The inlet remained open each summer after the dam filled in September 1992. This is reflected in recruitment of juveniles during the summer of 1992–3 (c) and thereafter, in subsequent summers (not illustrated). Figs. b, c and d clearly reflect growth of cohorts during each successive trip. Low sample size reflects low prawn density and difficulty in collecting animals (see Figure 7.15).**

1987; Reddering 1988; Whitfield & Bruton 1989). Ultimately, the estuary may close off from the sea completely.

The opening and closing of tidal inlets has direct implications for, *inter alia*, estuarine organisms that require a marine phase of development during their respective life cycles. Besides *Upogebia africana*, species identified as having planktonic larvae that require a marine phase of development include the saltmarsh crabs *Sesarma catenata* and *Paratylodiplax edwardsii* (T.W., unpublished data). In *Upogebia* the marine development phase is obligatory. *U. africana* larvae do not metamorphose if trapped in estuaries and recruitment ceases (Wooldridge 1994). Thus, mudprawn populations are directly affected by tidal inlet dynamics and their continued existence in temporary closed estuaries is dependent on the relationship between open and closed phases of the inlet. In extreme cases populations become locally extinct if inlets remain closed for extended periods.

The consequences of freshwater attenuation on tidal inlet dynamics and the subsequent impact on estuarine mudprawn populations is well illustrated in the Great Brak estuary. The completion in 1990 of the Wolwedans Dam (maximum capacity 24×10^6 m³) 5 km above the Great Brak estuary seriously aggravated freshwater supply and the tidal inlet opened infrequently over the following 28 months as the dam filled (Figure 7.13). Recruitment ceased during periods of mouth closure and this is reflected by the discontinuity of size classes in the Great Brak population (Figure 7.14a, b, c). Integration of the growth rate (Hanekom & Baird 1992) with the lower size limits of cohorts present in the estuary (Figure 7.14) provided information about when recruitment last took place. This occurred only when the inlet opened to the sea for periods exceeding one month. The age of the smallest size classes of *U. africana* sampled in January 1992 (Figure 7.14a) was about 17–21 months (carapace length

16–18 mm: Hanekom & Baird 1992). Recruitment therefore last took place in April 1990 after the inlet had remained open for seven months. Subsequent growth of prawns is reflected in Figure 7.14b (sampled late August 1992, Figure 7.13). This figure also shows that a small number of prawns colonised the estuary in November 1991 when the inlet remained open for 30 days. On the third and fourth sampling occasions (May and August 1993, Figure 7.14c and d respectively), relatively large numbers of prawns had recruited into the population. Thereafter, the estuary opened each summer (Fig 7.13) during which time successful recruitment took place. The population recovery rate is relatively slow and prawn density (expressed as number of holes per 0.25 m²) recovered to near predicted maximum only after four years (Figure 7.15). These data are compared with those in the adjacent Little Brak system which has remained permanently open at least since 1991.

The extent of larval dispersion away from parent estuaries is also an important aspect in the recruitment process. No data are presently available on marine dispersion patterns, but it is hypothesised that larvae remain in the general locality of parent estuaries and do not disperse widely in marine nearshore waters (Wooldridge 1994).

This is in contrast to the pattern described for other

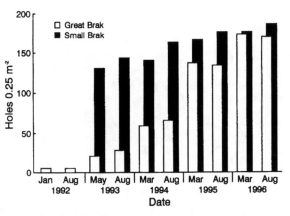

Figure 7.15. **Prawn density (expressed as the number of holes per 0.25 m²) in the Great Brak estuary compared with density in the Little Brak system (counts not conducted in 1992 in the latter estuary). No recruitment occurs in winter (non-breeding period) and this is reflected in similar hole counts at the end of summer and the following spring. Data are the mean of 15 random counts.**

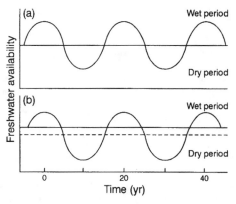

Figure 7.16. **Schematic representation of the cyclic wet and dry periods experienced by summer rainfall regions in South Africa (adapted from Figure 14, Davies *et al.* 1993). Reduced freshwater supply to estuaries in these areas will, from a practical point of view, result in estuaries experiencing drought conditions for longer periods (b) relative to the natural cycle (a). During extreme droughts, freshwater inflow may be insufficient to offset evaporative losses and prolonged hypersalinity may eventually become lethal to the estuarine biota. Although fluctuations in salinity occur naturally between wet and dry periods, average salinity values are being artificially shifted and maintained at higher levels for prolonged periods through excessive freshwater abstraction (e.g. Kromme estuary).**

estuarine organisms that have a marine phase of development (e.g. *Callinectes sapidus*: Little & Epifanio 1991). This implies that, as longshore distance between estuaries in southern Africa increases there is potentially a corresponding decline in the success of larval exchange between estuaries over a given period. Ultimately, recolonisation of an isolated estuary may only occur after a protracted period (years) should populations of *U. africana* become locally extinct. Although subjective, development of larvae in the general vicinity of parent estuaries is supported by recruitment patterns in the Great and Little Brak estuaries. Little or no recruitment occurred at times when the tidal inlet in the Great Brak opened for less than 30 days, while recruitment in the nearby Little Brak (9 km distant) was relatively high and uninterrupted over the same period. High numbers of newly-settled prawns (carapace length 4–5 mm) were present on all sampling occasions in the latter estuary (Figure 7.15). The recovery curve in the Great Brak also follows an exponential pattern. This lends support to the hypothesis that larvae develop in the general vicinity of parent estuaries: the higher the density of the breeding population in the estuary, the greater the potential for recruitment to the population.

General conclusions

As a result of anthropogenic perturbations, the heterogeneity of estuarine types in South Africa is artificially changing. Individual systems are becoming less dynamic

and there is a shift towards lower zooplankton biomass as freshwater inflow decreases. This is probably in response to reduced phytoplankton production. Along the southeast coast (no data are available for other regions), numerical abundance of *Pseudodiaptomus hessei* and biomass of estuarine endemic copepods in general are driven primarily by rainfall events within years, rather than by seasonal changes in water temperature. In the longer term, average copepod abundance between years would predictably follow interannual variations in rainfall (dry–wet cycles). These dry–wet cycles have a wavelength of about 20 years in summer rainfall areas (Davies *et al.* 1993). Consequently, estuaries will experience (from a practical point of view) a proportionally longer phase in the dry period due to freshwater attenuation (Figure 7.16). This will artificially shift more estuaries towards marine dominance and more stable salinity regimes that favour macrophytes such as *Zostera capensis* (Adams & Talbot 1992) rather than phytoplankton. Similarly, modification to river flood patterns deprives local estuaries of a key mechanism that regulates natural opening and closing of tidal inlets. If estuaries remain closed for extended periods, at least one important benthic species can potentially become locally extinct.

References

Adams, J.B. & Talbot, M.M.B. (1992). The influence of river impoundment on the estuarine seagrass *Zostera capensis* Setchell. *Botanica Marina*, **35**, 69–75.

Allanson, B.R. & Read, G.H.L. (1995). Further comment on the response of Eastern Cape Province estuaries to variable freshwater inflows. *Southern African Journal of Aquatic Sciences*, **21**, 56–70.

Alldredge, A.L. & King, J.M. (1980). Effects of moonlight on the vertical migration patterns of demersal zooplankton. *Journal of Experimental Marine Biology and Ecology*, **44**, 133–56.

Ambler, J.W., Cloern, J.E. & Hutchinson, A. (1985). Seasonal cycles of zooplankton from San Francisco Bay. *Hydrobiologia*, **129**, 177–97.

Bayly, I.A.E. (1965). Ecological studies on the planktonic Copepoda of the Brisbane River estuary, with special reference to *Gladioferens pectinatus* (Brady) (Calanoida). *Australian Journal of Marine and Freshwater Research*, **16**, 315–50.

Blaxter, J.H.S. (1964). Spectral sensitivity of the herring *Clupea harengus* L. *Journal of Experimental Biology*, **41**, 155–62.

Bobovich, M.A. (1976). Osmotic regulation in the brackish-water mysid *Neomysis integer* (Leach). *Soviet Journal of Ecology*, **7**, 368–70.

Bollens, S.M. & Frost, B.W. (1989). Predator-induced diel vertical migration in a planktonic copepod. *Journal of Plankton Research*, **11**, 1047–65.

Bollens, S.M. & Frost, B.W. (1991). Ovigerity, selective predation, and variable diel migration in *Euchaeta elongata* (Copepoda: Calanoida). *Oecologia*, **87**, 155–61.

Bollens, S.M., Frost, B.W., Thoreson, D.S. & Watts, S.J. (1992). Diel vertical migration in zooplankton: field evidence in support of the predator avoidance hypothesis. *Hydrobiologia*, **234**, 33–9.

Bosch, H.F. & Taylor, W.R. (1973). Diurnal vertical migration of an estuarine cladoceran, *Podon polyphemoides*, in the Chesapeake Bay. *Marine Biology*, **19**, 172–81.

Bowers, J.A. & Grossnickle, N.E. (1978). The herbivorous habits of *Mysis relicta* in Lake Michigan. *Limnology and Oceanography*, **23**, 767–76.

Branch, G. & Branch, M. (eds.) (1981). The living shores of southern Africa. Cape Town: C. Struik.

Branch, G.M., Griffiths, C.L., Branch, M.L. & Beckley, L.E. (eds.) (1994). *Two oceans*. Cape Town: David Philip Publishers.

Bremer, P. & Vijverberg, J. (1982). Production, population biology and diet of *Neomysis integer* (Leach) in a shallow Frisian Lake (The Netherlands). *Hydrobiologia*, **93**, 41–51.

Christy, J.H. & Stancyk, S.E. (1982). Timing of larval production and flux of invertebrate larvae in a well mixed estuary. In *Estuarine comparisons*, ed. V. Kennedy, pp. 489–503. New York: Academic Press.

Clutter, R.I. (1969). The microdistribution and social behaviour of some pelagic mysid shrimps. *Journal of Experimental Marine Biology and Ecology*, **3**, 125–55.

Connell, A.D. (1974). Mysidacea of the Mtentu River estuary, Transkei, South Africa. *Zoologica Africana*, **9**, 147–59.

Cooper, K.L., Hyatt, K.D. & Rankin, D.P. (1992). Life history and production of *Neomysis mercedis* in two British Columbia coastal lakes. *Hydrobiologia*, **230**, 9–30.

Cooper, S.D. & Goldman, C.R. (1980). Opossum shrimp (*Mysis relicta*) predation on zooplankton. *Canadian Journal of Fisheries and Aquatic Sciences*, **37**, 909–19.

Cronin, L.E., Daiber, J.C. & Hulbert, E.M. (1962). Quantitative seasonal aspects of zooplankton in the Delaware River estuary. *Chesapeake Science*, **3**, 63–93.

Cushing, D.H. (1951). The vertical migration of planktonic Crustacea. *Biological Reviews*, **26**, 158–92.

Davies, B.R., O'Keeffe, J.H. & Snaddon, C.D. (eds.) (1993). *A synthesis of the ecological functioning, conservation and management of South African River ecosystems*. WRC Report TT 62/93. Pretoria: Water Research Commission.

Day, J.H. (1981). The estuarine fauna. In *Estuarine ecology with particular reference to Southern Africa*, ed. J.H. Day, pp. 147–78. Cape Town: A.A. Balkema.

Day, J.W. (1989). Zooplankton, the drifting consumers. In *Estuarine ecology*, ed. J.W. Day Jr, C.A.S. Hall, W.M. Kemp & A. Yanez-Arancibia. New York: John Wiley.

Debus, L., Mehner, T. & Thiel, R. (1992). Spatial and diel patterns of migration for *Neomysis integer*. In *Taxonomy, biology and ecology of Baltic mysids*, ed. J. Köhn, M.B. Jones & A. Moffat, pp. 79–82. Rostock University Press.

DeCoursey, P.J. (1983). Biological timing. In *The biology of Crustacea*, Vol. 7, *Ecology and behaviour*, ed. D.E. Bliss, F.J. Vernberg & W.B. Vernberg, pp. 107–61. New York: Academic Press.

Department of Water Affairs (1986). *Management of the water resources of southern Africa*. Pretoria: Government Printers.

De Vries, M.C., Tankersley, R.A., Forward, R.B. Jr, Kirb-Smith, W.W. & Luettich, R.A. Jr (1994). Abundance of estuarine crab larvae is associated with tidal hydrolic variables. *Marine Biology*, **118**, 403–13.

Dittel, A.I. & Epifanio, C.E. (1990). Seasonal and tidal abundance of crab larvae in a tropical mangrove system, Gulf of Nicoya, Costa Rica. *Marine Ecology Progress Series*, **65**, 25–34.

Dittel, A.I., Epifanio, C.E. & Lizano, O. (1991). Flux of crab larvae in a mangrove creek in the Gulf of Nicoya, Costa Rica. *Estuarine, Coastal and Shelf Science*, **32**, 129–40.

Emmerson, W.D. (1986). The ecology of *Palaemon pacificus* (Stimpson) associated with *Zostera capensis* Setchell. *Transactions of the Royal Society of South Africa*, **46**, 79–97.

Epifanio, C.E. (1988). Transport of invertebrate larvae between estuaries and the continental shelf. In *Larval fish and shellfish transport through inlets*, ed. M.P. Weinstein. Proceedings of workshop, Ocean Springs, Mississippi. *American Fisheries Society Symposium*, **3**, 104–14.

Epifanio, C.E., Valenti, C.C. & Pembroke, A.E. (1984). Dispersal and recruitment of blue crab larvae in Delaware Bay, USA. *Estuarine, Coastal and Shelf Science*, **18**, 1–12.

Fancett, M.S. & Kimmerer, W.J. (1985). Vertical migration of the demersal copepod *Pseudodiaptomus* as a means of predator avoidance. *Journal of Experimental Marine Biology and Ecology*, **88**, 31–43.

Fenton, G.E. (1996a). Diet and predators of *Tenagomysis tasmaniae* Fenton, *Anisomysis mixta australis* (Zimmer) and *Paramesopodopsis rufa* Fenton from south-eastern Tasmania (Crustacea: Mysidacea). *Hydrobiologia*, **323**, 31–44.

Fenton, G.E. (1996b). Production and biomass of *Tenagomysis tasmaniae* Fenton, *Anisomysis mixta australis* (Zimmer) and *Paramesopodopsis rufa* Fenton from south-eastern Tasmania (Crustacea: Mysidacea). *Hydrobiologia*, **323**, 23–30.

Forbes, A.T. (1989). Mysid shrimps (Crustacea: Mysidacea) in the St Lucia narrows before and after cyclone Demoina. *Lammergeyer*, **40**, 21–9.

Forward, R.B. Jr (1987). Larval release rhythms of decapod crustaceans, an overview. *Bulletin of Marine Science*, **41**, 165–76.

Fulton, R.S. III. (1982). Predatory feeding of two marine mysids. *Marine Biology*, **72**, 183–91.

Gliwicz, M.Z. (1986). Predation and the evolution of vertical migration in zooplankton. *Nature*, **320**, 746–8.

Grange, N. & Allanson, B.R. (1995). The influence of freshwater inflow on nature, amount and distribution of seston in estuaries of the Eastern Cape, South Africa. *Estuarine, Coastal and Shelf Science*, **40**, 403–20.

Greenwood, J.G. (1981). Occurrences of congeneric pairs of *Acartia* and *Pseudodiaptomus* species (Copepoda, Calanoida) in Moreton Bay, Queensland. *Estuarine, Coastal and Shelf Science*, **13**, 591–6.

Greenwood, J.G., Jones, M.B. & Greenwood, J. (1989). Salinity effects on brood maturation of the mysid crustacean *Mesopodopsis slabberi*. *Journal of the Marine Biological Association of the UK*, **69**, 683–94.

Grice, G.D. & Marcus, N.H. (1981). Dormant eggs of marine copepods. *Oceanography and Marine Biology Annual Review*, **19**, 125–40.

Grindley, J.R. (1964). Effect of low-salinity water on the vertical migration of estuarine plankton. *Nature*, **203**, 781–2.

Grindley, J.R. (1972). The vertical migration behaviour of estuarine plankton. *Zoologia Africana*, **7**, 13–20.

Grindley, J.R. (1981). Estuarine plankton. In *Estuarine ecology with particular reference to southern Africa*, ed. J.H. Day, pp. 117–46. Cape Town: AA. Balkema.

Grindley, J.R. & Wooldridge, T.H. (1974). The plankton of Richards Bay. *Hydrobiological Bulletin*, **8**, 201–12.

Haigh, E.H. & Whitfield, A.K. (1993). Larval development of *Gilchristella aestuaria* (Gilchrist, 1914) (Pisces: Clupeidae) from southern Africa. *South African Journal of Zoology*, **28**, 168–72.

Hakala, I. (1978). Distribution, population dynamics and production of *Mysis relicta* (Lovén) in southern Finland. *Annales Zoologici Fennici*, **15**, 243–58.

Hanekom, N. & Baird, D. (1992). Growth, production and consumption of the thalassinid prawn *Upogebia africana* (Ortmann) in the Swartkops estuary. *South African Journal of Zoology*, **27**, 130–9.

Hanekom, N., Baird, D. & Erasmus, T. (1988). A quantitative study to assess standing biomasses of macrobenthos in soft substrata of the Swartkops estuary, South Africa. *South African Journal of Marine Science*, **6**, 163–74.

Harrison, T.D. & Whitfield, A.K. (1990). Composition, distribution and abundance of ichthyoplankton in the Sundays River estuary. *South African Journal of Zoology*, **25**, 161–8.

Hart, R.C. & Allanson, B.R. (1975). Preliminary estimates of production by a calanoid copepod in subtropical Lake Sibaya. *Verhandlungen des Internationale Vereinigung für Theoretische und Angewandte Limnologie*, **19**, 1434–41.

Hart, R.C. & Allanson, B.R. (1976). The distribution and diel vertical migration of *Pseudodiaptomus hessei* (Mrázek) (Calanoida: Copepoda) in a subtropical lake in southern Africa. *Freshwater Biology*, **6**, 183–98.

Heip, C.H.R., Goosen, N.K., Herman, P.M.J., Kromkamp, J., Middelburg, J.J. & Soetaert, K. (1995). Production and consumption of biological particles in temperate tidal estuaries. *Oceanography and Marine Biology Annual Review*, **33**, 1–149.

Heymans, J.J. & Baird, D. (1995). Energy flow in the Kromme estuarine ecosystem, St Francis Bay, South Africa. *Estuarine, Coastal and Shelf Science*, **41**, 39–59.

Hill, A.E. (1991). Vertical migration in tidal currents. *Marine Ecology Progress Series*, **75**, 39–54.

Hilmer, T. & Bate, G.C. (1990). Covariance analysis of chlorophyll distribution in the Sundays River estuary, eastern Cape. *Southern African Journal of Aquatic Sciences*, **16**, 37–59.

Hilmer, T., Talbot, M.M.B. & Bate, G.C. (1988). A synthesis of recent botanical research in the Swartkops estuary. In *The Swartkops estuary*, ed. D. Baird, J.F.K. Marais & A.P. Martin, pp. 25–40. South African National Scientific

Programmes Report No. 156. Pretoria: CSIR.

Hough, A.R. & Naylor, E. (1992). Distribution and position maintenance behaviour of the estuarine mysid *Neomysis integer*. *Journal of the Marine Biological Association of the UK*, **72**, 869–76.

Huizinga, P. (1987). *Hydrodynamic model studies of the Swartvlei estuary*. CSIR report T/SEA 8709. Stellenbosch: CSIR.

Ianora, A. & Santella, L. (1991). Diapause embryos in the neustonic copepod *Anomalocera patersoni*. *Marine Biology*, **108**, 387–94.

Jeffries, H.P. (1962). Succession of two *Acartia* species in estuaries. *Limnology and Oceanography*, **7**, 354–64.

Jerling, H.L. (1993). *Feeding ecology of mesozooplankton in the Sundays River estuary*. PhD thesis, University of Port Elizabeth.

Jerling, H.L. & Wooldridge, T.H. (1991). Population dynamics and estimates of production for the calanoid copepod *Pseudodiaptomus hessei* in a warm temperate estuary. *Estuarine, Coastal and Shelf Science*, **33**, 121–35.

Jerling, H.L. & Wooldridge, T.H. (1992). Lunar influence on distribution of a calanoid copepod in the water column of a shallow, temperate estuary. *Marine Biology*, **112**, 309–12.

Jerling, H.L. & Wooldridge, T.H. (1994). Comparative morphology of the feeding appendages of four mesozooplankton species in the Sundays River estuary. *South African Journal of Zoology*, **29**, 252–7.

Jerling, H.L. & Wooldridge, T.H. (1995a). Plankton distribution and abundance in the Sundays River estuary, South Africa, with comments on potential feeding interactions. *South African Journal of Marine Science*, **15**, 169–84.

Jerling, H.L. & Wooldridge, T.H. (1995b). Relatively negative δ¹³C ratios of mesozooplankton in the Sundays River estuary, comments on potential carbon sources. *Southern African Journal of Aquatic Sciences*, **21**, 71–7.

Jerling, H.L. & Wooldridge, T.H. (1995c). Feeding of two mysid species on plankton in a temperate South African estuary. *Journal of Experimental Marine Biology and Ecology*, **188**, 243–59.

Johnson, G.E. & Gonor, J.J. (1982). The tidal exchange of *Callianassa californiensis* (Crustacea, Decapoda) larvae between the ocean and the Salmon River estuary, Oregon. *Estuarine, Coastal and Shelf Science*, **14**, 501–16.

Johnston, N.T. & Lasenby, D.C. (1982). Diet and feeding of *Neomysis mercedis* Holmes (Crustacea, Mysidacea) from the Fraser River estuary, British Columbia. *Canadian Journal of Zoology*, **60**, 813–24.

Jury, M.R. & Levey, K. (1993). The eastern Cape drought. *Water SA*, **19**, 133–7.

Kimmerer, W.J. & McKinnon, A.D. (1987). Zooplankton in a marine bay. II. Vertical migration to maintain horizontal distributions. *Marine Ecology Progress Series*, **41**, 53–60.

Kleppel, G.S. (1993). On the diets of calanoid copepods. *Marine Ecology Progress Series*, **99**, 183–95.

Knatz, G. (1978). Succession of copepod species in a middle Atlantic estuary. *Estuaries*, **1**, 68–71.

Köpche, B. & Kausch, H. (1996). Distribution and variability in abundance of *Neomysis integer* and *Mesopodopsis slabberi* (Mysidacea, Crustacea) in relation to environmental factors in the Elbe estuary. *Archiv für Hydrobiologie, Supplement*, **110**, 263–82.

Lance, J. (1962). Effects of water of reduced salinity on the vertical migration of zooplankton. *Journal of the Marine Biological Association of the UK*, **42**, 131–54.

Lee, W.Y. & McAlice, B.J. (1979). Seasonal succession and breeding cycles of three species of *Acartia* (Copepoda: Calanoida) in a Maine estuary. *Estuaries*, **2**, 228–35.

Little, K.T. & Epifanio, C.E. (1991). Mechanism for the re-invasion of an estuary by two species of brachyuran megalopae. *Marine Ecology Progress Series*, **68**, 235–42.

McConaugha, J.R. (1988). Export and reinvasion of larvae as regulators of estuarine populations. In *Larval fish and shellfish transport through inlets*, ed. M.P. Weinstein. Proceedings of workshop. *American Fisheries Society Symposium*, **3**, 90–103.

MacKay, H.M. & Schumann, E.H. (1990). Mixing and circulation in the Sundays River estuary, South Africa. *Estuarine, Coastal and Shelf Science*, **31**, 203–16.

Mallin, M.A. (1991). Zooplankton abundance and community structure in a mesohaline North Carolina estuary. *Chesapeake Science*, **14**, 481–8.

Mallin, M.A., Paerl, H.W., Rudek, J. & Bates, P.W. (1993). Regulation of estuarine primary production by watershed rainfall and river flow. *Marine Ecology Progress Series*, **93**, 199–203.

Malone, T.C., Crocker, L.H., Pike, S.E. & Wendler, B.W. (1988). Influences of river flow on the dynamics of phytoplankton production in a partially stratified estuary. *Marine Ecology Progress Series*, **48**, 235–49.

Marcus, N.H. (1984). Recruitment of copepod nauplii into the plankton: importance of diapause eggs and benthic processes. *Marine Ecology Progress Series*, **15**, 47–54.

Marcus, N.H. (1990). Calanoid copepod, cladoceran, and rotifer eggs in sea-bottom sediments of northern Californian coastal waters: identification, occurrence and hatching. *Marine Biology*, **105**, 413–18.

Marcus, N.H. (1991). Planktonic copepods in a sub-tropical estuary: seasonal patterns in the abundance of adults, copepodids, nauplii, and eggs in the sea bed. *Biological Bulletin*, **181**, 269–74.

Mauchline, J. (1970). The biology of *Schistomysis ornata* (Crustacea, Mysidacea). *Journal of the Marine Biological Association of the UK*, **50**, 169–75.

Mauchline, J. (1980). The biology of mysids and euphausiids. In *Advances in Marine Biology*, **18**, 1–369. London: Academic Press.

Mees, J., Abdulkerim, Z. & Hamerlynck, O. (1994). Life history, growth and production of *Neomysis integer* in the Westerschelde estuary (SW Netherlands). *Marine Ecology Progress Series*, **109**, 43–57.

Mees, J., Cattrijsse, A. & Hamerlynck, O. (1993). Distribution and abundance of shallow-water hyperbenthic mysids (Crustacea, Mysidacea) and euphausids (Crustacea, Euphausiacea) in the Voordelta and the Westerschelde. *Netherlands Journal of Aquatic Ecology*, **27**, 359–76.

Mees, J., Fockedey, N. & Hamerlynck, O. (1995). Comparative study of the hyperbenthos of three European estuaries. *Hydrobiologia*, **311**, 153–74.

Miller, C.B. (1983). The zooplankton of estuaries. In *Ecosystems of the World*, Vol. 26, *Estuaries and enclosed seas*, ed. B.H. Ketchum. Amsterdam: Elsevier.

Miroshnichenko, M.P. & Vovk, F.I. (1973). A model of Mysidae production process as exemplified with *Paramysis intermedia* (Czern) from Tsimlyanskoye Reservoir. *Gidrobiologicheskii Zhurnal*, **9**, 36–44.

Moffat, A.M. & Jones, M.B. (1992). Bionomics of *Mesopodopsis slabberi* and *Neomysis integer* (Crustacea, Mysidacea) in the Tamar estuary. In *Taxonomy, biology and ecology of Baltic mysids*, ed. J. Köhn, M.B. Jones & A. Moffat, pp. 109–19. Rostock University Press.

Moffat, A.M. & Jones, M.B. (1993). Correlation of the distribution of *Mesopodopsis slabberi* (Crustacea, Mysidacea) with physico-chemical gradients in a partially-mixed estuary (Tamar, England). *Netherlands Journal of Aquatic Ecology*, **27**, 155–62.

Orsi, J.J. (1986). Interaction between diel vertical migration of a mysidacean shrimp and two-layered estuarine flow. *Hydrobiologia*, **137**, 79–87.

Pereyra-Lago, R. (1988). Phototactic behaviour and the nature of the shadow response in larvae of the estuarine crab *Sesarma catenata*. *South African Journal of Zoology*, **23**, 150–4.

Queiroga, H., Costlow, J.D. & Moreira, M.H. (1994). Larval abundance patterns of *Carcinus maenas* (Decapoda, Brachyura) in Canal de Mira (Ria de Aveiro, Portugal). *Marine Ecology Progress Series*, **111**, 63–72.

Read, G.H.L. & Whitfield, A.K. (1989). The response of *Grandidierella lignorum* (Barnard) (Crustacea: Amphipoda) to episodic flooding in three eastern Cape estuaries. *South African Journal of Zoology*, **24**, 99–105.

Reddering, J.S.V. (1988). Prediction of the effects of reduced river discharge on the estuaries of the south-eastern Cape Province, South Africa. *South African Journal of Science*, **84**, 726–30.

Reddering, J.S.V. & Esterhuysen, K. (1987). The effects of river floods on sediment dispersal in small estuaries: a case study from East London. *South African Journal of Geology*, **90**, 458–70.

Reddering, J.S.V. & Rust, I.C. (1990). Historical changes and sedimentary characteristics of Southern African estuaries. *South African Journal of Science*, **86**, 425–48.

Richards, S.W. & Riley, G.A. (1967). The benthic epifauna of Long Island Sound. *Bulletin of the Bingham Oceanographic Collection*, **19**, 89–129.

Rogers, H.M. (1940). Occurrence and retention of plankton within an estuary. *Journal of the Fisheries Research Board of Canada*, **5**, 164–71.

Rudstam, L.G., Hansson, S. & Larsson, U.

(1986). Abundance, species composition and production of mysid shrimps in a coastal area of the northern Baltic proper. *Ophelia, Supplement*, **4**, 225–38.

Sandifer, P.A. (1975). The role of pelagic larvae in recruitment to populations of adult decapod crustaceans in the York River estuary and adjacent lower Chesapeake Bay, Virginia. *Estuarine and Coastal Marine Science*, **3**, 269–79.

San Vicente, C. & Sorbe, J.C. (1995). Biology of the suprabenthic mysid *Schistomysis spiritus* (Norman, 1860) in the southeastern part of the Bay of Biscay. *Scientia Marina*, **59**, 71–86.

Schlacher, T.A. & Wooldridge, T.H. (1994). Tidal influence on distribution and behaviour of the opossum shrimp, *Gastrosaccus brevifissura*. In *Changes in fluxes in estuaries: implications from science to management*, ed. K.R. Dyer & R.J. Orth, pp. 307–12. Fredensborg, Denmark: Olsen & Olsen.

Schlacher, T.A. & Wooldridge, T.H. (1995). Small-scale distribution and variability of demersal zooplankton in a shallow, temperate estuary: tidal and depth effects on species-specific heterogeneity. *Cahiers de Biologie Marine*, **36**, 211–27.

Schlacher, T.A. & Wooldridge, T.H. (1996a). Patterns of selective predation by juvenile, benthivorous fish on estuarine macrofauna. *Marine Biology*, **125**, 241–7.

Schlacher, T.A. & Wooldridge, T.H. (1996b). Origin and trophic importance of detritus – evidence from stable isotopes in the benthos of a small, temperate estuary. *Oecologia*, **106**, 382–8.

Schlacher, T.A.& Wooldridge, T.H. (1996c). Ecological responses to reductions in freshwater supply and quality in South Africa's estuaries: lessons for management and conservation. *Journal of Coastal Conservation*, **2**, 115–30.

Schumann, E.H. & Pearce, M.W. (1997). Freshwater hydrology of the Gamtoos estuary, South Africa. *Estuaries* (in press).

Sell, D.W. (1982). Size-frequency estimates of secondary production by *Mysis relicta* in Lakes Michigan and Huron. *Hydrobiologia*, **93**, 69–78.

Shushkina, G.M. (1973). Evaluation of the production of tropical zooplankton. In *Life activity of pelagic communities in the Ocean tropics*, ed. M.E. Vinogradov, pp. 172–83.

Siegfried, C.A. & Kopache, M.E. (1980). Feeding of *Neomysis mercedis* (Holmes). *Biological Bulletin*, **159**, 193–205.

Siegfried, C.A., Kopache, M.E. & Knight, A.W. (1979). The distribution and abundance of *Neomysis mercedis* in relation to the entrapment zone in the western Sacramento–San Joaquin delta. *Transactions of the American Fisheries Society*, **108**, 262–8.

Stich, H.-B. & Lampert, W. (1981). Predator evasion as an explanation of diurnal vertical migration by zooplankton. *Nature*, **293**, 396–8.

Sullivan, B.K. & McManus, L.T. (1986). Factors controlling seasonal succession of the copepods *Acartia hudsonica* and *A. tonsa* in Narragansett Bay, Rhode Island: temperature and resting egg production. *Marine Ecology Progress Series*, **28**, 121–8.

Tackx, M., Irigoien, X., Daro, N., Castel, J., Zhu, L., Zhang, X. & Nijs, J. (1995). Copepod feeding in the Westerschelde and the Gironde. *Hydrobiologia*, **311**, 71–83.

Talbot, M.M.J-F. & Baird, D. (1985). Feeding of the estuarine round herring *Gilchristella aestuarius* (G & T) (Stolephoridae). *Journal of Experimental Marine Biology and Ecology*, **87**, 199–214.

Talbot, M.M.B. & Bate, G.C. (1987). The spatial dynamics of surf diatom patches in a medium energy cuspate beach. *Botanica Marina*, **30**, 459–65.

Talbot, M.M.B. & Bate, G.C. (1988). Distribution patterns of the surf diatom *Anaulus birostratus* in an exposed surfzone. *Estuarine, Coastal and Shelf Science*, **26**, 137–53.

Talbot, M.M.B., Bate, G.C. & Campbell, E.E. (1990). A review of the ecology of surf-zone diatoms, with special reference to *Anaulus australis*. *Oceanography and Marine Biology Annual Review*, **28**, 155–75.

Tattersall, W.M. & Tattersall, O.S. (1951). *The British Mysidacea*. London: Ray Society.

Uye, S. (1985). Resting egg production as a life-history strategy of marine planktonic copepods. *Bulletin of Marine Science*, **37**, 440–9.

Uye, S., Iwai, Y. & Kasahara, S. (1983). Growth and production of the inshore marine copepod *Pseudodiaptomus marinus* in the central part of the inland sea of Japan. *Marine Biology*, **73**, 91–8.

Uye, S. & Kasahara, S. (1983). Grazing of various development stages of *Pseudodiaptomus marinus* (Copepoda: Calanoida) on naturally occurring particles. *Bulletin of the Plankton Society of Japan*, **30**, 147–58.

Walter, T.C. (1986). The zoogeography of the genus *Pseudodiaptomus* (Calanoida: Pseudodiaptomidae). In *Proceedings of the Second International Conference on Copepoda*, ed. G. Schriever, H.K. Schminke, & C.-t. Shih, pp. 502–8. Ottawa, Canada.

Webb, P., Perissinotto, R. & Wooldridge, T.H. (1987). Feeding of *Mesopodopsis slabberi* (Crustacea, Mysidacea) on naturally occurring phytoplankton. *Marine Ecology Progress Series*, **38**, 115–23.

Webb, P. & Wooldridge, T.H. (1990). Diel horizontal migration of *Mesopodopsis slabberi* (Crustacea: Mysidacea) in Algoa Bay, southern Africa. *Marine Ecology Progress Series*, **62**, 73–7.

Webb, P., Wooldridge, T.H. & Schlacher, T. (1997). Osmoregulation and spatial distribution in four species of mysid shrimps. *Comparative Biochemistry and Physiology* (in press).

Webber, H.H. & Thurmann, H.V. (eds.) (1991). *Marine Biology*. New York: HarperCollins.

Wellerhaus, S. (1969). On the taxonomy of planktonic Copepoda in the Cochin Backwater (a south Indian estuary). *Veröffentlichungen des Instituts für Meeresforschung in Bremerhaven*, **11**, 245–86.

Whitfield, A.K. (1994). Abundance of larval and 0+ juvenile marine fishes in the lower reaches of three southern African estuaries with differing freshwater inputs. *Marine Ecology Progress Series*, **105**, 257–67.

Whitfield, A.K. & Bruton, M. (1989). Some biological implications of reduced freshwater inflow into eastern Cape estuaries: a preliminary assessment. *South African Journal of Science*, **87**, 192–7.

Winter, P.E.D. & Baird, D. (1988). Diversity, productivity, and ecological importance of macrobenthic invertebrates in selected eastern Cape estuaries. In *Towards an environmental plan for the eastern Cape*, ed. M.N. Bruton & F.W. Gess, pp. 149–54. Grahamstown: Grocott & Sherry.

Wittmann, K.J. (1992). Morphogeographic variations in the genus *Mesopodopsis* Czerniavsky with descriptions of three new species (Crustacea, Mysidacea). *Hydrobiologia*, **241**, 71–89.

Wooldridge, T.H. (1977). The zooplankton of Mgazana, a mangrove estuary in Transkei, southern Africa. *Zoologica Africana*, **12**, 307–22.

Wooldridge, T.H. (1983). Ecology of beach and surfzone mysid shrimps in the eastern Cape, South Africa. In *Sandy beaches as ecosystems*, ed. A. McLachlan & T. Erasmus, pp. 449–60. The Hague: Dr W. Junk.

Wooldridge, T.H. (1986). Distribution, population dynamics and estimates of production for the estuarine mysid *Rhopalophthalmus terranatalis*. *Estuarine, Coastal and Shelf Science*, **23**, 205–23.

Wooldridge, T.H. (1991). Exchange of two species of decapod larvae across an estuarine mouth inlet and implications of anthropogenic changes in the frequency and duration of mouth closure. *South African Journal of Science*, **87**, 519–25.

Wooldridge, T.H. (1994). The effect of periodic inlet closure on recruitment in the estuarine mudprawn, *Upogebia africana* (Ortmann). In *Changes in fluxes in estuaries: implications from science to management*, ed. K.R. Dyer & R.J. Orth, pp. 329–33. Fredensborg, Denmark: Olsen & Olsen.

Wooldridge, T.H. & Bailey, C. (1982). Euryhaline zooplankton of the Sundays estuary and notes on trophic relationships. *South African Journal of Zoology*, **17**, 151–63.

Wooldridge, T.H. & Erasmus, T. (1980). Utilization of tidal currents by estuarine zooplankton. *Estuarine, Coastal and Shelf Science*, **11**, 107–14.

Wooldridge, T.H. & Loubser, H. (1996). Larval release rhythms and tidal exchange in the estuarine mudprawn, *Upogebia africana*. *Hydrobiologia*, **337**, 113–21.

Wooldridge, T.H. & Melville-Smith, R. (1979). Copepod succession in two South African estuaries. *Journal of Plankton Research*, **1**, 329–41.

Wooldridge, T.H. & Webb, P. (1988). Predator–prey interactions between two species of estuarine mysid shrimps. *Marine Ecology Progress Series*, **50**, 21–8.

Zaret, T.M. & Suffern, J.S. (1976). Vertical migration in zooplankton as a predator avoidance mechanism. *Limnology and Oceanography*, **21**, 804–13.

8 Studies on estuarine macroinvertebrates

The macrobenthos

Casper de Villiers and Alan Hodgson

Nektonic invertebrates

Anthony Forbes

Olifants River estuary

THE MACROBENTHOS

Introduction

Estuaries are considered to be amongst the most productive ecosystems (Kokkinn & Allanson 1985), due largely to the richness of nutrients within them. Nutrients are derived from both freshwater and marine sources and therefore primary productivity in estuaries can be high (Day 1981a; Knox 1986). Many benthic invertebrates capitalise on this productivity either directly or indirectly as deposit feeders, suspension feeders, decomposers, predators or scavengers. In general, estuaries are ecosystems which have high densities and biomasses of invertebrates, but low species diversity (Day 1981b).

Prior to 1980, benthic invertebrate research in South African estuaries concentrated on documenting species diversity and distribution (often qualitatively only), as well as determining biomass and density of invertebrates in a few of the larger estuaries. Some workers studied the biology of selected species. This earlier research was summarised by Day (1981b, c) and Hill (1981). These works, along with the global synthesis of Wolff (1983), provided a foundation from which to develop more detailed studies. Since the publication of these reviews, the macrobenthic composition of a greater number of estuaries has been studied, and more detailed investigations have been carried out on the biology and ecology of estuarine benthic organisms. Perhaps of greater importance is the fact that researchers have turned their attention to determining the responses of macrobenthic invertebrates to environmental fluctuations, as well as studying macrobenthic productivity, consumption, turnover rates and the roles the benthos play in energy flow through estuarine ecosystems. The aim of this chapter is to summarize the more recent research developments on estuarine macrobenthos, although of necessity we will have to refer to earlier publications.

Benthic invertebrate species diversity, community composition and productivity – west to east coast; open vs. closed estuaries

Based largely on water temperatures, South African estuaries have been grouped into three broad biogeographic regions: cool temperate, warm temperate and subtropical (Day 1981c; Potter et al. 1990; Whitfield 1994). In the most recent of these classifications, Whitfield (1994) regarded cool temperate estuaries as extending from beyond Walvis Bay (22° 59' S, 14° 31' E) to Cape Point (34° 22' S, 18° 30' E); warm temperate from Cape Point to the Mbashe River estuary (32° 17' S, 28° 54' E); and estuaries north of the Mbashe River as subtropical (Figure 8.1). As estuaries are subjected to both terrestrial and marine influences (see Day 1981c for full discussion), the boundaries of the above biogeographic regions for these ecosystems cannot be precise. In addition, many estuarine species have a broad geographic distribution, and within each geographic region estuaries are structurally variable (Day 1981c). Nevertheless, the above biogeographic provinces serve as a convenient basis on which to discuss community composition.

Although detailed quantitative macrobenthic surveys have been carried out on estuaries in each of the biogeographic regions, there are still significant gaps in the data-

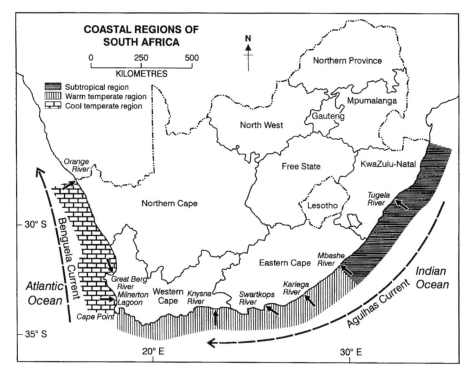

Figure 8.1. **Southern Africa showing biogeographical regions and some of the estuaries mentioned in the text (from Whitfield 1994).**

base. Quantitative data are available for only 13% (= 38 estuaries; Table 8.1) of the 289 South African estuaries.[1] In the Eastern Cape there have never been holistic quantitative macrobenthic surveys of the Bushmans River, Kowie River or Great Fish River estuaries, which are all relatively large systems. Furthermore, many estuaries were last surveyed over two decades ago. Since these surveys, considerable changes have often occurred to either the estuary itself (e.g. urban development) or the catchment (e.g. construction of dams). The Knysna River estuary was shown to have the richest macrobenthic species diversity (310 species) in South Africa (Day 1981c). Since the original surveys of the 1950s there has been considerable urban and industrial development around this estuary but which from recent surveys have not impacted significantly upon the diversity of the macrobenthos (B.R. Allanson pers. comm.)

Day (1964) noted that the fauna of South African estuaries depends more on the physical characteristics of the estuary (e.g. depth and permanence of the mouth, clarity of water, nature of bottom sediments, salinity regime) than geographic position. Thus it is very difficult to make unequivocal biogeographical comparisons of community structure, species diversity, etc. Rather than comparing

individual estuaries, Day (1974) compared the total number of macrobenthic species recorded from all surveyed estuaries within each of the biogeographic regions of South Africa. From this, Day noted that the warm temperate estuaries had a higher total species diversity (357) than either the subtropical estuaries (237) or the cool temperate estuaries (59). Since this synthesis, the macrobenthic community composition of a greater number of South African estuaries has been documented and studied in more detail. In this section we not only examine and describe geographic trends in estuarine macrobenthic community composition and species diversity but also the productivity of estuarine communities and their dominant invertebrates.

Cool temperate estuaries

The Atlantic coast of South Africa has few functional estuaries (Day 1981a), a result of low rainfall and high evaporation along this region of coastline. Of the classified estuaries of the west coast, only the Berg River estuary, Olifants River estuary and Milnerton are estuarine in nature and these three have been the subject of quantitative studies.

Results from these three cool temperate estuaries reveal that the macrobenthic communities are comprised of very few species. For example, only 25 species have been recorded from the Berg River estuary (Kalejta & Hockey 1991), 43 species (includes all invertebrates recorded) from the Olifants River estuary (Day 1981c) and 35 species from

[1] In Chapter 2 the geomorphologists have recorded 465 river outlets to the sea along the South African coastline. Not all of these have been visited or surveyed by estuarine research groups so that the actual number of estuaries, either temporary or permanent, which have been subject to some degree of surveillance is a good deal smaller.

Table 8.1. Number of invertebrate species (predominantly macrobenthic) recorded from some South African estuaries. O, estuaries always open to the sea; MO, estuaries mainly open to the sea; PO, estuaries periodically open to the sea; C, closed estuaries

Estuary	Location	Number of species	Open/closed	Reference
Cool temperate				
Berg	32°47' S, 18°10' E	25	O	Kalejta & Hockey 1991
Olifants	31°42' S, 18°11' E	43	O	Day 1981c
Milnerton	33°54' S, 18°28' E	35	MO	Grindley & Dudley 1988
Warm temperate				
Sandvlei	34°05' S, 18°28' E	22	O	Morant & Grindley 1982
Lourens	34°06' S, 18°49' E	6	MO	Cliff & Grindley 1982
Palmiet	34°20' S, 18°59' E	28	O	Branch & Day 1984
Bot	34°21' S, 19°07' E	25	C	De Decker & Bally 1985
Klein	34°24' S, 19°18' E	128	MO	De Decker 1989
Heuningnes	34°43' S, 20°07' E	18	C	Day 1981c
Breede	34°24' S, 20°51' E	149	O	Carter 1983
Great Brak	34°03' S, 22°46' E	90	MO; PO since 1990	Day 1981c
Swartvlei	34°01' S, 22°46' E	44	MO	Whitfield 1989
Knysna	34°04' S, 23°03' S	310	O	Day 1981c
Keurbooms	34°02' S, 23°23' E	42	O	Duvenhage & Morant 1984
Groot	33°59' S, 23°34' E	10	C	Morant & Bickerton 1983
Kromme	34°08' S, 24°52' E	56	O	Bickerton & Pierce 1988
Swartkops	33°51' S, 25°38' E	122	O	Baird et al. 1986
Sundays	33°42' S, 25°51' E	102	O	du Preez 1978
Bushmans	33°41' S, 26°42' E	80	O	Hodgson unpublished
Kariega	33°41' S, 26°41' E	107	O	Hodgson 1987
Kleinemond	33°33' S, 27°03' E	34	PO	Brown 1953
Keiskamma	33°17' S, 27°29' E	51	O	Branch & Grindley 1979
Subtropical				
Mbashee	32°16' E; 28°54' E	70	O	Branch & Grindley 1979
Mngazana	31°42' S, 29°25' E	209	O	Branch & Grindley 1979
Port St Johns area (Umgazi, Umzimvubu, Second Beach)	31°26' S, 29°34' E	75	O	Branch & Grindley 1979
Port Shepstone area (Umtanvuna, Umzimkulu, Untentweni, Ifafa)	30°44' S, 30°27' E	42	O	Branch & Grindley 1979
Umkomaas area (Umkomaas, Isipingo, Umzimbazi, Amalonga, Umgababa, Amazimtoti)	33°12' S, 30°47' E	83	O	Branch & Grindley 1979
Umlalazi	28°58' S, 31°48' E	84	O	Day 1981c
St Lucia	28°23' S, 32°26' E	84	O	Day 1981c

Milnerton estuary (Grindley & Dudley 1988) (Table 8.1). Day (1981b) suggests that the drastic changes which occur (often as much as 11 °C during upwelling) in west coast water temperatures have a major influence on reducing species composition within these estuaries. However, similar rapid declines in water temperature due to upwelling have been recorded in the warm temperate Kariega River (Taylor 1992) and Knysna River (B.R. Allanson, personal communication) estuaries, systems which have very rich macrobenthic faunas (Day 1967; Hodgson 1987). Thus the rapid temperature fluctuations caused by upwelling are unlikely to be directly responsible for the paucity of species in cool temperate estuaries.

Although most major phyla are represented in the macrobenthos of cool temperate estuaries, the recent study on the Berg River estuary revealed that gastropods and polychaetes dominated in terms of density, biomass and productivity (Kalejta & Hockey 1991). Three species, *Hydrobia* sp. (Gastropoda), *Ceratonereis erythraeensis* and *C. keiskama* (Polychaeta), are particularly abundant, with

Hydrobia sp. reaching densities of $182\,500$ m^{-2} (Kalejta & Hockey 1991), *C. erythraeensis* 7034 m^{-2} and *C. keiskama* 14897 m^{-2} (Kalejta 1992). *C. erythraeensis* and *Hydrobia* sp. were found to contribute on average 75% of the total invertebrate biomass, and at one site *C. erythraeensis* constituted 70% of the biomass (Kalejta & Hockey 1991). Kalejta and Hockey also determined that maximum biomass was reached in winter, early spring and autumn with a marked decrease in summer. Productivity of *C. erythraeensis* was found to range from 12.2 to 20.4 g m^{-2} year^{-1} and Kalejta (1992) calculated a P:B ratio of 1.90 for this species, within the range of other species of nereid polychaetes.

The work of Kalejta and Hockey (1991) is the only one which has attempted to determine the productivity of the macrobenthic community of a southern African cool temperate estuary. They calculated that the total annual production of the benthos in the Berg River estuary ranged from 66.7 to 146.9 g m^{-2} y^{-1} (average 87.6 g m^{-2} y^{-1}). This in turn gave a P:B ratio of 4.52, similar to that of Langebaan Lagoon (P:B 5) and San Francisco Bay (P:B 4.5) and much

Table 8.2. *Mean annual biomass (dry weights), production and P/B ratios of macrobenthic communities at some intertidal mudflats. Data from Kalejta and Hockey (1991) who give sources of original data*

Site	Latitude	Dominant invertebrates	Mean biomass (g m^{-2})	Production (g m^{-2} year^{-1})	P/B
South Africa					
Berg River estuary	32°S	Gastropoda/Polychaeta	19.35	87.58	4.52
Langebaan lagoon	33°S	Gastropoda/Amphipoda	19.21	95.25	4.96
Swartkops estuary	34°S	Decapoda	69.30	77.70	1.12
N. Hemisphere systems					
Forth estuary, Scotland	57°N	Oligochaeta/Bivalvia	10.5	12.90	1.23
Humber estuary, England	54°N	Bivalvia	24.00	27.00	1.12
Grevelingen estuary, Holland	53°N	Mollusca	25.70	41.30	1.60
Wadden Sea, Holland	53°N	Bivalvia/Polychaeta	27.00[a]	30.00[a]	1.10[a]
Lynher estuary, England	51°N	Bivalvia	13.24	13.31	1.01
Southampton Water, England	51°N	Bivalvia	140.00	188.50	1.35
Petpeswick inlet, Canada	44°N	Bivalvia/Gastropoda	16.22	21.20	1.31
San Francisco Bay, USA	37°N	Bivalvia	16.64[a]	74.93[a]	4.50[a]
Banc d'Arguin, Mauritania	11°N	Bivalvia/Polychaeta	3.20[a]	6.00–12.00[a]	3.30[a]
Other S. Hemisphere systems					
Westernport Bay, Australia	32°S		10.00–15.00[a]	43.00[a]	3.40[a]
U. Waitemata estuary, New Zealand	35°S		17.90	27.30	1.53

[a] = g AFDW m^{-2}.

higher than the P:B ratios of northern hemisphere cool temperate estuaries (P:B 1.01–1.60: Table 8.2). This high productivity enables these systems to support large bird populations (Kalejta & Hockey 1991).

The dominance of small polychaetes and gastropods in the Berg River estuary is in marked contrast to estuaries from other South African coastal regions, where crustaceans and bivalves are the most dominant taxa (see below).

Warm temperate estuaries

The number of species comprising the macrobenthic communities of warm temperate estuaries is very variable (Table 8.1). As a general rule, the warm temperate systems which occur between Cape Point and the Mbashe River estuary, and are mainly open to the sea, have on average about 80–100 macroinvertebrate species per estuary (Table 8.1) although there are some notable exceptions. The Palmiet River estuary, for example, which lies to the west of Cape Agulhas, has only 28 species of macroinvertebrate (Branch & Day 1984) and the small False Bay estuaries are even more impoverished, often a result of human activities (Morant 1991). In contrast to this, Knysna River estuary has 310 species (Day 1981c). The paucity of species in the Palmiet River estuary is probably a result of seasonal (winter) freshwater floods and strong tidal inflow into the estuary (Branch & Day 1984). The abundance of invertebrate species in the Knysna River estuary can be attributed to the large size of the estuary, its physical features (clear water, a relatively low freshwater inflow, extensive mudflats and a permanently open mouth) (Day 1974) and a diversity of habitats. Other warm temperate open estuaries

with a relatively low freshwater input and high clarity (e.g. Swartkops, Kariega & Sundays River estuaries) also have high species diversity (Table 8.1; see also Hodgson 1987).

Although the size of the estuary and volume of freshwater inflow influence the total number of species within an open estuary, perhaps the most significant factor which dictates species diversity is whether the mouth of the estuary remains mainly open or predominantly closed. Only 37 (12.8%) of South Africa's estuaries maintain permanent tidal inlets (Reddering & Rust 1990).[2] Those estuaries which are permanently or mainly open at the mouth tend to have richer macrobenthic communities (usually >60 species; Table 8.1), whereas estuaries in which the mouth is closed for lengthy periods are relatively poor in species (<35 species; Table 8.1). For example, within the Eastern Cape over 107 species have been recorded from the permanently open Kariega River estuary (Hodgson 1987), whereas the nearby Kleinmonde River estuary, the mouth of which is closed for long periods, has only 34 species (Brown 1953). Low species diversity is also a feature of closed estuaries in Western Australia (Platell & Potter 1996). The low species diversity of closed estuaries can be attributed largely to the limited recruitment of individuals from the sea (Day 1964; Platell & Potter 1996). Recent studies on the reproductive biology of several species of macrobenthic decapods have shown that they have an obligate (and therefore vital) marine larval phase (Emmerson 1983; Wooldridge 1991; Pereyra Lago 1993). Closed estuaries clearly restrict the movement

[2] It is noteworthy that although the majority of South Africa's estuaries are small systems with a closed mouth, most quantitative macrobenthic surveys have been on the larger open systems: see Table 8.1.

Table 8.3. Mean densities of *Upogebia africana* recorded from some South African estuaries

Estuary	Mean density m⁻² ± S.D.	Reference
Upogebia africana		
Olifants	500	Day 1981
Knysna	12–175	Cretchley 1996
Kromme	100 ± 70	Hanekom 1982
Swartkops	218 ± 166	Hanekom *et al*. 1988
Sundays	121 ± 79	du Preez 1978
Bushmans	120	Hodgson 1987
Kariega	120–500	Hodgson 1987
Kowie	up to 600	Day 1981
Mngazana	up to 90	Branch & Grindley 1979
Upogebia capensis		
Berg	1.6–9.4	Kalejta & Hockey 1991

and therefore survival and recruitment of such larvae. Despite the lack of information on the larval biology of South African estuarine polychaetes and bivalves, the absence of many species from closed systems suggests that they have obligate marine larvae.

Although warm temperate estuaries may contain a diverse array of species, only a few species of deposit-feeding and filter-feeding crustaceans constitute the greatest proportion of the invertebrate biomass. In the muddy intertidal regions of many open estuaries the filter-feeding mudprawn *Upogebia africana* is often the most abundant invertebrate (Table 8.3). In the Swartkops River estuary, *U. africana* constitutes about 80% of the total biomass of 75.4×10^3 kg (Hanekom *et al*.1988) and has a somatic production of 1093–1864 kJ m⁻² y⁻¹ (P:B 0.79–0.99) or 23 g C m⁻² y⁻¹ (Baird 1988; Hanekom & Baird 1992). *U. africana* also dominates the biomass in parts of the Kariega River estuary (Hodgson 1987). Brachyuran crabs are also extremely abundant, particularly amongst the estuarine vegetation (Hodgson 1987; Taylor 1988; Taylor & Allanson 1993). Within the *Zostera* of the Kariega River estuary, *Cleistostoma (Paratylodiplax) edwardsii* reaches densities of 600 m⁻² (biomass of 11.4 g m⁻² dry weight (DW): (Hodgson 1987) and, in the saltmarsh vegetation *Sesarma catenata* has densities of 47 m⁻² (= 8.7 g m⁻² acid free dry weight (AFDW): Taylor & Allanson 1993). These crabs play a significant role in the exchange of carbon across the surface of saltmarshes (Taylor & Allanson 1993) (see also Chapter 4).

In addition to crustaceans, filter-feeding bivalves can be very abundant in warm temperate estuaries. Within the Kariega River estuary, Hodgson (1987) determined that 50% of the macrobenthic biomass comprised filter-feeders, mainly in the form of *Solen cylindraceus* which reached densities of 400 m⁻² (biomass about 80 g m⁻² DW). Bivalves are also dominant in the Bot River estuary (*Arcuatula capensis*: de Decker & Bally 1985) and regions of the Swartvlei

estuary (*Loripes clausus* and *Macoma litoralis* >50% of invertebrate biomass in *Zostera*: Whitfield 1989).

In estuaries which are either closed or only periodically open, and in the sandflats of open estuaries, *U. africana* is replaced by the deposit/filter-feeding anomuran *Callianassa kraussi*. Like *U. africana*, *C. kraussi* can be extremely abundant and its success in closed estuaries is because, unlike *U. africana*, larval development of *C. kraussi* is abbreviated, being completed within the estuary (Forbes 1973). In the Palmiet River estuary and middle and upper regions of the Swartvlei River estuary, *C. kraussi* had a biomass of approximately 9.0 and 13.5–15.5 g m⁻² (DW), respectively (Branch & Day 1984; Whitfield 1989). For the Palmiet River estuary, Branch and Day (1984) calculated that there was a standing stock of 2070 kg (DW) of this crustacean. In many estuaries (e.g. Swartkops and Knysna) *C. kraussi* densities can be over 100 m⁻² (Hanekom *et al*. 1988). In the Swartkops River estuary, *C. kraussi* was found to be the second most important macrobenthic invertebrate, constituting 10% of the benthic biomass with a productivity of 1791 kJ m⁻² y⁻¹ (P:B 1.41) (Hanekom *et al*. 1988; Hanekom & Baird 1992). The importance of *C. kraussi* in organic turnover within estuaries is discussed later in this chapter.

There are few studies of macrobenthic community productivity of warm temperate estuaries, with the only available figures being calculated for the Swartkops River estuary (Kalejta & Hockey 1991). Although this estuary has a relatively high mean macrobenthic standing stock (= 69.30 g m⁻²), the productivity has been estimated at only 77.70 g m⁻² resulting in a P:B ratio of 1.12, a value which is similar to northern hemisphere cool temperate estuaries (Table 8.2).

Subtropical estuaries

As with the warm temperate estuaries, species diversity in subtropical estuaries is correlated with factors such as freshwater input and whether the estuary mouth is open or closed. Although the total number of species recorded from South African subtropical estuaries is relatively high (237: Day 1974) this is lower than total numbers for warm temperate estuaries. In addition, the number of species per open estuary is often fewer than 80, slightly lower than warm temperate counterparts (Table 8.1). Species diversity increases only in the tropical estuaries (e.g. Morrumbene, which has 378 species: Day 1974). Day attributes the lower species diversity of subtropical estuaries to the large silt loads carried by the rivers (often a result of poor land management in the catchments). In large silt-laden systems such as the Tugela River estuary, very few species have been recorded (Day 1981c). There are some notably species-rich estuaries such as the Mngazana which has over 200 species (Branch & Grindley 1979). The richness of the

Mngazana River estuary can be attributed to the favourable physical conditions within it. Unlike other subtropical estuaries, this estuary is permanently open at the mouth and the silt load of the river is minimal (Branch & Grindley 1979), making it one of the clearest of the subtropical systems.

The macrobenthos of only a few subtropical estuaries has been quantified in detail. Begg (1984a) noted that of the 73 estuaries in Natal, 51 had been overlooked in terms of any scientific study. Although Begg (1978, 1984a, b) carried out surveys of many of the estuaries of Natal, these did not include detailed macrobenthic work. Nevertheless those studies which have been forthcoming (Branch & Grindley 1979; Day 1981c) enable us to make some comments on the macrobenthos within subtropical estuaries.

Like warm temperate estuaries, the macrobenthos of subtropical estuaries is dominated by crustaceans. In these estuaries it is the brachyurans which dominate. In the Mngazana River estuary, Branch & Grindley (1979) found that between 80–100% of the biomass is composed of detritivorous brachyurans (mainly *Uca* spp. and *Sesarma* spp.), in contrast to the warm temperate estuaries which are dominated by filter- or deposit feeders. The high organic content and physical characteristics of the sediments of the subtropical estuaries presumably favour a deposit-feeding existence. In contrast, waters which are permanently silt-laden are probably unfavourable to filter-feeders, clogging filtering organs and appendages.

An exception to being dominated by detritivores is the St Lucia estuary which, after a period of stable salinities, was dominated by the filter-feeding bivalve *Solen cylindraceus* (maximum biomass recorded at Gilly's point 14.3 g m^{-2} DW, mean density 498 ± 150 m^{-2}) and the polychaete *Marphysa macintoshi* (maximum biomass recorded at Makakatana >4 g m^{-2} DW, mean density 8885 ± 1319 m^{-2}) (Blaber *et al.* 1983).

No studies on the productivity of the macrobenthos or macrobenthic communities of South African subtropical estuaries have been undertaken; this represents a serious gap in our knowledge of such systems.

Conclusions

A number of factors influence macrobenthic community structure in South African estuaries. These factors include the size of the estuary, freshwater input (and associated silt load), and whether the mouth of the estuary is permanently open or has phases in which it is closed. Estuaries in which the mouth is permanently or predominantly closed, have a poor macrobenthic diversity. Open estuaries which have a high input of freshwater (e.g. Palmiet River estuary) and silt (e.g. Tugela River estuary), have very low species diversity. Clear water estuaries with a relatively low freshwater input (e.g. Knysna, Swartkops, Kariega and Mngazana River estuaries) sustain a high number of species. Nevertheless, when the richness of macrobenthic species in permanently open estuaries is compared on a geographic basis, the trend is toward decreasing species abundance from the warm temperate to both cool temperate and subtropical estuaries. Although many species are not restricted to estuaries from one geographic region, there are some interesting faunistic differences between regions. The macrobenthic communities of warm temperate estuaries are dominated by filter-feeding and deposit-feeding crustaceans (Anomura and Brachyura) and bivalves. In the subtropics, deposit-feeding brachyurans are more prevalent. In cool temperate estuaries, small gastropods and polychaetes are most abundant. From the limited data available, the P:B ratio for the macrobenthic communities of open estuaries is higher in cool temperate (e.g. Berg River estuary) than warm temperate systems (e.g. Swartkops River estuary).

Factors influencing the distribution and abundance of benthic macroinvertebrates within South African estuaries

In contrast to the relative stability of the sea, waters of estuaries are characterised by sudden and often extensive change in physical conditions. The periodicity, rate and magnitude of these changes are determined by local geographic, hydrological and meteorological conditions (Newell 1976; 1979, Day 1981a; and other chapters in this book). South Africa is an arid country, and freshwater is arguably one of its most precious natural resources, its importance growing with ever increasing demand (Read & Whitfield 1989; Whitfield & Bruton 1989). Rainfall is generally very seasonal and erratic, and in many instances much of the annual total will fall over a short period of a few days. Erratic flow and floods are therefore a feature of many South African estuaries (e.g. MacNae 1957; Allanson & Read 1987). As estuaries in South Africa tend to be narrow, shallow and microtidal, they are subject to rapid changes in physical conditions (Day 1981a). Flooding results in a sudden and marked decrease in salinity which may persist for a few days to a few weeks. Droughts are also common in the catchments of many estuaries, and the consequent reduced runoff, together with impoundments and the abstraction of water, results in no freshwater input, with many estuarine systems (both open and closed) becoming hypersaline for prolonged periods (see Blaber *et al.* 1983; Whitfield & Bruton 1989). Furthermore, tidal exchange can result in short-term salinity fluctuations which may be quite steep.

Estuarine sediment composition is influenced by freshwater inflow. Water temperatures within South African estuaries fluctuate not only seasonally, but also under conditions of variable freshwater inflow. In addition, shorter term temperature fluctuations of 10–12 °C within a single tidal cycle occur during coastal upwelling (Day 1981*c*; Taylor 1992). The extent of temperature fluctuations, therefore, depends on the geographical location of the estuary, the volume of freshwater entering the system, magnitude of flooding, intensity of coastal upwelling, and meteorological conditions. Furthermore, catchment mismanagement and poor land usage, combined particularly with flood and drought conditions, can result in substantial variation within the seston along the length of the estuary (Grange & Allanson 1995).

The composition, distribution and abundance of estuarine macrofauna is a result of the combined effects of numerous factors. Successful occupation of estuarine habitats requires organisms to respond (physiologically and/or behaviourally) to both long and short-term changes in the environment. In the 1970s a number of researchers in South Africa determined the physiological tolerances of the dominant estuarine macrobenthic organisms in an attempt to explain the distribution and abundances of species both within and between estuaries. This earlier work was synthesised by Day (1981*b*) and Hill (1981). More recently, researchers have begun to examine the functional responses of some estuarine macrobenthic species to fluctuating physical conditions. In addition, more attention has been paid to human impact on estuarine species and community structure. Such studies are crucial for predicting the responses of estuarine communities to any alteration or disturbance to the estuarine environment.

Responses to salinity

Most biologically orientated definitions of estuaries emphasise salinity, or perhaps more accurately, variable salinity, as an essential feature of estuarine systems (Day 1981*a*, Knox 1986). Thus one of the principal characteristics of an estuarine organism is its ability to accommodate changes in salinity. Within South African estuaries, salinity changes may be erratic and extreme. Many estuaries experience short periods of virtually fresh water, through normal periods where salinity gradients are not pronounced and the tidal section is exposed to marine conditions, to potentially prolonged periods of hypersalinity during droughts (Forbes 1974, McLachlan & Erasmus 1974, Blaber *et al.* 1983; de Villiers & Allanson 1989; Whitfield & Bruton 1989).

Although the composition and distribution of the macrobenthos within an estuary are a result of the combined effects of numerous factors, with sediment composition probably having an overriding influence on infaunal distribution (Hill 1981), the salinity tolerance of a species is also important. Further, salinity tolerance is a convenient means of categorising and so, very broadly, separating

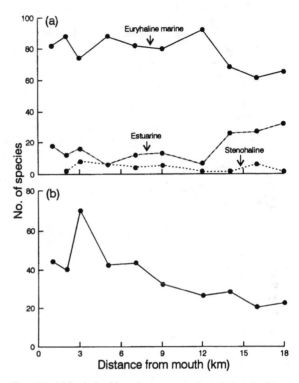

Figure 8.2. **(a) Analysis of faunal components along the length of the Kariega River estuary. Data expressed as a percentage of the total number of species found. (b) Change in total number of species found along the length of the Kariega River estuary. (Both redrawn from Hodgson 1987.)**

different estuarine faunal groups. Day (1981*b*) suggested the following components amongst the estuarine fauna based on their salinity tolerance ranges.

1. Stenohaline marine component (salinity tolerance 30–40).
2. Euryhaline marine component (salinity tolerance 15–60).
3. True estuarine component (salinity tolerance 2–60).
4. Migratory component (salinity tolerance variable to independent).
5. Terrestrial component (mostly independent).

Clearly, there is considerable overlap between some of these categories. Nevertheless, such broad groupings of the fauna as proposed by Day (1981*b*), together with the subsequent responses of the animals to changes in salinity, enables some explanation of, and predictions as to, their distribution and success within an estuarine environment.

In South Africa, most estuarine invertebrates fall into the euryhaline marine component, with only a small percentage which can be classed as truly estuarine (Figure 8.2*a*). The normal distribution pattern of species diversity within an estuary is for numerous species to occur at the mouth with a steady decrease toward the head

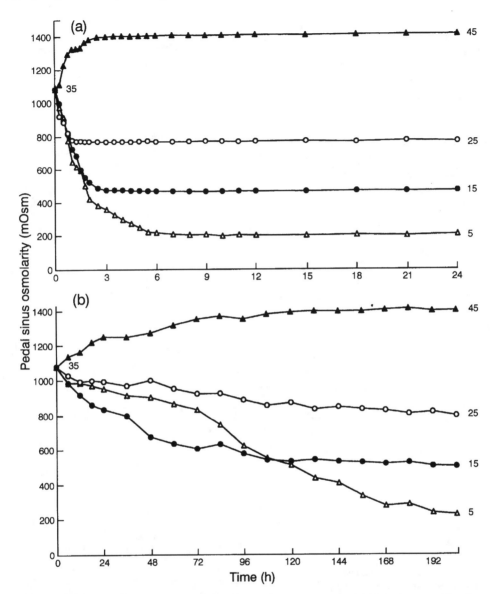

Figure 8.3. **Variation in pedal sinus osmolarity with time in *Solen cylindraceus* acclimated to a salinity of 35 (1074 mOsm) and exposed to salinities of 45 (1394 mOsm), 25 (754 mOsm), 15 (451 mOsm), and** 5 (167 mOsm). (a) Animals removed from burrows; (b) Animals retained in burrows. (After de Villiers and Allanson 1989.)

(Figure 8.2*b*) (e.g. Branch & Grindley 1979; Hodgson 1987). This observation may not necessarily hold for closed systems in which the invertebrate fauna often exhibit not only a low species diversity, but a more uniform distribution – most likely the result of calmer waters, little variability in sediment composition and prolonged periods of more uniform and stable salinities (e.g. Koop *et al.* 1983; Platell & Potter 1996). There is a reasonable amount of information on the salinity tolerance/osmoregulatory ability of species and their distribution within South African estuaries (Hill 1981). For most South African estuarine species a salinity range of 5 to 55 is usually non-lethal.

One of the few species in which the functional responses to salinity change have been investigated is the infaunal bivalve *Solen cylindraceus*. This species is an osmoconformer with a fairly wide salinity tolerance range of *c.* 10–65 (Day 1974; McLachlan & Erasmus 1974; de Villiers 1990). This animal effectively utilises its burrowing ability to buffer short-term effects of any variations in salinity of the overlying water. de Villiers and Allanson (1989) showed that there was a marked delay in attaining osmoconformity in those animals which, on exposure to different salinities, were within their burrows when compared to those that were removed (Figure 8.3*a*, *b*). This is considered

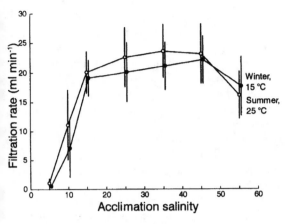

Figure 8.4. **Filtration rates (mean ± SD) of *Solen cylindraceus* collected during summer and winter and acclimated to different salinities. (After de Villiers *et al*. 1989*b*.)**

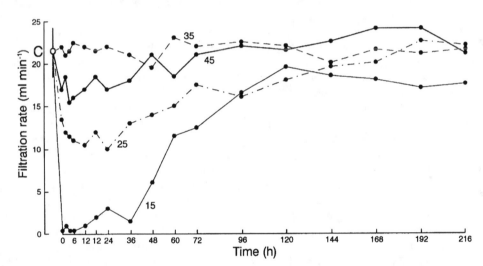

Figure 8.5. **Filtration rates of *Solen cylindraceus*, acclimated to a salinity of 35 at 25 °C, subjected to abrupt salinity changes (after de Villiers *et al*. 1989*b*). To avoid cluttering ± SD not shown: at no stage was a SD greater than ±4.5 ml min⁻¹ recorded. *C* = control value, overall mean ± SD of animals maintained at original acclimated conditions.**

to be attributable to a behavioural response by the animal to unfavourable salinity change, as it was noted that those exposed to a salinity of 5 had retreated to the base of their burrows where a more osmotically favourable 'microhabitat' is likely to be found. As interstitial salinities are reasonably stable for relatively long periods of time, despite salinity variations in the overlying water (Sanders *et al*. 1965; McLusky 1968; Kinne 1971), it is concluded that the observed behavioural osmoregulatory response by *S. cylindraceus* is linked to the stability of the interstitial salinity.

de Villiers *et al*. (1989*b*) carried out a detailed examination of the effect of exposure to different salinities on the filtration rate of *S. cylindraceus*. Optimal filtration rates (19–23 ml min⁻¹) were recorded over a range of salinities, 15–45 (Figure 8.4). When subjected to changes in salinity,

falling within this range, the animal is capable of complete acclimation. However, there was an initial inhibition of filtration rate and the subsequent rate of acclimation appears to be dictated, not only by the magnitude, but direction of salinity change (Figure 8.5). Exposure to increased salinity induced a lesser initial inhibition and subsequent shorter acclimation time than did exposure to lower salinities. Similar observations have been made for a number of other bivalves (Van Winkle 1972; Shumway 1977; Paparo & Dean 1984). These initial decreases in filtration rate may be explained by Shumway's (1977) proposal of decreased, and possibly arrested, ciliary activity in response to osmotic stress. In the case of *S. cylindraceus*, such variation in filtration rate, in response to a change in salinity, would serve to decrease the volume of water pumped by the animal. This, together with the behavioural osmoregulatory response associated with its infaunal habitat, effectively buffers the degree of osmotic shock and effects a greater time period for *Solen* to conform and acclimatise. Such a combination of physiological and behavioural responses which allows *Solen* successfully to occupy estuarine environments is probably typical of many other species.

Responses to temperature

The influence of temperature in defining distribution and abundance of South African estuarine macrobenthos has been known for some time (see Hill 1981 for review). Many of the warm temperate and subtropical infaunal species have been found to survive water temperatures of c. 35 °C and resist brief exposure to higher temperatures. The burrows of many species provide some protection against transient high temperatures. Temperatures above 35 °C are not uncommon in some Southern African estuaries, e.g. 39 °C in Kosi Bay (Blaber 1973), 38.5 °C in St Lucia (Millard & Broekhuysen 1970), 36 °C in Swartkops River estuary (MacNae 1957), 34 °C in Kariega River estuary (Taylor 1988). However, the lethal limits possessed by a species are rarely a restricting factor within its normal distribution and habitat and such data are of limited ecological value. Direct lethal effects are not the only manner in which temperature can influence survival and distribution. Sub-lethal effects, which reduce an animal's fitness or activity may prove indirectly lethal. The effect of moderate temperature change on growth, feeding, reproduction and other physiological and behavioural functions may combine to render a population non-viable. Estuarine water temperatures vary seasonally but, in addition, many estuaries experience large, short-term, fluctuations in temperature of 6–12 °C (Day 1981b; Taylor 1992), a result of the tidal movement of cold seawater into the estuary during periods of coastal upwelling. Although estuaries can be subjected to large and often rapid changes in temperature, very little attention has been paid to the effect of such on the macrobenthic invertebrates, even though they often comprise the largest proportion of the fauna in many southern African systems.

The role of temperature in restricting the spread of temperate fauna into sub-tropical and tropical areas, is unlikely to operate simply at the lethal temperature level, but rather through a range of sub-lethal effects which may influence population structure and viability of adults and/or larvae. Hanekom and Erasmus (1989) indicate that *Upogebia africana* in the Swartkops River estuary have a distinctive spring and summer breeding cycle and suggest that a change in temperature may be the environmental cue initiating the breeding. Hill (1977), too, found that the breeding season for *U. africana* was related to temperature, and noted that not only did warmer temperatures result in an extended breeding season, but they were also associ-

ated with smaller sized animals (*Upogebia* from Natal being smaller than those from the Cape) and, since the number of eggs carried is related to size, smaller females thus carry fewer eggs. A similar situation is reported for the crab *Hymensoma orbiculare*, which in Lake Sibaya breeds throughout the year, peaking in summer (Hill & Forbes 1979), while in the Cape estuaries, Broekhuysen (1955) reported winter breeding. In both areas breeding coincides with the rainy season, but salinity is not believed to be the trigger. *Hymensoma* has also been shown to breed at a smaller size and produce fewer eggs in the warmer areas (the northern end of its range).

More recently Bernard *et al.* (1988) examined the effect of temperature on embryonic development of the truly estuarine mytilid *Brachidontes virgiliae*. At 24 °C and a salinity of 9, embryonic development was rapid and spat formation occurred within 24 hours of fertilisation. Development slowed by a factor of 1.5 at 20 °C and 2.3 at 15 °C. Furthermore, whereas 18% of developing ova failed to develop to spat at 20–24 °C, 34% failed to develop at 15 °C. Thus for species with short larval phases, short-term temperature fluctuations within an estuary could have a significant effect on recruitment and therefore population structure.

The effect of temperature on filtration rate of bivalves is a well-known and recurrent feature in the literature (see Jorgensen 1990 for review). Despite this wealth of information, very few data are available on southern African species, particularly local infaunal estuarine bivalves and crustaceans. The only work to date which examines the effect of temperature on filtration rate in a local estuarine bivalve is that of de Villiers *et al.* (1989a) working on *Solen cylindraceus*. They described a thermal optimum for filtration (for acclimated animals) over the range 15–35 °C with a maximum rate of c. 23 ml min^{-1} recorded at 25 °C (Figure 8.6). In addition, they investigated the acute response of thermally acclimated animals to sudden temperature change and the subsequent adjustments in filtration rate (Figure 8.7). The initial response of animals, and the time taken to acclimate to different exposure temperatures varied depending upon the conditions to which the animals were previously acclimated. The general response, however, followed the three phases for non-genetic adaptation as described by Kinne (1963), i.e. immediate response to environmental change, stabilisation of this response, and a new steady state. In the results presented here (Figure 8.7), animals previously acclimated to 25 °C subsequently reacclimated to exposure temperatures in the range 15–35 °C, within 120–168 h (5–7 days). However, animals exposed to extreme temperatures of 10 °C exhibited no acclimatory response, and those at 40 °C displayed a gradual, limited acclimation. The reduced filtration rate at 40 °C is interpreted as a response to thermal stress. McLachlan and Erasmus (1974) have indicated an upper

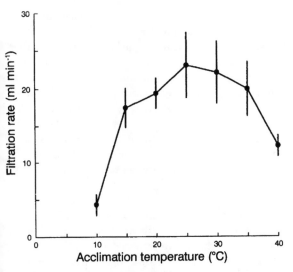

Figure 8.6. **Filtration rate (mean ± SD) of *Solen cylindraceus* acclimated to different temperatures. (After de Villiers *et al.* 1989*a*.)**

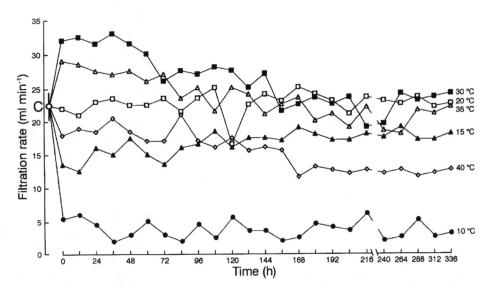

Figure 8.7. **Filtration re-acclimation rate of *S. cylindraceus* acclimated to conditions of 25 °C and a salinity of 35, and exposed to different temperatures. C, control value ±SD of animals maintained** at original acclimated conditions. Error bars are excluded to avoid cluttering. At no point is the SD greater than ±2.96. (After de Villiers *et al.* 1989*a*.)

lethal temperature under *in vitro* conditions of 44.5 °C for *S. cylindraceus*).

As an intertidal animal, it is unlikely that *S. cylindraceus* will be exposed to only one single temperature, but will respond rather to short-term fluctuations within the tidal period, reflecting the difference between estuarine and seawater temperatures. To this end, the acute response will probably reflect more accurately the filtering activity of the population. It is suggested that, for *Solen*, there is a temperature range within which complete acclimation (Precht Type 2) occurs, and beyond which only a partial

(Precht Type 3 for 40 °C) or no (Precht Type 4 for 10 °C) acclimatory response is recorded (see Newell & Branch 1980). Although *S. cylindraceus* appears incapable of adjusting its filtration rate sufficiently rapidly to accommodate short-term temperature fluctuations (within a single tidal cycle), it nevertheless exhibits an effective seasonal compensation. *Solen* is capable of complete acclimation over the thermal optimal range of 15–35 °C, provided conditions persist for a few days (Figure 8.4). Such plasticity would allow for optimal filtration throughout the year.

Not all estuarine invertebrates acclimate to seasonal

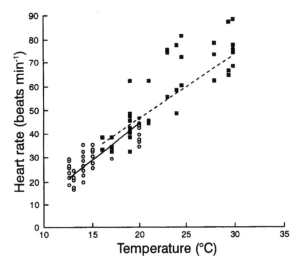

Figure 8.8. **Relationships between daytime heart rate of *Siphonaria oculus* and ambient temperature during summer (■) and winter (○) in the Kariega River estuary. (Redrawn from Marshall & McQuaid 1994.)**

changes in temperature. *In situ* measurements of heart rate of the estuarine pulmonate *Siphonaria oculus* (Figure 8.8) have shown that this species does not adjust its metabolic rate in winter (Marshall & McQuaid 1994). It is suggested that during the lower winter temperatures, *S. oculus* allows its metabolic rate to decline and thus reduce energy expenditure. Whether this is linked to reduced winter food availability for this grazing gastropod remains to be determined.

In conclusion, the sub-lethal effect of temperature on South African estuarine invertebrates has largely been overlooked and requires a great deal more investigation. Researchers should extend beyond the determination of lethal temperature limits, and adopt a more holistic, functional approach.

Temperature salinity interactions

Both temperature and salinity are important environmental variables, and an animals' success depends upon its ability to accommodate, tolerate and/or avoid unfavourable fluctuations in these two factors. In many situations changes in temperature and salinity do not occur singly but together. The south Western Cape, for example, is a winter rainfall area and thus estuarine animals in this region may often encounter simultaneous conditions of low salinity and low temperature. On the Natal coast, winter temperatures do not drop as low as those further south, and the rainy season is in summer. Estuaries in this region rarely experience conditions of low temperature and low salinity. Hill (1981) has argued that because of such varying conditions in the different regions, that this could prevent many northern warm water species inhabiting the more southerly estuaries.

Despite the realisation that both temperature and salinity are important physical variables which impact on an animal's success and, more importantly, that these two variables often act in concert (see Kinne 1964), most workers have tended to examine the effects singly. In South Africa, very little research has been done which investigates the combined effect of temperature and salinity on the macrobenthos.

Hill (1974) has described a marked temperature/salinity interaction for the first-stage zoeae of *Scylla serrata* in which 50% of the larvae survive exposure to a salinity of 17.5 at temperatures of 22.5–25 °C, while over 90% survive the same salinity at temperatures of 7–15 °C. Thus, there appears to be a greater tolerance to lower salinity at lower temperatures. A similar response has been reported by Avis (1988) for the hermit crab *Diogenes brevirostris*, which tolerates lower salinities at low temperatures (Figure 8.9*a*). These findings appear to contradict Lockwood's (1962) postulation that the tolerance of crustacea to low salinity increased at higher temperatures. However, Forbes and Hill (1969) have suggested that higher temperatures may be important in permitting both zoeae and adult *Hymenosoma orbiculare* to survive in freshwater. Their work on the effect of temperature on salinity tolerance by *H. orbiculare* (Figure 8.9*b*) concurs with Lockwood's hypothesis but, given the mortalities recorded at 28 °C, they suggest that this interaction is not necessarily positive throughout the entire temperature range (for *H. orbiculare* at least). Such apparently contradictory results as described above, highlight the need for further and more rigorous investigations.

In investigating the combined effects of temperature and salinity on an animal, it is important not only to consider survival (i.e. lethal limits), as both these factors may vary in the natural environment, yet remain within the tolerance range of a particular species. This may influence an animal's performance at a sub-lethal level. The only work of this nature on a South African species is that of de Villiers *et al.* (1989*b*) and de Villiers (1990) who examined the effect of simultaneous, abrupt, temperature and salinity changes on the filtration rate of the infaunal bivalve *Solen cylindraceus*. The response of summer (25 °C; salinity 35) acclimated animals to conditions of increased temperature and salinity (Figure 8.10) was an initial increase in filtration rate. This is interpreted as a response attributable primarily to increased temperature, since exposure to increased salinity on its own induced an initial decrease in filtration rate (see Figure 8.5), while exposure to increased temperature resulted in an increased rate (see Figure 8.7). Responses to conditions of lower temperature coupled with unchanged or decreased salinity was an immediate and rapid decrease in filtration rate (Figure 8.10), of a magnitude greater than that recorded for equivalent changes

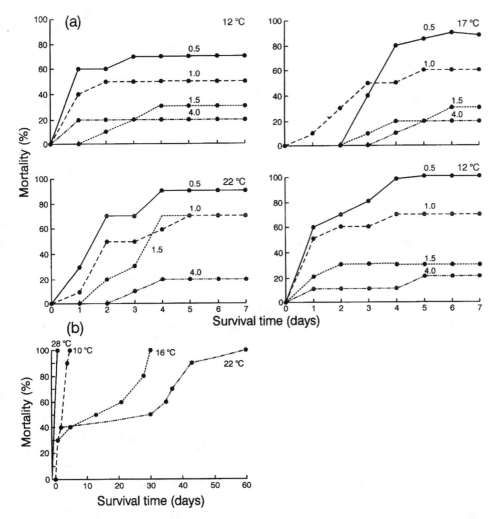

Figure 8.9. (a) Percentage mortality of *Diogenes brevirostris* at four different temperatures and salinities (redrawn from Avis 1988).

(b) Percentage mortality of *Hymenosoma orbiculare* at a salinity of 0.8 at different temperatures. (Redrawn from Forbes & Hill 1969.)

in temperature and salinity alone (Figures 8.5 and 8.7). It appears, then, under conditions of exposure to increased salinity and temperature, that temperature exerts an overriding effect. Under conditions of exposure to decreased salinity and temperature, both factors have a cumulative, acute effect, dictating the magnitude and subsequent time taken for the animal to re-acclimate.

Effects of flooding or reduced freshwater inflow

Many South African estuaries are subjected to periodic flooding (MacNae 1957; Hill 1971; Hanekom 1989). Within the Eastern Cape, for example, the Swartkops River estuary has probably flooded 15 times between 1951 and 1982 (Hanekom 1989). Thus flooding is part of the normal cycle of events. Allanson (1992) has pointed out that regular flooding is of long-term benefit to the health of many estuaries by providing or maintaining access to the sea. Many

plants and animals which live in estuaries are adapted to cope with episodic flood events. In South Africa, only a few studies have attempted to quantify the immediate effects of floods on the macrobenthos or to monitor recovery of populations from floods.

All studies to date have shown that major floods can result in short- to medium-term decreases in the abundance of some macrobenthic organisms. Hanekom (1989) reported 49% and 28% losses in the populations of *Upogebia africana* and *Callianassa kraussi*, respectively, following floods of the Swartkops River estuary in 1975 (Figure 8.11a, b). The decline in the bivalve *Solen cylindraceus* was even more spectacular, with 93% of the population disappearing (Hanekom 1989) (Figure 8.11c). Similar mortalities of this bivalve have been observed in other Eastern Cape systems, e.g. Kowie River estuary (Hill 1981), Kariega River estuary (A.N.H. and C.J. de V, personal observations).

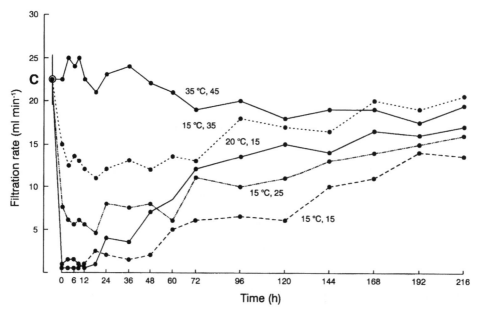

Figure 8.10. **Filtration rates of *Solen cylindraceus* acclimated to summer (25 °C, and salinity 35) conditions and subjected to simulta-** neous, abrupt, temperature and salinity changes. (After de Villiers *et al.* 1989*b*.)

Other species which have been shown to be affected deleteriously by floods in Eastern Cape estuaries include the gastropod *Nassarius kraussianus* (Palmer 1980), the echiuroid *Ochaetostoma capense* (MacNae 1957) and the portunid crab *Scylla serrata* (Hill 1975). Such mortalities are not necessarily due only to hypo-osmotic stress, as many benthic invertebrates exhibit physiological and behavioural responses which would allow them to survive periods of reduced salinity (Hill 1981, and previous discussion). Minor floods lasting a few days would be tolerated by most estuarine species. More severe floods of longer duration, however, would have a more detrimental effect, principally for three reasons.

1   Prolonged exposure to decreased salinities.
2   Physical scouring effects, which may displace much of the estuarine sediments and disrupt the benthic fauna or physically remove animals from their burrows.
3   Deposition of fine silt.

Flooding not only affects the abundance of individual macrobenthic organisms, but alters the community structure of the estuary as a whole. The effect of a severe flood on the benthic community in the Swartkops River estuary was described by McLachlan and Grindley (1974). Prior to flooding, the infaunal bivalves *Solen cylindraceus*, *Macoma litoralis* and *Dosinia hepatica* dominated the middle reaches of the estuary, whereas the anomuran *Upogebia africana* was most abundant in the lower reaches. Both bivalve and mudprawn numbers were severely depleted by the flood. As the macrobenthos recovered from the flood it was noted that the population of *U. africana* had spread further up the estuary. However, some 15 months after the flood, the bivalve populations were recovering and displacing *U. africana* from the middle reaches. McLachlan and Grindley (1974) argued that this shift back to the pre-flood situation was a result of recruitment and growth of bivalve spat, as the populations increased to pre-flood densities.

Within subtropical estuaries, the heavy rains caused by Cyclone Domoina in January 1984 resulted in a scouring of two river channels, and some species common to the lakes of St Lucia (e.g. *Apseudes digitalis*, *Scolelepis squamata*, *Solen cylindraceus*) were redistributed to the channel linking the lake to the sea (Forbes & Cyrus 1992).

Flooding can alter behaviour and benefit some species. The amphipod *Grandidierella lignorum* normally forms part of the benthic invertebrate community (Davies 1982). During flooding it becomes pelagic, the environmental stimulus which triggers this behavioural change being the drop in salinity (Read & Whitfield 1989) which accompanies higher river discharge. These animals remain in the water column until salinity starts increasing again, after which they resume a benthic existence (Figure 8.12). It is suggested by Read and Whitfield (1989) that such a planktonic existence allows this amphipod to exploit the increased total and particulate organic carbon resources that are associated with increased river discharge. Further, they suggest that in addition to increased food availability, the long-term effect could be to increase

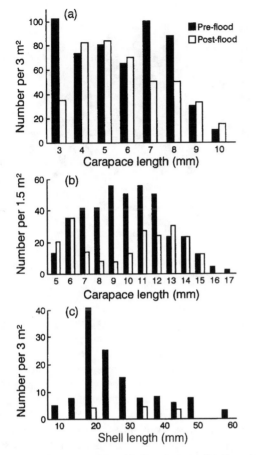

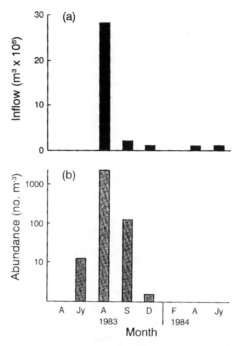

Figure 8.12. **Monthly river discharge (a) and abundance of Grandidierella lignorum (b) at one site in the Kariega River estuary. (Redrawn from Read & Whitfield 1989.)**

Figure 8.11. **Size composition of Callianassa kraussi (a), Upogebia africana (b) and Solen cylindraceus (c) before and after floods at three sites in the Swartkops River estuary. (Redrawn from Hanekom 1989.)**

population size, colonise new areas and prevent genetic isolation of populations. Whether short-term floods are of direct benefit to other estuarine invertebrates remains to be determined.

Very little is known about the direct effect of long periods of reduced freshwater inflow on South African estuarine macrobenthos. A reduced freshwater inflow results in an altered physical environment, characterised by three principal effects: (i) a change in the substrata, often in the form of increased size of sand shoals and a generally larger flood tide delta, which tend to constrict the channel of the lower estuary reducing tidal prism volume and possibly resulting in closure of the mouth; (ii) the development of hypersaline conditions, a consequence of high evaporation; and (iii) a reduction in nutrients making the system dependent upon tidal exchange to supply N and P (Allanson 1992; Allanson & Read 1995).

Hypersaline conditions have a dramatic effect on the benthic fauna. Day (1981c) noted that the high salinities

(52) in the northern lakes of St Lucia resulted in low macrobenthic species diversity. In the South Lake where salinities were lower (35–40), species diversity was greater. Boltt (1975) found that when salinities in the North Lake dropped to 20–25 a rapid re-colonisation from the South Lake occurred, mobile animals and those with planktonic larvae being the first to appear. However, Blaber et al. (1983) reported that after a period of elevated, stable salinities (31–44), there was a marked increase in animal diversity and biomass in this area (Table 8.4). The biomass of the benthic invertebrates was dominated by the filter-feeding bivalve Solen cylindraceus, which ten years previously had been recorded as rare (Boltt 1975). The success of this bivalve in the Kariega River estuary is also thought to relate to conditions of elevated, stable salinities (31–45), within the range that would allow Solen to filter optimally for most of the year (de Villiers et al. 1989b). It is also possible that the stable marine conditions also favour recruitment and survival of its larvae.

Where Man has exacerbated drought conditions by impoundment of flow upstream, the effects may be catastrophic. Whitfield and Bruton (1989) report a case study of the Seekoei River estuary in the Eastern Cape which, following the complete abstraction of all river flow by farm dams, closed and become hypersaline with salinities of 98 recorded! Such hypersaline conditions resulted in the loss of a high proportion of aquatic biota.

Table 8.4. Changes in relative biomass of some benthic invertebrates in South Lake, St Lucia between 1972–3 and 1981–2 following periods of stable salinities

Taxa with increased biomass	Taxa with decreased biomass
Solen cylindraceus	Assiminea sp.
Eumarcia paupercula	Prionospio sexoculata
Dosinia hepatica	Nematoda
Nemertea	Harpacticoida
Marphysa macintoshi	Nassarius kraussianus
Glycera convoluta	
Grandidierella lignorum	
Cumacea	

Source: Modified after Blaber et al. (1983).

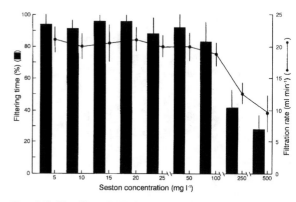

Figure 8.13. **The effect of different seston concentrations on filtration rate and time spent filtering by *Solen cylindraceus*. (Redrawn from de Villiers & Hodgson 1993.)**

Influence of human activities, excluding bait collecting

Although it is widely recognised that human activity can affect the fauna and flora of estuaries, there have been very few detailed studies on how such activities affect southern African estuarine macrobenthos. Activities commonly include the impoundment of freshwater, artificial breaching of closed estuaries, building of marinas and bridges, dredging, artificial stabilisation of the mouth by canalisation, macrophyte harvesting, and recreational activity such as power boating. Impoundment of freshwater not only affects the salinity regime of the estuary, but also often results in the closure of the mouth which, in turn, can have deleterious effects on some macrobenthic organisms. For example, the completion in 1990 of the Wolwedans Dam on the Great Brak river resulted in closure of the estuary mouth, a previously mainly open system. The consequence of this was no recruitment to the population of *Upogebia africana* and a subsequent shift in size-class distribution, with only large adults being present two years later (Wooldridge 1994).

In an assessment of the effects of the 1983 artificial breaching of the Bot River estuary mouth, de Decker (1987) found that the opening of the mouth resulted in a reduction in the number of species from 23 to 15. Most of the species loss was attributed to the collapse of the macrophyte *Ruppia maritima*, on which species such as *Arcuatula capensis*, *Melita zeylanica* and *Cyathura estuaria* were largely dependent. Artificial breaching, however, may to a certain extent redress previous man-made influences. For example, the breaching of the mouth of the Great Brak River enabled the larvae of *U. africana* to enter the estuary, and for the population of mudprawns to recover (B. Newman, personal communication).

The harvesting of macrophytes can have a deleterious effect on benthic invertebrates. Stewart and Davies (1986) showed that in Marina da Gama (Zandvlei estuary) areas where regular harvesting of *Potamogeton pectinatus* occurred regularly, invertebrate standing stocks were always far lower (c. four times) than in unharvested sites.

The pivotal importance of macrophytes as refugia for some species of macroinvertebrate is exemplified also by the natural die-back of *P. pectinatus* in Swartvlei in 1975. Prior to this the filter-feeding bivalve *Brachidontes virgiliae* constituted 90% of the invertebrate biomass in the system. After the die-back the bivalve population collapsed to under 5% (Stewart & Davies 1986). Any reduction (whether natural or man-made) in the filter-feeding community clearly has important ecological consequences. Filter-feeders are not only an important source of food for fish and birds, but also they must play a role in maintaining water quality.

Waves created as a result of power boating and water-skiing can have a dramatic effect on the intertidal areas of estuaries. These not only promote erosion of estuarine embankments (C.J. de V., personal observations) but cause resuspension of large amounts of surface sediments. de Villiers (1990) demonstrated this, recording a fivefold increase in total suspended material over the intertidal mudbanks of the Kariega River estuary (from 100 to 540 mg l^{-1} – the result of a single pass of boat and ski). What long-term effects this would have on the fauna is uncertain and requires investigation. de Villiers and Hodgson (1993) have demonstrated a marked reduction in not only the rate (50% reduction), but the time spent filtering (70% reduction) by *Solen cylindraceus*, under conditions in which the seston concentration was 500 mg l^{-1} (Figure 8.13). As filter-feeders are the dominant macrobenthos in the Kariega River estuary (Hodgson 1987), prolonged exposure to such suspension loads could possibly have deleterious effects, and may explain in part, why filter-feeders are reduced in silt-laden estuaries. Silt would also have a negative impact on macrophytes such as *Zostera capensis*, which in turn could affect macrobenthic species diversity and abundance.

Not all human activity is detrimental to estuaries. Baird *et al.* (1981) determined that the development of a marina near to the mouth of the Kromme River estuary provided an increased habitat area for benthic macro-invertebrates. This is also the case for the marina developed at the mouth of the Kowie River estuary (A.N.H., unpublished data). The colonisation of marinas by some estuarine invertebrates, however, is not always welcome. The development of Marina da Gama in Zandvlei provided a perfect habitat for the estuarine polychaete *Ficopomatus enigmaticus*, which achieved a standing stock of 2.88 ± 2.24 t (dry weight excluding tube) (Davies *et al.* 1989). This worm is perceived as a problem in the marina, as it forms dense encrustations on jetties and the concrete walls of canals, which can be hazardous to residents (due to its sharp calcareous tube), who make intense recreational use of the waterways. Despite this, removal is not an option, as Davies *et al.* (1989) have shown that *Ficopomatus* plays a fundamental role in the maintenance of water quality in this system by greatly reducing particle loads through its filter-feeding activity. It is estimated that this polychaete removes *c.* 130 kg (wet weight) of suspended material per hour.

Bait collecting

Angling is one of the most popular leisure activities carried out in estuaries. In addition, some estuaries (e.g. Knysna River estuary) support a subsistence fisheries (Cretchley 1996). Associated with fishing activities is the collection of bait. Except for some information on the collecting of *Upogebia africana* in the Swartkops River estuary (Martin 1991; Hanekom & Baird 1992), the only study on the impact of bait collecting on South African estuarine macrobenthic invertebrates is that of Cretchley (1996) in the Knysna River estuary, although a number of research workers had already suggested that it could be a problem (Hill 1967; Forbes 1973; Hanekom 1980; Gaigher 1987; Baird 1988). Of those species collected for bait, *Callianassa kraussi*, *Upogebia africana* and some polychaetes are most often taken by anglers. Although the work of Cretchley (1996) is the only thorough investigation on the impact of bait-collecting in an estuary, other studies have been carried out in Langebaan Lagoon (Wynberg & Branch 1991, 1994). The findings from the studies in the Knysna River estuary and Langebaan (which has a similar species composition to the lower reaches of many estuaries), we suggest, can be used to predict what might occur in other estuaries.

Wynberg and Branch (1991) determined that in Langebaan, bait collectors only removed about 0.01% of the sand- and mudprawn populations, although in a heavily utilised area the percentage rose to 3.2%. Whilst this represents the removal of over 1.2 million prawns, they concluded that in Langebaan the prawn populations were not under threat, but their data did suggest that in those regions where the prawns were collected the modal body sizes were smaller. Martin (1991) determined that licensed bait outlets and private fishermen collect approximately 4000 kg dry mass of *U. africana* per annum from the Swartkops River estuary. In the Knysna River estuary, Cretchley (1996) estimated that bait collecting removed about 700 kg dry mass of *U. africana* annually from six popular bait-collecting sites. This represented 8.5% of the mudprawn stocks at these sites, or an estimated 0.85% of the entire estuarine population. Cretchley (1996) also concluded that such a level of exploitation was not a threat to the mudprawn populations.

What is perhaps of greater significance is the secondary impact of bait collection. In Langebaan, bait-collecting was found to result in 1300 kg of macrofauna being disturbed annually, 80% of which was probably preyed upon by scavenging gulls (Wynberg & Branch 1991). Wynberg and Branch also undertook a series of experiments to determine the long-term effects of disturbance caused by bait digging on the biota of the intertidal sand flats. Their experiments showed that removal of *C. kraussi* and *U. africana* by digging or use of a prawn pump resulted in a large decline in population densities (*c.* 70%) in the disturbed areas, even though only 10% and 46%, respectively, of the total proportions of prawns were initially removed. Recovery of the prawn populations took 18 months. Wynberg and Branch (1994) suggested that disturbance and compaction of sediments have much greater effects than the actual removal of animals. Hanekom and Baird (1992), using the disturbance figures of Wynberg and Branch (1991), estimated that in the Swartkops River estuary there is a disturbance mortality of 192×10^6 kJ of *U. africana*. By contrast, Cretchley (1996) concluded that the methods of collection used by subsistence bait collectors in the Knysna River estuary (= 70% of collectors), who use prawn pushers or tin cans, caused minimal disturbance to other macrobenthic organisms.

In addition to the effects on the prawn populations, the experimental bait-digging of Wynberg and Branch (1994) affected the chlorophyll levels in the sediment, meiofaunal numbers, and macrofaunal numbers, biomass and species richness. One month after disturbance of *C. kraussi*, chlorophyll-*a* concentrations increased and remained elevated for 2–3 months. By contrast, removal of *U. africana* caused a nett decrease in chlorophyll-*a* for about one month. Meiofaunal numbers declined rapidly after disturbance, but this was followed by enormous increases and then a return to control numbers. Finally, their work on the macrofauna showed that disturbance caused a decrease in numbers, biomass and species richness and that this component of the benthos still had not recovered after 18 months.

The studies of Wynberg and Branch (1991, 1994) and Cretchley (1996) suggest that over-exploitation of prawns in Langebaan Lagoon and the Knysna River estuary is not yet a problem. However, the disturbance associated with bait-collecting is more problematical. The increasing accessibility of estuaries to the South African population, together with an increase in sport angling, will mean that there will be even greater pressure on the macrobenthos in many estuaries. Unlike the Knysna River estuary, other South African estuaries are relatively small and therefore could be very sensitive to exploitation.

Conclusions

A great deal is known about the physiological tolerances of many invertebrates which inhabit South African estuaries. It has become abundantly clear that although the distribution of many organisms can be correlated with salinity tolerances and therefore salinity gradients, these relationships are not necessarily causal. A wide salinity tolerance does not always imply success in estuarine environments and a number of additional factors play a role, e.g. temperature, nature of the substratum, turbidity, and water turbulence (flooding). Nevertheless, it is clear that an animal with a limited salinity tolerance would not be able to establish itself, or develop, far up most estuaries.

The studies on the responses of the macrobenthic species and communities to varying environmental conditions and perturbations (both natural and man-made) have provided a sound scientific basis on which to base future management decisions for estuaries and their catchments. It is now possible to give accurate predictions on the impact of freshwater abstraction/impoundment, breaching of the sandbars at estuary mouths, macrophyte harvesting and urban development (marinas) on not only the macrobenthos, but also estuarine ecosystems.

Of great importance are the recent studies of Wynberg and Branch (1991, 1994) and Cretchley (1996). Their results on the impact of bait collecting, combined with regular monitoring of bait stocks, will enable conservation authorities to allow exploitation of estuarine resources at sustainable levels. Such information is of paramount importance, particularly in view of the fact that estuaries are becoming increasingly utilised by man.

Macrobenthic invertebrates and estuarine food webs

Macrobenthos as consumers and their role in energy and carbon flow

The majority of macrobenthic invertebrates in estuaries are suspension feeders, deposit feeders or shredders of plant material. As we have shown earlier, South African warm temperate estuaries are dominated by suspension

and filter-feeding crustaceans and bivalves. It is only relatively recently that determinations have been made on consumption rates by some of these invertebrates and their role in energy and carbon flow.

Callianassa kraussi has traditionally been classified as a deposit-feeder, although this has recently been challenged by Schlacher and Wooldridge (1996a) whose work using stable isotopes suggest that it is a filter-feeder. Whatever its mode of feeding, *C. kraussi* has a considerable impact on the sediment and its associated organisms (Branch & Pringle 1987). Although the consumption rate of *C. kraussi* was not calculated, Branch and Pringle (1987) determined that the activities of this animal results in a bioturbation of about 12 kg m^{-2} per day. The faeces have an organic content (35%) and in the Palmiet River estuary, *Callianassa* faeces amounted to an annual turnover of 235 000 kg organic dry mass (Branch & Day 1984), a figure which is probably exceeded in the larger Swartvlei River estuary (Whitfield 1989). Whitfield (1989) also suggests that *Callianassa* faecal pellets are important for surface detritivores such as the amphipods *Grandidierella lignorum* and *Urothoe pulchella*. The burrowing activities of *C. kraussi* have a significant effect on bacteria in the sediment. Branch and Pringle (1987) found that bacterial numbers, concentrated around the linings of burrows, increased by 30 to 100% in the presence of sand-prawns. In addition burrowing plays a significant role in oxygenating the sediments.

Despite the fact that macrobenthic filter-feeders are abundant in estuaries in South Africa, relatively little work has been done on their feeding biology, and to date most of the work done has been devoted to investigating the zooplankton (e.g. Grange 1992; Jerling & Wooldridge 1994; Jerling 1994). de Villiers and Allanson (1988) have shown that *Solen cylindraceus*, in common with other bivalves, effectively retains particles over a wide spectrum of diameters (2–10 μm) (Figure 8.14). Particles in this size range can be, volumetrically, a significant fraction of the estuarine filtrate (Figure 8.15). Exploitation of such particles may mean that nanoplankton, microplankton, bacterial floc assemblages and a large proportion of organic detrital material contributes to the diet of *Solen* (de Villiers & Allanson 1989; Grange & Allanson 1995). Filtration rates for *Solen* vary minimally within the range of environmental conditions generally experienced by the bivalve (see previous discussion and Figures 8.4 and 8.6). Optimal filtration rates are consequently maintained through most of the year. From these data, it was calculated that in the Kariega River estuary, *Solen* filters c. 4% of the tidal volume per tide (de Villiers 1990). *Solen cylindraceus* shares the intertidal mudflats with the anomuran prawn *Upogebia africana*, which is also a filter-feeder. Despite the fact that this animal is often the most dominant macrobenthic species, nothing is known of its filtration rates and

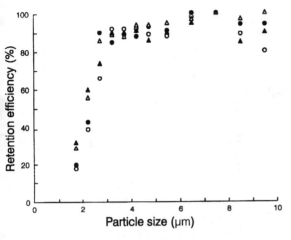

Figure 8.14. **Percentage retention of particles by *Solen cylindraceus* at different salinity/temperature combinations (▲, 35/25 °C; △, 15/25 °C; ●, 35/15 °C; ○, 15/15 °C). (After de Villiers & Allanson 1988.)**

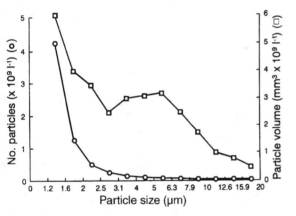

Figure 8.15. **Particle-size distribution of the Kariega River estuary in terms of both number and volume (redrawn from Grange 1992).**

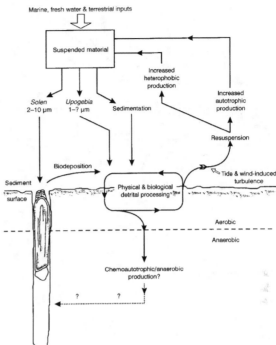

Figure 8.16. **Hypothetical model of consumption and recycling pathways of particulate organic matter by *Solen cylindraceus* and *Upogebia africana*. Illustrated is the physical and biological processing of detritus, and the tidally mediated resuspension and subsequent augmentation of this food resource by related increased autotrophic and heterotrophic production, and by biodeposition. A possible pathway for the contribution from chemoautotrophic and/or anaerobic production in the sediments, to the nutritional requirements of *S. cylindraceus* is also shown. (Modified from de Villiers 1990.)**

food selectivity. However, based on a scanning electron microscopy examination of the feeding appendages, Harris (1992) suggests that *U. africana* could retain particles at least as small as 1μm in diameter. If this is the case, *U. africana* would access an additional and significant volume of suspended particles (Figure 8.15) which is unavailable to other filter-feeders such as *Solen*. Hence competition for food being a driving force for their relative distribution in the estuary is unlikely. A hypothetical model showing the consumption and recycling pathways of particulate organic matter by *S. cylindraceus* and *U. africana* is shown in Figure 8.16.

The filter-feeding serpulid polychaete *Ficopomatus enigmaticus*, which has a maximum retention efficiency of particles ranging between 2 and 16 μm, is the dominant macrobenthic invertebrate in the marina on Zandvlei estuary, with a standing stock of *c.* 3 tons (Davies *et al.* 1989). Davies *et al.* calculated that these polychaetes filter 2.47 ×

10^7 l water per hour, which extrapolates to the entire volume of the marina being filtered in 26 hours! *Ficopomatus enigmaticus* is abundant also on the rocks of many estuaries where, together with large populations of other filter-feeders (e.g. *S. cylindraceus* and *U. africana*), they must process a significant proportion of the water volume each day, and thus play a major role in the turnover of particulate organic matter.

Equally important to determining filtration rates and retention efficiencies of such species, is the question of identifying the source of the particulate matter upon which they feed. The identification of food sources by gut analyses of filter-feeders provides, at best, only a snap-shot view of the animals' diet, and a very rough indication of the nature of the food ingested, with minimal information as regards its original source. The use of stable isotope analyses (^{13}C:^{12}C and ^{15}N:^{14}N ratios) provides for a greater resolution in food web studies. Conceptually simple, 'you are what you eat', there are numerous factors which limit the precision of this technique (for further details see e.g. De Niro and Epstein 1978; Rau *et al.* 1983; Fry & Sherr 1984; Knox 1986). Nevertheless, stable isotope analyses, if interpreted with caution, provide a useful tool in assisting in the

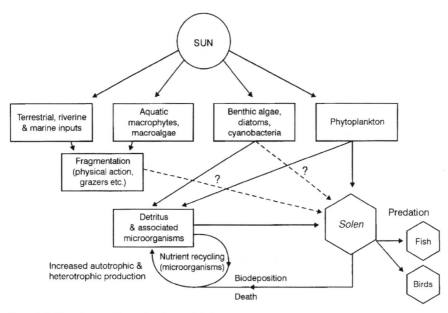

Figure 8.17. **Flow diagram illustrating the possible food sources and pathways for *Solen cylindraceus* in the Kariega River estuary. (Modified from de Villiers 1990.)**

identification of original food sources and their relative importance in the diet.

Investigations on macrobenthic invertebrates using stable isotope analyses have been done in three systems in South Africa: the Kariega River estuary (de Villiers 1990); Langebaan Lagoon (Harris 1992) and the Gamtoos River estuary (Schlacher & Wooldridge 1996a). In Langebaan Lagoon Harris (1992) found *Spartina* and *Zostera* to be the principal sources of carbon assimilated by *Upogebia*, whereas *Callianassa* appear to obtain carbon from a mixture of vascular plant detritus, diatoms and phytoplankton. But, as the values were skewed toward salt-marsh plants, it was suggested that *Callianassa* used vascular plant detritus in the sediment as its main carbon source. The situation in the Kariega and Gamtoos River estuaries is, however, remarkably different. Macrophytes are the most important source of primary production in several Eastern Cape estuaries (e.g. Allanson & Read 1995; Heymans & Baird 1995). In the Kariega River estuary *Zostera* and *Spartina* are abundant (Hodgson 1987) and *Phragmites australis* is the most abundant macrophyte in the upper Gamtoos River estuary (Schlacher & Wooldridge 1996a). None of these primary producers is implicated as an important source of carbon for the dominant filter-feeders (*Solen* and *Upogebia*) in the Kariega River estuary (de Villiers 1990) (Figure 8.17), nor in the Gamtoos River estuary (*C. kraussi* and *Grandidierella lignorum*) (Schlacher & Wooldridge 1996a). These species are thought to derive their carbon primarily from suspended material of phytoplankton and detritus origin (Figure 8.17). From

analyses undertaken in the Kariega River estuary, it appears that there is also a substantial contribution to the detrital pool from two allochthonous sources.

1   Freshwater derived imports – mainly terrestrial detritus and vegetation deposited during flood events.

2   Riparian litter – mainly indigenous terrestrial vegetation overhanging the intertidal areas. Many of the lower branches of trees and bushes are submerged during high tide (Hodgson 1987) and the intertidal banks are littered with debris originating from this riparian vegetation. Both flood-borne material and riparian litter would, due to the hydrodynamics of the Kariega River estuary (Allanson & Read 1987), remain in the estuary for prolonged periods and so be incorporated into the food web. Litterfall has, indeed, been shown to constitute a significant and often major source of organic material in aquatic ecosystems (see e.g. Cummins *et al.* 1973; Gasith & Hasler 1976).

It is very likely that few estuarine invertebrates consume macrophytes directly, although some amphipods play a vital role in macrophyte breakdown (Byren & Davies 1986; Whitfield 1989). Field experiments in Zandvlei showed that invertebrates (primarily the amphipods *Melita zeylanica* and *Austrochiltonia subtenuis*) reduced the biomass of the macrophyte *Potamogeton pectinatus*, which was housed in litter bags, by 1% per day (Byren & Davies 1986). Without these invertebrates, decomposition was reduced

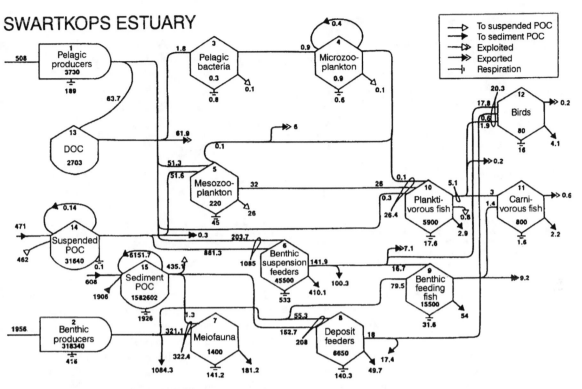

Figure 8.18. **Carbon flow network of the Swartkops River estuary.**
Biomass given in mg C m⁻² and flows in mg C m⁻² d⁻¹. Note feedback
loops in compartments 4, 14 and 15. (From Baird & Ulanowicz 1993.)

by 2.5 times (0.4% d⁻¹). Whitfield (1989) also concluded that in the Swartvlei River estuary, the amphipods *Orchestia ancheidos* and *O. rectipalma* assisted the breakdown of *Zostera capensis* wrack, although their precise role has still to be determined.

Subtropical estuaries often receive a large organic input in the form of mangrove leaves. Brachyuran crabs play a major role in the breakdown of this material. For example, about 74% of the diet of *Sesarma meinerti* is mangrove leaf litter (Steinke *et al.* 1993). Emmerson and McGwynne (1992) calculated that *S. meinerti* in the Mngazana River estuary consumed about 44% of the leaf fall, with individual crabs of 40 g body weight consuming 0.78 g m⁻² d⁻¹ DM (= 284 g m⁻² y⁻¹ or about 653 kJ m⁻² y⁻¹). In addition they found that the assimilation efficiency of the crabs was very high at just over 84%.

The construction of models of energy flow for South African estuaries has lagged behind such studies in marine ecosystems. The recent modelling work of Baird, *et al.* (1991), Baird and Ulanowicz (1993) and Heymans and Baird (1995) has shown that the relative importance of the suspension and deposit-feeding macroinvertebrates can be very different in some open estuaries. Baird and Ulanowicz (1993) calculated that in the Swartkops River estuary, suspension feeders had a biomass 6.8 times

greater than that of the deposit feeders (45 500 mg C m⁻² vs. 6650 mg C m⁻²) (Figure 8.18). In addition, the role of the suspension feeders in energy transfer was determined to be five times greater (1085 mg C m⁻² d⁻¹ vs. 208 mg C m⁻² d⁻¹). In contrast, a model of the Kromme River estuary revealed that the detritivores (brachyuran crabs), although having a smaller standing stock, play a greater role in energy transfer, consuming 2084 mg C m⁻² d⁻¹ compared with 507 mg C m⁻² d⁻¹ for suspension feeders (Heymans & Baird 1995; see also Chapter 11). The energy for the detritivores is derived from sediment bacteria, sediment POC and benthic microalgae. In both the Swartkops and Kromme River estuaries, the majority of the carbon from the suspension and deposit feeders ends up as particulate organic carbon.

The pivotal role of saltmarsh brachyurans in carbon transfer was earlier demonstrated by Taylor and Allanson (1993). By a combination of laboratory and field work it was shown that when crabs (primarily *Sesarma catenata* and *Cleistostoma* (*Paratylodiplax*) *edwardsii*) were absent from saltmarshes, the saltmarsh acted as a carbon sink. When crabs were present, the marsh acted as a carbon source. Clearly any disruption of the functioning of salt marshes and their crabs could have a considerable impact on the functioning of estuarine ecosystems.

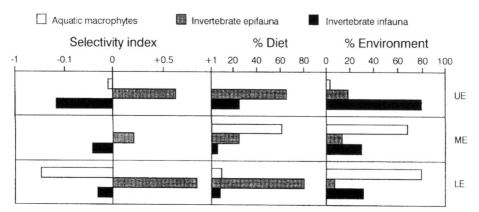

Figure 8.19. **The selectivity of fishes at three localities in the Swartvlei estuary for aquatic plants, epifauna and infauna. Note the negative selectivity of infaunal invertebrates and the positive selectivity of epifaunal invertebrates in all regions of the estuary. Linear** index of selection: +, positive selection; 0, random selection; −, negative selection. **UE, upper estuary; ME, middle estuary; LE, lower estuary. (Redrawn from Whitfield 1988.)**

There have been very few studies on the role of predatory invertebrates in South African estuaries. For example, nothing is known of consumption rates by predatory polychaetes, or the nemertean *Polybrachiorhyncus dayi* which is much prized by anglers as bait. The crab *Hymenosoma orbiculare* has also been classified as an estuarine predator (Branch & Branch 1981). The recent use of stable isotopes suggest that this animal is an omnivore, utilising both detritus and meiofauna (Schlacher & Wooldridge 1996a). A common predatory gastropod in the lower reaches of many estuaries is *Natica*. In a preliminary investigation into the feeding habits of *N. tecta* in Eastern Cape estuaries, Hodgson (unpublished data) found that this gastropod is a major predator of *Nassarius kraussianus*. In some regions of the estuary up to 80% of the empty shells of *N. kraussianus* sieved from sediments had been drilled. Unfortunately it could not be determined at what rate *N. kraussianus* was consumed in the estuaries. Laboratory experiments, however, revealed that *Natica tecta* would consume about four *Nassarius kraussianus* per week. Estimates from the Kromme River estuary (Heymans & Baird 1995) indicate that *Natica genuana* consumes about 10 mg C m^{-2} d^{-1} of filter-feeders.

Macrobenthic invertebrates as a food source for fish

South African estuaries are important habitats for some juvenile and adult fish (Day *et al.* 1981). According to Wallace *et al.* (1984), of the approximately 1500 species of continental shelf fishes recorded from South African waters fewer than 100 species make use of estuaries, although in a more recent analysis Whitfield (1994) increased this number to 142 species of fishes. Bennett and Branch (1990) summarised the theories that have been proposed for estuaries favouring both permanent and transient fish species. One theory is that estuaries provide fishes with a rich source of food.

There is now considerable information on the diets of adult and juvenile estuarine fishes (for literature review see Whitfield 1991). In general, juvenile (postlarval) fishes consume primarily zooplankton (Whitfield 1985; Bennett 1989), whereas the adult fish community has a much wider feeding spectrum. Fishes which feed primarily on benthic invertebrates include species from the Blenniidae, Gobiidae, Sillaginidae, Sparidae, Platycephalidae, Sciaenidae, Lethrinidae and Lutjanidae (Day *et al.* 1981).

Most publications on macrobenthivorous fish feeding have simply documented what the fish eat, given the relative importance of invertebrate species, estimated the amount consumed as a percentage of the body mass per day, or examined what factors influence fish feeding. In some instances the macrobenthic invertebrate species which is consumed is simply a consequence of prey abundance. In the Palmiet River estuary, for example, carnivorous fish consume the most abundant amphipod *Grandidierella bonnieroides*, whereas in the Bot River estuary the same fish consume the more numerous *Corophium triaenonyx* (Bennett 1989). However, availability of prey is also clearly important. Whitfield (1988) recently demonstrated a strong positive selectivity by fishes for epibenthic invertebrates but a negative selectivity for infaunal invertebrates (Figure 8.19). Furthermore, Whitfield (1988) suggested that the infaunal habits of some macrobenthic invertebrates in the Swartvlei River estuary enabled them to avoid fish predation. Thus the infaunal bivalves *Macoma litoralis* and *Loripes clausus*, which are found within eelgrass beds, escape consumption because they are screened from predators. Deeper burrowing species are also largely able to avoid most predatory fish, although both *Callianassa kraussi* and *Upogebia africana* are the most important dietary items of the spotted grunter (Hecht & van der Lingen 1992). The importance of these prey to the spotted

Table 8.5. The total production, consumption by fish and percentage of production consumed of some benthic invertebrates in the Bot River estuary. (From Bennett & Branch 1990.)

Invertebrate consumed	Total production (mg m^{-2} year^{-1})	Amount consumed (mg m^{-2} year^{-1})	% of production consumed
Polychaeta			
Capitella capitata	342	23	6.86
Ceratonereis erythraensis	528	2	0.40
Isopoda			
Cyathura estuaria	1594	24	1.50
Exosphaeroma hylecoetes	5762	1009	17.50
Amphipoda			
Corophium triaenonyx	912	669	73.36
Grandidierella bonnieroides	610	165	27.05
Melita zeylanica	1963	587	29.92
Decapoda			
Callianassa kraussi	1622	24	1.48
Hymenosoma orbiculare	404	347	85.90
Mollusca			
Hydrobia sp.	1319	551	41.78
Tomichia sp.	418	137	32.75

grunter *Pomadasys commersonnii* is, however, reduced in turbid estuaries (Hecht & van der Lingen 1992).

Selectivity may be biased towards one sex of prey. Schlacher and Wooldridge (1996b) showed that in the Gamtoos River estuary corophioid amphipods were the main prey of White steenbras *Lithognathus lithognathus* and Cape stumpnose *Rhabdosargus holubi*. The fish mainly consumed male *Grandidierella lignorum*, which the authors attributed to the fact that the males were more active on the surface than females, who rarely left their tubes. It was suggested that the low selectivity for juvenile *G. lignorum* was due to lower predator detection or capture efficiency.

There have been very few attempts to determine the rate by which invertebrates are consumed by fish. Bennett and Branch (1990) calculated daily consumption rates of macrobenthic invertebrates by fish in the Bot and Palmiet River estuaries (Table 8.5). From this they extrapolated the data to estimate annual consumption rates. In addition they estimated the impact of predation on the macrobenthos by calculating what percentage of invertebrate production was consumed by the fish.

Predation impact was highly variable. Some macrobenthic species such as *Hymenosoma orbiculare* and *Corophium triaenonyx* have over 70% of their annual production consumed by fish; in other species it is less than 1% (Table 8.5). Despite the fact that some species do have a substantial proportion of their production consumed by fish, Bennett and Branch (1990) concluded that fish predation was not having a significant effect on the overall prey resource in the Bot River estuary. Similarly, Hanekom and Baird (1992) have calculated that, in the Swartkops River estuary, fish only consume 5% of the annual somatic production of *U. africana*.

In models of carbon flow (Figure 8.18), Baird and Ulanowicz (1993) determined that benthic feeding fish in the Swartkops River estuary consumed 16.7 mg C m^{-2} d^{-1} of benthic suspension feeders and 9.4 mg C m^{-2} d^{-1} of benthic suspension and deposit feeders in the Kromme River estuary. These consumption rates represent only 0.03% and 0.02% of the macrobenthic standing stock of carbon in these estuaries, respectively.

Macrobenthic invertebrates as a food source for birds

One hundred and twenty-seven species of birds have been recorded from the estuaries of South Africa and of these, 68 species feed on invertebrates (Siegfried 1981). Although the diets of both resident and migratory birds appear to be well known, few studies have attempted to determine the impact of birds on their invertebrate prey or to calculate energy transfer between invertebrates and birds. This is presumably attributable to the inherent difficulties of such studies. Baird *et al.* (1985) reviewed the four main methods which have been used to estimate food consumption by birds, all of which have their drawbacks.

One of the first studies to examine energy consumption by waders in South Africa was by Puttick (1977, 1980) whose study area was Langebaan Lagoon. The results from the Langebaan studies could be atypical for South Africa estuaries as Langebaan Lagoon is a tidal marine inlet, and the density of waders in this seawater lagoon (e.g. 37000–50000 curlew sandpipers during the summer: Pringle & Cooper 1975) is far higher than those recorded for most true estuaries. However, it is interesting to note that Baird *et al.* (1985) calculated that the birds consume about 142 kJ m^{-1} y^{-1} (= 20%) of the total invertebrate production (705 kJ m^{-1} y^{-1}). Of this the curlew sandpipers consume 87 kJ m^{-1} y^{-1}, or 12% of the net annual production available to them.

Velásquez *et al.* (1991) used indirect methods to estimate and compare the energy intake of birds in the Berg and Swartkops River estuaries. Their calculations suggest that invertebrate feeders consume 2886–6982 kJ ha^{-1} d^{-1} and 5699–7633 kJ ha^{-1} d^{-1} in the Berg and Swartkops River estuaries, respectively. Baird and Ulanowicz (1993) in their model of carbon flow of the Swartkops River estuary (Figure 8.18) calculated that birds consume 17.8 mg C m^{-2} d^{-1} of benthic suspension feeders and 0.6 mg C m^{-2} d^{-1} of benthic deposit-feeders. This represents a small fraction only of the carbon flow from the macrobenthos.

The thalassinid prawn *Upogebia africana* is one of the dominant macrobenthic species in Eastern Cape estuaries, and it has been estimated that approximately 85% of its total somatic production (1077 × 10^6 kJ) occurs within the intertidal region (Hanekom & Baird 1992). Because of its preferred habitat, *U. africana* is vulnerable to predation by both fish and birds. Martin (1991) has calculated that the 11 most important bird species (≥80% of the bird numbers) in the Swartkops River estuary consume approximately 13% of the annual somatic production (= 136 × 10^6 kJ y^{-1}) of *U. africana*.

Meiobenthos

Although this chapter has been concerned with an examination of estuarine macrobenthos, meiofauna are an important component of estuarine sediments. They play a significant role in the biodegrative processes in estuaries, and therefore in trophic transfer. Studies on South African estuarine meiofauna are very limited, and results from early research were summarised by Dye and Furstenburg (1981). Since this work only a small amount of research has been carried out on the abundance and distribution of meiofauna in the Mngazana River estuary (Dye 1983a, b) and the Bot River estuary (De Decker & Bally 1985). In addition, the work of Wynberg and Branch (1994) in Langebaan Lagoon has provided some insights into the impact that sediment disturbance can have on meiofaunal abundance. All studies of meiofauna are still confined at the level of the major meiofaunal taxa, as the taxonomy of southern African meiofauna is poorly understood (Dye 1983b).

Dye (1983a) found that the mean meiofaunal densities (2.7 × 10^6 m^{-2}) in the mangroves of the Mngazana River estuary were higher than those of warm temperate South Africa estuaries (e.g. 0.8 and 0.08 × 10^6 m^{-2}). Most of the meiofauna was located within the top 10 cm of the sediment (although animals could be found as far down as the water table), and was dominated by nematodes (80% of meiofaunal community). Dye (1983b) further showed that there was some seasonal fluctuation, with a tendency for densities to increase in summer, and that in the Mngazana River estuary the production from the standing crop was 4.34 g C m^{-2} y^{-1}. De Decker and Bally (1985) determined that the meiofauna of the Bot River estuary was impoverished, which they attributed to the sediment character-

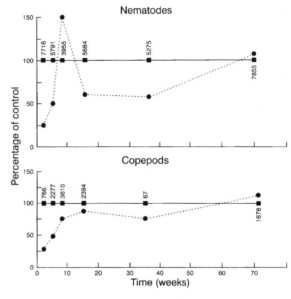

Figure 8.20. **Responses of two types of meiofauna in control (■) and dug (●) areas in relation to the time after disturbance of *Callianassa kraussi* in Langebaan lagoon. Control values are numbers per 100 g dry sediment. (Redrawn from Wynberg & Branch 1994.)**

istics of this estuary. Experiments by Wynberg and Branch (1994) have shown that meiofauna numbers can be significantly affected by sediment disturbances. Immediately after digging, meiofaunal numbers declined, but this decline was followed by an explosive increase and then a return to control levels (Figure 8.20). These results suggest that meiofauna are remarkably resilient.

Some work has been forthcoming on the role of meiofauna in energy transfer in the Kromme and Swartkops River estuaries. In the Kromme, it has been estimated that the meiofauna consumed about 4% of the benthic microalgal production (Heymans & Baird 1995). However, data for meiofaunal biomass and productivity are very limited. Such a lack of information restricts successful modelling of energy flow in estuarine systems. The role of the meiofauna in estuarine ecosystems will thus remain poorly understood until more detailed investigations are undertaken.

Conclusions

Very few studies have examined the role of the macrobenthos in energy and carbon flow in South African estuaries. Recent models of warm temperate systems, however, have shown that in some estuaries more carbon flows through the suspension feeders, while in others the deposit-feeding pathway is more important. In addition, relatively little of this carbon flow is to fish and birds. The use of stable carbon isotopes has provided a good indication as to the initial source of energy for the macrobenthos. Although macrophytes may be the main primary producers in many estuaries, the suspension and deposit-feeding macrobenthos may receive the majority of their carbon from phytoplankton. The macrobenthic suspension feeders, which filter considerable volumes of estuarine water, must play a pivotal role in the turnover of particulate organic matter in warm temperate systems.

Within many warm temperate and subtropical estuaries, Crustacea play an essential role in carbon transfer in saltmarshes and between the saltmarsh and the estuary. They are also vital for initiating the breakdown of macrophyte wrack, thus making it available to other consumers. How much carbon is made available in this way is not clear.

Although we have shown that there is a greater understanding of the role of macrobenthic invertebrates in South Africa's estuaries, all models of energy and carbon flow are for warm temperate, open estuaries. Clearly these models cannot be applied to other estuarine systems. The majority (87.2%) of estuaries in South Africa have periods during which they are closed to the sea. Very little is known about the trophic pathways and energy flow in these systems. As most closed estuaries are small, and are the sites of urban development, an understanding of how they function is vital to their management. Of particular importance is investigating whether the pathways of energy flow in closed systems change during the open phases.

NEKTONIC INVERTEBRATES

Introduction

Under the category nektonic invertebrates of South African estuaries are included those larger species which are capable of active swimming and which characteristically migrate between sheltered nursery habitats, usually estuaries, and offshore spawning grounds occupied by the larvae and adults. The groups or species falling into this category comprise, in rough order of ecological and economic significance, the penaeid prawns belonging to the genera *Penaeus* and *Metapenaeus*, the portunid crab *Scylla serrata* Forskal, and the palaemonid shrimps *Palaemon peringueyi* (Stimpson) (MacPherson 1990) and *Nematopalaemon tenuipes* (Henderson). Of these, the penaeids (de Freitas 1980) and *S. serrata* (Barnard 1950) are tropical Indo-Pacific species extending down the east coast and becoming progressively less common southwards, while *P. peringueyi* (Emmerson 1983) is a more temperate species and more common on the southeast and south coasts. *N. tenuipes* has an Indo-Pacific distribution and is common in India and the Philippines (Holthuis 1980). It is poorly known in the western Indian Ocean, although unpublished data indicate that it occurs within South African limits in KwaZulu-Natal where it appears to be a migrant and is accordingly included here.

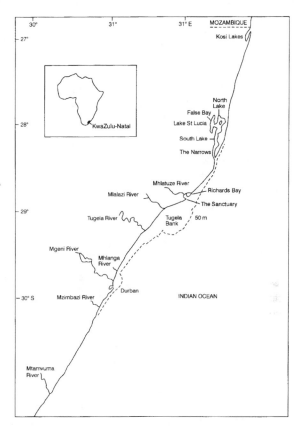

Figure 8.21. **The coast of KwaZulu-Natal showing the areas mentioned in the text. The extent of the Tugela Bank is indicated by the 50 m contour which runs close to the shore both north and south of the demarcated area.**

The palaemonid genus *Macrobrachium* includes seven species (Kensley 1972) which occur in brackish and fresh waters and do not appear to have a marine phase, but in order to cover the larger invertebrates are included here.

The penaeid prawns

Spatial and temporal distribution

The penaeid prawn fauna associated with South African estuaries and shallow inshore waters is dominated by five species, namely (in order of abundance): the white prawn *Penaeus indicus* Milne Edwards; the brown prawn *Metapenaeus monoceros* (Fabricius) and the tiger prawn *P. monodon* Fabricius; followed by, in much smaller numbers, the ginger or bamboo prawn *P. japonicus* Bate and the green prawn *P. semisulcatus* de Haan. These are the most common species of the Western Indian Ocean and all occur in the subtropical estuaries of KwaZulu-Natal (Figure 8.21) but decline in abundance and regularity of occurrence in the more temperate waters of the Eastern and Western Cape

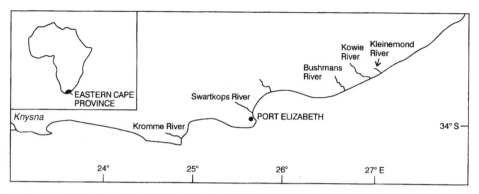

Figure 8.22. **The coast of the southeastern Cape Province showing the areas mentioned in the text.**

Provinces (Figure 8.22). *Penaeus semisulcatus* and *M. monoceros* have been recorded as far south as Durban Bay (Figure 8.21), *P. monodon* as far as Port Elizabeth, and *P. indicus* and *P. japonicus* as far as Knysna on the Western Cape south coast (Figure 8.22) (de Freitas 1980).

Distribution within estuaries can be influenced by salinity and to some extent temperature although penaeid prawns and particularly the juvenile stages, i.e. the estuarine phases, show relatively broad tolerances to both temperature and salinity extremes. Juveniles appear to survive a temperature range from *c.* 9 to 39 °C with low temperature tolerances being greater in temperate species such as *P. japonicus*. Tolerance of salinity extremes is more variable. In the South African context the abundance of *P. indicus* in the St Lucia system declines at salinities below 10 (A.T.F., personal observations) but *P. monodon* can tolerate freshwater (Motoh 1981) and *M. monoceros* was the only species which was regularly recorded in freshwater above the limit of tidal influence in Mozambique rivers (de Freitas 1980). Upper tolerance limits appear to be in the range of 50–60.

Distinct benthic habitat preferences are characteristic of the shallow water penaeids both as juveniles in estuaries and as adults in offshore environments. In the southeast African region this is best shown in the large, mangrove-dominated estuaries and coastal bays of Mozambique (de Freitas 1986) but is also apparent in the generally smaller South African systems. Juvenile *P. japonicus* are typical of bare intertidal sand or muddy sandflats in which they bury themselves during the day. In South Africa they are best known from Kosi Bay and Durban Bay (Figure 8.21), both of which have extensive sandbanks. The latter habitat has, however, been negatively impacted by harbour developments and catches of this species have declined drastically (unpublished Natal Parks Board records). The sandy areas along the eastern shores in the South and North Lake areas of the St Lucia system in KwaZulu-Natal appear to provide a suitable habitat, but

are separated from the sea by the muddy 20 km of the Narrows (Figure 8.21). This muddy stretch appears to be a barrier to the recruitment of postlarvae of this species into the upper reaches of the system and it has always been extremely rare in the bait fishery catches. Attempts to locate populations of juveniles of *P. japonicus* in sandy areas of Richards Bay (Figure 8.21) by beam trawling were unsuccessful (A.T.F., unpublished data).

Both *P. indicus* and *P. monodon* in estuaries are typical of turbid water and muddy areas within mangrove swamps. *P. indicus* is invariably more common and, while this has been attributed to difficulties of sampling amongst mangrove pneumatophores (Hughes 1966), the disproportion extends through to later stages in the life cycle when prawns are no longer associated with pneumatophores. It also appears that nowhere in its range does *P. monodon* form a major component of any prawn fishery, although it remains a prime and valuable target in all areas of occurrence because of its relatively large size.

P. semisulcatus is one of the least common species in South African estuaries, a situation probably arising from its well-established association with submerged macrophytes. This habitat is either rare or only intermittently available in the major KwaZulu-Natal estuarine nursery areas of Richards Bay and St Lucia.

M. monoceros in Mozambique was recorded in a greater diversity of habitats than the other species (de Freitas 1986) but in South African nursery areas it typically occurs in the same muddy habitats as *P. indicus* and *P. monodon*. These overlapping preferences extend to the adult habitat where all three species are taken in the offshore fishery.

The abundance of all the penaeid prawns in South African estuaries is seasonal, being most common in summer and declining in early autumn following emigration to the marine breeding grounds. This is most apparent in the case of the white prawn *P. indicus*, which is the most common species both as a juvenile in the estuaries (Demetriades 1990; Forbes & Benfield 1986a) and also as an adult on the offshore fishing grounds on the Tugela Banks (Figure 8.21) (Demetriades & Forbes 1993).

Emigration from the estuarine nursery grounds by this and the other species taken in the fishery may be associated with decreases in salinity, and this would be an overriding excluding factor. Estuarine catches invariably decline following flood events but other cues triggering emigration also appear to be involved. Catches in the offshore fishery (Demetriades 1990) include small individuals of *P. indicus*, which appear to begin their emigration from the estuaries without any salinity cue and at a size well below the maximum (*c.* 25 mm carapace length) attained in the nursery grounds. *M. monoceros* also begins to emigrate from the estuaries well before maturity but *P. monodon* remains until virtually adult, as shown by the size of individuals taken in the estuarine catches and the absence of this species from all but the largest size categories in the offshore catches. This greater size in the estuaries appears to reflect the well-known faster growth rates, which have made it the species of choice in prawn culture programmes, rather than necessarily a longer stay in the nursery grounds.

Life histories and recruitment patterns

The life histories of the shallow water penaeid prawns have generally been well described (Garcia & LeReste 1981; Dall *et al.* 1990). The eggs are spawned into the water column where they float for 24–36 hours before hatching. Based on the rearing of several *Penaeus* species, there are eleven larval stages comprising five nauplius, three protozoea and three mysis stages before the first postlarval stage is reached. The postlarva has the general adult appearance but the rostral tooth formula is incomplete and this holds for several early stages each of which is characterised by a particular rostral formula. After achievement of the adult rostral formula it is referred to as a juvenile. When the external sexual organs (petasma in the male, thelycum in the female) are completely formed, the prawn is described as sub-adult. Adulthood is reached when the prawn is able to reproduce. Growth during the juvenile stage varies between 30 and 60 mm per month in total length and the whole life cycle can be completed in one year.

Spawning and larval development occur at sea, the larvae and early postlarvae being planktonic. The postlarvae congregate in inshore areas and the mouths of estuaries from where they are transported upstream by flood-tidal currents. In most cases this occurs at night and predominantly during new moon spring tides (Garcia & Le Reste 1981). In the only studies on the east coast of Africa postlarval recruitment was investigated at Kosi Bay, St Lucia, Richards Bay and Durban Bay at various times between 1982 and 1992 (Forbes & Benfield 1986a, b; Forbes & Cyrus 1991; Forbes *et al.* 1994). The species composition of the immigrating postlarvae included the three major commercial species as well as *P. japonicus* and *P. semisulcatus*. All species showed geographical, seasonal and temporal variations in abundance which appeared to be species specific.

There was remarkable consistency in the size of the recruiting postlarvae (Forbes & Benfield 1986b). More than 80% of the *P. indicus*, *P. japonicus* and *P. semisulcatus* postlarvae had carapace lengths (CL) between 1.5 and 1.99 mm. *P. monodon* was slightly larger but equally consistent, with more than 90% of postlarvae falling between 2 and 2.99 mm CL. *M. monoceros* was more variable than the other species; postlarval carapace lengths varied from 1.5 to 3.5 mm CL.

Recruitment into the most northerly of the east coast estuaries, Kosi Bay on the Mozambique border, was totally dominated (>95%) by *P. japonicus* which occurred at densities of *c.* 30 per 100 m³ during the main recruitment periods. A seasonal pattern was not clear although three of the four lowest densities were recorded in summer between November and January (Forbes & Cyrus 1991). Species ratios in the other three systems sampled, i.e. St Lucia, Richards Bay and Durban Bay, showed that *P. japonicus* either dominated, although not to the same extent as at Kosi Bay, or was at least as numerous as the next most common species, *P. indicus*. All other species were represented (Forbes *et al.* 1994). Not all areas could be sampled at the same time and consequently comparisons are problematic, but some general trends were apparent. Abundance of *P. japonicus* postlarvae appeared to be bimodal with peaks in late summer/early autumn and in spring, and the same pattern appeared in *P. indicus*. This bimodality has been recorded in other Indo-Pacific species (Garcia 1985; Staples 1985) and attributed to spawning patterns (Crocos & Kerr 1983). The two species did not follow the same year-to-year abundance patterns. Abundance of *P. japonicus* peaked at St Lucia in 1983 at 120–160 postlarvae per 100 m³ while the highest densities of *P. indicus*, 100–140 per 100 m³, were recorded at St Lucia in 1988. *P. monodon*, *P. semisulcatus* and *M. monoceros* rarely exceeded densities of 5 per 100 m³, which made the recognition of any seasonal patterns difficult, although the two *Penaeus* species appeared to be more common in summer (Forbes & Benfield 1986a). No pattern could be discerned in the case of *M. monoceros*.

It is noteworthy that despite the ubiquity of *P. japonicus* postlarvae in all the systems sampled, juveniles appeared to survive in any numbers only in Kosi Bay and Durban Bay. It is apparent therefore that while the mechanisms exist in prawn postlarvae for location and recruitment into estuaries, subsequent survival depends on the ability of the migrants to locate suitable habitats within the estuary.

Sampling of postlarvae during day and night flood and ebb tides at St Lucia between 1982 and 1984 (Forbes &

Benfield 1986b) indicated a contrast in behaviour between the two commonest species, *P. indicus* and *P. japonicus*. *P. japonicus* was significantly more abundant in bottom samples than in surface samples; this difference was less apparent in the case of *P. indicus*. *P. japonicus* was more abundant over night flood tides but *P. indicus* was found in similar densities during day and night flood tides. The abundance of *P. indicus* in the water column during ebb tides was typically much lower than during floods and was generally consistent with the use of flood tides to bring about a net transport of postlarvae into and up the estuary. The earliest investigations of the mechanisms involved in this process were made on *P. duorarum* in the Gulf of Mexico (Hughes 1969). These suggested a response to changing salinity whereby postlarvae moved off the bottom following exposure to increasing salinities during flood tides but remained on the bottom during periods of decreasing salinities associated with ebbing tides. More recent work (Forbes & Benfield 1986b; Rothlisberg *et al.* 1995) has not discarded the salinity response hypothesis but has suggested that responses to tidally induced pressure changes may be a more reliable cue to enhance the transport of postlarvae by tidal currents. The proposed mechanism involves a similar response to that described by Hughes (1969) for changing salinity, with increasing or decreasing pressures acting, respectively, as stimuli or inhibitors of movement into the water column.

The relatively high levels of activity of *P. japonicus* postlarvae over ebb tides (Forbes & Benfield 1986b) recorded at St Lucia, which does not accord with net transport into the system, can probably be accounted for by the unsuitability of the sediments in the St Lucia Narrows for this species, which is strongly associated with sandy substrata. Activity during ebb tides would thus be associated with an avoidance response to an unsuitable substratum.

Role in trophic dynamics

The significance of the penaeid prawns as consumers or prey items in South African estuaries is little known. Unpublished studies by N.T. Demetriades (personal communication) in which she investigated and compared the foregut contents of juvenile *P. indicus*, *P. monodon* and *M. monoceros* from St Lucia and Richards Bay, indicated that all were basically benthic omnivores. In all three species 65–70% of the diet consisted of crustaceans, particularly the small crab *Paratylodiplax blephariskios*, and detrital floc. The rest of the diet was composed of plant material, copepods, foraminiferans and fish. *P. monodon* was the only consumer of gastropod and bivalve molluscs and also ate a greater proportion of plant material than the other two species which, unlike *P. monodon*, took copepods. Detrital floc was most significant in *M. monoceros*. These differences probably relate partly to the size difference, *P. monodon* in

the estuaries generally being much larger than the other species and therefore better able to handle hard-shelled molluscs.

The impact of penaeid prawn predation on benthic food species and biomass has not been quantified. At the same time there are no benthic production figures available to give a measure of carrying capacity or to allow comparison of different estuaries.

Predation on prawns by fish in the St Lucia system was noted by Whitfield and Blaber (1978) who found that penaeid prawns were 'a common food item of piscivorous fish in St Lucia lake' and specifically mentioned *Elops machnata* and *Argyrosomus hololepidotus*, which constituted 91% of the piscivorous fish in the system at the time. In winter, when prawns occur in minimal numbers in the system, they comprised less than 10% of the diets of both fish species but this increased to 70% in summer when prawn populations peaked. Penaeids formed a significant component of the diet in four out of six common KwaZulu-Natal juvenile carangid fish (Blaber & Cyrus 1983). In the Mlalazi estuary (Figure 8.21) one of the commonest species of the benthic fish community, *Rhabdosargus sarba*, ate 21% Penaeidae by dry weight; this figure rose to 79.3% in the Kosi estuary (Blaber 1984). The full extent and significance of fish predation on prawns in South African estuaries is still unknown. It is noteworthy in this regard that Salini *et al.* (1990) found that 37 out of a total of 77 fish species recorded in a tropical Australian estuary preyed on prawns, although they did not provide any data on total consumption.

The only non-fish predator on prawns of any significance appears to be the water mongoose *Atilax paludinosus* (Whitfield & Blaber 1980). Prawn remains were recorded in 39% of 31 mongoose scats collected at St Lucia in 1975 and 1976.

The fluctuating and highly seasonal nature of prawn populations in estuaries in South Africa and elsewhere presumably precludes any predator from specialising on this resource although it is clear that substantial density-dependent predation on penaeid prawns does occur. It has been estimated that consumption of penaeids by demersal fish in the Arabian Gulf is four times the commercial catch (Pauly & Palomares 1987).

Economic importance

The penaeid prawns of KwaZulu-Natal support both an offshore fishery on the Tugela Banks and estuarine fisheries in the two major estuarine nursery grounds of St Lucia and Richards Bay. Both fisheries are small by global standards, producing about 100 tonnes per annum from the offshore fishery and a combined catch until 1995 of about 20 tonnes from the two estuaries. The retail value of the offshore catch is about R10 million p.a. and from the estuaries

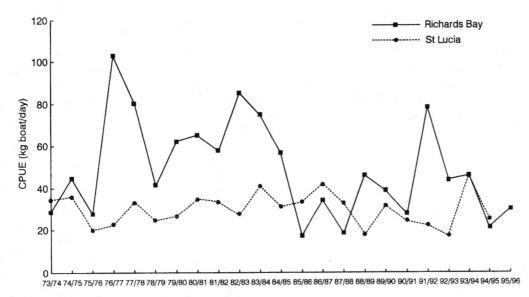

Figure 8.23. **Average annual (June to May) catch per unit effort (Catch per boat day) of the penaeid prawn bait fisheries in St Lucia and Richards Bay.**

about R500 000. The offshore catch has always been taken by commercial trawler operators since the inception of the fishery in about 1966. The estuarine operations have a much longer history, going back to the 1930s although catch data are available for only the last 25 years. The prawn fisheries developed to supply bait for anglers who historically had access to these areas despite their protected nature reserve status. The operations were controlled until the 1990s by the regional conservation body, the Natal Parks Board. The Richards Bay fishery, which operates in the mangrove-lined channels to the west of the main harbour area, has now (1993) been taken over by private enterprise although it is still operated under the same constraints as previously, i.e. trawling is restricted to two dinghies driven by 30 hp outboard motors pulling a gate net or beam trawl 4.9 m × 1.1 m with a 25.4 mm stretch mesh bag. The operation at St Lucia employed the same catching methods, using a maximum of three boats with all netting being done in the 20 km of the system known as the Narrows.

The future of both estuarine fisheries is now uncertain. The St Lucia operation in its present form has been phased out (1996), presumably partly because of the acceptance of the incongruity of such a fishery in a proclaimed nature reserve. At the same time it can be accepted that there will be increasing pressure for the development of an artisanal fishery along those sections of shoreline abutting on traditionally owned land. The future of the Richards Bay fishery, which operates within the broad harbour limits, will depend on the nature of harbour development plans, which could annihilate the present nursery grounds,

although these proposals are being assessed regarding their ecological impact. The Sanctuary area which comprises the Mhlatuzi estuary (Figure 8.21) and was part of the original Richards Bay, but which was separated from the harbour during the construction period, has supported an unofficial artisanal fishery for several years. Attempts are presently being made to establish a programme to monitor the catches taken in this fishery.

In both of the areas where the bait fisheries operate the habitat consists of soft, muddy substrata with associated turbid water. The catch is accordingly dominated by *P. indicus*, which typically contributes in excess of 80%, followed by *M. monoceros* and *P. monodon* (Demetriades 1990). The bulk of the catch is taken in mid to late summer (January–March). The size of the catch has fluctuated from year to year, although the CPUE (catch per boat day) has remained remarkably consistent at St Lucia in comparison with Richards Bay (Figure 8.23). The catch is also influenced by demand and as the prime use is for bait by anglers, fishing conditions are a significant factor.

The CPUE at Richards Bay, despite the development of the bay as a harbour, has frequently been higher than at St Lucia (Figure 8.23) while the total catches have been similar despite the very different sizes of the areas involved (<1 km² vs. c. 8 km²) and the difference in the numbers of boats (Demetriades 1990). Available data on recruitment of postlarvae to these areas (Forbes & Benfield 1986a; Forbes & Cyrus 1991; Forbes et al. 1994) do not cover the same periods and consequently it is difficult to relate these differences to contrasting levels of immigration. Richards Bay is closer to the major offshore concentrations and consequently one would expect greater input of post-larvae to occur, although this remains to be tested. The possibility of a greater carrying capacity in the Richards Bay area exists but the benthic fauna in the two areas is similar in species composition, both being dominated by the small ocypodid crab *Paratylodiplax blephariskios* (Hay 1985). There are no productivity figures for the benthos of either area.

Scylla serrata

Spatial and temporal distribution

This large portunid swimming crab ostensibly has a very extensive range in estuaries and inshore shallow waters from the western Pacific Ocean as far south as northern New Zealand, across to the western Indian Ocean and down the east coast of Africa (Barnard 1950) as far as Knysna (Figure 8.22) (du Plessis 1971). Although there have not been any suggestions that more than one species occurs within South African limits, doubts have been expressed (Estampador 1949; Quinn & Kojis 1987) about the validity of the single-species status over the entire range.

The abundance of *S. serrata* in South Africa becomes more erratic at the southern limit of its range (Robertson 1996). This is probably a reflection of variations in recruitment and the influence of estuarine salinities in combination with steadily decreasing average temperatures with increasing latitude. This combined latter effect was initially suggested by Hill (1981) on the basis of his observations at the Kleinemond estuary (Figure 8.22) in the Eastern Cape Province (33° 30′ S) (Hill 1975) where, following 'severe flooding', 'dead crabs were found on the beach at the mouth of the estuary'. This contrasted with the situation at St Lucia (28° S), where no mortalities were recorded following flooding associated with Cyclone Domoina in January 1984 and follow-up trap catches did not show any impact on the population. This would probably have been related to St Lucia temperatures in the range 25–27 °C while the Kleinemond temperatures would have been between 12 and 16 °C (Forbes & Hay 1988). The effect of such low temperatures was demonstrated by Hill (1980) who showed that emergence time and movement at 12 °C were 24% and 33%, respectively, of their levels at 25 °C and feeding activity stopped.

Density estimates of crabs in one Eastern Cape and three KwaZulu-Natal estuaries exist but there are no long-term records and the different systems were not sampled at the same time, so comparisons should be made with caution. From south to north the results were as follows. In the Kleinemond Estuary (33° 31′ S) in the Eastern Cape

(Figure 8.22), Hill (1975) estimated a density of 0.806 crabs per 100 m^2 in March 1972 based on crabs above 100 mm carapace width (CW). Robertson and Piper (1991), working in two small KwaZulu-Natal estuaries, the Mzimbazi (30° 08′ S) and the Mhlanga (29° 42′ S) in December 1988 (Figure 8.21), estimated densities of crabs above 100 mm CW at 0.282 and 0.280 per 100 m^2, respectively. In January 1979, Hill (1979a) estimated crab densities in three areas of the large (320 km^2) St Lucia Lake system (28° S), namely the Narrows, South Lake and North Lake, at 0.339, 0.156 and 0.037 per 100 m^2 respectively. These areas are progressively further from the sea and therefore from tidal effects and the possibilities of transport or migration of megalopae. Density estimates in three areas of Queensland, Australia, fell in the range 0.02–0.11 per 100 m^2 (Hill 1982) but these referred to crabs above 150 mm CW. On the basis of the above data there does not seem to be any consistent latitudinal gradient of crab densities in South African estuaries nor any distinct difference between South African crab densities and those in the slightly lower latitudes of southern Queensland.

The only other data on comparative temporal and spatial abundance relate to 18 estuaries spread along c. 450 km of coastline, from the Mtamvuna River (31° S) on the southern KwaZulu-Natal border (Figure 8.21) to the Bushmans River (33° 40′ S) east of Port Elizabeth (Figure 8.22), which were sampled by Robertson (1996) between 1991 and 1994. She found that the CPUE and size composition varied considerably amongst estuaries as well as from year to year within the same estuary and attributed this to erratic larval recruitment into the different systems.

Distribution of crabs within estuaries appears to change with age and size (Robertson 1987). Juvenile crabs (<30 mm CW) preferred shallow, protected habitats and tend to remain in intertidal areas during low tide. Larger crabs occurred in deeper waters but moved into the intertidal zone during high tide periods. The nature of diel activity patterns is uncertain as trapping at Richards Bay and in the Mlalazi estuary (Robertson 1989) indicated no difference between day and night trap catches and therefore between day and night activity. This conflicted with results obtained by Hill (1976, 1978) in the laboratory and the field which suggested that crabs were more active at night. Robertson (1989) suggested that her results were caused by the highly turbid conditions in the sampling areas which restricted light penetration and thereby removed any light-based inhibition of crab activity.

The best indications of activity of *S. serrata* under natural conditions were obtained by Hill (1978), who attached ultrasound transmitters to six adult crabs in the Kowie estuary (Figure 8.22) in the Eastern Cape, where they were tracked continuously for 24 hour periods. The crabs were active for about 13 h out of 24, mainly at night, and moved a mean distance of 461 m (range 219–910 m) although their mean displacement was only 62 m (range 1–126 m). Similar observations of relatively restricted foraging areas were made by Robertson & Piper (1991) during mark–recapture studies in two small KwaZulu-Natal estuaries, the Mhlanga and Mzimbazi, north and south of Durban respectively. Hill (1978) categorised the movements as (i) restricted movements around a more-or-less permanent home site; (ii) free-ranging movements during which crabs foraged over extensive distances and did not return to a fixed place each day; and (iii) migration associated with reproduction. The first two categories would involve both males and females and do not distinguish between foraging and movement associated with other activities such as mate location. The third category would be restricted to the females which, unlike the males, migrate to the marine spawning grounds.

Life history and recruitment patterns

S. serrata is a migrant species in which the females typically spawn at sea, the eggs hatch after *c.* 12–17 days depending on temperature (du Plessis 1971) and the larvae pass through five zoeal stages in about 20 days before moulting to megalopae. The megalopae then migrate to estuarine nursery grounds. The only investigation of megalopa recruitment into an estuary was carried out at St Lucia in 1983 and 1984 (Forbes & Hay 1988). Sampling of the water column during night spring flood tides produced both megalopae and first-stage crabs. Densities of megalopae during peak recruitment periods were *c.* 10 per 100 m³. There was an extended recruitment period, with numbers building up from spring (October) and only declining in early winter (June/July). At this stage the source of megalopae recruiting into South African estuaries is unknown, i.e. there is no indication as to how far offshore impregnated females move nor whether megalopae recruiting into South African estuaries are derived from local South African populations or Mozambican populations. There are no offshore fisheries to pinpoint breeding populations and the work of Hill (1974) on egg and larval development suggests that offshore bottom temperatures towards the southern limits of the range of this species might well impair egg development. The source of megalopae then becomes an open question and challenges the nature of existing conservation measures of estuarine populations which are based on the idea of self-sustaining populations.

The abundance of juveniles (<30 mm carapace width) in three KwaZulu-Natal estuaries, Richards Bay, Mlalazi and Mgeni (Figure 8.21) was greatest between May and December (Robertson 1987) and presumably reflects the pattern of extended summer recruitment found by Forbes

& Hay (1988) at St Lucia. Robertson and Kruger (1990) calculated that a CW of about 50 mm was attained after one year, 125 mm after two years and a maximum of about 160 mm after three years in KwaZulu-Natal estuaries. Hill (1975), working in the Kleinemonde and Kowie estuaries in the eastern Cape Province, estimated a CW of 80–160 mm after 12–15 months, after which growth slowed and a CW of 140–180 mm was attained after three years.

These data are inadequate to demonstrate any latitudinal effect on growth rates but there does appear to be an effect on size of both sexes at sexual maturity. If this is based on the size at which 50% of the population is mature, there appears to be a latitudinal increase in females (Robertson & Kruger 1994; Robertson 1996) from *c.* 123 mm CW in KwaZulu-Natal to *c.* 140 mm at the southern limits of distribution in the eastern and southern Cape Province.

Maturity in males is less easily determined and while 50% of males were producing sperm at a CW of 92 mm, the smallest male in which mating was observed in KwaZulu-Natal was in the 115–119 mm size class (Robertson & Kruger 1994) while mating males in the Kleinemond and Kowie estuaries fell into the larger size range, 141–166 mm (Hill 1975). Mating takes place in the estuaries and occurs throughout the year in KwaZulu-Natal but becomes progressively more of a summer phenomenon in the south of the range. Spawning follows a similar pattern.

Following the attainment of maturity and mating the females emigrate from the estuaries to the marine breeding grounds at an age of about two years while the males remain in the estuaries. Hill (1974) has related this breeding migration to the tolerance of the first-stage zoeae to temperature–salinity combinations. Temperatures above 25 °C and salinities below 17.5 caused 'considerable' mortality, indicating that this stage in the life cycle is unsuited to estuarine environments. Low temperature effects on egg development do not appear to be known but first-stage zoeae were inactive below 10 °C in seawater. In South Africa this would limit the distance females could move offshore to about 10 km because of thermoclines and cool, deeper water.

Role in trophic dynamics

Hill (1976) analysed the foregut contents of *S. serrata* from the Kowie and Kleinemond estuaries and compared them with crabs from southern Queensland in Australia. In both areas 50% had mollusc remains, but whereas in South African crabs, gastropods were most common (47% with gastropods, 15% with bivalves), in Australian crabs, bivalves, chiefly mytilids predominated (30% with bivalves, 20% with gastropods). In South African crabs the most common gastropod was *Nassarius kraussianus*. Bivalve genera included *Musculus*, *Lamya*, *Modiolus*, *Pitaria* and

Solen. Crustaceans were found in 22.5% and 20% of South African and Australian crabs, respectively, and in both cases small grapsid and hermit crabs predominated. Fish remains were found in only three Australian crabs (out of 67) and two South African crabs (out of 40), indicating the lack of importance of this particular component in the diet.

The interaction(s) or inter-relations between penaeid prawns and *S. serrata* have not been fully clarified despite the co-occurrence of the crabs and the commonest penaeid species in South African subtropical estuaries. In an experiment (Hill 1976) where prawns were kept in tanks with fed and unfed crabs as well as on their own, there were no significant differences in prawn mortality rates between any of the tanks although no dead prawns were found in tanks with crabs, indicating that crabs would eat dead or incapacitated prawns. *S. serrata* does not appear to catch live prawns. From a comparative feeding point of view, the larger and much more strongly built crabs would be able to deal with more heavily shelled prey than the prawns, such as the larger bivalves and thick-shelled gastropods. There is, however, scope for overlap as regards more soft-shelled, smaller prey, and as bivalves are generally uncommon in South African estuaries (Hill 1976), the possibility of a competitive feeding interaction between prawns and crabs remains. A comparison of the results of gut content studies of prawns by Demetriades (personal communication) with those of Hill (1976) indicates that the crabs are able to handle larger, tougher items than the prawns and therefore should not compete, but in the absence of bivalves greater overlap of food types is likely and therefore competition for food may occur.

Investigations of food and feeding strategies in *S. serrata* by Hill (1976, 1979*a, b*) in the laboratory, in the Kowie River in South Africa and in Moreton Bay in Australia, showed that crushing of bivalve, crustacean or fish tissue induced activity in the crabs. Food location was mainly by contact chemoreception via the dactyls of the walking legs. In soft substrata the dactyls would be pushed into the substratum and this would be followed by use of the chelae to extract burrowed bivalves. Pebbles with attached bivalves were explored with the dactyls which were then used to guide the chelipeds to the prey. Mobile prey such as small crabs were trapped by the legs and pulled forward to the chelae. Tracking investigations in which crab activity and frequency of occurrence in the Kowie River were investigated in relation to food availability showed that crabs spent more time in those areas where prey organisms were more abundant. In terms of potential prey items, burrowing and attached bivalves were some 30 times more abundant than the two species of small crabs, *Cleistostoma algoense* (now *Paratylodiplax algoensis* Manning & Holthuis 1981) and *Hymenosoma orbiculare*. Electivity indices indicated no

preference for nor avoidance of the bivalves but a strong positive selection of the small crabs. Although the crabs never dominated the total food item intake by mass, the energy contribution, because of the threefold greater energy content of live decapod tissue relative to live bivalves, contributed nearly one and a half times as much energy as the bivalves.

Predation on *S. serrata* in South African estuaries is poorly documented. Analysis of pellets regurgitated by grey herons *Ardea cinerea* (8) and the great white egret *Egretta alba* (200) did not show any sign of *S. serrata* (Hill 1979*a, b*). Bass *et al.* (1973) found one crab in the gut contents of 99 *Carcharinus leucas,* the only large shark to occur in southern African estuaries. The major recorded predator on *S. serrata* was the water mongoose *Atilax paludinosus* in the St Lucia system. Of 31 scats collected at St Lucia during 1975 and 1976 (Whitfield & Blaber 1980), 81% contained brachyuran remains although these included the freshwater crab *Potamonautes* sp. Observations of foraging behaviour showed that mongooses carried crabs ashore where they were placed dorsal side down, the claws removed but not eaten and the sternum opened. The remains of large crabs, *c.* 150 mm CW, which were found with chelipeds attached were suggested by Whitfield & Blaber (1980) to have been moribund or dead and to have been scavenged by the mongooses which did not need to remove the otherwise potentially dangerous chelipeds. Crab mortality at St Lucia was most prevalent in July 1976 when the salinity in South Lake dropped to 2.3 and temperatures as low as 12 °C and dead crabs were found washed up on the shores.

Economic importance

To date *S. serrata* has not assumed any great economic importance in South Africa, despite the crab density in those estuaries which have been investigated falling in the range of densities found in Queensland, Australia, where this species is the basis of a major fishery (Hill 1982). The single, small-scale experimental fishery at Richards Bay does not seem to have developed (A.T.F., personal observations) and Robertson (1996) considers that the abundance of crabs in Eastern Cape estuaries is insufficient and too erratic to support a commercial fishery. She acknowledges that there may be scope for recreational, artisanal and subsistence fisheries in years when crabs are abundant but points out that the present size limits of 114–115 mm CW are below the minimum size at which males are capable of mating and that only *c.* 16% of females are mature at this size. Careful management would therefore be required as heavy fishing pressure may drastically reduce the breeding population (Robertson & Kruger 1994). This statement would nevertheless have to be considered in the light of the absence of any information on the source of megalopae to South African estuaries. The long-term potential of an estuarine fishery for this species is likely to be limited, apart from the above factors, by the relatively small estuarine area in South Africa, the general degradation or loss of this habitat through a variety of human influences and the fact that the St Lucia Lake system which is the only estuarine area of any size in the country is a declared Ramsar site and a nature reserve which is part of the Greater St Lucia Wetland Park.

The Caridean prawns

Palaemon peringueyi

Spatial and temporal distribution

Under its former status as *P. pacificus* this species was described as a widespread caridean shrimp occurring in the Indo-Pacific from the Red Sea to Japan and in South Africa from Walvis Bay on the west coast to Kosi Bay on the east (Barnard 1950). Confusion about its presence or absence along the East African coast north of Kosi Bay and the relationship with populations in the northern Indian Ocean have been resolved by the recognition of the southern African population as a separate species, *P. peringueyi* (MacPherson 1990).

Known salinity and temperature tolerance ranges (Robertson 1984) suit this species to life in most South African estuaries. It remains active and is capable of moulting and growth at salinities between 10 and 50 and temperatures of 10 to 30 °C. Greater extremes of salinity and lower temperatures may be tolerated for limited periods but temperatures from 30 °C upwards are rapidly lethal. The relatively low upper temperature limit raises questions regarding the ability of this species to survive north of Kosi Bay as well as its links with populations in the northern Indian Ocean. Tolerance, albeit limited, of subadults to temperatures of 4 °C and salinities down to 5 (Robertson 1984) would be of benefit to these shrimps during flood conditions in local estuaries, particularly at their southerly limits of distribution.

In South Africa this species has a marine adult phase and a juvenile phase which, unlike the strongly estuary-associated species described thus far, may be completed in the marine environment through the use of intertidal rock pools (Emmerson 1985a).

Life history and recruitment patterns

Palaemon peringueyi reaches a maximum size of c. 65 mm total length corresponding to an age of c. 3 years although a full life cycle may be completed in one year (Emmerson 1985b). Females produce between 400 and 1600 eggs,

depending on size. Ovigerous females appear in winter (July) and numbers peak between August and November. In the natural situation ovigerous females have only been recorded in the marine environment and not in estuaries. Development of eggs and larvae in these situations has not been described but in the laboratory the hatching period was 30 days at 20 °C while larval development through six zoeal stages at the same temperature required a further 27–29 days. Development of the final zoeal stage may be delayed or extended, a feature which Emmerson (1985b) described as an adaptation to assist in location of a suitable juvenile habitat.

Movement of *P. peringueyi* between the marine and estuarine environments has been monitored only in the Swartkops estuary (Figure 8.22) in Eastern Cape Province (Emmerson 1983). Abundance of final-stage larvae (zoea 6) in the estuary mouth peaked during night flood tides, especially over spring tides, and although zoeae were also present during ebb tides there was a net influx into the estuary. Retention was apparently aided by lateral movements to the edges of the channel during ebb tides.

Recruitment into Eastern Cape estuaries occurred in summer, beginning in the more southerly systems, the Kromme and Swartkops, and being followed towards the end of summer by the Sundays (Figure 8.22). Size at recruitment was 8–10 mm total length, occasionally 6–8 mm. Size at emigration varied from 30 to 40 mm total length which corresponded to a residence time of nine months to one year in the estuary. Male:female ratios in the estuaries typically deviated from 1:1 and approached 1:2 in most cases although there were significant differences between different estuaries.

Abundance of shrimps in the estuaries investigated was strongly correlated with the presence and abundance of the seagrass *Zostera capensis* Setchell. Loss of the *Zostera* beds in the Kromme estuary by siltation during a flood was followed by a period of decreased numbers of *P. peringueyi* (Emmerson 1986). This sequence was also found in the Bushmans River (Robertson 1984). Abundance of shrimps in the *Zostera* beds in the three estuaries (Emmerson 1986) peaked in summer at 200–400 m^{-2}, declining to 20–100 m^{-2} in winter–spring. This corresponded to a decline in dry biomass from 3–6 g m^{-2} to 0.3–1.5 g m^{-2}. The mean annual somatic production varied from 12.42 g m^{-2} in the Swartkops to 15.3 g m^{-2} in the Sundays River where up to four cohorts in the population could be recognised. The mean P:B ratio was 5.71.

Distribution of the shrimps in the *Zostera* beds was correlated with size and water depth. Large individuals were typically absent from shallow water while small shrimp, although not excluded from deeper water, tended to prefer shallow water. Emmerson (1986) attributed this to a matting effect of the *Zostera* fronds at low tide which produced a physical barrier to larger shrimp. In the case of subtidal *Zostera* this pattern of distribution was less noticeable.

In a study of the migration of *P. peringueyi* into and out of a *Zostera* bed in the Swartkops estuary over a tidal cycle, the shrimps moved into the *Zostera* during the flood tide and this occurred to a greater degree during the day than at night (Emmerson 1987).

Role in trophic dynamics

The feeding behaviour of *P. peringueyi* was investigated by Emmerson (1987) in the Swartkops estuary in relation to the tidal movement into and out of *Zostera* and *Spartina maritima* beds. A gut fullness index exhibited day and night peaks coinciding with high tide periods. The nature of the gut contents changed with time. During movement into the *Zostera* beds the gut contents were dominated by the epiphytic diatom *Cocconeis* sp. along with naviculoid diatoms and detritus. As the shrimps moved through the *Zostera* and into the *Spartina*, *Cocconeis* sp. declined and the proportion of naviculoids and detritus increased. On the return to the *Zostera* with the ebbing tide the gut contents consisted mainly of detritus with epiphytes and low numbers of two species of nematode, *Helicotylenchus californicus* and *Tylenchus* sp. There was an increase in the amount of *Cocconeis* sp. and a decrease in detritus and naviculoids once the shrimps returned to the channel at low tide.

Total *P. peringueyi* production in the three estuaries studied ranged from 0.15 tonnes in the Sundays to 1.12 t in the Swartkops and 1.17 t in the Kromme (Emmerson 1986). Known fish predators of this species include *Argyrosomus hololepidotus*, *Lithognathus lithognathus*, *Pomatomus saltatrix*, *Lichia amia*, *Epinephelus guaza* and *Pomadasys commersonnii*, all of which are common in Eastern Cape estuaries (Emmerson 1986). The relative contribution of *P. peringueyi* to the diets of these species is, however, unknown.

Economic importance

This species has no economic importance in the South African region.

Nematopalaemon tenuipes

Spatial and temporal distribution

This is the least known of any of the species falling into the category of estuarine nektonic invertebrates. Its range extends through the Indo-Pacific from New Zealand to the coast of KwaZulu-Natal (Holthuis 1980).

Life history and recruitment patterns

No published data exist. The only available information

comes from unpublished data collected incidentally during monitoring of postlarval penaeid prawn recruitment into the St Lucia system during 1982–4 and 1987–8. Specimens were obtained at the southern end of the St Lucia Narrows only during the third year of the programme during reduced salinity periods which followed the Cyclone Domoina floods in January–February 1984. Relatively low salinity conditions also existed during 1987–8 and *N. tenuipes* was again present. The largest catches were typically taken in the stationary plankton net in bottom samples during the initial stages of either the flood or ebb tide. Both periods of occurrence were seasonal, the prawns being present from January to July, i.e. summer to early winter, with peaks in May.

The sizes of specimens caught ranged from 5–13 mm carapace length in males and 5–18 mm in females. Ovigerous females were recorded whenever the species was present and were most abundant in May when they made up 25% of the mature females. The male:female ratio was approximately 2:1.

The absence of the species from the St Lucia Narrows in some years, combined with the seasonal occurrence and the absence of small individuals, implies a migratory pattern associated with irregular favourable conditions in the Narrows. The presence of ovigerous females of an apparently seasonal, migratory species in the estuary is unusual as this is contrary to more typical patterns of breeding migrations of mobile estuarine species whereby mature females would generally move out to marine breeding grounds.

Role in trophic dynamics

Unknown, but as the species is rare in South African estuaries any role is likely to be negligible.

Economic importance

This species has no economic importance in the South African region.

The genus *Macrobrachium*

The approximately 125 species in this genus occur in brackish and fresh waters in tropical and subtropical areas (Holthuis 1980). Only one species, *M. intermedium* from Australia, is fully marine (Williamson 1972) although several species appear to require brackish conditions for larval development. Apart from the general but dated taxonomic and distributional information provided by Barnard (1950), information on the occurrence and biology of species in this genus in South Africa is limited to the St Lucia system (Figure 8.21) of KZN (Bickerton 1989) and the Keiskamma estuary in the Eastern Cape Province (Figure 8.22) (Read 1983, 1985*a*, *b*).

Spatial and temporal distribution

Although Kensley (1972) recognised seven species in South Africa, Bickerton (1989) recorded only three, i.e. *M. equidens*, *M. rude* and *M. scabriculum*, in the St Lucia system while Read (1983) found only *M. petersi* further south in the Keiskamma River and estuary in the Eastern Cape. The decline in the number of species towards the more temperate areas corresponds with the tropical and subtropical affinities of the genus.

Read (1983) showed experimentally that although the larvae of the freshwater *M. petersi* required saline water and could tolerate marine salinities, this tolerance was adversely affected by low temperatures. Colonisation of estuaries via a marine larval dispersal phase in more temperate areas would therefore be restricted and this was suggested by Read (1983) to be the mechanism whereby the southerly limits of distribution were determined. It appears that the broad geographical distributions characteristic of some species are associated with greater degrees of salt tolerance in either larvae or adults than is found in strictly freshwater, more localised species.

Distribution of *Macrobrachium* species in both the St Lucia (Bickerton 1989) and Keiskamma systems (Read 1985*a*) was strongly correlated with salinities. The St Lucia studies (Bickerton 1989) were precipitated by the appearance of *Macrobrachium* spp. in the penaeid bait prawn fishery following a series of relatively wet years and associated general reductions in salinities during the mid 1970s. During subsequent drier years there was a general decline in the abundance of this genus and a reversion to the dominance of penaeid prawns in the fishery. During the periods of greatest abundance of the three *Macrobrachium* species in the system, *M. scabriculum* tended to be localised around the mouths of streams and points of freshwater inflow into the lake. *M. equidens* occurred in the Narrows (Figure 8.21) where its appearance coincided with the disappearance of *M. rude* and the restriction of the latter species to the lake. Rising salinities in the lake and Narrows appear to force the various *Macrobrachium* species into the tributary rivers of the system although the dynamics of this process, the patterns of movement and recruitment and the interactions of the species within the genus and with other comparably sized benthic invertebrates such as the penaeid prawns and the crab *Scylla serrata* remain unclear.

Subtleties in the effects of salinity on distribution were noted by Read (1985*a*) in the Keiskamma estuary in the case of *M. petersi* where a low salinity preference in the adults was overridden by the salinity requirements of the larvae such that adults migrated downstream, thereby ensuring the proximity of the larval stages to more saline waters which favoured growth and development. This was not detected in the St Lucia system.

Life histories and recruitment patterns

Ovigerous females of all the *Macrobrachium* species were recorded in either the St Lucia (Bickerton 1989) or Keiskamma (Read 1983) systems. Egg production occurred primarily during summer. Egg development times in local species are unknown. Larval development in *M. petersi* involved nine stages (Read 1985*b*) and lasted *c.* 28 days at 25 °C (Read 1983). Details of the other species are unknown. Available information suggests an absence of any distinct marine phase in these species and Read (1985*b*) proposed that behavioural adaptations to reduce larval losses to the marine environment were more the norm despite the possible significance of larval phases in colonising or re-establishing populations in other estuaries.

Development appears to be slower and life spans considerably longer than in the penaeid prawns. On the basis of laboratory observations that growth virtually stopped at temperatures of 20–22 °C, and because temperatures in St Lucia dropped to this level for five months of the year, Bickerton (1989) suggested a life span of at least 3–4 years in the case of *M. rude* in this area. Sexual dimorphism, especially as regards the enlarged chelipeds in the males, is a major feature of all species in the genus. Unpublished observations on mating pairs of *M. rude* indicated that the chelae are used to enclose females during courtship and simultaneously as barriers to deter rival males.

Role in trophic dynamics

The local species of this genus appear to be omnivorous benthic feeders (Bickerton 1989) but their impact on the benthos or their róle as prey items for larger predators is unknown.

Economic importance

The population of *Macrobrachium* spp. in the St Lucia Narrows contributed to the bait prawn fishery for *c.* 3 years during the mid 1970s but apart from this there has never been any exploitation of estuarine populations of this genus.

Acknowledgements

The author gratefully acknowledges the cooperation of Ms Wendy Robertson and Professor Winston Emmerson in making available the results of their research, both published and unpublished, on *Scylla serrata* and *Palaemon* (*pacificus*) *peringueyi* respectively. Professor Emmerson's advice on taxonomic aspects was also extremely useful.

References

Allanson, B.R. (1992). *An assessment of the SANCOR Estuaries Research Programme from 1980 to 1989.* Committee for Marine Science Occasional Report No.1. Pretoria: Foundation for Research Development.

Allanson, B.R. & Read, G.H.L. (1987). *The response of estuaries along the south eastern coast of southern Africa to marked variation in freshwater inflow.* Institute for Freshwater Studies Special Report No. 2/87. Grahamstown: Rhodes University.

Allanson, B.R. & Read, G.H.L. (1995). Further comment on the response of eastern Cape province estuaries to variable freshwater inflows. *South African Journal of Aquatic Science*, **21**, 56–70.

Avis, A.M. (1988). Temperature and salinity tolerance of adult hermit crabs, *Diogenes brevirostris* Stimpson (Crustacea: Decapoda: Anomura). *South African Journal of Zoology*, **23**, 351–5.

Baird, D. (1988). Synthesis of ecological research in the Swartkops Estuary. In *The Swartkops Estuary*, Proceedings of a symposium held on 14 and 15 September 1987 at the University of Port Elizabeth, ed. D. Baird, J.F.K. Marais & A.P. Martin, pp. 41–56. South African National Scientific Programmes Report No. 156.

Baird, D., Evans, P.R., Milne, H., & Pienkowski, M.W. (1985). Utilization by shorebirds of benthic invertebrate production in intertidal areas. *Oceanography and Marine Biology Annual Review*, **23**, 573–98.

Baird, D., Hanekom, N. & Grindley, J.R. (1986). *Estuaries of the Cape. Part II. Synopses of available information on individual systems*, ed. A.E.F. Heydorn & J.R. Grindley. *Report No. 23: Swartkops (CSE3).* CSIR Research Report No. 422.

Baird, D., McGlade, J.M. & Ulanowicz, R.E. (1991). The comparative ecology of six marine ecosystems. *Philosophical Transactions of the Royal Society, London, Series B*, **333**, 15–29.

Baird, D., Marais, J.F.K. & Wooldridge, T. (1981). The influence of a marina canal system on the ecology of the Kromme estuary, St Francis Bay. *South African Journal of Zoology*, **16**, 21–34.

Baird, D. & Ulanowicz, R.E. (1993). Comparative study on the trophic structure, cycling and ecosystem properties of four tidal estuaries. *Marine Ecology Progress Series*, **99**, 221–37.

Barnard, K.H. (1950). Descriptive catalogue of South African Decapod Crustacea (crabs and shrimps). *Annals of the South African Museum*, **38**, 1–837.

Bass, A.J., D'Aubrey, J.D. & Kistnasamy, N. (1973). *Sharks of the east coast of Southern Africa.* Oceanographic Research Institute Investigational Report No. 33.

Begg, G.W. (1978). *The estuaries of Natal.* Natal Town and Regional Planning Report 41.

Begg, G.W. (1984a). *The estuaries of Natal Part 2.* Natal Town and Regional Planning Report 55.

Begg, G.W. (1984b). *The comparative ecology of Natal's smaller estuaries.* Natal Town and Regional Planning Report 62.

Bennett, B.A. (1989). The diets of fish in three southwestern Cape estuarine systems. *South African Journal of Zoology*, **24**, 163–77.

Bennett, B.A. & Branch G.M. (1990). Relationships between production and consumption of prey species by resident fish in the Bot, a cool temperate South African estuary. *Estuarine, Coastal and Shelf Science*, **31**, 139–55.

Bernard, R.T.F., Davies, B.R. & Hodgson, A.N. (1988). Reproduction in a brackish-water mytilid: Gametogenesis and embryonic development. *The Veliger*, **30**, 278–90.

Bickerton, I.B. (1989). Aspects of the biology of the genus *Macrobrachium* (Decapoda: Caridea: Palaemonidae) in the St Lucia system. MSc thesis, University of Natal, Durban. Published as CSIR Research Report 684.

Bickerton I.B. & Pierce, S.M. (1988). *Estuaries of the Cape, Part II: Synopses of available information on individual systems*, ed. A.E.F. Heydorn & P.D. Morant. *Report No. 33: Krom (CMS 45), Seekoei (CMS 46) and Kabeljous (CMS 47).* CSIR Research Report 432. Stellenbosch: CSIR.

Blaber, S.J.M. (1973). Temperature and salinity tolerance of juvenile *Rhabdosargus holubi* (Steindachner) (Teleostei: Sparidae). *Journal of Fish Biology*, **5**, 593–8.

Blaber, S.J.M. (1984). The diet, food selectivity and niche of *Rhabdosargus sarba* (Teleostei: Sparidae) in Natal estuaries. *South African Journal of Zoology* **19**, 241–6.

Blaber, S.J.M. & Cyrus, D.P. (1983). The biology of Carangidae (Teleostei) in Natal estuaries. *Journal of Fish Biology*, **22**, 173–88.

Blaber, S.J.M., Kure, N.F., Jackson, S. & Cyrus, D.P. (1983). The benthos of South Lake, St Lucia following a period of stable salinities. *South African Journal of Zoology*, **18**, 311–19.

Boltt, R.E. (1975). The benthos of some southern African lakes. Part 5. The recovery of the benthic fauna of St Lucia following a period of excessively high salinity. *Transactions of the Royal Society of South Africa*, **41**, 295–323.

Branch, G. & Branch, M. (1981). *The living shores of southern Africa.* Cape Town: C. Struik.

Branch, G.M. & Day, J.A. (1984). Ecology of southern African estuaries. Part XIII. The Palmiet river estuary in the south-western Cape. *South African Journal of Zoology*, **19**, 63–77.

Branch, G.M. & Grindley, J.R. (1979). Ecology of southern African estuaries. Part XI. Mngazana: a mangrove estuary in Transkei. *South African Journal of Zoology*, **14**, 149–70.

Branch, G.M. & Pringle, A. (1987). The impact of the sand prawn *Callianassa kraussi* Stebbing on sediment turnover and on bacteria, meiofauna, and benthic microfauna. *Journal of Experimental Marine Biology and Ecology*, **107**, 219–35.

Broekhuysen, G.J. (1955). The breeding and growth of *Hymenosoma orbiculare* Desm. (Crustacea, Brachyura). *Annals of the South African Museum*, **41**, 313–43.

Brown, A.C. (1953). *A preliminary investigation of the ecology of the larger Kleinmond River estuary, Bathurst District.* MSc thesis, Rhodes University, Grahamstown.

Byren, B.A. & Davies, B.R. (1986). The influence of invertebrates on the breakdown of *Potamogeton pectinatus* L. in a coastal marina (Zandvlei, South Africa). *Hydrobiologia*, **137**, 141–51.

Carter, R.A. (1983). *Estuaries of the Cape. Part II. Synopses of available information on individual systems*, ed. A.E.F. Heydorn &

J.R Grindley. *Report No. 21: Bree (CSW 22).* CSIR Research Report 420. Stellenbosch: CSIR.

Cliff, S. & Grindley, J.R. (1982). *Estuaries of the Cape. Part II. Synopses of available information on individual systems*, ed. A.E.F. Heydorn & J.R. Grindley. *Report No. 17: Lourens (CSW 7).* CSIR Research Report 416. Stellenbosch: CSIR.

Cretchley, R. (1996). *Exploitation of the bait organism* Upogebia africana *(Crustacea: Anomura) in the Knysna estuary.* MSc thesis, Rhodes University, Grahamstown.

Crocos, P.J. & Kerr, J.D. (1983). Maturation and spawning of the banana prawn, *Penaeus merguiensis*, in the Gulf of Carpentaria. *Journal of Experimental Marine Biology and Ecology*, **69**, 37–59.

Cummins, K.W., Petersen, R.C., Howard, F.O., Wuycheck, J.C. & Holt, V.I. (1973). The utilization of leaf litter by stream detritivores. *Ecology*, **54**, 336–45.

Dall, W.D., Hill B.J., Rothlisberg, P.C. & Staples, D.J. (1990). The biology of Penaeidae. *Advances in Marine Biology*, **27**, 1–489.

Davies, B.R. (1982). Studies on the zoobenthos of some southern Cape coastal lakes. Spatial and temporal changes in the benthos of Swartvlei, South Africa, in relation to changes in the submerged littoral macrophyte community. *Journal of the Limnological Society of Southern Africa*, **8**, 33–45.

Davies, B.R., Stuart, V. & de Villiers, M. (1989). The filtration activity of a serpulid polychaete population (*Ficopomatus enigmaticus* (Fauvel)) and its effects on water quality in a coastal marina. *Estuarine, Coastal and Shelf Science*, **29**, 613–20.

Day, J.H. (1964). The origin and distribution of estuarine animals in South Africa. In *Ecological studies in southern Africa*, ed. D.H.S. Davis, pp. 159–73. The Hague: Dr W. Junk.

Day, J.H. (1967). The biology of Knysna estuary, South Africa. In *Estuaries*, ed. G.H. Lauff, pp. 397–407. Washington, DC: American Association for the Advancement of Science.

Day, J.H. (1974). The ecology of Morrumbene estuary, Moçambique. *Transactions of the Royal Society of South Africa*, **41**, 43–97.

Day, J.H. (ed.) (1981a). *Estuarine ecology with particular reference to Southern Africa.* Cape Town: A.A. Balkema.

Day, J.H. (1981b). The estuarine fauna. In *Estuarine ecology with particular reference to Southern Africa*, ed. J.H. Day, pp. 147–78. Cape Town: A.A. Balkema.

Day, J.H. (1981c). Summaries of current knowledge of 43 estuaries in southern Africa. In *Estuarine ecology with particular reference to Southern Africa*, ed. J.H. Day, pp. 251–329. Cape Town: A.A. Balkema.

Day, J.H., Blaber, S.J.M., & Wallace, J.H. (1981). Estuarine fishes. In *Estuarine ecology with particular reference to Southern Africa*, ed. J.H. Day, pp. 197–222. Cape Town: A.A. Balkema.

De Decker, H.P. (1987). Breaching the mouth of the Bot river estuary, South Africa: impact on its benthic macrofaunal communities. *Transactions of the Royal Society of South Africa*, **46**, 231–50.

De Decker, H.P. (1989). *Estuaries of the Cape. Part II. Synopses of available information on individual systems*, ed. A.E.F. Heydorn & P.D. Morant. *Report No. 40: Klein (CSW 12).* CSIR Research Report 439. Stellenbosch: CSIR.

De Decker, H.P. & Bally, R. (1985). The benthic macrofauna of the Bot River estuary, South Africa, with a note on its meiofauna. *Transactions of the Royal Society of South Africa*, **45**, 379–96.

de Freitas, A.J.O. (1980). *The Penaeoidea of southeast Africa.* PhD thesis, University of the Witwatersrand, Johannesburg.

de Freitas, A.J.O. (1986). Selection of nursery areas by six southeast African Penaeidae. *Estuarine, Coastal and Shelf Science*, **23**, 901–8.

Demetriades, N.T. (1990). *Some aspects of the population dynamics of the commercially important Natal penaeid prawns.* MSc thesis, University of Natal, Durban.

Demetriades N.T. & Forbes A.T. (1993). Seasonal changes in the species composition of the penaeid prawn catch on the Tugela Bank, Natal, South Africa. *South African Journal of Marine Science*, **13**, 317–22.

De Niro, M.J. & Epstein, S. (1978). Influence of diet on the distribution of carbon isotopes in animals. *Geochimica et Cosmochimica Acta*, **42**, 495–506.

de Villiers, C.J. (1990). *Aspects of the biology of the infaunal bivalve mollusc* Solen cylindraceus *(Hanley) in the Kariega estuary.* PhD thesis, Rhodes University, Grahamstown.

de Villiers, C.J. & Allanson, B.R. (1988). Efficiency of particle retention in *Solen cylindraceus* (Hanley) (Mollusca: Bivalvia). *Estuarine, Coastal and Shelf Science*, **26**, 421–8.

de Villiers, C.J. & Allanson, B.R. (1989). Osmotic properties of an infaunal estuarine bivalve *Solen cylindraceus* (Hanley). *Journal of Molluscan Studies*, **55**, 45–51.

de Villiers, C.J., Allanson, B.R. & Hodgson, A.N. (1989a). The effect of temperature on the filtration rate of *Solen cylindraceus* (Hanley) (Mollusca: Bivalvia). *South African Journal of Zoology*, **24**, 11–17.

de Villiers, C.J. & Hodgson, A.N. (1993). The filtration and feeding physiology of the infaunal estuarine bivalve *Solen cylindraceus* (Hanley 1843). *Journal of Experimental Marine Biology and Ecology*, **167**, 127–42.

de Villiers, C.J., Hodgson, A.N. & Allanson, B.R. (1989b). Effect of salinity and temperature on the filtration rate and distribution of *Solen cylindraceus* (Hanley). In *Reproduction, genetics and distribution of marine organisms*, ed. J.S. Ryland & P.A. Tyler, pp. 459–65. Denmark: Olsen and Olsen.

du Plessis, A. (1971). *A preliminary investigation into the morphological characteristics, feeding, growth, reproduction and larval rearing of* Scylla serrata *Forskal (Decapoda, Portunidae) held in captivity.* Unpublished report of the Fisheries Development Corporation of South Africa.

du Preez, H.H. (1978). *Aspects of intertidal macrofauna ecology of Sundays River estuary.* Third Year Project, University of Port Elizabeth.

Duvenhage, I.R. & Morant, P.D. (1984). *Estuaries of the Cape. Part II. Synopses of available information on individual systems*, ed. A.E.F. Heydorn & P.D. Morant. *Report No. 31: Keurbooms–Bitou system (CMS 19), Piesang (CMS 18).* CSIR Research Report 430. Stellenbosch: CSIR.

Dye, A.H. (1983a). Composition and seasonal fluctuations of meiofauna in a southern African mangrove estuary. *Marine Biology*, **73**, 165–70.

Dye, A.H. (1983b). Vertical and horizontal distribution of meiofauna in mangrove sediments in Transkei, southern Africa. *Estuarine, Coastal and Shelf Science*, **16**, 591–8.

Dye, A.H. & Furstenburg, J.P. (1981). Estuarine meiofauna. In *Estuarine ecology with particular reference to southern Africa*, ed. J.H. Day, pp. 179–86. Cape Town: A.A. Balkema.

Emmerson, W.D. (1983). Tidal exchange of two decapod larvae, *Palaemon pacificus*

(Caridea) and *Upogebia africana* (Thalassinidae) between the Swartkops river estuary and adjacent coastal waters. *South African Journal of Zoology*, **18**, 326–30.

Emmerson, W.D. (1985*a*). Seasonal abundance, growth and production of *Palaemon pacificus* (Stimpson) in eastern Cape tidal pools. *South African Journal of Zoology*, **20**, 221–31.

Emmerson, W.D. (1985*b*). Fecundity, larval rearing and laboratory growth of *Palaemon pacificus* (Stimpson) (Decapoda, Palaemonidae). *Crustaceana*, **49**, 277–89.

Emmerson, W.D. (1986). The ecology of *Palaemon pacificus* (Stimpson) associated with *Zostera capensis* Setchell. *Transactions of the Royal Society of South Africa*, **46**, 79–97.

Emmerson, W.D. (1987). Tidal migration and feeding of the shrimp *Palaemon pacificus* (Stimpson). *South African Journal of Science*, **83**, 413–16.

Emmerson, W.D. & McGwynne, L.E. (1992). Feeding and assimilation of mangrove leaves by the crab *Sesarma meinerti* de Man in relation to leaf-litter production in Mgazana, a warm-temperate southern African mangrove swamp. *Journal of Experimental Marine Biology and Ecology*, **157**, 41–53.

Estampador, E.P. (1949). Studies on *Scylla* (Crustacea, Portunidae). I. Revision of the genus. *Philippine Journal of Science*, **78**, 95–108.

Forbes, A.T. (1973). An unusual abbreviated larval life in the estuarine burrowing prawn *Callianassa kraussi* (Crustacea: Decapoda: Thalassinidea). *Marine Biology*, **22**, 361–5.

Forbes, A.T. (1974). Osmotic and ionic regulation in *Callianassa kraussi* Stebbing (Crustacea: Decapoda: Thalassinidea). *Journal of Experimental Marine Biology and Ecology*, **16**, 301–11.

Forbes, A.T & Benfield, M.C. (1986*a*). Penaeid prawns in the St Lucia Lake system: postlarval recruitment and the bait fishery. *South African Journal of Zoology*, **21**, 224–8.

Forbes, A.T. & Benfield, M.C. (1986*b*). Tidal behaviour of postlarval penaeid prawns (Crustacea: Decapoda: Penaeidae) in a southeast African estuary. *Journal of Experimental Marine Biology and Ecology*, **102**, 23–34.

Forbes, A.T. & Cyrus, D.P. (1991). Recruitment and origin of penaeid prawn postlarvae in two southeast

African estuaries. *Estuarine, Coastal and Shelf Science*, **33**, 281–9.

Forbes, A.T. & Cyrus, D.P. (1992). Impact of a major cyclone on a Southeast African estuarine lake system. *Netherlands Journal of Sea Research*, **30**, 265–72.

Forbes, A.T. & Hay, D.G. (1988). Effects of a major cyclone on the abundance and larval recruitment of the portunid crab *Scylla serrata* (Forskal) in the St Lucia estuary. *South African Journal of Marine Science*, **7**, 219–25.

Forbes, A.T. & Hill, B.J. (1969). The physiological ability of a marine crab *Hymenosoma orbiculare* Desm., to live in a subtropical freshwater lake. *Transactions of the Royal Society of South Africa*, **38**, 271–83.

Forbes, A.T., Niedinger, S. & Demetriades, N.T. (1994). Recruitment and utilisation of nursery grounds by penaeid prawn postlarvae in Natal, South Africa. In *Changes in Fluxes in estuaries*, ed. K.R. Dyer & R.J. Orth. Denmark: Olsen & Olsen.

Fry, B. & Sherr, E.B. (1984). $\delta^{13}C$ measurements as indicators of carbon flow in marine and freshwater ecosystems. *Contributions in Marine Science*, **27**, 13–47.

Gaigher, C.M. (1987). *The distribution and status of burrowing bait organisms and the management of the estuarine environment of the Cape Province*. Unpublished report, Chief Directorate of the Cape Province, Nature and Environmental Conservation, Cape Provincial Administration.

Garcia, S. (1985). Reproduction, stock assessment models and population parameters in exploited penaeid shrimp populations. In *Second Australian National prawn seminar*, ed. P.C. Rothlisberg, B.J. Hill & D.J. Staples, pp. 139–58. Cleveland, Australia: NPS2.

Garcia, S. & Le Reste, L. (1981). *Life cycles, dynamics, exploitation and management of coastal penaeid shrimp stocks*. FAO Fisheries Technical Paper No. 203.

Gasith, A. & Hasler, A.D. (1976). Airborne litterfall as a source of organic matter in lakes. *Limnology and Oceanography*, **21**, 253–8.

Grange, N. (1992). *The influence of contrasting freshwater inflows on the feeding ecology and food resources of zooplankton in two eastern Cape estuaries, South Africa*. PhD thesis, Rhodes University, Grahamstown.

Grange, N. & Allanson, B.R. (1995). The influence of freshwater inflow on the nature, amount and distribution of seston in estuaries of the eastern Cape, South Africa. *Estuarine, Coastal and Shelf Science*, **40**, 403–20.

Grindley, J.R. & Dudley, S. (1988). *Estuaries of the Cape. Part 2. Synopses of available information on individual systems*, ed. A.E.F. Heydorn & P.D. Morant. *Report Number 28: Reitvlei (CW 24) and Diep (CW 25)*. CSIR Research Report No. 427. Stellenbosch: CSIR.

Hanekom, N. (1980). *A study of two thalassinid prawns in the non-Spartina regions of the Swartkops estuary*. PhD thesis, University of Port Elizabeth.

Hanekom, N. (1989). A note on the effects of a flood of medium intensity on macrobenthos of soft substrata in the Swartkops estuary, South Africa. *South African Journal of Marine Science*, **8**, 349–55.

Hanekom, N. & Baird, D. (1992). Growth, production and consumption of the thalassinid prawn *Upogebia africana* (Ortmann) in the Swartkops estuary. *South African Journal of Zoology*, **27**, 130–9.

Hanekom, N. & Erasmus, T. (1989). Determination of the reproductive output of populations of a thalassinid prawn *Upogebia africana* (Ortmann) in the Swartkops estuary. *South African Journal of Zoology*, **24**, 244–50.

Hanekom, N., Baird, D. & Erasmus, T. (1988). A quantitative study to assess standing biomasses of macrobenthos in soft substrata of the Swartkops estuary, South Africa. *South African Journal of Marine Science*, **6**, 163–74.

Harris, J.M. (1992). *Relationships between invertebrate detritivores and gut bacteria in marine systems*. PhD thesis, University of Cape Town.

Hay, D.G. (1985). *The macrobenthos of the St Lucia Narrows*. MSc thesis, University of Natal, Pietermaritzburg.

Hecht, T. & van der Lingen, C.D. (1992). Turbidity-induced changes in feeding strategies of fish in estuaries. *South African Journal of Zoology*, **27**, 95–107.

Heymans, J.J. & Baird, D. (1995). Energy flow in the Kromme estuarine ecosystem, St Francis Bay, South Africa. *Estuarine, Coastal and Shelf Science*, **41**, 39–59.

Hill, B.J. (1967). *Contributions to the ecology of the anomuran mud prawn Upogebia africana (Ortmann)*. PhD thesis, Rhodes University, Grahamstown.

Hill, B.J. (1971). Osmoregulation by an estuarine and a marine species of *Upogebia* (Anomura, Crustacea). *Zoologica Africana*, **6**, 229–36.

Hill, B.J. (1974). Salinity and temperature tolerance of zoeae of the portunid crab *Scylla serrata*. *Marine Biology*, **25**, 21–4.

Hill, B.J. (1975). Abundance, breeding and growth of the crab *Scylla serrata* in two South African estuaries. *Marine Biology*, **32**, 119–26.

Hill, B.J. (1976). Natural food, foregut clearance rate and activity of the crab *Scylla serrata* in an estuary. *Marine Biology*, **34**, 109–16.

Hill, B.J. (1977). The effect of heated effluent on egg production in the estuarine prawn *Upogebia africana* (Ortmann). *Journal of Experimental Marine Biology and Ecology*, **29**, 291–302.

Hill, B.J. (1978). Activity, track and speed of movement of the crab *Scylla serrata* in an estuary. *Marine Biology*, **47**, 135–41.

Hill, B.J. (1979a). Biology of the crab *Scylla serrata* (Forskal) in the St Lucia system. *Transactions of the Royal Society of South Africa*, **44**, 55–62.

Hill, B.J. (1979b). Aspects of the feeding strategy of the predatory crab *Scylla serrata*. *Marine Biology*, **55**, 209–14.

Hill, B.J. (1980). Effects of temperature on feeding and activity in the crab *Scylla serrata*. *Marine Biology*, **59**, 189–92.

Hill, B.J. (1981). Adaptations to temperature and salinity stress in southern African estuaries. In *Estuarine Ecology with particular reference to Southern Africa*, ed. J.H. Day, pp. 187–96. Cape Town: A.A. Balkema.

Hill, B.J. (ed.) (1982). *The Queensland mud crab fishery*. Queensland Fisheries Information Series FI 8201.

Hill, B.J. & Forbes, A.T. (1979). Biology of *Hymenosoma orbiculare* Desm in Lake Sibaya. *South African Journal of Zoology*, **14**, 75–9.

Hodgson, A.N. (1987). Distribution and abundance of the macrobenthic fauna of the Kariega estuary. *South African Journal of Zoology*, **22**, 153–62.

Holthuis, L.B. (1980). *FAO species catalogue*. Vol. 1. *Shrimps and prawns of the world. An annotated catalogue of species of interest to fisheries*. FAO Fisheries Synopses 125.

Hughes, D.A. (1966). Investigations of the nursery areas and habitat preferences of juvenile penaeid prawns in Moçambique. *Journal of Applied Ecology*, **3**, 349–54.

Hughes, D.A. (1969). Responses to salinity change as a tidal transport mechanism of pink shrimp, *Penaeus duorarum*. *Biological Bulletin*, **136**, 43–53.

Jerling, H.L. (1994). *Feeding ecology of mesozooplankton in the Sundays River estuary*. PhD thesis, University of Port Elizabeth.

Jerling, H.L. & Wooldridge, T.H. (1994). Comparative morphology of the feeding appendages of four zooplankton species in the Sundays River estuary. *South African Journal of Zoology*, **29**, 252–7.

Jorgensen, C.B. (1990). *Bivalve filter feeding: hydrodynamics, bioenergetics, physiology and ecology*. Denmark: Olsen and Olsen.

Kalejta, B. (1992). Distribution, biomass and production of *Ceratonereis erythraeensis* (Fauvel) and *Ceratonereis keiskama* (Day) at the Berg River estuary, South Africa. *South African Journal of Zoology*, **27**, 121–9.

Kalejta, B. & Hockey, P.A.R. (1991). Distribution, abundance and productivity of benthic invertebrates at the Berg River estuary, South Africa. *Estuarine, Coastal and Shelf Science*, **33**, 175–91.

Kensley, B. (1972). *Shrimps and Prawns of Southern Africa*. South African Museum, Cape Town.

Kinne, O. (1963). The effects of temperature and salinity on marine and brackish water animals. I. Temperature. *Oceanography and Marine Biology Annual Review*, **1**, 301–40.

Kinne, O. (1964). The effects of temperature and salinity on marine and brackish water animals. II. Salinity and temperature, salinity combinations. *Oceanography and Marine Biology Annual Review*, **2**, 281–339.

Kinne, O. (1971). Salinity and animals – Invertebrates. In *Marine ecology*, Vol. 1, Part 2, *Environmental factors*, pp. 821–995. London: Wiley Interscience.

Knox, G.A. (1986). *Estuarine ecosystems: a systems approach*, Vols. 1 and 2. Florida: CRC Press.

Kokkinn, M.J. & Allanson, B.R. (1985). On the flux of organic carbon in a tidal salt marsh, Kowie river estuary, Port Alfred, South Africa. *South African Journal of Science*, **81**, 613–17.

Koop, K., Bally, R. & McQuaid, C.D. (1983). The ecology of South African estuaries. Part XII. The Bot river, a closed estuary in the south-western Cape. *South African Journal of Zoology*, **18**, 1–10.

Lockwood, A.P.M. (1962). The osmoregulation of Crustacea. *Biological Reviews*, **37**, 257–305.

McLachlan, A. & Erasmus, T. (1974). The temperature tolerances and osmoregulation in some estuarine bivalves. *Zoologica Africana*, **9**, 1–13.

McLachlan, A. & Grindley, J.R. (1974). Distribution of macrobenthic fauna of soft substrata in the Swartkops estuary with observations on the effects of floods. *Zoologica Africana*, **9**, 147–60.

McLusky, D.S. (1968). Some effects of salinity on the distribution and abundance of *Corophium volutator* in the Ythan estuary. *Journal of the Marine Biological Association of the UK*, **48**, 443–54.

MacNae, W. (1957). The ecology of the plants and animals in the intertidal regions of the Zwartkops estuary near Port Elizabeth, South Africa. Part 1. *Journal of Ecology*, **45**, 113–31.

MacPherson, E. (1990). The validation of *Palaemon peringueyi* (Stebbing 1915) from Southern African waters and its relationship with *Palaemon pacificus* (Stimpson 1860) (Decapoda, Palaemonidae). *Journal of Natural History*, **24**, 627–34.

Manning, R.B. & Holthuis, L.B. (1981). West African brachyuran crabs (Crustacea: Decapoda). *Smithsonian Contributions to Zoology*, **306**, 1–379.

Marshall, D.J. & McQuaid, C.D. (1994). Seasonal and diel variations of *in situ* heart rate of the intertidal pulmonate limpet *Siphonaria oculus* Kr. (Pulmonata). *Journal of Experimental Marine Biology and Ecology*, **179**, 1–9.

Martin, A.P. (1991). *Feeding ecology of birds on the Swartkops estuary, South Africa*. PhD thesis, University of Port Elizabeth.

Millard, N.A.H. & Broekhuysen, G.J. (1970). The ecology of South African estuaries. Part X. St Lucia: a second report. *Zoologica Africana*, **5**, 277–307.

Morant, P.D. (1991). The estuaries of False Bay. *Transactions of the Royal Society of South Africa*, **47**, 629–40.

Morant, P.D. & Bickerton, I.B. (1983). *Estuaries of the Cape. Part II. Synopses of available information on individual systems*, ed. A.E.F. Heydorn & J.R. Grindley. *Report No. 19: Groot (Wes) (CMS 23) and Sout (CMS 22)*. CSIR Research Report 418. Stellenbosch: CSIR.

Morant, P.D. & Grindley, J.R. (1982). *Estuaries of the Cape*. Part II. *Synopses of available information on individual systems*, ed. A.E.F. Heydorn & J.R. Grindley. *Report No. 14: Sand (CSW 4)*. CSIR Research Report 413. Stellenbosch: CSIR.

Motoh, H. (1981). *Studies on the fisheries biology of the giant tiger prawn*, Penaeus monodon, *in the Philippines*. Aquaculture Department, SEAFDEC, Technical Report No.7.

Newell, R.C. (ed.) (1976). *Adaptation to environment*. London: Butterworths.

Newell, R.C. (1979). *Biology of intertidal animals*. Marine Ecological Surveys Ltd. Kent. Cape Town: Rustica Press.

Newell, R.C. & Branch, G.M. (1980). The influence of temperature on the maintenance of metabolic energy balance in marine invertebrates. *Advances in Marine Biology*, **17**, 329–96.

Palmer, C.G. (1980). *Some aspects of the biology of* Nassarius kraussianus *(Dunker) (Gastropoda: Prosobranchia: Nassariidae), in the Bushmans River estuary with particular reference to recolonisation after floods*. MSc thesis, Rhodes University, Grahamstown.

Paparo, A.A. & Dean, R.C. (1984). Activity of the lateral cilia of the oyster *Crassostrea virginica* Gmelin: response to changes in salinity and to changes in potassium and magnesium concentration. *Marine Behaviour and Physiology*, **11**, 111–30.

Pauly, D. & Palomares, M.L. (1987). Shrimp consumption by fish in Kuwait waters: a methodology, preliminary results and their implications for management and research. *Kuwait Bulletin of Marine Science*, **9**, 101–25.

Pereyra Lago, R. (1993). Tidal exchange of larvae of *Sesarma catenata* (Decapoda, Brachyura) in the Swartkops estuary, South Africa. *South African Journal of Zoology*, **28**, 182–91.

Platell, M.E. & Potter, I.C. (1996). Influence of water depth, season, habitat and estuary location on the macrobenthic fauna of a seasonally closed estuary. *Journal of the Marine Biological Association of the UK*, **76**, 1–21.

Potter, I.C., Beckley, L.E., Whitfield, A.K. & Lenanton, R.C.J. (1990). Comparisons between the roles played by estuaries in the life cycles of fishes in temperate western Australia and southern Africa. *Environmental Biology of Fishes*, **28**, 143–78.

Pringle, J.S. & Cooper, J. (1975). The palaearctic wader population of Langebaan lagoon. *Ostrich*, **46**, 213–18.

Puttick, G.M. (1977). Spatial and temporal variations in intertidal animal distribution in Langebaan lagoon, South Africa. *Transactions of the Royal Society of South Africa*, **42**, 403–40.

Puttick, G.M. (1980). Energy budgets of curlew sandpipers at Langebaan lagoon, South Africa. *Estuarine and Coastal Marine Science*, **11**, 207–15.

Quinn, N.J. & Kojis, B.L. (1987). Reproductive biology of *Scylla* spp. (Crustacea: Portunidae) from the Labu estuary in Papua New Guinea. *Bulletin of Marine Science*, **41**, 234–41.

Rau, G.H., Mearns, A.J., Young, D.R., Olsen, R.J., Schafer, H.A. & Kaplan, I.R. (1983). Animal $^{13}C/^{12}C$ correlates with trophic level in pelagic food webs. *Ecology*, **64**, 1314–18.

Read, G.H.L. (1983). Possible influence of high salinity and low temperature on the distribution of *Macrobrachium petersi* (Hilgendorf) (Crustacea: Caridea) along the south-east coast of South Africa. *Transactions of the Royal Society of South Africa*, **45**, 35–43.

Read, G.H.L. (1985a). Factors affecting the distribution and abundance of *Macrobrachium petersi* (Hilgendorf) in the Keiskamma River and estuary, South Africa. *Estuarine, Coastal and Shelf Science*, **21**, 313–24.

Read, G.H.L. (1985b). Aspects of larval, post-larval and juvenile ecology of *Macrobrachium petersi* (Hilgendorf) in the Keiskamma estuary, South Africa. *Estuarine, Coastal and Shelf Science*, **21**, 501–10.

Read, G.H.L. & Whitfield, A.K. (1989). The response of *Grandidierella lignorum* (Barnard) (Crustacea: Amphipoda) to episodic flooding in three eastern Cape estuaries. *South African Journal of Zoology*, **24**, 99–105.

Reddering, J.S.V. & Rust, I.C. (1990). Historical changes and sedimentary characteristics of Southern African estuaries. *South African Journal of Science*, **86**, 425–08.

Robertson, W.D. (1984). *Aspects of the ecology of the shrimp* Palaemon pacificus *(Stimpson) (Decapoda, Palaemonidae) in the Bushmans River estuary*. MSc thesis, Rhodes University, Grahamstown.

Robertson, W.D. (1987). *Biology of the mangrove crab*, Scylla serrata. Oceanographic Research Institute, Poster Series No. 8.

Robertson, W.D. (1989). Factors affecting catches of the crab *Scylla serrata* (Forskal) (Decapoda: Portunidae) in baited traps: soak time, time of day and accessibility of the bait. *Estuarine, Coastal and Shelf Science*, **29**, 161–70.

Robertson, W.D. (1996). Abundance, population structure and size at maturity of *Scylla serrata* (Forskal) (Decapoda: Portunidae) in Eastern Cape estuaries, South Africa. *South African Journal of Zoology*, **31**, 177–85.

Robertson, W.D. & Kruger, A. (1990). *A growth model for the mangrove crab*, Scylla serrata. Oceanographic Research Institute, Poster Series No. 21.

Robertson, W.D. & Kruger, A. (1994). Size at maturity, mating and spawning in the portunid crab *Scylla serrata* (Forskal) in Natal, South Africa. *Estuarine, Coastal and Shelf Science*, **39**, 185–200.

Robertson, W.D. & Piper, S.E. (1991). Population estimates of the crab *Scylla serrata* (Forskal, 1755) (Decapoda: Portunidae) in two closed estuaries in Natal, South Africa, from mark–recapture methods. *South African Journal of Marine Science*, **11**, 193–202.

Rothlisberg, P.C., Church, J.A. & Fandry, C.B. (1995). A mechanism for near-shore concentration and estuarine recruitment of post-larval *Penaeus plebejus* Hess (Decapoda, Penaeidae). *Estuarine, Coastal and Shelf Science*, **40**, 115–38.

Salini, J.P., Blaber, S.J.M. & Brewer, D.T. (1990). Diets of piscivorous fishes in a tropical Australian estuary, with special reference to predation on penaeid prawns. *Marine Biology*, **105**, 363–74.

Sanders, H.L., Mangelsdorf, P.C. & Hampson, G.R. (1965). Salinity and faunal distribution in the Pocasset river, Massachussetts. *Limnology and Oceanography*, **10** Supplement, 216–29.

Schlacher, T.A. & Wooldridge, T.H. (1996a). Origin and trophic importance of detritus – evidence from stable isotopes in the benthos of a small, temperate estuary. *Oecologia*, **106**, 382–8.

Schlacher, T.A. & Wooldridge, T.H. (1996b). Patterns of selective predation by juvenile, benthivorous fish on estuarine macrofauna. *Marine Biology*, **125**, 241–7.

Shumway, S.E. (1977). Effect of salinity fluctuation on the osmotic pressure and Na$^+$, Ca^{++} and Mg^{++} ion concentrations in the haemolymph of bivalve molluscs. *Marine Biology*, **41**, 153–77.

Siegfried, W.R. (1981). The estuarine avifauna of southern Africa. In *Estuarine ecology with particular reference to Southern Africa*, ed. J.H. Day, pp. 223–50. Cape Town: A.A. Balkema.

Staples, D.J. (1985). Modelling the recruitment processes of the banana prawn, *Penaeus merguiensis*, in the southeastern Gulf of Carpentaria, Australia. In *Second Australian National prawn seminar*, ed. P.C. Rothlisberg, B.J. Hill & D.J. Staples, pp. 175–84. Cleveland, Australia: NPS2.

Steinke, T.D., Rajh, A. & Holland, A.J. (1993). The feeding behaviour of the red mangrove crab *Sesarma meinerti* de Man 1887 (Crustacea: Decapoda: Grapsidae) and its effect on the degradation of mangrove leaf litter. *South African Journal of Marine Science*, **13**, 151–60.

Stewart, B.A. & Davies, B.R. (1986). Effects of macrophyte harvesting on invertebrates associated with *Potamogeton pectinatus* L. in the Marina da Gama, Zandvlei, Western Cape. *Transactions of the Royal Society of South Africa*, **46**, 35–49.

Taylor, D.I. (1988). *Tidal exchanges of carbon, nitrogen and phosphorus between* Sarcocornia *salt-marsh and the Kariega estuary, and the role of salt-marsh brachyura in this transfer*. PhD thesis, Rhodes University, Grahamstown.

Taylor, D.I. (1992). The influence of upwelling and short-term changes in concentrations of nutrients in the water column on fluxes across the surface of a salt marsh. *Estuaries*, **15**, 68–74.

Taylor, D.I. & Allanson, B.R. (1993). Impacts of dense crab populations on carbon exchanges across the surface of a salt

marsh. *Marine Ecology Progress Series*, **101**, 119–29.

Van Winkle, W. (1972). Ciliary activity and oxygen consumption of excised bivalve gill tissue. *Comparative Biochemistry and Physiology*, **42A**, 472–95.

Velásquez, C.R., Kalejta, B. & Hockey, P.A.R. (1991). Seasonal abundance, habitat selection and energy consumption of waterbirds at the Berg River estuary, South Africa. *Ostrich*, **62**, 109–23.

Wallace, J.H., Kok, H.M., Beckley, L.E., Bennett, B., Blaber, S.J.M. & Whitfield, A.K. (1984). South African estuaries and their importance to fishes. *South African Journal of Science*, **80**, 203–7.

Whitfield, A.K. (1985). The role of zooplankton in the feeding ecology of fish fry from some southern African estuaries. *South African Journal of Zoology*, **20**, 166–71.

Whitfield, A.K. (1988). The fish community of the Swartvlei estuary and the influence of food availability on resource utilization. *Estuaries*, **11**, 160–70.

Whitfield, A.K. (1989). The benthic invertebrate community of a southern Cape estuary: structure and possible food resources. *Transactions of the Royal Society of South Africa*, **47**, 159–79.

Whitfield, A.K. (1991). *A bibliography of southern African estuarine research (1950–1990)*. J.L.B. Smith Institute of Ichthyology Investigational Report No. 36.

Whitfield, A.K. (1994). An estuary-association classification for the fishes of southern Africa. *South African Journal of Science*, **90**, 411–17.

Whitfield, A.K. & Blaber, S.J.M. (1978). Food and feeding ecology of piscivorous fishes at Lake St Lucia, Zululand. *Journal of Fisheries Biology*, **13**, 675–91.

Whitfield, A.K. & Blaber, S.J.M. (1980). The

diet of *Atilax paludinosus* (water mongoose) at St Lucia, South Africa. *Mammalia*, **44**, 315–80.

Whitfield, A.K. & Bruton, M.N. (1989). Some biological implications of reduced freshwater inflow into eastern Cape estuaries: a preliminary assessment. *South African Journal of Science*, **85**, 691–4.

Williamson, D.I. (1972). Larval development in a marine and freshwater species of *Macrobrachium* (Decapoda: Palaemonidae). *Crustaceana*, **23**, 282–98.

Wolff, W.J. (1983). Estuarine benthos. In *Ecosystems of the World 26. Estuaries and enclosed seas*, ed. B.H. Ketchum. Amsterdam: Elsevier.

Wooldridge, T.H. (1991). Exchange of two species of decapod larvae across an estuarine mouth inlet and implications of anthropogenic changes in the frequency and duration of mouth closure. *South African Journal of Science*, **87**, 519–25.

Wooldridge, T.H. (1994). The effect of periodic inlet closure on recruitment in the estuarine mudprawn, *Upogebia africana* (Ortmann). In *Changes in fluxes in estuaries: implications from science to management*, ed. K.R. Dyer & R.J. Orth, pp. 329–33. Fredensborg, Denmark: Olsen & Olsen.

Wynberg, R.P. & Branch, G.M. (1991). An assessment of bait-collecting for *Callianassa kraussi* Stebbing in Langebaan lagoon, Western Cape and of associated avian predation. *South African Journal of Marine Science*, **11**, 141–52.

Wynberg, R.P. & Branch, G.M. (1994). Disturbance associated with bait-collection for sandprawns (*Callianassa kraussi*) and mudprawns (*Upogebia africana*): long-term effects on the biota of intertidal sandflats. *Journal of Marine Research*, **52**, 523–8.

9 The ichthyofauna

Ichthyoplankton diversity, recruitment and dynamics

Alan Whitfield

The estuarine ichthyofauna

Hannes Marais

Orange River estuary

ICHTHYOPLANKTON DIVERSITY, RECRUITMENT AND DYNAMICS

Introduction

Estuaries in South Africa and around the world serve as nursery areas for a wide variety of fish species (Kennish 1990). Although the juvenile and adult life stages of southern African estuarine-associated species have been well studied (e.g. Wallace & van der Elst 1975; Beckley 1984; Bennett 1989; Marais 1988), information on the larval biology and ecology of most taxa is generally lacking. Specific studies on the ichthyoplankton component were initiated in the warm temperate Eastern and Western Cape estuaries (e.g. Melville-Smith & Baird 1980; Beckley 1985; Whitfield 1989a; Harrison & Whitfield 1990), with the first studies in subtropical KwaZulu-Natal being conducted in the St Lucia and Kosi systems (Harris & Cyrus 1994, 1995; Harris et al. 1995). No published information is available on the fish larvae or post-larvae from cool temperate estuaries in the Western or Northern Cape. In this chapter, the information on ichthyoplankton and ichthyonekton in subtropical and warm temperate estuaries is collated and placed in context with the life-history styles of fishes associated with these systems.

Although numerous studies have been conducted on the distribution, seasonality, habitat requirements, feeding and growth of juvenile marine fishes in both northern and southern hemisphere estuaries (Yanez-Aráncibia 1985), information on specific factors influencing immigration into these systems is generally lacking (Norcross & Shaw 1984). In recent years, the movement of fish larvae from marine spawning grounds to estuarine nursery areas has become the focus of increasing attention (Miller et al. 1984), the importance of passive and active transport mechanisms for several species having

already been determined (Norcross 1991). Although the routes followed by certain species are slowly being unravelled (Shaw et al. 1988), the cues which govern postlarval fish immigration behaviour are poorly known (Boehlert & Mundy 1988). An attempt is made in this chapter to identify the mechanisms and cues involved in larval and juvenile fish migration between southern African marine spawning grounds and estuarine nursery areas. In addition, the movements of the embryos and larvae of certain estuarine spawners are examined and related to the life-history styles of these particular species.

Spawning and early life history of estuarine-associated species

The fishes associated with South African estuaries can be divided into two broad categories according to whether they spawn in estuarine systems or the sea. The former group are referred to here as estuarine and the latter group marine. The main feature of the life cycle of most marine species utilising South African estuaries is a division into a juvenile period that is predominantly estuarine and an adult stage that is primarily marine (Wallace 1975a). Although some species may attain sexual maturity within the estuarine environment, spawning always occurs in the sea (Wallace 1975b) where the relatively stable marine environment is more suitable for the survival of the egg, embryonic and larval stages.

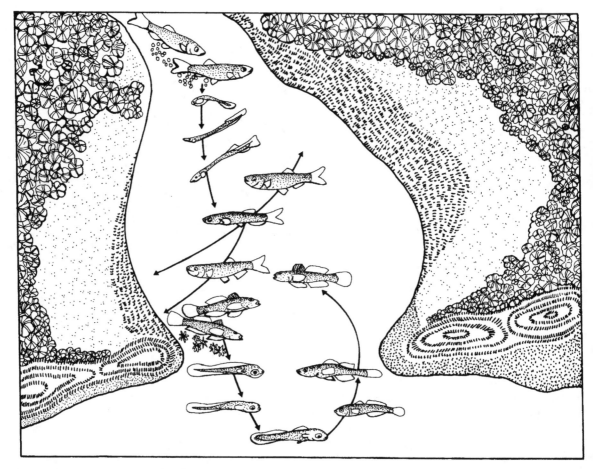

Figure 9.1. **Diagrammatic representation of the predominant life cycles of estuarine spawners. The two examples shown here are the** estuarine roundherring **Gilchristella aestuaria** and Knysna sandgoby **Psammogobius knysnaensis.**

There are relatively few fish taxa which can complete their entire life cycle within South African estuaries and these are invariably small species which mature at less than 70 mm standard length (SL) (Whitfield 1990). Wallace (1975b) has suggested that the small size of estuarine species would reduce their physical ability to undertake migrations to and from the sea, when compared with large species. Predation by abundant piscivorous fish populations in the sea may also mitigate against mass spawning migrations by these small species in the marine environment.

Some estuarine species (e.g. Knysna sandgoby *Psammogobius knysnaensis*, prison goby *Caffrogobius multifasciatus* and kappie blenny *Omobranchus woodi*) do have a marine larval phase which is achieved by a synchronised hatching of the embryos at high tides, thus facilitating a mass export to the sea during the ebb tide (Figure 9.1). Most estuarine species have reproductive specialisations which facilitate the retention of eggs, embryos and larvae within the estuarine environment. For example, the male longsnout pipefish *Syngnathus acus* has a brood pouch in which the offspring

are carried until they have reached an advanced developmental stage, thus avoiding the rigours of the estuary and increasing the chances of remaining within aquatic plant beds (Melville-Smith & Baird 1980). Other species such as estuary klipfish *Clinus spatulatus* are viviparous and their young exceed 15 mm in length when released into the estuarine environment (Bennett 1983). The Cape silverside *Atherina breviceps* has relatively large eggs with glutinous filaments which are used for attachment to aquatic plants and other objects (Neira *et al*. 1988), thus aiding their retention in the estuarine environment. Although the eggs of the estuarine roundherring *Gilchristella aestuaria* are freefloating, this species spawns in the upper reaches of open estuaries (Talbot 1982), thereby reducing losses to the sea (Figure 9.1). Very few *G. aestuaria* eggs were recorded in the lower half of the Sundays estuary when compared with the upper half (Wooldridge & Bailey 1982).

In contrast to the small estuarine spawners, marine species utilising estuaries usually mature above 200 mm SL (Whitfield 1990). Evidence from both KwaZulu-Natal and

Table 9.1. Some possible processes and behaviour influencing marine fish immigration to South African estuarine nursery areas

	Habitat			
	Nearshore marine	**Surf zone**	**Estuary mouth**	**Estuary**
Physical process	Wind/coastal current driven surface drift	Longshore transport	Tidal flux transport	Estuarine
Fish development	Egg/embryo/larval stages	Larval/juvenile stages	Larval/juvenile stages	Juvenile stage
Fish movement	Mainly passive	Mainly active	Active and passive	Mainly active
Orientation to estuarine cues	No	Yes	Yes	Yes

Cape waters suggests that the spawning of most marine species occurs close inshore, often in the vicinity of estuary mouths (Wallace 1975b; Lasiak 1983). Inshore currents along the KwaZulu-Natal coast retain eggs, embryos and larvae in the region (Wallace 1975b), thus reducing the distance between breeding and nursery areas. According to Heydorn et al. (1978), the retention of these early life stages inshore prior to migration into KwaZulu-Natal estuaries is favoured by the slow overall movement of the water mass between the Agulhas Current and the coast, frequent current reversals parallel with the shore, and the occurrence of onshore components. A similar situation applies to the Eastern Cape Province where shear edge eddies on the inside of the Agulhas Current feed into the coastal shelf area where current reversals are commonplace (Schumann & Brink 1990). Within Algoa Bay and other bays along the Eastern Cape coast, cyclonic within-bay circulation occurs (Harris 1978). It has been suggested that in Algoa Bay the embryos and larvae become entrained within this cyclonic circulation and, upon completion of larval development, juveniles would not have to swim great distances to recruit into coastal nursery areas (Beckley 1986). Drift card analyses indicate that in the southern Cape there is a greater frequency of surface onshore flow during the summer than in winter (Shannon & Chapman 1983), with the path taken by oil from the Venpet/Venoil tanker collision supporting this conclusion (Moldan et al. 1979). Onshore water currents, particularly during the spring and summer, would assist the movement of estuarine-associated fish larvae and post-larvae towards the coast.

Some marine species can spawn in estuaries when conditions are suitable. The estuarine bream Acanthopagrus berda has been recorded spawning in the mouth region of the Kosi estuary at night and the eggs are transported out to sea during peak ebb tides (Garratt 1993). The blackhand sole Solea bleekeri breeds in specific parts of Lake St Lucia when conditions within this system are suitable (Cyrus 1991) but has not been recorded spawning in other KwaZulu-Natal estuaries. The white seacatfish Galeichthys feliceps is a mouthbrooder (Tilney & Hecht 1993) and adults have been recorded carrying eggs and young in Eastern Cape estuaries (Marais & Venter 1991), thus providing juve-

niles with direct access to the estuarine environment. In contrast, the eggs of other marine species are mainly pelagic (Brownell 1979) and postlarvae need to locate and enter estuaries without parental assistance. Perhaps the most arduous routes to estuarine nursery areas are undertaken by larvae and juveniles of the strepie Sarpa salpa and leervis Lichia amia. Spawning of these two species occurs along the KwaZulu-Natal coast (Joubert 1981; van der Elst 1988) with the eggs, embryos and larvae probably being transported southward on the inside of the Agulhas Current. Consequently, juveniles of S. salpa and L. amia frequent Eastern and Western Cape estuaries (Beckley 1983, 1984; Whitfield & Kok 1992) but are normally absent from KwaZulu-Natal estuaries and inshore reefs (Joubert 1981; van der Elst 1988).

Recruitment mechanisms and possible cues

As has been suggested earlier, the egg, embryonic and larval stages of marine species are retained within the nearshore environment by cyclonic circulation patterns and predominantly onshore surface current components. Once the postlarval stage of development has been reached (i.e. attainment of complete fin ray counts and beginning of squamation), fish movement becomes active and the individuals then enter the surf zone which may be used as an interim nursery area before seeking out an estuary (Table 9.1, Figure 9.2). More than 85% of the larvae and/or postlarvae found in the Swartvlei Bay surf zone were estuarine-associated marine species (Whitfield 1989b). The increase in postlarval fish densities towards the Swartvlei estuary mouth suggests that there is an active onshore and longshore movement towards the estuary mouth. In addition, concentrations of the ichthyoplankton in the surf zone were highest during the summer when most species migrate into Cape estuaries. Although there was no significant difference between day and night densities, tidal phase may well be important in governing ichthyoplankton abundance within the shallow surf zone (Whitfield 1989b).

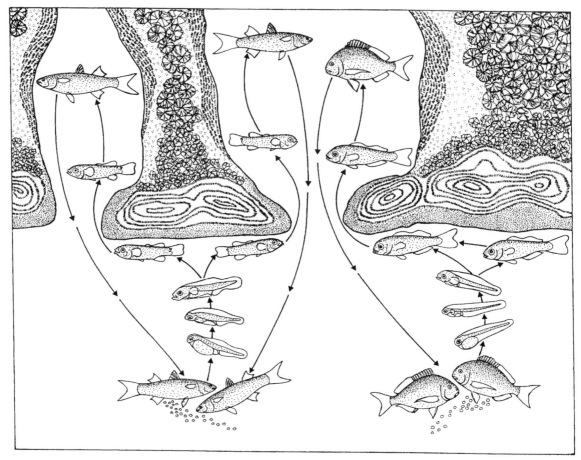

Figure 9.2. **Diagrammatic representation of the predominant life cycles of estuarine-associated marine spawners. The two examples** shown here are the flathead mullet *Mugil cephalus* and Cape stump-nose *Rhabdosargus holubi*.

How do the postlarvae locate the estuary mouth once they have entered the surf zone? Blaber & Blaber (1980) have suggested that turbidity gradients in the marine environment may aid juvenile fishes to locate estuarine nursery areas but the importance of this factor has yet to be quantified. Whitfield (1992a) has shown that immigration of postlarvae occurs regularly during oceanic over-topping of temporarily closed estuary sandbars. The absence of linked estuarine/marine turbidity gradients during such extreme high tide events mitigates against the use of this particular cue in locating estuarine nursery areas. Whitfield (1994a) compared the recruitment of larval and juvenile fishes into three Eastern Cape estuaries with differing turbidities and found that both the Sundays (average turbidity = 10 Nephelometric Turbidity Units) and Great Fish (average turbidity = 65 NTU) had similar ichthyonekton abundance. These results suggest that in Eastern Cape estuaries turbidity gradients do not play a major role in attracting larvae and postlarvae towards these systems.

Whitfield (1994a) found that the abundance of newly-recruited marine fishes into Eastern Cape estuaries showed a significant positive correlation with longitudinal salinity gradients within these systems. However, in the absence of detectable salinity gradients within the marine environment, he suggested that it was riverine and estuarine olfactory cues associated with these gradients that were attracting the postlarvae into estuaries and not salinity gradients *per se*. Indirect evidence to support this hypothesis comes from the observation that marine postlarvae are able to enter temporarily closed estuaries when marine overtopping of the mouth sandbar occurs (Whitfield 1992a) and no salinity gradient between the estuary and sea is present. Olfactory cues associated with seepage of estuarine water through relatively narrow sandbars could provide the key to locating the mouth area of temporarily closed estuaries, and thus facilitate the migration of postlarvae over these barriers during oceanic overtopping events.

The immigration of larvae and postlarvae into large, well-flushed estuaries of the northern hemisphere mainly takes place using passive and/or selective tidal transport for both entry to and retention within these systems (e.g. Weinstein *et al.* 1980, Fortier & Leggett 1982; Boehlert & Mundy 1988). In contrast, in poorly-flushed estuaries in South Africa, Australia and New Zealand, where for much of the year the two-layered circulation pattern is less pronounced or absent, the larvae and juveniles of some marine species enter these systems on the flood tide and are retained by rapidly settling along the banks or on the bottom where water movements are reduced (Beckley 1985; Roper 1986; Whitfield 1989c; Neira & Potter 1994). Motile juveniles can, however, easily enter estuaries on the ebb tide by keeping to the margins where current speeds are attenuated (Whitfield 1989c). In the Zotsha estuary on the KwaZulu-Natal south coast, postlarval fish were recorded moving into the system through the bottom of standing waves, swimming upstream in a series of steps (Harrison & Cooper 1991). Yet another recruitment mechanism, already mentioned above, involves the larvae and juveniles entering temporarily closed or closing estuaries during oceanic overtopping of the sandbar at the mouth (Whitfield 1992a). Preliminary indications are that large–scale ichthyoplankton migration into temporarily closed estuaries may occur in this manner.

Once the larvae and juveniles have entered an estuary, several cues are available to assist orientation towards specific nursery areas. Salinity gradients are perhaps the most obvious but have been largely discounted by Blaber (1987) on the basis of experimental and field evidence. Temperature is an unlikely cue, as thermal gradients within estuaries are irregular and highly variable, depending upon cloud cover, tidal regime, river flow, oceanic upwelling, etc. However, the juveniles of a number of fish species are attracted to warm littoral areas, so the use of water temperatures, possibly in conjunction with other cues, cannot be discounted. Turbidity gradients are usually strongly developed within estuaries, and Blaber (1987) is of the opinion that both vertical and horizontal gradients may be important to certain species. Studies by Cyrus and Blaber (1987a, b, c) have shown that the juvenile marine fishes of KwaZulu-Natal estuaries can be divided into five main groups according to their occurrence in various turbidities. The above field and laboratory studies indicate that turbidity plays a significant role, either singly or in combination with other variables, in determining the distribution of juvenile marine fishes in estuaries. The role of olfactory cues within estuaries, including those released by fishes, should be investigated since these 'markers' hold great potential in 'fine tuning' juvenile movements within the estuarine environment.

In contrast to the marine migrants, larvae of species such as *Gilchristella aestuaria* do not have to enter the estuary. Instead, they need to remain within the system during the vulnerable embryonic and larval stages. Melville-Smith *et al.* (1981) have described how *G. aestuaria* larvae in the Sundays estuary utilise tidal transport in order to avoid being swept out to sea. There is evidence to suggest that these larvae remain in the middle and upper reaches of the estuary where zooplanktonic food resources are most abundant, and avoid the more marine areas near the mouth until at least the juvenile stage has been attained. The cues that assist these larvae in monitoring their position in the upper half of the estuary are unknown.

Spatial and temporal patterns

The migration of 0+ marine juveniles into KwaZulu-Natal estuaries occurs mainly in late winter and spring when river flow is often at a minimum (Wallace & van der Elst 1975). However, many of the smaller systems along the KwaZulu-Natal coast are closed during the winter and only open after spring rains in September/October (Whitfield 1980a). Recruitment into these temporarily closed estuaries is therefore only possible when increased water flow forces open the mouths of the above systems. The prolonged period of juvenile immigration (Figure 9.3), which is a function of the extended spawning season of most species, may be regarded as a strategy against unseasonal floods which could open blind estuaries prematurely, and droughts which would delay the opening of these systems until midsummer.

In Eastern and Western Cape estuaries, a similar prolonged recruitment pattern is evident, but in this case the peak immigration phase occurs in late spring and early summer when primary and secondary production within estuaries is reaching a maximum. Thus, in both KwaZulu-Natal and Cape estuaries, juvenile fishes are able to exploit the abundant summer food resources and warm temperatures to grow rapidly before the onset of winter. Refuges in the form of submerged aquatic vegetation are also most prolific in summer (Whitfield 1984) and higher turbidities due to increased river flow would aid predator avoidance by juvenile fish (Cyrus & Blaber 1987a).

Spawning by estuarine fish taxa occurs mainly during spring, and the embryos, larvae and juveniles are particularly abundant during summer (Figure 9.4). Some of these fish species grow very rapidly in the warm highly productive waters, and with the onset of winter have already attained sexual maturity: for example, *Gilchristella aestuaria* consume approximately 12% of body mass per day in summer and mature within 7 months of hatching (Talbot & Baird 1985). Daily food consumption then declines to

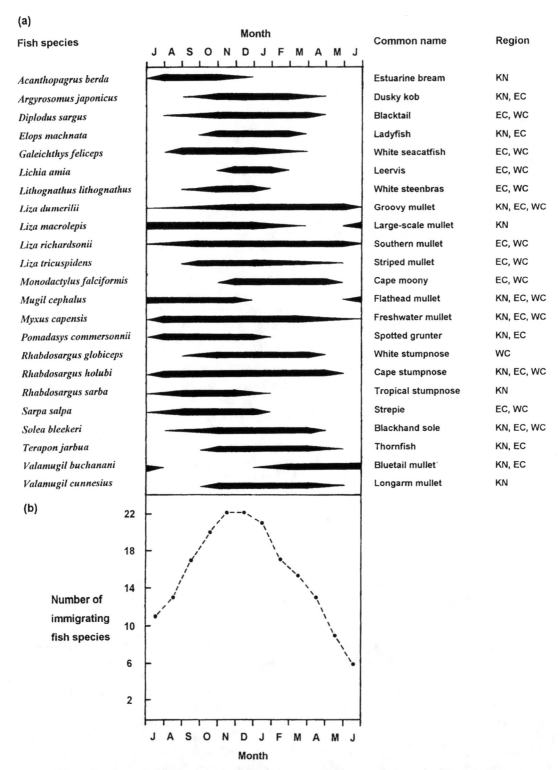

Figure 9.3. (a) Diagrammatic representation of the immigration periods of estuarine-associated fish larvae and juveniles into South African systems. Regional information used: KN, KwaZulu-Natal; EC, Eastern Cape; WC, Western Cape. Data from Wallace & van der Elst (1975), Bok (1979), Melville-Smith & Baird (1980), Whitfield (1990), Whitfield & Kok (1992) and Harris & Cyrus (1995). (b) Monthly variation in the numbers of species entering southern African estuaries, as derived from bar representations shown in (a).

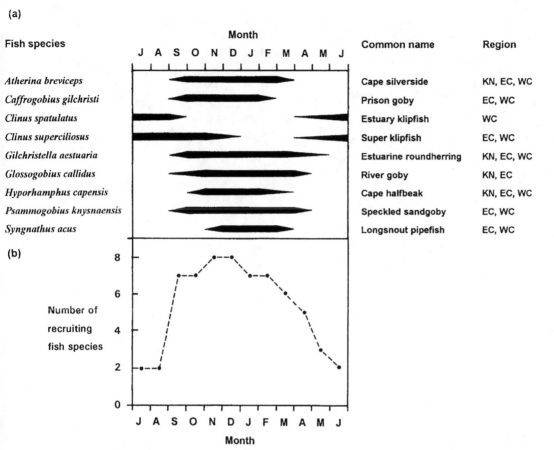

Figure 9.4. (a) Diagrammatic representation of the periods of abundance of larvae spawned within South African aestuaries. Regional information used: KN, KwaZulu-Natal; EC, Eastern Cape; WC, Western Cape. Data from Melville-Smith & Baird (1980), Bennett (1983), Whitfield (1990) and Harris & Cyrus (1995). (b) Monthly variation in the numbers of species represented by larvae in southern African estuaries, as derived from bar representations shown in (a).

less than 2% of body mass in winter when water temperatures decline and zooplankton resources become more scarce.

A major factor influencing the composition of ichthyoplankton in South African estuaries is sea temperature, which is often linked to latitude. Species diversity generally declines from the subtropical systems in the northeast to the cool temperate systems in the southwest (Whitfield 1994a). The subtraction of tropical fish species is clearly reflected in Table 9.2 which shows the decline in family representation between the Kosi and Swartvlei estuaries. Approximately half the fish families represented in the St Lucia estuary were present in Eastern and Western Cape systems. The decline becomes even more apparent if the Kosi estuary diversity is compared with that of the warm temperate estuaries (Table 9.2). In both KwaZulu-Natal and Cape systems, ichthyoplankton densities fluctuate widely and are generally much higher in summer than in winter (Melville-Smith & Baird 1980; Whitfield 1989a; Harrison & Whitfield 1990; Harris & Cyrus 1995). In the Great Fish, Kariega, Sundays and St Lucia estuaries, larval and postlarval fish densities showed a strong positive correlation with temperature (Whitfield 1994b; Harris & Cyrus 1995).

Few studies have quantified ichthyoplankton interchange in South African estuaries (Beckley 1985; Whitfield 1989c). Table 9.3 shows the net gain or loss of the most common species moving across a transect in the Swartvlei estuary mouth over 24 h tidal cycles. Results indicated that there was a net gain of marine migrant species (e.g. Cape stumpnose *Rhabdosargus holubi*) and a net loss of the embryos and larvae of certain estuarine spawners (e.g. *Psammogobius knysnaensis*) which subsequently returned to the estuary as postlarvae. Other estuarine spawners (e.g. *Syngnathus acus*) lost newly-released postlarvae to the marine environment and these were not recorded returning to the estuary on subsequent flood tides. Movements of embryos, larvae and postlarvae between the estuary and sea occurred mainly during twilight and nocturnal hours when predation rates would probably be lower than during the day.

Table 9.2. Family composition (+ = presence) of ichthyoplankton (excluding oceanic taxa) sampled in six South African estuaries

Family	Kosi Subtropical	St Lucia Subtropical	Great Fish Warm temperate	Sundays Warm temperate	Swartkops Warm temperate	Swartvlei Warm temperate
Elopidae	+	+	+		+	
Megalopidae	+	+				
Anguillidae		+	+			
Congridae	+					
Muraenidae		+				
Opichthidae	+	+	+			
Muraenesocidae		+				
Clupeidae	+	+	+	+	+	+
Engraulidae	+	+	+	+	+	+
Chanidae	+	+				
Gobiesocidae	+	+	+		+	+
Notocheiridae	+					
Atherinidae		+	+	+	+	+
Hemiramphidae	+	+		+	+	+
Syngnathidae	+	+	+	+	+	+
Scorpaenidae	+	+				
Platycephalidae	+	+		+	+	
Triglidae		+				
Ambassidae	+	+	+	+		
Serranidae		+				
Teraponidae	+	+	+	+		
Apogonidae	+					
Pomatomidae			+	+		
Haemulidae	+	+	+	+	+	
Lujanidae	+					
Sparidae	+	+	+	+	+	+
Lethrinidae	+					
Scorpididae	+					
Coracinidae					+	
Monodactylidae	+	+		+	+	+
Gerreidae	+	+		+		
Mullidae	+					
Sillaginidae	+	+				
Sciaenidae	+	+	+	+	+	+
Leiognathidae	+	+				
Chaetodontidae					+	
Carangidae	+		+		+	
Pempheridae	+	+				
Labridae	+	+				
Scaridae	+	+				
Mugilidae	+	+	+	+	+	+
Sphyraenidae	+	+				
Trichonotidae	+					
Creediidae	+					
Mugiloididae	+					
Blenniidae	+	+	+	+	+	+
Tripterygiidae	+	+				
Clinidae	+		+	+	+	+
Callionymidae	+					
Schindleriidae	+					
Gobiidae	+	+	+	+	+	+
Eleotridae	+	+				
Trichiuridae		+				
Siganidae		+				
Bothidae	+					
Pleuronectidae	+					
Cynoglossidae	+	+				
Soleidae		+	+	+	+	+
Paralichthyidae	+					
Monacanthidae	+					
Ostraciidae		+				
Tetraodontidae		+	+	+	+	+
Total	48	41	21	20	21	15

Source: Information from Melville-Smith & Baird (1980), Wooldridge & Bailey (1982), Beckley (1985), Whitfield (1989*a*), Whitfield (1989*c*), Harrison & Whitfield (1990), Whitfield (1994b), Harris & Cyrus (1994) and Harris *et al.* (1995).

Table 9.3. Net gain or loss of selected ichthyoplankton species moving into (+) and out of (−) the Swartvlei estuary mouth over three 24 h periods (n.r. = not recorded). Numbers of larvae/postlarvae are calculated from the actual densities (individuals m⁻³) for each species, in conjunction with the estimated volumes (m³) of water entering and leaving the estuary (see Whitfield 1989c for details)

Fish species	Month		
	October	January	March
Marine migrants			
Liza ?richardsonii	+18 108	+2849	+131 511
Heteromycteris capensis	+3666	+7172	+161 890
Monodactylus falciformis	+4050	+2352	+10 254
Rhabdosargus holubi	+5211	n.r.	+2292
Argyrosomus japonicus	+5814	+3306	n.r.
Solea bleekeri	n.r.	+26 725	+9657
Estuarine spawners			
Psammogobius knysnaensis	−548 533	−6903	+135 560
Caffrogobius ?multifasciatus	−511 330	+84 540	+32 862
Omobranchus woodi	−1922	−118 101	+3281
Syngnathus acus	−65 376	n.r.	−7838
Atherina breviceps	−3905	n.r.	n.r.
Hippocampus capensis	−4091	n.r.	n.r.

Dependence of fish larvae and postlarvae on estuaries

Although the juveniles of a number of marine fish species are dependent on estuaries as nursery areas (Whitfield 1994a), the larvae of these species are normally absent from estuaries. As has already been mentioned, spawning occurs at sea, the embryo and larval stages of development being completed in the marine environment. The available evidence suggests that abiotic constraints to the survival of egg, embryonic and larval stages, under fluctuating estuarine salinity, temperature and dissolved oxygen regimes, could be problematical. For example, eggs of the flathead mullet Mugil cephalus cannot survive salinities less than 28 for more than 24 h (Sylvester et al. 1975). In addition, tidal fluctuations within estuaries could easily transport embryos and larvae into unfavourable reaches of an estuary, or result in their deposition on intertidal sand and mudflats at low tide.

The biotic environment within estuaries may also present problems for the survival of large numbers of ichthyoplanktonic organisms. Larval fishes feed mainly on microzooplankton, whereas postlarval fishes utilise both micro- and macrozooplankton (Whitfield 1985). According to Vernberg & Vernberg (1972) zooplankton are the dominant primary consumers in oceanic waters and therefore fish larvae are more likely to find a suitably sized and reliable planktonic food supply in this environment. Carter (1978) found that mean zooplankton biomass off the KwaZulu-Natal coast was higher over the continental shelf than in offshore areas. The highest biomass values were recorded within 5 km of the coast, a region usually occupied by the fish larvae of estuarine-associated marine species (Heydorn et al. 1978). Research on zooplankton in South African estuaries has revealed that this resource is highly variable, the east coast estuaries tending to have a higher biomass (1–1200 mg m⁻³) than south coast estuaries (1–112 mg m⁻³) (Grindley 1981). The fluctuating nature of estuarine zooplankton, even in those systems with characteristically abundant stocks, is a further mitigating factor against the use of this environment by the larvae of marine species.

Virtually all of the marine fish larvae and postlarvae which enter South African estuaries feed on zooplankton, regardless of their ultimate juvenile or adult diet (Whitfield 1985). Calanoid and cyclopoid copepods are particularly important prey items to newly-recruited ichthyoplankton. The switch from a predominantly zooplanktonic diet to one dominated by zoobenthos (e.g. evenfin pursemouth Gerres methueni), detritus (e.g. large-scale mullet Liza macrolepis) or aquatic plants (e.g. Rhabdosargus holubi) is rapid and usually occurs between 10 and 30 mm SL. Since most species first enter estuaries between 10 and 20 mm SL, the transition in diet occurs within the estuarine environment. Blaber & Whitfield (1977) have shown that mullet postlarvae in estuaries feed initially on zooplankton, then vertically migrating zooplankton and meiobenthos, and finally microbenthos. The role of vertically migrating zooplankton in initiating dietary switches by the postlarvae may be significant. In addition, it is tempting to suggest that the scarcity of zooplankton in certain South African estuaries may force a switch in the diet of postlarvae earlier than would otherwise have occurred.

The seasonal abundance of ichthyoplankton in estuaries is usually positively correlated with copepod abundance (Harrison & Whitfield 1990), thereby increasing the potential growth and survival of these individuals. The spring and summer peak in zooplanktonic productivity is particularly important to those estuarine fish species whose larvae do not develop within the marine environment. There is increasing evidence to suggest that estuaries with large zooplankton stocks also support high densities of resident planktivores such as Gilchristella aestuaria. Swartvlei is an estuarine lake which has poor zooplankton resources (Coetzee 1981) and G. aestuaria larvae averaged 26 individuals per 100 m³ during 1986–7 (Whitfield 1989a). In contrast, the Sundays estuary has a rich zooplanktonic fauna (Wooldridge & Bailey 1982) and G. aestuaria larvae averaged 204 individuals per 100 m³ during 1986–7 (Harrison & Whitfield 1990). In addition, larval G. aestuaria from the Swartvlei system are narrower bodied than those from the Sundays estuary (Haigh & Whitfield 1993), thus reinforcing the suggestion by Blaber et al. (1981) that zooplanktonic prey abundance may influence the morphometrics of this species. The thinner larvae from Swartvlei would presumably be more susceptible to starvation and associated mortalities than those in the Sundays estuary.

Future research directions

Studies on the early life stages of fishes in South African estuaries have progressed considerably within the last two decades, largely as a result of the efforts of a handful of scientists. However, it would appear that ichthyoplankton research is still in its infancy, and studies on a wide variety of topics are needed to fill some of the major gaps which exist. For example, there is very little or no information on the breeding behaviour, spawning localities and larval development of most estuarine-associated species. Baseline studies on the ichthyoplankton from many regions are lacking, and the cool temperate Western Cape estuaries have been totally neglected. Field and laboratory studies on the cues which govern larval and postlarval immigration into estuaries would assist in identifying the possible role freshwater plays in influencing fish movements into and within estuaries. Finally, information on the physiological tolerances, diet, growth and survival of fish larvae and postlarvae in both the marine and estuarine environment would be most useful in understanding the ecology of these poorly known early life stages.

Acknowledgements

I thank Ray Vogel for drawing the illustrations used in Figures 9.1 and 9.2, and Angus Paterson for commenting on an initial draft of the manuscript.

THE ESTUARINE ICHTHYOFAUNA

Introduction

Because of the extensive utilisation of estuaries by man, interest in the ichthyofauna of these specialised environments has increased. This interest is despite the fact that South African estuaries are characterised by a relatively low ichthyofaunal diversity but high abundance of individual taxa (Whitfield 1994c). Blaber (1973) showed that even small estuaries may have single-species populations ranging from 10 000–50 000 individuals while the densities in these systems usually exceed that of the adjacent marine or freshwater environments (Whitfield 1993).

Many factors, both biotic and abiotic, directly or indirectly influence the utilisation of estuaries by fishes and research has been conducted on many naturally occurring factors as well as anthropogenic effects that cause changes in the abundance and diversity of fishes in estuaries. These will form the central theme of this chapter.

The utilisation of estuaries by fishes

South African estuaries can be classified into permanently open, temporarily open or closed, estuarine bays, estuarine lakes and river mouths (Whitfield 1992b). According to Reddering and Rust (1990) they are characterised by the following features.

1   The majority are small, their tidal prisms being 10^6 m^3 or less.
2   Most occupy drowned river valleys and only a few have developed on coastal plains.
3   The tidal inlets are constricted and or periodically blocked by sandbars.
4   Flood-tidal deltas are well developed whereas ebb-tidal deltas are poorly developed.

Whitfield (1992b) gives estimates of the size of tidal prisms in the different estuarine types, ranging from $>10 \times 10^6$ m^3 to $<1 \times 10^6$ m^3 per spring tide cycle. Average mean salinity varies between 1 and >35.

Estuaries are food-rich, sheltered ecosystems where fewer than 100 of the more than 3000 species occurring around our coast, are regularly found (Wallace et al. 1984). The fish communities that use estuaries have been variously classified: Day et al. (1981) divided the communities into six groups using breeding and migratory habits; Heydorn (1989) on their dependence on estuarine or marine habitats; and Beckley (1985) and Whitfield (1989c) into three and four groups respectively on the basis of larval and postlarval recruitment. However, the most useful is that of Loneragan and Potter (1990) which distinguishes between five categories of fish that utilise estuaries and is incorporated in the classification system proposed by Whitfield (1994a) for South African estuaries.

1   Estuarine species which breed in southern African estuaries and further subdivided into:
    (a) Resident species which have not been recorded breeding in the marine or freshwater environment.
    (b) Resident species which also have marine or freshwater breeding populations.
2   Euryhaline marine species which usually breed at sea, with the juveniles showing varying degrees of

dependence on southern African estuaries, are further subdivided into:

(a) Juveniles dependent on estuaries as nursery areas.

(b) Juveniles occur mainly in estuaries, but are also found at sea.

(c) Juveniles occur in estuaries but are usually more abundant at sea.

3 Marine species which occur in estuaries in small numbers but are not dependent on these systems.

4 Euryhaline freshwater species, whose penetration into estuaries is determined primarily by salinity tolerance. This category includes some species which may breed in both freshwater and estuarine systems.

5 Catadromous species which use estuaries as transit routes between the marine and freshwater environments but may also occupy estuaries in certain regions. Further subdivided into:

(a) Obligate catadromous species which require a freshwater phase in their development.

(b) Facultative catadromous species which do not require a freshwater phase in their development.

From the discussion of reproductive behaviour of estuarine-associated fishes (Whitfield, this volume), the life cycles of estuarine fishes are such that, except for the estuarine spawners (although certain estuarine spawners have a marine larval component to their life cycle), they frequently migrate between estuaries and the sea. Migratory behaviour, which includes the influx of juveniles and larvae from the marine environment, assisted by larvae from estuarine spawners, has a major effect on their numbers in estuaries. Any attempt to determine the extent of utilisation of any estuary, or to compare different systems, should take cognisance of this fact. Therefore, abundance and catch composition of fishes in estuaries should be determined over extended periods using different gear that overlaps in terms of selectivity. Such a sampling protocol would ensure the collection of representative data from which reliable conclusions can be drawn.

Mouth condition, that is, whether open or closed, will determine the movement of juveniles in and out of estuaries and thus the ichthyofaunal composition (Bennett *et al.* 1985; Kok & Whitfield 1986). South African estuaries represent of a wide range of rainfall and climatic regimes, ranging from the semi-arid west coast where estuaries are closed for most of the year to the subtropical east coast estuaries characterised by summer thunderstorms, resulting in floods and silt-laden waters (Heydorn 1989). In between are the southwest coast estuaries, subject to summer droughts and winter floods and the south coast estuaries, where a bimodal rainfall prevails and estuarine opening or closure is more erratic depending on rainfall and wave action along the coast.

Within a particular estuary, a number of environmental variables will determine its suitability or attractiveness for its inhabitants and thus determine their abundance. This will differ among species (Kinne 1964; Whitfield *et al.* 1981; Whitfield 1996) and will be reviewed below.

Factors influencing fish abundance and diversity in estuaries

Salinity

Day (1981) pointed out that the main characteristic of an estuarine animal is its ability to tolerate salinity changes. According to Panikkar (1960), the essential adaptation by fish penetrating estuaries is an adjustment to salinity fluctuations. In this ever changing environment salinity has an important role, not only in determining the distribution of fishes within an estuary, but also the abundance and diversity of ichthyofauna. Salinities in South African estuaries range from freshwater conditions during floods (Marais 1982, 1983*b*) or below 10 for more than two years after floods (Cyrus 1988*b*), to 100 or more during extended drought periods (van der Elst *et al.* 1976; Whitfield & Bruton 1989).

The physical effects of reduced freshwater discharge into estuaries are that sand shoals situated in the lower reaches of many mature Eastern Cape estuaries increase in size (Reddering 1988). This larger flood tide delta may cause a reduction in the tidal prism of an estuary and take a particular flood longer to scour because there is more sediment to be removed. Reduced freshwater inflow may cause the development of hypersaline conditions (>40), as is often found in the upper reaches of estuaries such as the Kromme, Bushmans and Kariega. Salinities in Eastern Cape estuaries are usually inversely related to the amount of freshwater entering these systems (Whitfield & Bruton 1989).

The indirect effect of reduced freshwater inflow into an estuary appears to be a reduction in the abundance of the larger fish component (sampled by gill net). Analysis of gill-net catches sampling this component from 14 Eastern Cape estuaries, showed a positive correlation between fish numbers or biomass and decreasing salinity (Marais 1988). A similar relationship was found by Allen and Barker (1990) for North American estuaries. It seems, however, that it is the difference in longitudinal salinity between the head and mouth of the estuary, or the salinity

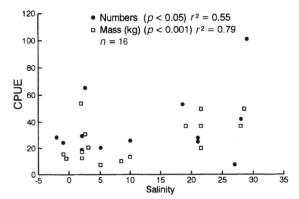

Figure 9.5. **Correlation between axial salinity gradient (difference between the head and mouth) and CPUE of fish, determined from gill-net catches in 14 Eastern Cape estuaries (after Marais 1988).**

gradient, rather than low salinity throughout the estuary that is important. Figure 9.5 illustrates that the catch per unit effort (CPUE) for both number and mass was correlated with increasing salinity range. Further supporting evidence is provided by the earlier studies of van der Elst *et al.* (1976), who found a doubling in Lake St Lucia gill-net catches in water salinities below 20 over those above 50 in Lake St Lucia.

A consequence of this response is that estuaries with regular freshwater input, especially irrigation return flows, had much larger catches than freshwater-deprived systems. The former estuaries had a particularly large mullet (e.g. *Mugil cephalus* and *Myxus capensis*) and carp (*Cyprinus carpio*) component in the upper reaches as well as predatory kob *Argyrosomus japonicus* (Marais 1981, 1982, 1983b; Plumstead 1984). Recent and as yet unpublished data, obtained from ten Eastern Province estuaries during a 4-year study, have provided further support for these findings. They indicate that juveniles, sampled by seine net, exhibit a negative correlation between bottom water salinity and numbers. A similar relationship has also been suggested between bottom water salinity and species diversity within the small fish community. We recognise that the proximate factor(s) operational in this correlation may have less to do with salinity than with higher production and carrying capacity of those reaches of an estuary with lowered salinity. Whitfield *et al.* (1994) and Whitfield (1996) suggested that the greater abundance of the larger size groups of detritivorous and iliophagous species may be linked to the large riverine-derived organic inputs into these systems.

Russell (1996) suggested that inputs of freshwater could provide strong cues for recruiting euryhaline fishes into the Swartvlei Lake system. However, river floods have quite the opposite effect. He reported (Russell 1994) the indirect impact of severe flooding and breaching of the Swartvlei estuary during February 1991, and concluded that the loss

of some 3000 specimens of the Knysna seahorse *Hippocampus capensis*, a resident endangered species of the estuarine fish community, was a result of a marked rise in water temperature (32 °C) in the shallow littoral, which because of the rapid flood flow, prevented the cooling effect of the tides.

The Mtata estuary, sampled as one of the 14 estuaries described by Marais (1988), exhibited low fish numbers. Artificial flood conditions were experienced at the time because of freshwater being released from the hydro-electric scheme at Umtata Dam. Marais (1983b) also found a negative correlation between water flow and catch rate in the Gamtoos estuary after prolonged flood conditions. Under such conditions current velocities, especially in the middle and upper reaches of estuaries, probably require fish to expend extra energy to swim upstream or even to maintain position. In the Sundays estuary high suspensoid loads, possibly associated with reduced oxygen levels in the water, resulted in clogging of the gills of fishes, and caused mass fish mortality after a flash flood (Whitfield & Paterson 1995).

The relationship between fish abundance and post-flood conditions may not always be simple. Whereas Remane and Schlieper (1971) attributed an impoverished fish fauna to salinities of 5–7, Marais (1982) found reduced fish numbers (especially mullet) in the channel-like Sundays estuary after floods, but enhanced numbers in the adjacent Swartkops estuary. In the Sundays estuary silt and organic material were removed by the flood, whereas silt and mud were deposited on the extensive intertidal banks of the Swartkops estuary where *Liza dumerilii* and *L. richardsonii* normally feed. In Lake St Lucia, Cyrus (1988a) could not find a change in *Solea bleekeri* abundance after cyclonic floods removed bivalves on which they prey.

Catchment size

Catchment size is normally an indication of estuary size, although this is not invariably so, e.g. the Knysna river estuary. In Eastern Cape estuaries, Marais (1988) found both higher fish numbers and biomass in larger systems. Figure 9.6, constructed from the catch returns and catchment sizes presented by Marais (1988), shows significant positive correlations between fish abundance in terms of both numbers and mass, and catchment size. The relationship (Marais 1992) is probably an indirect one because of high salinity gradients found in larger systems being correlated with a high fish abundance (Figure 9.5), as well as increased transparency which was negatively correlated with fish abundance (Figure 9.7). A further factor to be considered is the tendency towards higher nutrient and organic loading of estuaries with larger catchments and hence the potential for elevated primary and secondary productivity within these systems (Allanson & Read 1995).

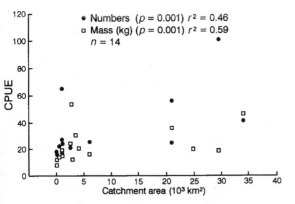

Figure 9.6. **Correlation between size of catchment (km²) and CPUE of fish, determined from gill-net catches in 14 Eastern Cape estuaries (after Marais 1988).**

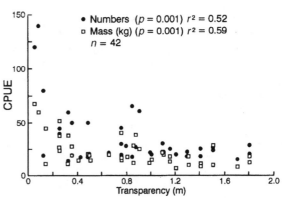

Figure 9.7. **Correlation between water transparency (m) and CPUE of fish, determined from gill-net catches in 14 Eastern Cape estuaries (after Marais 1988).**

Turbidity

Blaber & Blaber (1980) expressed the opinion that protection against predators afforded by turbid water may, in addition to calm water and abundant food, be one of the main reasons why juvenile fish enter Indo-Pacific estuaries. Gill-net catches from 14 Eastern Cape estuaries (Marais 1988) showed turbidity to be positively correlated with catch size in terms of both fish numbers and mass (Figure 9.7). Hecht & van der Lingen (1992) found that increased turbidity caused decreased predation rates by visual predators and changed feeding strategies. Low turbidity may, according to Whitfield et al. (1994), account for the presence of certain marine transients in the Kowie estuary resulting in higher species richness.

Blaber (1981) suggested that many southern African estuarine-associated fishes are essentially 'turbid-water' species and have evolved in turbid areas. Research conducted on juvenile marine fishes by Cyrus & Blaber (1987a, b, c) has shown that 80% of the species studied were turbid water taxa (>10 Nephelometric Turbidity Units), whereas only four (20%) could be classified as truly clear water (<10 NTU). This suggests that the protective isolation created by turbidity was advantageous to the survival of juvenile fish. Turbid estuaries often have a higher freshwater input, associated with more nutrients and organic material, and this input into the food web provides rich feeding conditions, which promotes growth. The fact that shallow turbid areas are usually absent along the South African coast, enhances the importance of estuaries which provide these conditions (Whitfield 1996), with Blaber (1981) showing that where these conditions are present outside estuaries in the Indo-Pacific region, the same juvenile fishes are present.

Cyrus & Blaber (1987c) remark that the protective isolation against bird and teleost predation created by turbidity, coupled with the low number of predators in calm, sheltered, shallow estuarine waters with an abundance of food, produces conditions advantageous to the survival

and growth of estuarine fishes. Various factors determine turbid and clear conditions in KwaZulu-Natal estuaries, which may occur near each other in the same estuary and provide a greater diversity of habitat to estuarine species (Cyrus 1988b, c). Cyrus & Blaber (1988), however, warn that increased sedimentation by fine particles in St Lucia estuary as a result of injudicious dredging and the introduction of water from the Mfolozi River could, in the long run, be detrimental to the fauna of St Lucia estuary and possibly St Lucia Lake as well. This is in agreement with Bruton (1985) who stated that the effects of suspensoids on fish could be either beneficial or detrimental, depending on the concentration of suspended particles, and the timing and duration of exposure.

Estuary size and mouth condition

Larger systems tend to support a larger, more diverse and more abundant fish component than smaller systems. Whitfield (1980b) has shown that large systems such as Kosi and St Lucia have more than 100 species whereas the small Sodwana estuary has fewer than 50 species. Dundas (1994) found the number of species caught by gill nets in small, periodically open estuaries, to be more a reflection of effort (number of nets used) than of size or mouth condition. A comparison of the gill-net catches of Marais (1988) and Dundas (1994) does, however, show a lower fish abundance in small periodically closed estuaries in terms of mass when compared with large permanently open systems. Seine nets, which sample the smaller fish component very effectively when performed under comparable conditions in the lower reaches of large and small estuaries, revealed little difference in abundance in terms of mass or numbers of smaller sized fish.

Bennett (1989) presented information on fish abundance and species diversity of three small south Western Cape estuaries in the winter rainfall area of South Africa. The data showed that differences in number, composition and abundance of species can be related to whether or not each estuary had been closed and, if so, for how long. Thus the permanently open Palmiet estuary had the highest percentage of marine migrant species and the closed Bot the lowest, with resident estuarine species following an opposite trend. Migrants contributed 1% and residents 99% of numbers caught in the Bot River lagoon, whereas the two groups occurred in approximately equal numbers in the Palmiet River estuary.

Whitfield & Kok (1992) in their study of the recruitment of juvenile marine fishes into the Knysna and Swartvlei estuaries over a 30 month period found that the dominant species included southern mullet *Liza richardsonii*, Cape stumpnose *Rhabdosargus holubi*, white steenbras *Lithognathus lithognathus*, blacktail *Diplodus sargus*, strepie *Sarpa salpa*, Cape moony *Monodactylus falciformis*, groovy mullet *Liza dumerilii*, striped mullet *L. tricuspidens*, flathead mullet *Mugil cephalus*, freshwater mullet *Myxis capensis* and leervis *Lichia amia*. The relatively higher densities of most species in the Knysna system are attributed very largely to the deep permanently open mouth and a strong marine influence. The sandbar blocking the entrance to the Swartvlei system effectively blocks migration into and out of the estuary for much of the year, thus directly influencing the composition of the fish community. As regards juveniles, Kok & Whitfield (1986) have shown, as has Dundas (1994), that mouth closure resulted in a marked decline of juvenile fish <50 mm TL but not of larger size groups. Hall *et al.* (1987) attributed the prolonged closed mouth phase and shallowness of interconnecting channels to the low diversity of fish fauna in the upper Wilderness lakes compared with other South African estuarine lake systems. Whitfield (1996) does, however, stress the benefits of mouth closure in increasing nursery areas as water level rises.

Dundas (1994) found that the greatest influence of mouth closure in periodically closed estuaries was a breakdown in recruitment of juvenile marine migrant species. She ascertained that immigration of fish into the Seekoei, Kabeljous and van Stadens estuaries was dominated by the Southern mullet *L. richardsonii*. On the other hand, Harrison and Whitfield (1995) in a detailed study of three temporarily closed estuaries in KwaZulu-Natal established that breaching of the bar results in a temporary slump in food resources due to the scouring action of the floods and the exposure of the previously inundated benthos. Competition between the resident fish species and marine migrants increases during the open phase due to limited food resources and nursery habitats. The large-scale recruitment of estuarine-dependent marine species during spring and summer accounts for the increasing dominance of the migrating species. Once the estuary mouth closes during late summer or autumn, there is a redistribution of fish groups within the system (Harrison

& Whitfield 1995). Weinstein (1985) suggests that warm temperate and tropical estuaries are highly programmed systems which allow for the spatial and temporal separation of fish species in order to more completely utilise the available resources and contribute towards a higher survival rate for otherwise competing species.

Russell (1996) reasoned that the artificial breaching of estuaries in the Swartvlei and Wilderness lake systems during optimum fish recruiting periods in 1991 and 1992 may have contributed to the increases in abundance during 1993. However, a longer tidal phase in the Swartvlei system during 1992 and 1993, compared with the Wilderness lake system, did not result in greater abundance of fish sampled. Russell concluded that there appears to be no justification for the artificial maintenance of permanently tidal conditions in the Swartvlei and Wilderness estuaries on the grounds of benefit for the maintenance of viable fish communities. Whitfield & Kok (1992) maintain that artificial breaching of the Swartvlei mouth in winter has generally led to premature closure of the system and loss of the 'head' of water needed for the spring/summer opening when maximum recruitment of most fish species occurs. On the other hand, if the timing of the breach coincided with peak immigration of marine migrants and there was sufficient head of water to scour sediments from the mouth, the breach was beneficial.

Where estuaries experience increasing periods of mouth closure, as for example the Bot River lagoon, De Decker (1987) reported that opening of the lagoon to the sea caused lowered macrofaunal abundance. Branch *et al.* (1985) had reservations about the artificial breaching of the lagoon and recommended either that it should never be breached, or that breaching should be done on a controlled basis in a manner determined by sound ecological principles, of which the most important are probably whether the opening coincides with the period of maximum recruitment and an extended open phase. The Bot River lagoon is regarded by Koop *et al.* (1983) as a temporarily closed estuary in an intermediate evolutionary stage between open estuary and coastal lake. Input of seawater is confined to occasional overtopping of the dune barrier via washover fans. Whitfield (1992b) has actually observed fish larvae introduced into closed estuaries by this mechanism. The lagoon had salinities of under 10 which in 1981 fell below 3 causing mass mortalities of marine as well as estuarine species (Koop *et al.* 1983). This is the opposite of what had been recorded in the estuary of the Seekoei River where elevated salinities as a result of severe drought caused mass fish mortalities.

Habitat

The habitat of an estuary will, to a large extent, determine its inhabitants. As might be expected, estuaries with a

Table 9.4. Simple correlation coefficients between fish abundance and diversity against selected estuarine characteristics for fish caught by seine net in ten Eastern Cape estuaries

Water/substratum characteristics		d.f.	Abundance		Richness	Diversity		Evenness mass	Dominance mass
			no.	mass		no.	mass		
Depth		80		−0.213*	0.264**		0.472**	0.287**	−0.438**
Temperature (°C)	(surface)	80			0.327**				
	(bottom)	80			0.266**				
Salinity	(bottom)	80	−0.227*	0.271**		−0.202*			
Zostera		80			0.228*				

* $p < 0.05$, ** $p < 0.01$.

wide range of substrata and littoral plant growth normally have a higher species diversity than uniform systems, as was shown by Whitfield (1983). The higher species richness found in the Kowie River estuary compared to the nearby Great Fish River estuary was attributed to the availability of a wider range of habitats, and hence an increased variety of food sources, shelter, etc. in the Kowie River estuary (Whitfield *et al.* 1994). However, abundance of ichthyofauna in the two systems was not linked to habitat diversity and the authors found a 3:1 ratio between fish abundance in the Great Fish and Kowie estuaries. They linked the higher abundance in the former estuary to large fluvatile organic and nutrient inputs into the system.

Harrison and Whitfield (1995) in their review of the fish community structure in three temporary closed estuaries in KwaZulu-Natal correlated species number, biomass, richness, evenness and dominance with a number of ecological variables. A similar approach has been used in the analysis of an extensive and intensive new data set from ten estuaries in the Eastern Province. The correlation analyses are given in Table 9.4. Only those coefficients which are significant at $p < 0.05$ and $p < 0.01$ are reported. In view of the richness of the data set the degrees of freedom are high (>70). Abundance and dominance in terms of mass are negatively correlated with depth, while richness, diversity, evenness and dominance are positively correlated. Richness appears to be a close function of bottom and surface temperature. Numerical abundance is negatively correlated with salinity or some function thereof, while mass is positively correlated. The positive correlation between eelgrass beds and richness is a frequent observation in the estuaries of the Eastern Cape and Western Cape south coast. Eelgrass beds expand where freshwater inflow is minimal or has ceased altogether (Adams & Talbot 1992) thus increasing this type of nursery area for fishes. The importance of eelgrass beds as a nursery habitat is stressed by Whitfield *et al.* (1989) who showed that *Zostera* beds are mainly colonised by small species and the juveniles of marine species. Senescence of

macrophyte beds in the Swartvlei estuary caused a decrease of 92% in CPUE of *Monodactylus falciformis* and 91% in *Rhabdosargus holubi* (Whitfield 1984). According to this author, macrophyte beds provide refuge from predators and an assured food supply in the form of a large biomass of invertebrates and epiphytes.

Studies by Beckley (1983) in the Swartkops estuary supported the view that *Zostera* beds contribute to a more abundant and diverse fish fauna than non-vegetated areas. Hanekom and Baird (1984) found only two species, *Monodactylus falciformis* and *Rhabdosargus holubi*, in significantly higher numbers in *Zostera* areas compared with non-*Zostera* areas in the Kromme estuary. However, contrary to most other studies, community analyses comparing numbers, species richness and diversity revealed no significant differences between the catches from the two habitats. A possible explanation forwarded by Hanekom and Baird (1984) was that the sampling sites of Beckley (1983) were dissimilar as regards substrata and water current strength whereas their sites were similar, i.e. factors other than macrophyte cover could have caused the differences.

Catch composition in the Swartkops and Kromme River estuaries showed strong similarities with both systems being dominated by a few species (Beckley 1983; Hanekom & Baird 1984). In samples of the Swartkops, 97% of the catch consisted of only ten species, namely: *Atherina breviceps* (46%); members of the family Mugilidae (*Mugil cephalus*, *Liza dumerilii*, *L. richardsonii* and *L. tricuspidens* − together 20%); *Rhabdosargus holubi* (12%); *Gilchristella aestuaria* (9%); *Diplodus sargus* (6%) and two species of Gobidae, *Psammogobius knysnaensis* and *Caffrogobius multifasciatus* (4%) (Beckley 1983). In the Kromme River estuary samples, all the above species (with the exception of *D. sargus*) were dominant, and as a group (together with *Glossogobius callidus*) formed 96% of the total catch (Hanekom & Baird 1984). Most of the above species also dominated seine net catches from the lower reaches of other Eastern Cape estuaries.

Although herbivores normally comprise a small percentage of estuarine fishes (18% in Swartvlei), detritivores accounted for 49% of catch biomass in the same estuary. This general trend has been documented in many estuaries and Whitfield (1983) concluded that detritus forms the food base of these estuaries, and that they act as detritus traps. Whitfield 1988 has shown that detritus is transported by tidal flow to macrophyte-free areas of the Swartvlei estuary where it enters the detrital food web.

A recent study by Paterson and Whitfield (1996) found that a saltmarsh creek in the Kariega River estuary was utilised by a wide cross-section of fish species associated with estuaries; and among this array, the euryhaline marine category was numerically and gravimetrically the most dominant. The dominance of this eurytopic fish community reflected the marked daily fluctuations in the abiotic features of the saltmarsh. Ter Morshuizen et al. (1996) have shown that this dominance also occurs in the ebb and flow region of the Great Fish estuary. The existence of several marine species (e.g. *Rhabdosargus holubi*, spotted grunter *Pomadasys commersonnii*, white steenbras *Lithognathus lithognathus* and *Liza* spp.) which have seldom, if ever, been recorded from freshwater may have to do with the high conductivity values (150–625 mS m^{-1}) that characterise the lower Great Fish River.

Temperature

No single factor operates in isolation and this is well demonstrated by the studies of Martin (1988). His investigations into the temperature tolerance ranges of three *Ambassis* species suggest that the osmoregulatory capability of *A. productus* in reduced salinities (<10) increases while that of *A. gymnocephalus* decreases sharply in salinities below 20. *A. natalensis*, which is endemic to the southeast coast of Africa, is adapted to a wide range of salinity conditions. Distribution and abundance of the three *Ambassis* species, within and between estuaries, is determined by interaction between temperature and salinity tolerances, which can also be inferred from the article by Whitfield (1995) on mass mortalities of estuarine-associated fish species in South Africa.

The sea has a buffering effect on estuarine temperatures in the middle and lower reaches of estuaries (Branch & Grindley 1979). But water temperatures show much wider seasonal fluctuations in the head regions of estuaries (Marais & Baird 1980a; Marais 1981, 1983a, b). This factor, in conjunction with lowered salinity and increased turbidity, may account for decreased diversity and abundance of estuarine ichthyofauna in the upper reaches of estuaries. Loneragan & Potter (1990) sampled different reaches of ten Australian shallow water estuaries and found that both species number and total density of fishes were higher during summer and autumn when temperature and salinity were at a maximum. This was true for both marine stragglers and marine estuarine opportunists. Marais (1994) has shown that surface and bottom temperature were positively correlated with species richness of fishes sampled by seine netting. This is probably more a reflection of seasonal spawning and migration than of temperature tolerance.

Depth

Water has to be of a certain depth to permit free movement of fish and allow a certain degree of protection, especially against bird predation. The negative correlation of water depth with fish abundance indicates that as water got deeper CPUE in terms of mass decreased (Table 9.4). This could be as a result of species such as *Gilchristella aestuaria*, which was numerous in the catches made in the shallow areas near the mouths of the ten Eastern Cape estuaries (Marais 1994), providing a strong bias in the correlation. However, depth was positively correlated with species richness, diversity (in terms of mass) and evenness of distribution and negatively with dominance (Table 9.4).

Other factors

Several factors that were either not discussed or just mentioned in passing in this section directly or indirectly influence the utilisation, and thus the abundance and diversity, of fishes in South African estuaries. Many of these are discussed in the reviews of Whitfield (1983, 1994c, 1996), Blaber (1985) and Marais (1988). Some of the most pertinent are latitude (Wallace 1975a; Blaber 1981), seasonality (Wallace 1975b; Branch & Grindley 1979), nearshore marine conditions (Whitfield 1989b; Potter et al. 1990), mouth depth and degree of marine influence (Whitfield & Kok 1992), physical constrictions within estuarine systems (Hall et al. 1987), intertidal saltmarsh creeks (Paterson & Whitfield 1996), trapping of organic material by impoundments (Plumstead 1990), dissolved oxygen levels (Russell 1994), predation (Blaber 1973; Whitfield & Blaber 1978a), competition (Whitfield & Blaber 1978b), reproductive condition (Wallace 1975b; Marais & Venter 1991), parasite loads (Whitfield & Heeg 1977; Schramm 1991; Whitfield 1996), and habitat degradation as a result of pollution (Blaber et al. 1984) or dredging (Cyrus & Blaber 1988).

The functional role of fish in estuaries

Fish are generally regarded as opportunists and will consume whatever nutritious material is readily available. Estuarine fish communities, however, have been divided into different categories based on structural and functional characteristics of the habitat. Whitfield (1988) examined the stomach contents of 1648 fishes in the Swartvlei estuary and found that the detritivore group comprised 49% of the total catch biomass, zoobenthivores 25%, herbivores 18%, piscivores 6%, and epifauna/zooplanktivores 2%. The linear index for selection indicated a strong positive selection by fishes for epifaunal invertebrates and poor utilisation of infauna and plants in the estuary. A low diversity of herbivorous fishes was recorded (only three species in Swartvlei), despite the abundance of seagrass beds in this system (Whitfield 1988). A similar situation was recorded in the Swartkops estuary (Hanekom & Baird 1984) and in Florida (Carr & Adams 1973).

Food chains in several South African estuaries are very largely based on detritus (Whitfield 1980b, c, 1982; Marais 1984; Heymans & Baird 1995). Notwithstanding the importance attached to *Zostera capensis* as a source of detritus in the Kromme River estuary, Heymans & Baird (1995) determined that marsh halophytes, particularly *Spartina maritima*, were responsible for 78% of primary production in that estuary of which 90% was first broken down to detritus before entering the food web.

One of the reasons why juvenile marine migrants as well as estuarine residents utilise estuaries successfully, is a food supply which is richer and more predictable than in the open sea. In a study of production and consumption of prey species in the Bot estuary, Bennett & Branch (1990) found that consumption of invertebrates by resident fishes accounted for 17% of secondary production, or 30% of production by prey species. This suggests that predation by fish was not likely to have a major impact on the prey resource and that food was not in short supply. Most of the fish consumed a wide variety of prey but no two species shared their most important prey category. This was also found for predatory marine migrants in Eastern Cape estuaries (Marais 1984). Bennett and Branch (1990) concluded that dietary niche width for the different species was significantly correlated with their abundance but not with their average overlap with the other species, implying that the most successful species are those with the broadest niches and that specialisation is a dubious means of reducing competition.

From a study of food utilisation by fishes in the St Lucia Lake system, Whitfield (1980b) describes different ways in which estuarine food resources may be segregated. For example, certain piscivorous fish species seldom enter shallow weedy areas (*Argyrosomus japonicus* and *Elops machnata*), whereas juveniles (mainly mullet) remain in shallow marginal areas thus avoiding large predators. Different foraging methods and prey size selection by various bird species, seasonal variation in the feeding habits of both fish and birds, variation in recruitment and migration of fishes and invertebrates to and from the system and prevailing salinity regime all contributed to the maintenance of a dynamic food web. He felt that St Lucia could not be regarded as a closed system since freshwater fishes, birds, crocodiles and hippopotamuses move between the lake and adjacent freshwater habitats, thus continually exchanging energy and nutrients between the different systems.

After studying the feeding ecology of major carnivorous fishes in four Eastern Cape estuaries, Marais (1984) commented on the diversity of prey as well as partitioning of prey by size. The food webs in these estuaries were characterised by a comparatively small number of energy pathways but high energy flow rates per pathway. The main pathway in the group containing most piscivores (*A. japonicus, L. amia* and *E. machnata*) was via algae, of which a varying amount was derived from phytoplankton and benthic microalgae to mullet and/or via copepoda to *G. aestuaria* (Schlacher & Wooldridge 1996). The main pathway to the group containing mostly benthivores (*P. commersonnii, G. feliceps*) was via algae and the mudprawn, *Upogebia africana*. At the top of the estuarine trophic chain, although not quantified, were birds, otters and human anglers (Marais 1984).

McLachlan and Bate (1985), Romer (1986) and Romer & McLachlan (1986) described three main pathways for the transfer of energy from primary foods, phytoplankton and detritus, to the fish community in an Eastern Cape beach/surf-zone ecosystem.

1   Via mullet grazing directly on diatoms (benthic microalgae) and sediment bacteria to piscivorous fish.
2   Via suspended particulate organic carbon (POC) and sediment POC pools to macrofauna and benthic-feeding fish, which in turn are consumed by piscivorous fish.
3   Via phytoplankton and suspended detritus, zooplankton and zooplanktivorous fish to piscivorous fish.

These pathways are similar to those of Eastern Cape estuarine systems. However, Baird & Heymans (1996) found that after the construction of the Mpofu dam in 1984, which severely reduced freshwater inflow into the Kromme River estuary, the first and second pathways became more prominent due mainly to reduced phytoplankton production. They (Baird and Heymans *op. cit.*) also noted a concomitant decline in the standing stocks of zooplankton in the Kromme estuary, and attributed that to reduced water column primary production. Baird and Ulanowicz (1993), in a comparative study of several estuaries, showed the phytoplankton–zooplankton–zooplanktivorous links to be more pronounced in systems with higher phytoplankton production and strong salinity gradients (e.g. the Swartkops estuary) than in those with low primary production and weak salinity gradients, e.g. the Kromme estuary.

Various categories of fish – invertebrate detritivores, planktivores and benthic feeders – occurred in the diet of piscivores in the Swartkops estuary. This was also found in comparative stomach content analyses performed on piscivores from these two estuaries (Marais 1984).

The changes that occurred in the carbon chains, after reduction of freshwater inflow in the Kromme estuary, were also highlighted by Allanson and Read (1995) in their comparison of the Kariega River and Great Fish River estuaries. They found a three orders of magnitude higher standing stock of zooplankton in the Great Fish River estuary, which is subject to a sustained freshwater inflow, emphasising the vital role such flows play in determining the structure and functioning of their communities. They argued that the generation of pycnoclines during the tidal cycle and the interfaces or fronts which develop, coupled with the phenomenon of hydrodynamic trapping and suspensoid flocculation, provide the stimulus for phytoplankton growth and the generation of POC. It can be accepted that it is the persistence of these water column processes, coupled with sufficient nutrients, that determines the richness of the phyto- and zooplankton standing stock.

This explains the higher planktonic productivity in estuaries with more regular freshwater inputs, a feature highlighted by Whitfield & Wooldridge (1994). The shift from a suspensoid–microalgal based energy source to an autochthonous, detritus base as was found in the Kariega estuary, is largely controlled by the ratio of river flow to tidal prism volumes (Allanson & Read 1995). This decline in planktonic productivity was also observed in the Kromme estuary (Baird & Heymans 1996) and, probably, the Bushmans River estuary. The Kariega and Bushmans estuaries had by far the lowest CPUE in terms of numbers of any of the ten estuaries sampled by Marais (1988). CPUE in terms of mass and the species diversity was equal to the other estuaries, although the larger fish species component was poorer in both number and biomass. Because of the decreased freshwater inflow, estuaries like the Kariega, Kromme and Bushmans have entered a marine-dominated phase where the benthic community of plants and animals expand which still afford a rich nursery area for largely benthic-feeding fishes.

Exploitation of estuarine fishes

The marine resources presently exploited can be grouped under a formal sector (i.e. commercial fishing and harvesting) and an informal sector (i.e. recreational angling and harvesting as well as subsistence angling). In 1990 it was estimated that the turnover value of recreational fishing and its associated tourism and tackle manufacture industries in South Africa is in the order of R150 million per annum (van der Elst & Adkin 1991). In the Eastern Cape Province these activities amounted to about R17 million in 1984 (Smale & Buxton 1985). It has been estimated that recreational fishing in South Africa is growing at a rate of about 6% per annum. Whether or not this rate of increase is sustainable is uncertain.

Estuarine angling involves boat traffic and bait collecting (digging for prawns, etc.), with potential negative effects on ecologically sensitive areas such as marshes and intertidal mud- and sandbanks (Baird et al. 1996). A crude estimate (van der Elst 1989) was that 50 000 anglers operated from light tackle boats in estuaries in 1987, thus underlining the extent of the potential problem.

Wallace et al. (1984) considered the value of estuaries as recreational angling facilities as one of the main reasons for their conservation. They pointed out that activities other than those associated with angling may be detrimental to fish stocks if they alter the natural functioning of estuaries and reduce their value as nurseries and fishing grounds.

An activity associated with angling is the removal of benthic organisms for bait, which can lead to the depletion of available prey for fish (Daniel 1994). Bait digging can, apart from the reduction in population size of the bait organisms (Baird et al. 1981), cause a reduction in the population size of non-bait species (Jackson & James 1979; van den Heilegenberg 1987), and the temporary disruption and redistribution of the bait species population (McLusky et al. 1983). Bait digging may also have a negative effect on bird and fish populations within estuarine systems (Baird & Martin 1987; van den Heilegenberg 1987; Hanekom et al. 1988; Martin 1988).

With regard to the effect of angling on fish stocks, Gulland (1987) pointed out that while modifying the structure of fish communities, the size of the effect is difficult to measure because this is easily confounded with environmental effects. Pinpointing the absolute causes of the depletion of fish stocks is further complicated by the migratory nature of the large fish component. Perturbations that occur elsewhere, e.g. in the marine environment or catchment, may negatively impact on fish populations in estuaries.

The negative effects of angling on estuarine-associated fish species are of great concern because fishing in South Africa is growing at an ever increasing rate (Coetzee & Baird 1981). This is particularly disturbing in the light of the conclusion by Bennett and Atwood (1993) that there is a clear relationship between CPUE and degree of exploitation as is evident from angling catch data from around southern Africa. Most fisheries tend to target one species of fish with other species being caught incidentally (Gulland 1987). This seems to be the case for the recreational line fishery in the Swartkops estuary where *Pomadasys commersonnii* dominated anglers' catches (Marais & Baird 1980a). Whether this is an artefact of the dominance of this species in the estuary (Marais & Baird 1980b) or its greater catchability (Marais 1985) is of little consequence as it inevitably renders spotted grunter vulnerable to over-exploitation.

A study, using both gill net and anglers CPUE, was performed to reassess the relative abundance of, especially, angling species in the Swartkops and Sundays estuaries (Baird et al. 1996) and to compare these results with those obtained earlier for the two respective estuaries (Marais 1981; Marais & Baird 1980 a, b). The study revealed that a mean of 148 and 43 anglers use the Swartkops and Sundays estuaries respectively over weekends. It was found that over a period of nearly 20 years the composition of angler's catches in the two Eastern Cape estuaries showed little change. Gill-net catches, although not significantly different in terms of total CPUE, indicated a decline in the abundance of *P. commersonnii* in both estuaries. This is not surprising because the questionnaire survey conducted by Baird et al. (1996) revealed that 77% of respondents targeted spotted grunter followed by dusky kob. These two species together comprised 87% and 90% of anglers' catches in the Swartkops and Sundays estuaries, respectively. It was concluded that present management and conservation policies in Eastern Cape estuaries are ill defined and that the philosophy of ecosystem preservation, rather than individual species conservation, should be seriously considered. In the Durban Bay estuary, where *P. commersonnii* also dominated in anglers' catches, catch rates have declined over a period of 16 years and Guastella (1994) recommended careful management and policing of the harbour as essential prerequisites to ensure sustainable future catches.

The utilisation of highly sought after species such as white steenbras *Lithognathus lithognathus* has led to reduced abundance in both estuaries and the marine environment. *L. lithognathus* formed an important component of the Swartkops estuary seine net catches conducted by Gilchrist (1918) earlier this century. In the 1970s this species comprised only 3% of Swartkops anglers' catches (Marais & Baird 1980a) and in the latest study they were absent from gill-net catches and occurred in low numbers in anglers' catches (Baird et al. 1996). Over-exploitation is blamed by Bennett (1993) for the decrease of both commercial catches (mean annual catch for the period 1983–91 was only 14% of that for the years 1897–1906) and recreational catches (a decline of 90% since the 1970s) of *L. lithognathus* off the Eastern Cape coast.

Another species that all but disappeared from anglers' catches in the Swartkops estuary (Marais & Baird 1980a) but was once as abundant in catches as *P. commersonnii*, is the elf *Pomatomus saltatrix*. This species was still the most

important of the recreational rock-angling fishery on the southeast coast of South Africa in the mid-1980s (Clarke & Buxton 1989) despite the fact that there were already signs of over-exploitation (Coetzee et al. 1989). Although reductions in the abundance of angling fish species cannot be attributed to angling pressure alone it is undoubtedly an important factor, and a reduction in angling pressure could serve to increase present abundance of these species (Baird et al. 1996). A case in point is that of *P. saltatrix*, which has increased in numbers along the coast of KwaZulu-Natal and the east coast of the Eastern Province following increased protection (van der Elst & De Freitas 1987; Garrett & van der Elst 1990).

Marinas

Conflicts often arise between developers and ecologists when the construction of new marinas is being considered. This is because the building of marinas in estuaries, regardless of their location, can seriously degrade coastal resources (Lindall & Trent 1975). The study of Cloete (1993) supported the earlier findings of Baird et al. (1981) that, provided certain conditions are met, such as the maintenance of adequate connections to ensure acceptable flushing rates and water quality, marinas can increase the habitat for typical estuarine fauna, i.e. marinas may not have an adverse effect on the ecology of parent estuaries.

The small fish component, sampled by seine net, in the canals of the Royal Alfred Marina on the Kowie estuary, at Marina Glades in the Kromme estuary and in Marina Martinique, a man-made coastal lagoon with a tenuous artificial link to the sea, was shown by Cloete (1993) to exhibit seasonal fluctuations characteristic of connecting or nearby estuaries. Species composition was similar but abundance lower than in the open estuary to which they were connected. The composition of the large fish component, sampled by gill net, was also similar to that in the adjacent estuaries, but the abundance was considerably lower (Baird et al. 1981; Cloete 1993). The main reason for this is the shallowness of the marinas investigated. Higher ichthyoplankton numbers in the marina canals was attributed to reduced predation rates and lower current velocities enabling fish larvae to maintain their position more readily. Cloete (1993) concluded that marinas associated with estuaries offer an extension of the estuarine environment for the ichthyoplankton and small fish but not for large fish.

Conclusions

South African estuaries provide a unique, dynamic environment dominated by a limited number of food chains, which are predominantly dependent on detritus, especially where phytoplankton production is reduced as a result of lack of freshwater pulses. Of the many factors which influence the diversity of the ichthyofaunal structure and its utilisation of local estuaries, riverine pulses probably have the most important impact on species distribution, composition and abundance. Estuaries with decreased or zero freshwater inputs may pass into a marine-dominated mode with a diverse and abundant small fish component, especially where extensive macrophytic plant cover occurs, but an impoverished large fish component.

Man-made structures such as marinas can be regarded as extensions of the estuary with an ichthyofaunal component similar to the adjacent estuary, provided good circulation patterns exist. Whether periodically closed estuaries should be opened or not should not be determined haphazardly by local residents because of discomfort caused by rising water levels, but should be based on sound ecological principles.

Thus, with the increasing exploitation of estuarine resources not only by recreational and commercial anglers and subsistence fisherfolk, but also by other human activities, the need for a whole ecosystem approach to protection rather than measures to conserve individual species is urgently needed.

Acknowledgements

I would like to thank Alan Whitfield for critically reading the manuscript of the second part of this chapter, and Sheila Heymans for suggestions in the section on the functional role of fish in estuaries.

References

Adams, J.B. & Talbot, M.M. (1992). The influence of river impoundment on the estuarine eelgrass *Zostera capensis* Setchell. *Botanica Marina*, **35**, 69–75.

Allanson, B.R. & Read, G.H.L. (1995). Further comment on the responses of Eastern Cape Province estuaries to variable freshwater inflows. *Southern African Journal of Aquatic Sciences*, **21**, 56–70.

Allen, D.M. & Barker, D.L. (1990). Interannual variations in larval fish recruitment to estuarine epibenthic habitats. *Marine Ecology Progress Series*, **63**, 113–25.

Baird, D. & Heymans, J.J. (1996). Assessment of ecosystem changes in response to freshwater inflow in the Kromme river estuary, St Francis Bay, South Africa: a network analysis approach. *Water SA*, **22**, 1–10.

Baird, D. & Martin, A.P. (1987). A preliminary energy budget for birds in the Swartkops estuary. 6th National Oceanographic Symposium, Stellenbosch, abstracts.

Baird, D. & Ulanowicz, R.E. (1993). Comparative study on the trophic structure, cycling and ecosystem properties of four tidal estuaries. *Marine Ecology Progress Series*, **99**, 221–37.

Baird, D., Marais, J.F.K. & Daniel, C. (1996). Exploitation and conservation of angling fish species in selected South African estuaries. *Aquatic Conservation:*

Marine and fresh water ecosystems. (In press.)

Baird, D., Marais, J.F.K. & Wooldridge, T. (1981). The influence of a marina canal system on the ecology of the Kromme estuary, St Francis Bay. *South African Journal of Zoology*, **16**, 21–34.

Beckley, L.E. (1983). The ichthyofauna associated with *Zostera capensis* Setchell in the Swartkops estuary, South Africa. *South African Journal of Zoology*, **18**, 15–24.

Beckley, L.E. (1984). The ichthyofauna of the Sundays estuary, South Africa, with particular reference to the juvenile marine component. *Estuaries*, **7**, 248–50.

Beckley, L.E. (1985). Tidal exchange of ichthyoplankton in the Swartkops estuary mouth, South Africa. *South African Journal of Zoology*, **20**, 15–20.

Beckley, L.E. (1986). The ichthyoplankton assemblage of the Algoa Bay nearshore region in relation to coastal zone utilization by juvenile fish. *South African Journal of Zoology*, **21**, 244–52.

Bennett, B.A. (1983). *Clinus spatulatus*, a new species of clinid fish (Perciformes: Blenniodei) from South Africa, with a modified definition of the genus *Clinus*. J.L.B. Smith Institute of Ichthyology Special Publication No. 29.

Bennett, B.A. (1989). A comparison of the fish communities in nearby permanently open, seasonally open and normally closed estuaries in the

south-western Cape, South Africa. *South African Journal of Marine Science*, **8**, 43–55.

Bennett, B.A. (1993). The fishery for white steenbras *Lithognathus lithognathus* off the Cape coast, South Africa, with some considerations for its management. *South African Journal of Marine Science*, **13**, 1–14.

Bennett, B.A. & Atwood, C.G. (1993). Shore-angling catches in the De Hoop nature reserve, South Africa, and further evidence for the protective value of marine reserves. *South African Journal of Marine Science*, **13**, 213–22.

Bennett, B.A. & Branch, G.M. (1990). Relationship between production and consumption of prey species by resident fish in the Bot, a temperate South African estuary. *Estuarine, Coastal and Shelf Science*, **31**, 139–55.

Bennett, B.A., Hamman, K.C.D., Branch, G.M. & Thorne, S.C. (1985). Changes in the fish fauna of the Bot estuary in relation to opening and closure of the estuary mouth. *Transactions of the Royal Society of South Africa*, **45**, 449–64.

Blaber, S.J.M. (1973). Population size and mortality of the marine teleost *Rhabdosargus holubi* (Pisces: Sparidae) in a closed estuary. *Marine Biology*, **21**, 219–25.

Blaber, S.J.M. (1981). The zoogeographical affinities of estuarine fishes in south-east Africa. *South African Journal of Science*, **77**, 305–7.

Blaber, S.J.M. (1985). The ecology of fishes of estuaries and lagoons of the Indo-Pacific with particular reference to south-east Africa. In *Fish community ecology in estuaries and coastal lagoons*, ed. A. Yanez-Aráncibia, pp. 247–66. Mexico City: Universidad Nacional Autonoma de México.

Blaber, S.J.M. (1987). Factors affecting recruitment and survival of mugilids in estuaries and coastal waters of southeastern Africa. *American Fisheries Society Symposium*, **1**, 507–18.

Blaber, S.J.M. & Blaber, T.G. (1980). Factors affecting the distribution of juvenile estuarine and inshore fish. *Journal of Fish Biology*, **17**, 143–62.

Blaber, S.J.M., Cyrus, D.P. & Whitfield, A.K. (1981). The influence of zooplankton food resources on the morphology of the estuarine clupeid *Gilchristella aestuarius* (Gilchrist, 1914). *Environmental Biology of Fishes*, **6**, 351–5.

Blaber, S.J.M., Hay, D.G., Cyrus, D.P. & Martin, T.J. (1984). The ecology of two degraded estuaries on the north coast of Natal, South Africa. *South African Journal of Zoology*, **19**, 224–40.

Blaber, S.J.M. & Whitfield, A.K. (1977). The feeding of juvenile mullet (Mugilidae) in south-east African estuaries. *Biological Journal of the Linnean Society*, **9**, 277–84.

Boehlert, G.W. & Mundy, B.C. (1988). Roles of behavioral and physical factors in larval and juvenile fish recruitment to estuarine nursery areas. *American Fisheries Society Symposium*, **3**, 51–67.

Branch, G.M. & Grindley, J.R. (1979). Ecology of southern African estuaries. Part XI. Mngazana: a mangrove estuary in Transkei. *South African Journal of Zoology*, **14**, 149–70.

Branch, G.M., Bally, R., Bennett, B.A., De Decker, H.P., Fromme, A.W., Heyl, C.W. & Willis, J.P. (1985). Synopsis of the impact of artificially opening the mouth of the Bot River estuary: implications for management. *Transactions of the Royal Society of South Africa*, **45**, 465–83.

Brownell, C.L. (1979). *Stages in the early development of 40 marine fish species with pelagic eggs from the Cape of Good Hope*. Ichthyological Bulletin of the J.L.B. Smith Institute of Ichthyology No. 40.

Bruton, M.N. (1985). The effects of suspensoids on fish. *Hydrobiologica*, **125**, 221–41.

Butterworth, D.S. (1983). *Some aspects of the scientific information requirements for line-fish management decisions*. Internal SANCOR Line-fish Programme Document.

Carr, W.E.S. & Adams, C.A. (1973). Food habits of juvenile marine fishes occupying seagrass beds in the estuarine zone near Crystal River, Florida. *Transactions of the American Fisheries Society*, **102**, 511–40.

Carter, R.A. (1978). *The distribution of calanoid copepoda in the Agulhas Current system off Natal, South Africa*. CSIR Research Report No. 363. Pretoria: CSIR.

Clarke, J.R. & Buxton, C.D. (1989). A survey of the recreational rock-angling fishery at Port Elizabeth, on the South-East coast of Africa. *South African Journal of Marine Science*, **8**, 183–94.

Cloete, A.E. (1993). *Comparative assessment of ichthyofauna in three marinas*. MSc thesis, University of Port Elizabeth.

Coetzee, D.J. (1981). Zooplankton distribution in relation to environmental conditions in the Swartvlei system, southern Cape. *Journal of the Limnological Society of Southern Africa*, **7**, 5–12.

Coetzee, P.S. & Baird, D. (1981). Catch composition and catch per unit effort of anglers' catches off St Croix island, Algoa Bay. *South African Journal of Wildlife Research*, **11**, 14–19.

Coetzee, P.S., Baird, D. & Tregoning, C. (1989). Catch statistics and trends in the shore angling fishery of the east coast, South Africa, for the period 1959–1982. *South African Journal of Marine Science*, **8**, 155–71.

Cyrus, D.P. (1988a). Episodic events in estuaries: effects of cyclonic flushing on the benthic fauna and diet of *Solea bleekeri* (Teleostei) in Lake St Lucia on the south-eastern coast of Africa. *Journal of Fish Biology*, **33**, 1–7.

Cyrus, D.P. (1988b). Turbidity and other physical factors in Natal estuarine systems. Part 1: selected estuaries. *Journal of the Limnological Society of Southern Africa*, **14**, 60–71.

Cyrus, D.P. (1988c). Turbidity and other physical factors in Natal estuarine systems. Part 2: estuarine lakes. *Journal of the Limnological Society of Southern Africa*, **14**, 72–81.

Cyrus, D.P. (1991). The reproductive biology of *Solea bleekeri* (Teleostei) in Lake St Lucia on the south-east coast of Africa. *South African Journal of Marine Science*, **10**, 45–51.

Cyrus, D.P. & Blaber, S.J.M. (1987a). The influence of turbidity on juvenile marine fish in the estuaries of Natal, South Africa. *Continental Shelf Research*, **7**, 1411–16.

Cyrus, D.P. & Blaber, S.J.M. (1987b). The influence of turbidity on juvenile marine fishes in estuaries. Part 1. Field studies at Lake St Lucia on the southeastern coast of Africa. *Journal of Experimental Marine Biology and Ecology*, **109**, 53–70.

Cyrus, D.P. & Blaber, S.J.M. (1987c). The influence of turbidity on juvenile marine fishes in estuaries. Part 2. Laboratory studies, comparisons with field data and conclusions. *Journal of Experimental Marine Biology and Ecology*, **109**, 71–91.

Cyrus, D.P. & Blaber, S.J.M. (1988). The potential effects of dredging activities and increased silt load on the St Lucia system, with special reference to turbidity and estuarine fauna. *Water SA*, **14**, 43–7.

Daniel, C. (1994). *A comparative assessment of gill net and anglers' catches in the Swartkops and Sundays estuaries, eastern Cape*. MSc thesis, University of Port Elizabeth.

Day, J.H. (1981). Summaries of present knowledge of 43 estuaries in southern Africa. In *Estuarine ecology with particular reference to Southern Africa*, ed. J.H. Day, pp. 251–329. Cape Town: A.A. Balkema.

Day, J.H., Blaber, S.J.M. & Wallace, J.H. (1981). Estuarine fishes. In *Estuarine Ecology with particular reference to Southern Africa*, ed. J.H. Day, pp. 197–221. Cape Town: A.A. Balkema.

De Decker, H.P. (1987). Breaching the mouth of the Bot River estuary, South Africa: impact on its benthic macrofaunal communities. *Transactions of the Royal Society of South Africa*, **46**, 231–50.

Dundas, A. (1994). *Comparative analysis of fish abundance and diversity in three semi-closed estuaries in the eastern Cape*. MSc thesis, University of Port Elizabeth.

Fortier, L. & Leggett, W. (1982). Fickian transport and the dispersal of fish larvae in estuaries. *Canadian Journal of Fisheries and Aquatic Sciences*, **39**, 1150–63.

Garratt, P.A. (1993). Spawning of riverbream, *Acanthopagrus berda*, in Kosi estuary. *South African Journal of Zoology*, **28**, 26–31.

Garratt, P.A. & van der Elst, R.P. (1990). Status of the fishery in Natal and

Transkei. In *Marine recreational fishing. Resource usage, management and research*, pp. 27–31. South African National Science Programmes Report No. 167. Pretoria: CSIR.

Gilchrist, J.D.F. (1918). Report on netting in the Swartkops River. *Marine Biological Report*, **4**, 54–72.

Grindley, J.R. (1981). Estuarine plankton. In *Estuarine ecology with particular reference to Southern Africa*, ed. J.H. Day, pp. 117–46. Cape Town: A.A. Balkema.

Guastella, L.A.-M. (1994). A quantitative assessment of recreational angling in Durban harbour, South Africa. *South African Journal of Marine Science*, **14**, 187–203.

Gulland, J.A. (1987). The effect of fishing on community structure. In *The Benguela and comparable ecosystems. South African Journal of Marine Science*, **5**, 839–49.

Haigh, E.H. & Whitfield, A.K. (1993). Larval development of *Gilchristella aestuaria* (Gilchrist, 1914) (Pisces: Clupeidae) from southern Africa. *South African Journal of Zoology*, **28**, 168–72.

Hall, C.M., Whitfield, A.K. & Allanson, B.R. (1987). Recruitment, diversity and the influence of constrictions on the distribution of fishes in the Wilderness lakes system, South Africa. *South African Journal of Zoology*, **22**, 163–9.

Hanekom, N. & Baird, D. (1984). Fish community structures in *Zostera* and non-*Zostera* regions of the Kromme estuary, St Francis Bay. *South African Journal of Zoology*, **19**, 295–301.

Hanekom, N., Baird, D. & Erasmus, T. (1988). A quantitative study to access the standing biomass of macrobenthos in the soft substrate of the Swartkops estuary, South Africa. *South African Journal of Marine Science*, **6**, 163–74.

Harris, S.A. & Cyrus, D.P. (1994). Utilization of the St Lucia estuary by larval fish. In *Proceedings of the Fourth Indo-Pacific Fish Conference, Faculty of Fisheries, Kasetart University, Bangkok, Thailand*, pp. 410–25.

Harris, S.A. & Cyrus, D.P. (1995). Recruitment of larval fish in the St Lucia estuary, KwaZulu-Natal, South Africa. *South African Journal of Marine Science*, **16**, 333–50.

Harris, S.A., Cyrus, D.P. & Forbes, A.T. (1995). Composition of ichthyoplankton in the mouth of the Kosi estuary, KwaZulu-Natal, South Africa. *South African Journal of Marine Science*, **16**, 351–64.

Harris, T.F.W. (1978). *Review of coastal currents in southern African waters*. South African National Scientific Programmes Report No. 30. Pretoria: CSIR.

Harrison, T.D. & Cooper, J.A.G. (1991). Active migration of juvenile mullet (Teleostei: Mugilidae) into a small lagoonal system on the Natal coast. *South African Journal of Science*, **87**, 395–6.

Harrison, T.D. & Whitfield, A.K. (1990). Composition, distribution and abundance of ichthyoplankton in the Sundays River estuary. *South African Journal of Zoology*, **25**, 161–8.

Harrison, T.D. & Whitfield, A.K. (1995). *Fish community structure in three temporarily open closed estuaries on the Natal coast*. Ichthyological Bulletin of the J.L.B. Smith Institute of Ichthyology No. 64.

Hecht, T. & van der Lingen, C.D. (1992). Turbidity induced changes in feeding strategies in fish in estuaries. *South African Journal of Zoology*, **27**, 95–107.

Heydorn, A.E.F. (1989). Estuaries and the open sea. In *Oceans of life off Southern Africa*, ed. A.I.L. Payne & R.J.M. Crafford, pp. 4–11. Cape Town: Vlaeberg Publishers.

Heydorn, A.E.F., Bang, N.D., Pearce, A.F., Flemming, B.W., Carter, R.A., Schleyer, M.H., Berry, P.F., Hughes, G.R., Bass, A.J., Wallace, J.H., van der Elst, R.P., Crawford, R.J.M. & Shelton, P.A. (1978). Ecology of the Agulhas Current region: an assessment of biological responses to environmental parameters in the south-west Indian Ocean. *Transactions of the Royal Society of South Africa*, **43**, 151–90.

Heymans, J.J. (1992). *Energy flow model and network analysis of the Kromme estuary, St Francis Bay, South Africa*. MSc thesis, University of Port Elizabeth.

Heymans, J.J. & Baird, D. (1995). Energy flow in the Kromme estuarine ecosystem, St Francis Bay, South Africa. *Estuarine, Coastal and Shelf Science*, **41**, 38–59.

Hilmer, T. & Bate, G.C. (1991). Vertical migration of a flagellate-dominated bloom in a shallow South African estuary. *Botanica Marina*, **34**, 113–21.

Jackson, M.J. & James, R. (1979). The influence of bait digging on cockle *Cerastoderma edule* populations in North Norfolk. *Journal of Applied Ecology*, **6**, 671–9.

Joubert, C.S.W. (1981). *Aspects of the biology of five species of inshore reef fishes on the Natal coast, South Africa*. Oceanographic Research Institute Investigational Report No. 51.

Kennish, M.J. (1990). Biological aspects: fishes. In *Ecology of Estuaries*, Vol. 2, pp. 291–350. Boca Raton, Florida: CRC Press.

Kinne, O. (1964). The effects of temperature and salinity on brackish water animals. II. Salinity and temperature, salinity combinations. *Oceanographic Marine and Biological Review A*, **2**, 281–339.

Kok, H.M. & Whitfield, A.K. (1986). The influence of open and closed mouth phases on the marine fish fauna of the Swartvlei estuary. *South African Journal of Zoology*, **12**, 309–15.

Koop, K., Bally, R. & McQuaid, C.D. (1983). The biology of South African estuaries. Part 12: The Bot River, a closed estuary in the south-western Cape. *South African Journal of Zoology*, **18**, 1–10.

Lasiak, T.A. (1983). Aspects of the reproductive biology of the southern mullet, *Liza richardsoni*, from Algoa Bay, South Africa. *South African Journal of Zoology*, **18**, 89–95.

Lindall, W.N. Jr & Trent, L. (1975). Housing developement canals in the coastal zone of the gulf of Mexico: ecological consequences, regulations and recommendations. *Marine Fisheries Review*, **37**, 19–24.

Loneragan, N.R. & Potter, I.C. (1990). Factors influencing community structure and distribution of different life-cycle categories of fishes in shallow waters of a large Australian estuary. *Marine Biology*, **106**, 25–37.

McLachlan, A. & Bate, G. (1985). Carbon budget for a high energy surf-zone. *Vie Mileu*, **35**, 67–77.

McLusky, D.S., Anderson, F.E. & Wolfe-Murphy, S. (1983). Distribution and population recovery of *Arenicola marina* and other benthic fauna after bait digging. *Marine Ecology Progress Series*, **11**, 173–9.

Marais, J.F.K. (1981). Seasonal abundance, distribution and catch per unit effort using gill-nets of fishes in the Sundays estuary. *South African Journal of Zoology*, **16**, 144–50.

Marais, J.F.K. (1982). The effects of river flooding on the fish populations of two Eastern Cape estuaries. *South African Journal of Zoology*, **17**, 96–104.

Marais, J.F.K. (1983a). Seasonal abundance distribution and catch per unit effort of fishes in the Krom estuary, South Africa. *South African Journal of Zoology*, **18**, 96–102.

Marais, J.F.K. (1983b). Fish abundance and distribution in the Gamtoos estuary with notes on the effect of floods. *South African Journal of Zoology*, **18**, 103–9.

Marais, J.F.K. (1984). Feeding ecology of major carnivorous fish from four eastern Cape estuaries. *South African Journal of Zoology*, **19**, 210–23.

Marais, J.F.K. (1985). Some factors influencing the size of fishes caught in gill nets in eastern Cape estuaries. *Fisheries Research*, **3**, 251–61.

Marais, J.F.K. (1988). Some factors that influence fish abundance in South African estuaries. *South African Journal of Marine Science*, **6**, 67–77.

Marais, J.F.K. (1992). Factors that influence fish abundance and diversity in South African estuaries. *World Fisheries Congress*, 14–19 May, Athens, Greece.

Marais, J.F.K. (1994). Characteristics of South African estuaries with special reference to abundance and diversity. *Conference during inauguration of Istanbul University*, 3–8 October, Istanbul, Turkey.

Marais, J.F.K. & Baird, D. (1980a). Seasonal abundance, distribution and catch per unit effort of fishes in the Swartkops estuary. *South African Journal of Zoology*, **15**, 66–71.

Marais, J.F.K. & Baird, D. (1980b). Analysis of anglers' catches in the Swartkops estuary. *South African Journal of Zoology*, **15**, 61–5.

Marais, J.F.K. & Venter, D. (1991). Changes in body composition associated with growth and reproduction in *Galeichthys feliceps* (Teleostei: Ariidae). *South African Journal of Marine Science*, **10**, 149–57.

Martin, T.J. (1988). Interaction of salinity and temperature as a mechanism for spatial separation of three co-existing species of Ambassidae (Cuvier) (Teleostei) in estuaries on the south-east coast of Africa. *Journal of Fish Biology*, **33**, 9–15.

Melville-Smith, R. & Baird, D. (1980). Abundance, distribution and species composition of fish larvae in the Swartkops estuary. *South African Journal of Zoology*, **15**, 72–8.

Melville-Smith, R., Baird, D. & Wooldridge, T. (1981). The utilization of tidal currents by the larvae of estuarine fish. *South African Journal of Zoology*, **16**, 10–13.

Miller, J.M., Reed, J.P. & Pietrafesa, L.J. (1984). Patterns, mechanisms and approaches to the study of migrations of estuarine-dependent fish larvae and juveniles. In *Mechanisms of migration in fishes*, ed. J.D. McCleave, pp. 209–25. New York: Plenum.

Moldan, A., Chapman, P. & Fourie, H.O. (1979). Some ecological effects of the Venpet-Venoil collision. *Marine Pollution Bulletin*, **10**, 60–3.

Neira, F.J., Beckley, L.E. & Whitfield, A.K. (1988). Larval development of the Cape silverside, *Atherina breviceps* Cuv. & Val., 1835 (Teleostei, Atherinidae) from southern Africa. *South African Journal of Zoology*, **23**, 176–83.

Neira, F.J. & Potter, I.C. (1994). The larval fish assemblage of the Nornalup-Walpole estuary, a permanently open estuary on the southern coast of Western Australia. *Australian Journal of Marine and Freshwater Research*, **45**, 1193–207.

Norcross, B.L. (1991). Estuarine recruitment mechanisms of larval Atlantic croakers. *Transactions of the American Fisheries Society*, **120**, 673–83.

Norcross, B.L. & Shaw, R.F. (1984). Oceanic and estuarine transport of fish eggs and larvae: a review. *Transactions of the American Fisheries Society*, **113**, 153–65.

Panikkar, N.K. (1960). Physiological aspects of adaptation to estuarine conditions. *Australian Fisheries Council Proceedings*, **32**, 168–75.

Paterson, A.W. & Whitfield, A.K. (1996). The fishes associated with an intertidal saltmarsh creek in the Kariega estuary, South Africa. *Transactions of the Royal Society of South Africa*, **51**, 195–218.

Plumstead, E.E. (1984). *The occurrence and distribution of fishes in selected Transkei estuaries*. MSc thesis, University of Transkei, Umtata.

Plumstead, E.E. (1990). Changes in ichthyofaunal diversity and abundance within the Mbashe estuary, Transkei, following construction of a river barrage. *South African Journal of Marine Science*, **9**, 399–407.

Potter, I.C., Beckley, L.E., Whitfield, A.K. & Lenanton, R.C.J. (1990). Comparisons between the roles played by estuaries in the life cycles of fishes in temperate

western Australian and southern Africa. In *Environmental biology of fishes*, **28**, ed. B.N. Bruton, pp. 143–78. Dordrecht: Kluwer Academic Press.

Reddering, J.S. (1988). Coastal and catchment basin controls on estuary morphology of the south-eastern Cape coast. *South African Journal of Science*, **84**, 154–7.

Reddering, J.S. & Rust, I.C. (1990). Historical changes and sedimentation characteristics of southern African estuaries. *South African Journal of Science*, **86**, 425–8.

Remane, A. & Schlieper, C. (1971). *Biology of brackish water*. New York: Wiley.

Romer, G. (1986). *Faunal assemblages and food chains associated with surf-zone phytoplankton blooms*. MSc thesis, University of Port Elizabeth.

Romer, G. & McLachlan, A. (1986). Mullet grazing on surf diatom accumulations. *Journal of Fish Biology*, **28**, 93–104.

Roper, D.S. (1986). Occurrence and recruitment of fish larvae in a northern New Zealand estuary. *Estuarine, Coastal and Shelf Science*, **22**, 705–17.

Russell, I.A. (1994). Mass mortality of marine and estuarine fish in the Swartflei and Wilderness lake systems, southern Cape. *Southern African Journal of Aquatic Sciences*, **20**, 93–6.

Russell, I.A. (1996). Fish abundance in the Wilderness and Swartvlei lake systems: changes relative to environmental factors. *South African Journal of Zoology*, **31**, 1–9.

Schlacher, T. & Wooldridge, T.H. (1996). Origin and tropic importance of detritus – evidence from stable isotopes in the benthos of a small, temperate estuary. *Oecologia*, **106**, 382–8.

Schramm, M. (1991). *Grillotia perelica* (Cestoda: Trypanoryncha) plerocercoids in mullet (Pisces: Mugilidae) from estuaries in Transkei, southern Africa. *South African Journal of Marine Science*, **11**, 169–78.

Schumann, E.H. & Brink, K.H. (1990). Coastal-trapped waves off the coast of South Africa: generation, propagation and current structures. *Journal of Physical Oceanography*, **20**, 1206–18.

Shannon, L.V. & Chapman, P. (1983). Suggested mechanism for the chronic pollution by oil of beaches east of Cape Agulhas, South Africa. *South*

African Journal of Marine Science, **1**, 231–44.

Shaw, R.F., Rogers, B.D., Cowan, J.H. & Herke, W.H. (1988). Ocean–estuary coupling of ichthyoplankton and nekton in the northern Gulf of Mexico. *American Fisheries Society Symposium*, **3**, 77–89.

Shelton, P.A. & Kriel, F. (1980). Surface drift and the distribution of pelagic-fish eggs and larvae off the south-east coast of South Africa, November and December 1976. *Fisheries Bulletin South Africa*, **13**, 107–9.

Smale, M.J. & Buxton, C.D. (1985). Aspects of the recreational ski-boat industry off the eastern Cape, South Africa. *South African Journal of Marine Science*, **3**, 131–44.

Sylvester, J.R., Nash, C.E. & Emberson, C.R. (1975). Salinity and oxygen tolerances of eggs and larvae of Hawaiian striped mullet, *Mugil cephalus* L. *Journal of Fish Biology*, **7**, 621–9.

Talbot, M.M.J. (1982). *Aspects of the ecology and biology of* Gilchristella aestuarius *(G & T) (Pisces: Clupeidae) in the Swartkops estuary, Port Elizabeth.* MSc thesis, University of Port Elizabeth.

Talbot, M.M.J. & Baird, D. (1985). Feeding of the estuarine round herring *Gilchristella aestuarius* (G & T) (Stolephoridae). *Journal of Experimental Marine Biology and Ecology*, **87**, 199–214.

Ter Morshuizen, L.D., Whitfield, A.K. & Paterson, A.W. (1996). Distribution patterns of fishes in an Eastern Cape estuary and river with particular emphasis on the ebb and flow region. *Transactions of the Royal Society of South Africa*, **51**, 257–80.

Tilney, R.L. & Hecht, T. (1993). Early ontogeny of *Galeichthys feliceps* from the south east coast of South Africa. *Journal of Fish Biology*, **43**, 171–93.

van den Heilegenberg, T. (1987). Effects of mechanical and manual harvesting of lugworms *A. marina* L. on the benthic fauna of tidal flats in the Dutch Wadden Sea. *Biologica Conserva*, **39**, 165–77.

van der Elst, R.P. (1988). *A guide to the common sea fishes of Southern Africa.* Cape Town: C. Struik Publishers.

van der Elst, R.P. (1989). Marine recreational angling in South Africa. In *Oceans of life off Southern Africa*, ed. A.I.L. Payne & R.J.M. Crawford, pp. 164–76. Cape Town: Vlaeberg Publishers.

van der Elst, R.P. & Akin, F. (1987). *Marine*

linefish. Priority species and research objectives in southern Africa. Oceanographic Research Institute Special Publication No. 1.

van der Elst, R.P., Blaber, S.J.M., Wallace, J.H. & Whitfield, A.K. (1976). The fish fauna of Lake St Lucia under different salinity regimes. In *St Lucia Scientific Advisory Workshop, Charters Creek, February 1976*, ed. A.E.F. Heydorn, pp. 1–10. Pietermaritzburg: Natal Parks Board.

van der Elst, R.P. & De Freitas, A.J. (1987). Long term trends in Natal marine fisheries. In *Long term data series relating to southern Africa's renewable natural resources*, ed. I.A.W. Macdonald & R.J.M. Crawford. *South African Journal of Marine Science*, **167**, 68–70.

Vernberg, W.B. & Vernberg, F.J. (1972). *Environmental physiology of marine animals.* Berlin: Springer-Verlag.

Wallace, J.H. (1975a). *The estuarine fishes of the east coast of South Africa. Part I. Species composition and length distribution in the estuarine and marine environments. Part II. Seasonal abundance and migrations.* Oceanographic Research Institute Investigational Report No. 40.

Wallace, J.H. (1975b). *The estuarine fishes of the east coast of South Africa. Part III. Reproduction.* Oceanographic Research Institute Investigational Report No. 41.

Wallace, J.H. & van der Elst, R.P. (1975). *The estuarine fishes of the east coast of South Africa. Part 4. Occurrence of juveniles in estuaries. Part 5. Ecology, estuarine dependence and status.* Oceanographic Research Institute Investigational Report No. 42.

Wallace, J.H., Kok, H.M., Beckley, L.E., Bennett, B., Blaber, S.J.M. & Whitfield, A.K. (1984). South African estuaries and their importance to fishes. *South African Journal of Science*, **85**, 203–7.

Weinstein, M.P. (1985). Distributional ecology of fishes inhabiting warm temperate and tropical estuaries: community relationships and implications. In *Fish community ecology in estuaries & coastal lagoons: towards an ecosystem integration*, ed. A. Yanez-Aráncibia. Mexico: Universidad Nacional Autonoma de México.

Weinstein, M.P., Weiss, S.L., Hodson, R.G. & Gerry, L.R. (1980). Retention of three taxa of postlarval fishes in an

intensively flushed tidal estuary, Cape Fear River, North Carolina. *Fishery Bulletin U.S.*, **78**, 419–36.

Whitfield, A.K. (1980a). Factors influencing the recruitment of juvenile fishes into the Mhlanga estuary. *South African Journal of Zoology*, **15**, 166–9.

Whitfield, A.K. (1980b). Food chains in Lake St Lucia. In *Studies on the ecology of Maputuland*, ed. M.N. Bruton & K.H. Cooper. Grahamstown: Rhodes University.

Whitfield, A.K. (1980c). A quantitative study of the tropic relations within the fishing community of Mhlanga estuary, South Africa. *Estuarine and Coastal Marine Science*, **10**, 417–35.

Whitfield, A.K. (1982). *Trophic relationships and resource utilization within the fish community of the Mhlanga and Swartvlei estuarine systems.* PhD thesis, University of Natal, Pietermaritzburg.

Whitfield, A.K. (1983). *Factors influencing the utilization of southern African estuaries by fishes.* Investigational Report of the Oceanogaphic Research Institute No. 79.

Whitfield, A.K. (1984). The effects of prolonged aquatic macrophyte senescence on the biology of the dominant fish species in a southern African coastal lake. *Estuarine, Coastal and Shelf Science*, **18**, 315–29.

Whitfield, A.K. (1985). The role of zooplankton in the feeding ecology of fish fry from some southern African estuaries. *South African Journal of Zoology*, **20**, 166–71.

Whitfield, A.K. (1988). The fish community of the Swartvlei estuary and the influence of food availability on resource utilization. *Estuaries*, **11**, 160–70.

Whitfield, A.K. (1989a). Fish larval composition, abundance and seasonality in a southern African estuarine lake. *South African Journal of Zoology*, **24**, 217–44.

Whitfield, A.K. (1989b). Ichthyoplankton in a southern African surf zone: nursery area for the postlarvae of estuarine associated fish species? *Estuarine, Coastal and Shelf Science*, **29**, 533–47.

Whitfield, A.K. (1989c). Ichthyoplankton interchange in the mouth region of a southern African estuary. *Marine Ecology Progress Series*, **54**, 25–33.

Whitfield, A.K. (1990). Life-history styles of fishes in South African estuaries.

Environmental Biology of Fishes, **28**, 295–308.

Whitfield, A.K. (1992a). Juvenile fish recruitment over an estuarine sand bar. *Ichthos*, **36**, 23.

Whitfield, A.K. (1992b). A characterisation of southern African estuarine systems. *Southern African Journal of Aquatic Sciences*, **12**, 89–103.

Whitfield, A.K. (1993). Fish biomass estimates from the littoral zone of an estuarine coastal lake. *Estuaries*, **16**, 280–9.

Whitfield, A.K. (1994a). An estuary-association classification for the fishes of southern Africa. *South African Journal of Science*, **90**, 411–17.

Whitfield, A.K. (1994b). Abundance of larval and 0+ juvenile marine fishes in the lower reaches of three southern African estuaries with differing freshwater inputs. *Marine Ecology Progress Series*, **105**, 257–67.

Whitfield, A.K. (1994c). A review of ichthyofaunal biodiversity in Southern African estuaries. *Annales Musée royale Afrique Centrale, Zoologie*, **275**, 149–63.

Whitfield, A.K. (1995). Mass mortalities of fish in South African estuaries. *Southern African Journal of Aquatic Sciences*, **21**, 29–34.

Whitfield, A.K. (1996). A review of factors influencing fish utilization of South African estuaries. *Transactions of the Royal Society of South Africa*, **51**, 115–37.

Whitfield, A.K., Beckley, L.E., Bennett, B.A., Branch, G.M., Kok, H., Potter, I.C. & Van der Elst, R.P. (1989). Composition, species richness and similarity of ichthyofauna in eelgrass *Zostera capensis* beds of Southern Africa. *South African Journal of Marine Science*, **8**, 251–9.

Whitfield, A.K. & Blaber, S.J.M. (1978a). Food and feeding ecology of piscivorous fishes at Lake St Lucia, Zululand. *Journal of Fish Biology*, **13**, 675–91.

Whitfield, A.K. & Blaber, S.J.M. (1978b). Resource segregation amongst iliophagus fish in Lake St Lucia, Zululand. *Environmental Biology of Fishes*, **3**, 293–6.

Whitfield, A.K., Blaber, S.J.M. & Cyrus, D.P. (1981). Salinity ranges of some southern African fish species occurring in estuaries. *South African Journal of Zoology*, **16**, 151–5.

Whitfield, A.K. & Bruton, M.N. (1989). Some biological implications of reduced freshwater inflow into eastern Cape estuaries: a preliminary assessment. *South African Journal of Science*, **85**, 691–4.

Whitfield, A.K. & Heeg, J. (1977). On the life cycles of the cestode *Ptychobothrium belones* and nematodes of the genus *Contracaecum* from Lake St Lucia, Zululand. *South African Journal of Science*, **73**, 121–2.

Whitfield, A.K. & Kok, H.M. (1992). *Recruitment of juvenile marine fishes into permanently open and seasonally open estuarine systems on the southern coast of South Africa*. Ichthyological Bulletin of the J.L.B. Smith Institute of Ichthyology No. 57.

Whitfield, A.K. & Paterson, A.W. (1995). Flood associated mass mortality of fishes in the Sundays estuary. *Water SA*, **21**, 385–9.

Whitfield, A.K., Paterson, A.W., Bok, A.H. & Kok, H.M. (1994). A comparison of the ichthyofaunas of two permanently open eastern Cape estuaries. *South African Journal of Zoology*, **29**, 175–85.

Whitfield, A.K. & Wooldridge, T.H. (1994). Changes in freshwater supplies to southern African estuaries: some theoretical and practical considerations. In *Changes in fluxes in estuaries*, ed. K.R. Dyer & R.J. Orth, pp. 41–50. International Symposium Series. Fredensborg, Denmark: Olsen & Olsen.

Wooldridge, T.H. (1991). Exchange of two species of decapod larvae across an estuarine mouth inlet and implications of anthropogenic changes in the frequency and duration of mouth closure. *South African Journal of Science*, **87**, 519–25.

Wooldridge, T. & Bailey, C. (1982). Euryhaline zooplankton of the Sundays estuary and notes on trophic relations. *South African Journal of Zoology*, **17**, 151–63.

Yanez-Aráncibia, A. (1985). *Fish community ecology in estuaries and coastal lagoons: towards an ecosystems integration*. México: Universidad Nacional Autónoma de México.

10 Estuarine birds in South Africa

Philip Hockey and Jane Turpie

Mntafufu River estuary

Introduction

Birds are a highly visible and frequently diverse component of the fauna of estuaries. Many species form large and dense foraging aggregations and thus have the potential to play key roles in estuarine ecosystem dynamics.

Until the 1970s, surprisingly little research had been carried out on estuarine birds in South Africa. This situation has greatly improved over the past 25 years. The distributional databases now available are better than for any other country in Africa and several studies have addressed ecological interactions between bird predators and their prey, although rather little attention has been paid to the roles of herbivorous birds in estuaries.

Most research effort has been invested at four sites: the Berg River estuary, Langebaan Lagoon, the Swartkops River estuary and Lake St Lucia. Fortuitously, these four sites span the full geographic range, and hence climatic range, of the South African coast.

This chapter addresses firstly aspects of broad-scale estuarine bird distribution, both in biogeographic terms and in relation to taxonomy and the degree of estuarine dependence of different species. We then progress to look in some detail at trophic partitioning and the ways in which estuarine birds obtain their food and minimise competition between species. We also consider ways in which feedback loops operate between predators and prey.

The ecological roles of birds in estuarine energy flows are assessed in terms of both the rate of prey removal by birds (birds as energy sinks) and the rates at which birds recycle energy within estuaries. This theme is developed into a discussion of whether estuarine birds in South Africa are food-limited and how competition influences broad-scale distribution patterns. The implications of the above for a national conservation strategy for estuarine

birds are evaluated and the priority sites for their conservation are discussed.

Bird diversity, abundance and estuarine dependence

Taxonomic diversity

One hundred and sixty-two bird species occur regularly in South African estuaries (listed with scientific names in Appendix 10.1). These species are taxonomically, functionally and morphologically diverse, ranging from piscivores weighing several kilograms, such as pelicans, to tiny insectivores, such as cisticolas, weighing only a few grams. Representatives of 13 of the 26 Orders of birds that occur in South Africa are found at estuaries (Table 10.1). This high taxonomic diversity can largely be attributed to the variety of habitat types represented in estuarine ecosystems. These habitats, which include shallow water, deep open water, intertidal mud- and sandflats, saltmarshes, reedbeds and mangroves, are further diversified by the existence of a strong salinity gradient which influences, *inter alia*, the spatial occurrence of plant and invertebrate species (Day 1981a; Kalejta & Hockey 1991; Velásquez & Hockey 1992). Because this diversity of habitats exists within a restricted area, estuaries attract species which have a range of habitat requirements for breeding, feeding and roosting. In six avian Orders, more than two-thirds of their species occur regularly in estuaries. These Orders typically are associated with aquatic habitats, although not necessarily exclusively or predominantly with estuaries. These are the Podicipediformes (grebes), Pelecaniformes

Table 10.1. *Number of species of each avian Order represented in South Africa, compared with number of species of each of these orders that occur on South African estuaries. Vagrants are excluded from both columns*

Order	Number of species		
	South Africa	Estuaries	%
Struthioniformes	1	0	0
Sphenisciformes	1	0	0
Podicipediformes	3	3	100
Procellariiformes	19	0	0
Pelecaniformes	9	7	78
Ciconiiformes	31	26	84
Phoenicopteriformes	2	2	100
Anseriformes	17	12	71
Falconiformes	58	8	14
Galliformes	17	0	0
Gruiformes	29	8	28
Charadriiformes	66	45	68
Pterocliformes	4	0	0
Columbiformes	14	0	0
Psittaciformes	5	0	0
Musophagiformes	4	0	0
Cuculiformes	15	1	7
Strigiformes	12	1	8
Caprimulgiformes	7	2	29
Apodiformes	10	0	0
Coliiformes	3	0	0
Trogoniformes	1	0	0
Alcediniformes	16	5	31
Coraciiformes	14	0	0
Piciformes	20	0	0
Passeriformes	316	41	13

(pelicans), Ciconiiformes (herons and storks), Phoenicopteriformes (flamingos), Anseriformes (ducks) and Charadriiformes (waders, gulls and terns) (Table 10.1). By contrast, only 13% of the Passeriformes (passerines), a characteristically terrestrial group, occur regularly at estuaries. However, Passeriformes are the most speciose bird Order in South Africa and account for a greater diversity of 'estuarine' birds than any other Order except the Charadriiformes (Table 10.1).

Usage and abundance

All but three of the species listed in Appendix 10.1 feed in estuaries. Of the three, the Southern Crowned Crane occurs commonly at only one estuary in South Africa, the Nxaxo, and uses the estuary only as a night roost. Of the 162 estuarine species, 96 (59%) breed in estuaries. Most of the breeding species are resident in estuaries, but a few migrate inland or northwards to the tropics in their nonbreeding seasons. Sixty-six species visit estuaries only to forage and/or roost. Some of these (mostly waders: Charadrii) are long-distance migrants from Palearctic breeding grounds, whereas others, such as flamingos and the Mangrove Kingfisher, are intra-African migrants (Appendix 10.1).

An accurate assessment of the total number of birds

that use South African estuaries is not possible, but estuaries support at least 345000 non-passerine birds during summer. Of these, 225000 are Charadriiformes, of which 150000 are waders (Underhill & Cooper 1984; Ryan & Cooper 1985; Ryan et al. 1986, 1988; Martin & Baird 1987; Underhill 1987a; Hockey 1993; Tree & Martin 1993). Curlew Sandpipers alone make up half of the wader numbers. There are several reasons why an accurate estimation of estuarine bird numbers is difficult. The potential for major inaccuracies arises because the abundance and composition of estuarine bird populations are dynamic over time scales as short as hours. Terns and gulls forage on the coast or at sea, but enter estuaries to roost, sometimes in very large numbers. Additionally, skulking species such as bitterns, rails and crakes, and nocturnal species such as owls, night herons and nightjars are frequently overlooked. Passerines, which spend much time in structurally complex habitats such as reedbeds, are usually excluded from waterbird counts.

The above figure of 345000 non-passerines is based largely on one-off counts made during the summer. During multiple counts at the lower Berg River estuary in summer, the non-passerine population averaged slightly over 12000 (Velásquez et al. 1991). Based on maximum counts of individual species, however, the total number of birds using the lower estuary during the year was estimated to be at least 37000. Extrapolating this ratio to the national summer total (which excludes estuaries between Coffee Bay and Port Edward), gives a figure of c. 1000000 non-passerines using South African estuaries during the year.

The most significant changes in estuarine avifaunas are seasonal. There is an influx of hundreds of thousands of migrants during the austral summer, a substantial proportion of which are Palearctic-breeding waders. The first of these arrive in August, and most have left by the end of May to return to their northern breeding grounds. Peak numbers are present between late November and March (Hockey & Douie 1995). The juveniles of most of the migratory waders do not return north with the adults in their first year and remain at estuaries throughout the winter, migrating north to breed for the first time at two years old (Martin & Baird 1988). Young of some of the larger species may delay northward migration until their third year (Hockey & Douie 1995). The numbers that remain behind vary from year to year, depending on the success of the previous breeding season, which in turn is linked to Arctic lemming cycles. In years when lemmings are scarce, Arctic predators switch their attention to birds, and bird breeding success is low. Lemming populations normally fluctuate on a three to four year cycle. The proportion of juvenile Curlew Sandpipers at Langebaan Lagoon in the austral summer is positively correlated with lemming abundance on the Taimyr Peninsula in the previous boreal summer

(Underhill 1987b). At three-year intervals, when lemmings are scarce, few juvenile birds recruit into the population. Underhill et al. (1989) found the same, synchronous demographic patterns in Knot populations at Langebaan Lagoon and used these patterns to deduce that these birds also breed on the Taimyr Peninsula. Year-to-year variations in breeding success lead to year-to-year variations in population size, and hence to the predatory impact that these birds have in estuaries.

As the Palearctic-breeding birds leave in autumn, other groups of birds move into South African estuaries. At the Berg River estuary, which lies in the winter-rainfall region, the avifauna changes from being dominated by migrant waders in summer to being dominated by resident waders and waterfowl in winter (Velásquez et al. 1991). At Langebaan Lagoon, flamingo numbers increase in winter to the point where these birds are the most important avian consumers in the system (Underhill 1987a). At the Swartkops estuary, winter brings an influx of cormorants, gulls, egrets and spoonbills (Martin & Baird 1987). This seasonal influx is not linked to high fish abundance during winter (Winter 1979; Marais & Baird 1980) but is more likely to be caused by an exodus of these species from the estuary during their breeding season (Martin & Baird 1987).

Estuarine dependence of birds

Because several estuarine habitat types, at least in the broad sense, can be found in non-estuarine situations, estuaries share a large proportion of their avifauna with other habitats such as freshwater wetlands, rivers, lagoons, sandy and rocky shores and terrestrial biotopes. Consequently, the majority of bird species that occur in estuaries are not, as species, dependent on them (Table 10.1, Appendix 10.1). Only 15 bird species can be classified as highly dependent on estuaries in South Africa (Appendix 10.1), their national populations occurring predominantly or entirely within estuaries for at least part of the year. Only three of these species, the Pinkbacked Pelican, Caspian Tern and Pinkthroated Longclaw, breed at estuaries. Both the Pinkbacked Pelican and Pinkthroated Longclaw are tropical species restricted in range in South Africa to the northern coast of KwaZulu-Natal, and both breed, within South Africa, only at Lake St Lucia (Brooke 1984). Caspian Terns breed at inland as well as coastal localities; their most important estuarine breeding colonies are at Lake St Lucia and at saltpans adjacent to the Berg and Swartkops River estuaries (Brooke 1984; Martin & Randall 1987, Velásquez et al. 1990). Half of the highly estuary-dependent species are Palearctic-breeding waders which migrate more than 10 000 km from their breeding grounds to spend the austral summer in South Africa. A local, intra-African migrant, the Mangrove Kingfisher, is also dependent on estuaries during the winter.

Table 10.2. Degree of regional residency of estuarine birds as a function of their level of estuarine dependence

Degree of estuarine dependence	Resident/ intra-African migrant		Inter- continental migrant	
	n spp.	%	n spp.	%
Low	98	73.1	8	29.6
Moderate + High	36	26.9	29	70.4
Totals	134	100	27	100

Thirty-nine species are partially dependent on estuaries, in that they also utilise habitats found in other wetlands and intertidal areas as well as in estuaries, but a relatively large proportion of their South African populations are thought to occur in estuaries. Of these, 21 species, mainly Ciconiiformes and Anseriformes, breed in estuaries (Appendix 10.1). The remainder, most of which are migrants, use estuaries for feeding; nine of them – six waders, two terns and the Osprey – are visitors from the Palearctic; some others are intra-African migrants. Among the intra-African migrants, Yellowbilled Storks and Damara Terns breed in and to the north of South Africa. Lesser Crested Terns are both intra-African and intercontinental migrants. Some migrate to the South African coast from the Horn of Africa, but others are thought to migrate from the Middle East and the Indian subcontinent (Harrison 1983).

Two-thirds of the species listed in Appendix 10.1 as occurring at estuaries have low estuarine dependence. Nearly all of the Passeriformes and Gruiformes (cranes and rails) fall into this category. The level of estuarine dependence is much lower among resident species (27%) than among intercontinental migrants (70%) ($X^2_1 = 19.01$, $p < 0.001$, Table 10.2).

Breeding seasonality

Geographical variation in the seasonality of breeding activity was analysed for all species recorded as breeding at estuaries (from Appendix 10.1). Timing of breeding was extracted from the nest record cards of the Southern African Ornithological Society for three different climatic regions of South Africa, namely the winter-rainfall region of the west and south coasts (termed the south Western Cape), the summer-rainfall region of KwaZulu-Natal, and the area in between, where rainfall is aseasonal (Eastern Cape). A species was recorded as breeding only in those months in which the number of breeding records was equal to or greater than 10% of the total number of breeding records for that species. The seasonality of breeding in the three regions was compared with seasonal rainfall patterns in Cape Town, Port Elizabeth and Durban, extracted from Pearce & Smith (1984).

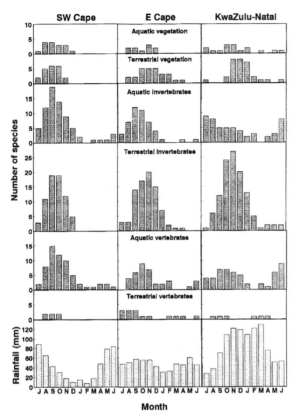

Figure 10.1. **Breeding seasonality of six dietary categories of estuarine birds in three rainfall regions of South Africa compared with
average monthly rainfall at Cape Town, Port Elizabeth and Durban.
(Climate data from Pearce & Smith 1984.)**

Breeding of all estuarine species in the south Western
Cape is highly seasonal (Figure 10.1). The breeding season
starts towards the end of the winter rains and peaks
between August and October. With a handful of exceptions
among aquatic invertebrate-feeders and piscivores, there
is no breeding activity between February and June.

A broadly similar seasonal pattern of breeding occurs
in the Eastern Cape, although the peak in breeding activity
is one to two months later than in the south Western Cape.
Among herbivores and invertebrate-feeders, there is a tendency for predominantly terrestrial species to breed
slightly later than their aquatic counterparts, with peak
activity in summer rather than in spring. The reasons for
this difference are not known, but may be temperature –
(and therefore production) – linked rather than rainfall-
linked. In general, the breeding season in the Eastern Cape
is longer than in the south Western Cape.

In KwaZulu-Natal, there is more variation in the breeding seasons of different guilds than exists in the more
temperate regions to the west. The pattern of later breeding by terrestrial than aquatic species, especially invertebrate-feeders, is more pronounced than in the Eastern

Cape. Breeding of terrestrial invertebrate-feeders peaks
during the warm, wet summer months whereas aquatic
invertebrate-feeders breed mostly during the cool, dry
winter months when water levels are low. Breeding activity of piscivores in the south Western and Eastern Cape is
unimodal, peaking in spring and early summer, but in
KwaZulu-Natal they have a bimodal breeding pattern with
peaks in early winter and again in spring.

Biogeography of estuarine birds

A biogeographical classification of the South African coast
was first proposed by Stephenson & Stephenson (1972) and
subsequently reviewed by Brown & Jarman (1978).
Distribution patterns of algae, invertebrates, fishes and
birds have all been used to try and delimit these regions
(e.g. Siegfried 1981; Emanuel et al. 1992). The general
pattern identified has been one of a Cool Temperate West
Coast Province extending from the Orange River south to
somewhere between Cape Point and Cape Agulhas; a
Warm Temperate South Coast Province extending east to
approximately Port St Johns; and a Subtropical East Coast
Province from Port St Johns north into Mozambique. Based
on his analysis of the distribution patterns of estuarine
birds, Siegfried (1981) did not consider that the South
Coast Province warranted recognition: this conclusion
was supported by Hockey et al. (1983) based on analysis of
the sandy beach avifauna. Both studies considered this
region as a zone of overlap between Cool Temperate and
Subtropical Provinces.

For comparative purposes, we divided the South African
coast into the same thirteen 200 km sections used by
Siegfried (1981) and Hockey et al. (1983) and extracted
distributional data for birds from Sinclair et al. (1993). This
approach was adopted in preference to analysing count
data for two reasons: firstly, counts exclude passerines, and
secondly, almost all count data are based on one-off
summer counts. They therefore cannot account for seasonal patterns of estuarine occupancy and they almost certainly contain a high proportion of false negatives.
Considering species richness alone (Figure 10.2), the subtropical subtraction effect is clearly evident. In the extreme
eastern sections, species richness averages 145, decreasing
rapidly to an average of 123 between Durban and Port
Elizabeth, and an average of 111 between Port Elizabeth and
St Helena Bay. North of St Helena Bay, the average species
richness per 200 km section is only 85, 57 of which are
species which occur around the entire South African coast.
In the same way as overall species richness decreases by 41%
from east to west, the number of estuarine-dependent
species (categories 2 and 3 in Appendix 10.1) also decreases

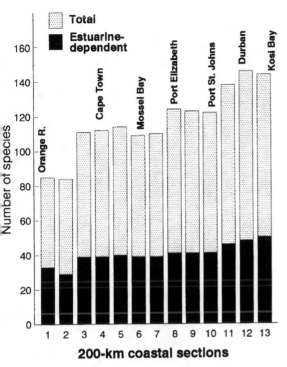

Figure 10.2. **Estuarine bird species richness in thirteen 200-km sections of the South African coast between the Orange River (section 1) and Ponta do Ouro (section 13). (Data extracted from Sinclair et al. 1993.)**

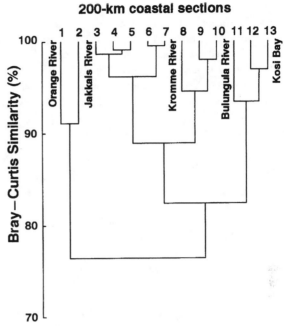

Figure 10.3. **Cluster analysis of species presence/absence in thirteen 200–km sections of the South African coast between the Orange River (section 1) and Ponta do Ouro (section 13). (Data extracted from Sinclair et al. 1993.)**

by c. 38%). However, the proportion of estuarine-dependent species in each 200 km section remains remarkably constant, ranging from 33 to 39% (mean = 34.7 ± 1.7%).

Presence/absence data (from Sinclair et al. 1993) were subjected to cluster analysis using PRIMER (Plymouth Routines in Multivariate Ecological Research). Similarity between all sections is high (>75% – Figure 10.3), but at the 85% level of similarity, the coast is divided into three zones. Zone 1 extends from the Orange River to Lamberts Bay, Zone 2 from Lamberts Bay to Port St Johns, and Zone 3 from Port St Johns northwards to the Mozambique border. These findings differ from those of Siegfried (1981), who found a very high similarity in the species complement of all sections between the Orange River and Mossel Bay. However, Siegfried did not use passerines in his analysis, and based his biogeographical conclusions on an analysis of resident species only. The dramatic reduction in species richness between sections 3 and 2 suggests that estuaries north of Lamberts Bay support a species assemblage very different from that of estuaries further south. This difference is due entirely to species subtraction rather than species replacement, the complement in sections 1 and 2 being a subset of the species complement further south. This might, however, be a consequence of the paucity of estuaries in this area: the only perennial estuaries are those of the Orange and Olifants Rivers.

Our analysis, based on 162 species, as distinct from the 90 used by Siegfried (1981), still does not allow any firm conclusion as to whether the south coast (sections 3–10) is a true biogeographic zone (for birds) or not. Only two of the 162 estuarine bird species included in the analysis have coastal distributions restricted to this area. These are the Great Crested Grebe and Levaillant's Cisticola, both of which have very low levels of estuarine dependence. There is an isolated population of Damara Terns on the south coast, but this species also occurs on the coasts of the Northern Cape Province and north into Namibia and southern Angola. Furthermore, the distributions of several predominantly tropical species extend well into the south coast 'region'; these include Goliath and Greenbacked Herons, Dwarf Bittern, Yellowbilled Stork, African Jacana and African Pied Wagtail. Others, including Pinkbacked Pelican, Black Egret, Squacco Heron, Saddlebilled Stork, Lesser Jacana, Wiretailed Swallow and Redfaced Cisticola have distributions which are truncated in southern KwaZulu-Natal. If an identical analysis is repeated using only estuary-dependent species (categories 2 and 3, Appendix 10.1), the picture is even less clear, with all sections between the Spoeg and Kwenxura Rivers clustering at similarities greater than 80%.

The analysis by Emanuel et al. (1992) of rocky shore invertebrates provides convincing evidence for the existence of a

South Coast Province, as does Prochazka's (1994) analysis of rocky intertidal fish distribution: at least 11 species of fish which occur intertidally on rocky shores are restricted or nearly restricted to the south coast. However, the avifaunas of both estuaries and sandy beaches tend to support the idea that the south coast is a zone of overlap and not a discrete province. Among estuarine fish, only one species, *Clinus spatulatus*, is restricted to the south coast (K. Prochazka, personal communication). The reason why this overlap zone should be so large for birds is not clear, but may be linked to the lack of latitudinal and climatic variation in this area resulting in an attenuated subtropical species subtraction. There are few examples of 'west coast' species which extend their distributions east of Cape Agulhas. This pattern is shown by the African Black Oystercatcher, Damara Tern and Yellowrumped Widow but, as with Great Crested Grebe and Levaillant's Cisticola, these species are not estuary-dependent. It is doubtful whether additional biogeographic analyses based on estuarine birds will shed further light on this problem. There are probably two main reasons why the patterns are unclear. Firstly, birds, relative to most other estuarine taxa (except possibly fish), are highly dispersive and regularly occur far from the cores of their ranges. Secondly, river catchments vary in their geology such that estuaries of a very different nature can occur close to one another. Among estuarine birds, there are several examples of species which occur in all putative biogeographical regions, but only at a small number of estuaries within each (Hockey & Douie 1995).

Diet, foraging behaviour and ecological segregation

Diet

Estuarine bird species fall into six major dietary categories, reflecting the diversity of habitats available in which to forage. These broadly comprise plant, invertebrate and vertebrate diets, in aquatic habitats on the one hand and terrestrial habitats on the other. Within each of the 13 estuarine bird Orders, species usually fall into one or two of these categories in terms of the main components of their diets (Figure 10.4, Appendix 10.1). Invertebrate feeders predominate in estuaries, comprising 90 species in 11 Orders. There are 48 species of vertebrate feeders in eight Orders and 22 species of herbivores in five Orders. Although aquatic and terrestrial feeders both have representatives from nine Orders, the former group has 94 species, as opposed to 66 terrestrial feeders. Aquatic feeders are, to a large extent, Charadriiformes and Ciconiiformes which eat mainly invertebrates and vertebrates respectively. Most herbivores belong to the Anseri-

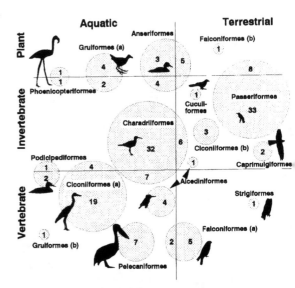

Figure 10.4. **Representation of the South African estuarine bird fauna in terms of diet, e.g. all seven Pelecaniformes prey mainly on aquatic vertebrates and 32 of the 45 species of Charadriiformes prey mainly on aquatic invertebrates. (Dietary data from Maclean 1993.)**

formes and Gruiformes, but several species of both orders also feed on aquatic invertebrates. Most terrestrially-feeding birds are Passeriformes.

The highest proportions of partially and highly estuarine-dependent species are aquatic foragers (Table 10.3). Ten of the 44 species which prey predominantly on aquatic invertebrates are highly dependent on estuaries in South Africa. The terrestrial feeding categories, on the other hand, are characterised by species with low estuarine dependence. Among these categories, only the Palmnut Vulture, Mangrove Kingfisher and Pinkthroated Longclaw are highly dependent on estuaries (Table 10.3). The dependence of Pinkthroated Longclaw is probably an artefact of recent loss of non-estuarine lowland wetlands in KwaZulu-Natal (Brooke 1984). This species' estuarine dependence is limited to South Africa and Mozambique: in Namibia, Botswana and Zimbabwe it occurs at inland wetlands. Outside South Africa, Palmnut Vultures also occur extensively away from estuaries. The Mangrove Kingfisher is confined to estuaries and mangroves during the non-breeding season throughout its range on the Indian Ocean coast of Africa (Fry *et al.* 1988).

Foraging behaviour

In terms of energy flow and the role of birds in the ecological functioning of estuaries, three of the six dietary categories identified above are of particular importance and include those birds which eat aquatic vegetation, aquatic invertebrates, or fish and amphibians. Relative to species in other dietary categories, these birds exhibit a high degree of estuarine dependence (Table 10.3).

Table 10.3. Number of species of estuarine birds of three degrees of estuarine dependence (1 = low, 2 = moderate, 3 = high) in each of six different dietary categories

Main dietary category	Degree of estuarine dependence		
	1	2	3
Aquatic vegetation	5	2	1
Aquatic invertebrates	25	10	9
Aquatic vertebrates	20	19	3
Terrestrial vegetation	12	1	1
Terrestrial invertebrates	40	4	2
Terrestrial vertebrates	5	1	0

Within these dietary categories, there is considerable variation both in foraging behaviour and methods of prey capture. Some herbivores, for example, obtain their food from the water surface while swimming (e.g. Yellowbilled Duck and Redknobbed Coot), although the same species will also dive for food. Purple Gallinules pull up aquatic macrophytes such as *Typha* and hold the plants in their feet while eating the corms. Lesser Flamingos have mouthpart adaptations that are analogues of those of the baleen whales, with specially adapted lamellae to filter microscopic algae, in particular the blue-green algae *Spirulina*, from the water column.

Many species, especially waders, prey on aquatic estuarine invertebrates. The majority of waders obtain their prey at or close to the surface of intertidal mudflats, by pecking or shallow probing. Most of these prey are located visually, either directly or by using secondary cues such as the surface casts of polychaete worms or the movements of prawns irrigating their burrows (e.g. Smith 1975). The range of foraging habitats available to these wading birds is limited by their morphology, as is the range of prey they are able to capture and handle. Short-legged waders, such as Little Stints, are constrained to foraging on exposed mudbanks, among saltmarshes or at the very edge of the water. Little Stints are the smallest of South Africa's estuarine waders, and their light weight (<25 g) allows them to forage on floating algal mats, a habitat unavailable to heavier waders. Among the scolopacid waders (sandpipers, curlews, godwits, etc.), there is a general tendency for large body size to be correlated with a long bill and long legs. The larger birds can thus forage in deeper water than small birds, and can also probe deeper into the mud for food. Whilst even the deep probers undoubtedly use visual cues for much of the time, they also use (as do some smaller waders) chemoreception and mechanoreception to help in the location and recognition of prey (Heppleston 1970; Gerritsen *et al.* 1983; Van Heezik *et al.* 1983; Gerritsen & Meiboom 1986). Plovers, on the other hand, are exclusively visual foragers. Large plovers have longer legs than small plovers, but all plovers are relatively short-billed and

do not habitually wade in water to obtain their prey. Larger species are able to handle larger prey than small species and, because of their long legs and greater visual radius, to locate their prey over greater distances (Pienkowski 1983).

Few invertebrate-feeding species obtain their prey by swimming or diving. Some waders, notably Ruff, Curlew Sandpiper and Avocet, do occasionally forage while swimming, and this is the normal foraging technique used by Cape and Hottentot Teals and Cape Shovellers. Blacknecked Grebes dive from the surface for invertebrates and Mangrove Kingfishers plunge-dive for large invertebrate prey. The only obligate filter-feeder is the crustacean-eating Greater Flamingo. Some aerial insects, notably chironomid midges, have aquatic larval stages. The larvae are preyed on by waders, and the adults are eaten by flock-foraging Whitewinged Terns and Redwinged Pratincoles, as well as by several species of swallows and martins.

The majority of piscivores use one of three foraging techniques. Herons and storks, as well as the Hamerkop, either wade steadily through the water or adopt a 'sit-and-wait' strategy, stabbing through the water at prey within range. Cormorants, darters, African Finfoots and grebes dive from the surface, whereas terns, kingfishers, African Fish Eagles and Ospreys plunge into the water from the air and are thus unable to capture their prey in such deep water as are the surface divers. Pelicans catch their prey close to the water surface while swimming. The African Skimmer has an unusual foraging technique in which the elongated lower mandible is trawled through the water while the bird is flying, the bill automatically snapping shut when a fish is encountered. This species, however, no longer occurs in South African estuaries, having bred for the last time at Lake St Lucia in 1943 (Berruti 1980).

Ecological segregation among estuarine bird species
High bird diversity and abundance in South African estuaries create the potential for interspecific competition for food. It is possible to identify mechanisms, behavioural or otherwise, which may serve to reduce the likelihood or extent of competition between species for food: options for segregation include where to feed, when to feed, how to feed and what to eat.

Differences in foraging habitat may exist at the macroscale between species within diet categories. At the Berg River estuary, most waders preferentially forage on mudflats, but Ruffs favour saltmarshes, as do herbivorous Redbilled Teals (Velásquez *et al.* 1991). Differences between wader foraging habitats also exist at the meso-scale, between different mudflat types within individual estuaries. Ringed Plovers, for example, favour areas of soft, fine mud, Whitefronted Plovers prefer sandy areas, and Terek Sandpipers aggregate in areas where the substratum is overlaid with Eelgrass *Zostera*. Wading piscivores at Lake St Lucia segregate spatially by foraging in water of different depths. The average wading depth of Little Egrets is 100 mm; the larger Great White Egrets and Grey Herons forage in depths of 160 mm and 190 mm respectively, while the world's largest heron, the Goliath Heron, forages in an average water depth of 325 mm (Whitfield & Blaber 1979*a*). As a parallel, intertidal mudflats are also exploited in three dimensions. Short-billed birds such as plovers are restricted to taking prey from or close to the surface, whereas long-billed probers, such as the Bartailed Godwit, can obtain prey from several centimetres depth.

Temporal segregation of foraging activity is a means of reducing interference competition, although it is not necessarily effective in reducing exploitative competition. Exploitative competition occurs when the numbers of birds result in the depletion of prey resources available to individuals to the point where availability is limited. Interference competition occurs when birds directly hinder one another's access to food, either by affecting prey behaviour or by restricting the space available in which to gain access to resources.

The extent to which temporal segregation of foraging is an option in an estuarine setting is limited for many species by the tidal rise and fall that occurs in open estuaries, severely constraining the foraging time available for those species which obtain their food intertidally. It is tidal rise and fall, and the associated short-term expansion and contraction of habitats, that account for the complex predation dynamics in estuaries. Because of restricted daylight foraging time, many bird species which forage intertidally at low tide are forced to forage by both day and night (Kalejta 1992; Turpie & Hockey 1993). Piscivorous species do not suffer from this time constraint, being able to forage at both low and high tide. Most estuarine piscivores are predominantly daytime foragers, spending the night at communal roosts. Goliath and Grey Herons, Yellowbilled Storks and African Spoonbills forage at night (as well as by day) at Lake Turkana, Kenya (Fasola & Canova 1993), Grey Herons forage by night and day at Lake St Lucia (Whitfield & Blaber 1978) and Little Egrets occasionally forage on moonlit nights at the Swartkops estuary (Martin 1991). Only two species, the Blackcrowned and Whitebacked Night Herons, are exclusively nocturnal foragers.

The greater potential for temporal segregation in foraging among piscivores than among invertebrate-feeders is illustrated in Table 10.4. Comparing absolute daily foraging times, piscivores feed for 5.1 ± 2.2 hours per day and invertebrate-feeders for 6.9 ± 1.3 hours. This difference is not significant (Mann–Whitney $Z_{7,8} = 1.3$, $p > 0.1$). In terms of the percentage of available foraging time used, however, piscivores use far less ($21.1 \pm 9.1\%$) than invertebrate-

Table 10.4. *Estimates of time spent foraging (h) per 24 h by estuarine birds at three sites in South Africa. Percentage of available foraging time is in parentheses*

Piscivores	Lake St Lucia		Swartkops estuary	
Eastern White Pelican	3.0[b]	(12.5)		
Whitebreasted Cormorant	2.3[b]	(9.6)		
Reed Cormorant	3.1[b]	(12.9)		
Grey Heron	6.7[a]	(27.9)		
Goliath Heron	6.4[a]	(26.7)		
Great White Egret	7.6[a]	(31.7)		
Little Egret	7.8[a]	(32.5)	5.0[c]	(20.8)
Invertebrate-feeders	Berg River estuary		Swartkops estuary	
Sacred Ibis			5.8[c]	(48.3)
African Black Oystercatcher			6.8[c]	(56.7)
Grey Plover	9.2[d]	(76.7)	7.0[c],5.9[e]	(58.3,49.2)
Turnstone			7.7[c]	(64.2)
Greenshank	4.8[d]	(40.0)	8.3[c]	(69.2)
Curlew Sandpiper	9.7[d]	(80.8)	9.4[c]	(78.3)
Curlew			5.8[c]	(48.3)
Whimbrel			6.8[c],5.8[e]	(56.7,48.3)

Data from Whitfield & Blaber 1978[a], 1979[b], Martin 1991[c], Kalejta 1992[d] and Turpie & Hockey 1993[e].

feeders ($58.2 \pm 10.3\%$). This difference is highly significant ($Z = 3.2$, $p < 0.001$). This calculation assumes that piscivores can forage at night, which may not be realistic for some species, especially plunge-divers. Even if this is the case, however, one important conclusion remains: most invertebrate-feeders cannot satisfy their energy demands during one low tide cycle and are thus forced to forage for some time at night. Diurnally-foraging piscivores on the other hand can easily satisfy their energy demands during daylight hours, spending on average only 42% of daylight hours foraging. The reciprocal line of evidence supporting this conclusion is that three piscivores – Blackcrowned and Whitebacked Night Herons and the (rarely estuarine) Pel's Fishing Owl *Scotopelia peli* – obtain their food entirely at night. No estuarine birds which eat mudflat-dwelling invertebrates forage exclusively at night. Seasonal variation in day length (sunrise to sunset) is not as extreme on the South African coast as at north temperate latitudes where comparable studies have been made. On the northwest coast (Port Nolloth), day length varies from 10.2 h to 14.0 h. In the south of the country (Port Elizabeth), the range is from 9.7 h to 14.3 h.

The third major axis along which foraging segregation can occur is the axis of prey choice. Potential competition can be lessened by reducing dietary overlap in prey species taken. However, the nature of estuarine prey resources limits the extent to which this option can be used. Whilst both invertebrate prey resources and estuarine fish resources may be diverse (Day 1981*b*), typically only a few species are abundant, and these dominate the diets of birds. By concentrating their predation on numerically dominant species, estuarine birds may contribute to the maintenance of diversity in invertebrate and fish communities (e.g. Schneider 1978).

In the lower reaches of the Berg River the Southern Mullet *Liza richardsonii* is an order of magnitude more abundant than all other fish species combined (Bennett 1994). At the same estuary, the main prey of most waders are two species of *Ceratonereis* polychaete. These dominate the mudflat macroinvertebrate biomass, and the entire mudflat macroinvertebrate fauna comprises only 25 species (Kalejta & Hockey 1991). Further east, at the Swartkops River estuary, the mudprawn *Upogebia africana* makes up 82% of aquatic invertebrate biomass (Hanekom et al. 1988) and is the principal prey of many species of waders and other waterbirds (Martin 1991). The ichthyofauna of subtropical Lake St Lucia is more diverse than that of either the Berg or Swartkops River estuaries (Millard & Broekhuysen 1970), but three taxa, Small Kob *Johnius dussumieri*, Mozambique Tilapia *Oreochromis mossambicus* and mullets (Mugilidae) make up the bulk of the fish consumed by birds (Whitfield & Blaber 1978, 1979a, b).

When the diversity of abundant prey is low, the principal axis of dietary segregation becomes size selection. For any species, the upper prey size limit is determined by morphological constraints on catching, handling or eating the prey. The lower size limit is probably set most often either by the ability to detect small prey, the dexterity to handle them or the profitability of accepting them. Size-selective predation is the norm. Among waders, size-selective predation occurs on prey which range widely in their maximum sizes, from mussels and limpets (Hockey & Underhill 1984), to amphipods (Howard & Lowe 1984) and even smaller prey. Grey Phalaropes *Phalaropus fulicarius* are at one extreme of the scale, exhibiting size-selective predation of zooplankton in Arctic tundra pools (Dodson & Egger 1980).

The degree of selectivity for prey size is linked to foraging mode. At the Berg River, Kalejta (1993a) found that obligate visual foragers (e.g. Ringed Plovers) were more size selective than facultative tactile foragers (e.g. Curlew Sandpipers).

Studies of differences in diet and foraging behaviour within waterbird assemblages in South African estuaries are few, and are restricted entirely to invertebrate-feeders and piscivores. Although there have been a few studies of the diets of individual herbivores, e.g. Redknobbed Coot at the Swartvlei and Bot River estuaries (Fairall 1981; Stewart & Bally 1985), none has compared diets of an assemblage of herbivores at the same site.

Case studies of ecological segregation

Diets of invertebrate-feeders have been studied at the Berg River estuary (Kalejta 1993a) and at the Swartkops River estuary (Martin 1991). The avifaunas of the two sites differ considerably: the Berg River estuary is dominated by small species, especially Curlew Sandpipers, which eat mostly small polychaete prey. Large waders, mainly Whimbrels and Grey Plovers, form a much greater proportion of the invertebrate-eating species assemblage at the Swartkops estuary, where the main prey is the large mudprawn *Upogebia africana* (Martin & Baird 1987; Velásquez et al. 1991).

Invertebrate-feeders of the Berg River estuary

Kalejta (1993a) studied the diets of four plover species (Grey, Blacksmith, Whitefronted and Kittlitz's), as well as Greenshanks and Curlew Sandpipers, in the lower Berg River estuary. Polychaete worms numerically dominated the diets of all species except Greenshank, whose diet comprised mostly crabs *Hymenosoma orbiculare*. Fish were a significant energy source for Greenshanks and Grey Plovers. The most important polychaetes in wader diets were *Ceratonereis erythraeensis* and *C. keiskama*. The abundance of these polychaetes increases upstream, concomitant with a switch from 100% *C. erythraeensis* in the lowest reaches to an even species mix in the mid-reaches and 70% dominance by the smaller *C. keiskama* in the upper reaches (Kalejta & Hockey 1991).

Of the 25 invertebrate species present in the mudflats of the Berg River estuary, only 14 (56%) have been recorded in wader diets. Crabs, although uncommon relative to polychaetes, contribute significantly to the energy intake of all species except Whitefronted and Kittlitz's Plovers.

There are seasonal variations in wader diets at the Berg River which are linked to the population dynamics of their dominant prey. For example, following the spring reproduction of nereid worms, their importance in the diets of most species increases. A local population crash of *C. erythraeensis* in the mid to upper reaches of the estuary was reflected in the diets of both Curlew Sandpipers and Grey Plovers. In the case of the latter, between December 1988 and April 1989, the numerical proportion of nereids in the diet decreased by 88%, with a parallel increase in the proportion of small, unidentified prey, probably crustaceans and gastropods.

Although diets of waders at the Berg River estuary overlap considerably in terms of their species complements (Table 10.5), there is marked segregation in the sizes of prey taken by different species. Plovers select large nereids irrespective of their relative abundance in the population. In autumn, the modal size of nereids in the substratum is c. 20 mm, and the modal size selected by Grey Plovers is between 35 and 45 mm. In summer, when the modal size of nereids is only 15 mm, Grey Plovers take mostly nereids in the 50–60 mm size class. Blacksmith Plovers are predominantly dryland foragers, and are less selective than Grey Plovers when foraging on mudflats,

Table 10.5. Summary of the diets of five wader species at the Berg River estuary. + + = main prey type, + = occasional prey type

	Whitefronted Plover	Kittlitz's Plover	Grey Plover	Blacksmith Plover	Curlew Sandpiper
Ceratonereis keiskama			++	++	++
C. erythraeensis	++		++	++	++
Perinereis nuntia	+				++
Boccardia sp.					+
Capitella capitata					
Exosphaeroma hyloecetes			++	++	+
Melita zeylanica					+
Orchestia sp.					+
Grandidierella lustosa					+
Hymenosoma orbiculare	+		++	+	+
Tabanid larvae				++	+
Coleoptera	+			++	+
Pupae	+				+
Indet. larvae					+
Hydrobia sp.				++	++
Seeds		+		++	+

Source: From Kalejta (1993*a*).

taking mostly nereids only slightly larger than the modal available size (Kalejta 1993*a*). Kittlitz's Plovers are analogues of Grey Plovers in selecting the largest worms, whereas Whitefronted Plovers are less selective. All of the above plover species are exclusively visual foragers. Curlew Sandpipers, by contrast, are facultative tactile foragers and show a very different pattern of prey size selection. At most times of year their predation is concentrated on the most abundant size classes of nereids, except in spring and early summer when these are very small (10 mm or less) and may be difficult for the sandpipers to detect.

Many waders are sexually dimorphic in both body size and bill length, females being heavier and longer-billed than males. This has been correlated with inter-sexual dietary differences in species foraging on both soft (Puttick 1981) and hard substrata (Hockey & Underhill 1984). Puttick's (1981) study demonstrated sexual differences in the diets of Curlew Sandpipers at Langebaan Lagoon, but only 40 km away, at the Berg River, no such differences have been detected (Kalejta 1993*a*).

Because of their different foraging modes and differing degrees of prey selectivity, the distribution of different wader species within an estuary is influenced by different prey-base attributes. This was demonstrated by comparing correlates of dispersion patterns of Curlew Sandpipers and Grey Plovers within the Berg River estuary (Kalejta & Hockey 1994). Both species' diets are dominated by polychaetes, but the same prey characteristics cannot be used to explain the birds' distributions. Foraging densities of Curlew Sandpipers, which do not show marked preference for large polychaetes, are strongly correlated with the numerical abundance of nereids in the substratum but not with nereid biomass. The dispersion of the strongly size-selective Grey Plovers, on the other hand, is best explained by nereid biomass (itself influenced by the presence of large polychaetes), rather than abundance.

Differences in the tidal foraging rhythms of waders at the Berg River estuary are neither seasonally consistent nor pronounced (Kalejta 1992). Consistent patterns of foraging activity are not evident even for Greenshanks, which spend only 44% of the low tide period foraging by day and 31% by night. Part of the reason why these tidal patterns are obscure may be related to the tidal lag of more than two hours between the lower and upper estuary: as the lower reaches become inundated on the flooding tide, birds move upstream to continue foraging. Such movements cannot be accounted for by point observations.

Invertebrate-feeders of the Swartkops River estuary

Foraging behaviour and diets of waders, as well as Kelp Gulls, Common Terns, Sacred Ibises and Little Egrets have been documented for the Swartkops estuary by Martin (1991). As at the Berg River, diets of all species are dominated by only a few prey types (Table 10.6). For most species the dominant prey is the estuarine mudprawn *Upogebia africana*. This one prey taxon accounts for more than 75% of the energy intake of Sacred Ibis, Grey Plover, Turnstone, Greenshank, Curlew Sandpiper, Curlew, Whimbrel, Kelp Gull and Common Tern. The diet of Little Egret is dominated by mugilid fish, and African Black Oystercatchers eat mostly Pencil Bait *Solen capensis*. Small *Cleistostoma* crabs are regular prey for most species, providing a maximum of 16% of energy intake (for Curlew Sandpipers). A study of the comparative diurnal and nocturnal foraging behaviour of Grey Plovers and Whimbrels indicated some change in diet between day and night (Turpie & Hockey 1993). Small crabs contribute less to the diets by night than by day, and juvenile Whimbrels eat no crabs at all at night.

Table 10.6. Summary of the diets of six wader species at the Swartkops estuary. ++ = main prey type; + = occasional prey type

	Grey Plover	Turnstone	Greenshank	Curlew Sandpiper	Curlew	Whimbrel
Upogebia africana	++	++	++	++	++	++
Sesarma catenata	++		++			++
Cleistostoma spp.	+	+		+		
Thaumastoplax spiralis	+					
Assiminea bifasciata	++	+		++		
Nassarius kraussianus					+	
Coleoptera		+		++		
Hymenoptera		+				
Diptera	+	+	++		+	
Indet. fish		+	++			
Sarcocornia perennis	+	++		++		+
Plant fragments	++	++	++	++		++

Source: From Martin (1991).

The foraging environment for waders at the Swartkops estuary is very different from that at the Berg River estuary. The high densities of large waders at the former site are facilitated not only by the dominance of one large crustacean (*Upogebia*) but also by the mudprawns' behaviour. The majority of mudprawns remain deep in their burrows during the low tide period, but a small proportion come to the surface. Mudprawns are vulnerable to bird predators when on the sand/mud surface because of their limited mobility, and most surfacing mudprawns are eaten by birds. This 'suicidal' behaviour is thought to be mediated by trematode and cestode parasites which induce surfacing behaviour as a means of bettering their chances of transmission to their final bird hosts (Hockey & Douie 1995): this cause/effect relationship has not, however, been demonstrated empirically.

The availability of surfacing mudprawns is lowest during the winter months, the time of year when most migratory birds are absent at their northern breeding grounds. However, many one-year-old Palearctic-breeding waders do not migrate north with the adults, remaining on the estuary for 16–18 months. These young birds respond to the reduced availability of *Upogebia* by changing their diets. In the case of Whimbrels, the numerical proportion of *Upogebia* in the diet falls from 59% in summer (September–April) to only 28% between May and August, when crabs and polychaetes become more important in the diet. The diet of Grey Plovers becomes numerically dominated by polychaetes at this time of the year. Among the resident species, Kelp Gulls become much more reliant on small prey than on *Upogebia* during the winter.

The majority of *Upogebia* eaten by all birds except Sacred Ibises and Curlews are caught while on the surface; their size can therefore readily be assessed by these predators. Surfacing mudprawns range in size from 10 to 80 mm, and the full size spectrum is consumed by birds. The mudprawns themselves are fairly fragile and most

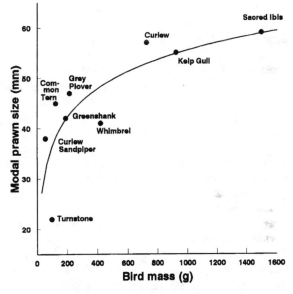

Figure 10.5. **The relationship between bird mass and the modal size of mudprawns *Upogebia africana* consumed by nine bird species at the Swartkops River estuary. (Data from Martin 1991.)**

birds are capable of breaking them up before eating them (although long handling time carries with it the risk of kleptoparasitism of the prey by other birds). Under these conditions, it might be predicted that size selectivity would not be marked, but this is not the case. All species locate all or most prawns visually and there is clear size selectivity linked to bird body size (Figure 10.5). The largest prawns (modal length 59 mm) are taken by Sacred Ibises (weight 1500 g), and the smallest (22 mm) by Turnstones (99 g). The relationship between bird size and prey size is not entirely consistent. Curlew Sandpipers are smaller than Turnstones (57 g), but their *Upogebia* prey are considerably larger (modal size 38 mm). This is in contrast to the situation on rocky shores where Turnstones eat significantly larger prey than Curlew Sandpipers (authors' unpublished data). On rocky shores, however, neither species eats prey of the size of *Upogebia*.

The sizes of *Upogebia* preyed on by Common Terns vary during the tidal cycle. Common Terns start to forage before the intertidal mudflats are exposed, when waders are not foraging. At this time (±3 h before low tide), they take fairly large prawns (c. 55 mm median length). As the low tide progresses, however, they take smaller and smaller prey: by the time the mudflats are inundated again, median prawn size has fallen to 35 mm. When the mudflats are exposed, Kelp Gulls and most waders select large prawns. Martin (1991) hypothesises that the tidal change in the diet of Common Terns serves the primary function of reducing the risk of the birds losing their prey to heterospecific kleptoparasites. The pattern of selection for decreasing prawn size during the low tide period is not linked to patterns of prey size availability. The largest prawns are available early and late in the low tide cycle (Martin 1991). Away from estuaries, Kelp Gulls also select small prey items when the risk of kleptoparasitism, usually by conspecifics, is high (Hockey & Steele 1990). The advantage in selecting small prey under these situations is that small prey have a shorter handling time and are thus vulnerable to kleptoparasites for the minimum time possible (Steele & Hockey 1995).

Tidal foraging rhythms at the Swartkops estuary are much more pronounced than at the Berg River estuary. Grey Plovers and Whimbrels, which eat *Upogebia* of very similar sizes (Turpie & Hockey 1993), have different peak foraging times. The former concentrate their foraging on the falling tide whereas the foraging activity of Whimbrels is symmetrical around low tide (Martin 1991). A trend of maximal foraging effort on the falling tide is also evident for Sacred Ibises, Turnstones, Greenshanks, Kelp Gulls and Common Terns. Curlew Sandpipers have similar foraging rhythms to Whimbrels, while Curlews and Little Egrets concentrate their efforts on the rising tide.

Detailed analyses of species' distribution patterns at the Swartkops estuary have not been made, but across the estuary there is a strong correlation ($r_s = 0.79$, $p < 0.01$) between low tide bird density and the standing stock biomass of *Upogebia*. This further highlights strong links between the distributions of several bird species and the distribution of a single prey taxon.

The piscivores of Lake St Lucia

The ichthyofauna of Lake St Lucia is far more diverse than the benthic invertebrate fauna of either the Berg or the Swartkops River estuaries. More than 190 fish species occur in the system (Millard & Broekhuysen 1970) as compared with 25 benthic invertebrate species at the Berg River (Kalejta & Hockey 1991) and 62 species at the Swartkops estuary (MacNae 1957). As in the above cases of

predation of benthic invertebrates by birds in estuaries, only a small proportion of the potential prey diversity at Lake St Lucia is utilised by piscivores. Only 28 fish species (14.6%) are known to be eaten by birds (Table 10.7). Of these, only Riverbream *Acanthopagrus berda*, Flathead Mullet *Mugil cephalus*, Spotted Grunter *Pomadasys commersonnii* and Mozambique Tilapia *Oreochromis mossambicus* occur in the diet of eight or more of the 10 piscivores studied by Whitfield & Blaber (1978, 1979a, b). *Oreochromis mossambicus* makes up 30% or more by numbers of the diets of six species: Reed Cormorant, Eastern White Pelican, Great White Egret, Grey and Goliath Herons, and Pied Kingfisher. Thus, as for invertebrate-feeders, it is the abundance and availability of only a few prey taxa, rather than the diversity of potential prey species present, that is important to the predator assemblage.

Dietary similarities between piscivores, in terms of prey species, are determined largely by foraging mode. Species which forage either by aerial plunge-diving or by diving from the surface capture a significantly greater diversity of prey species (mean = 11.6 ± 3.8 SD) than do species which swim or wade to capture prey (mean = 6.2 ± 2.7; Mann–Whitney $Z_{5,5} = 2.1$, $p < 0.05$) (Table 10.7). This difference reflects the greater foraging mobility of the former group, either through their ability to search a large area (plunge-divers) or to exploit a wide range of water depths (surface divers).

Based on a presence/absence analysis of prey taxa (Figure 10.6), plunge-divers (Pied Kingfisher and Caspian Tern) cluster together, as do surface divers (cormorants) and the larger wading herons. The two species with the most similar diets (in terms of prey species), Great White Egret and Goliath Heron, eat prey with an order of magnitude difference in their maximum and modal sizes. The three large herons use predominantly the same 'sit-and-wait' hunting strategy, but segregate their foraging locations on the basis of water depth. An order of magnitude difference in prey size also exists between the two surface-diving cormorants and between the two plunge-divers. The largest prey are taken by African Fish Eagles and Eastern White Pelicans. The eagles eat prey of at least 4–5 times the size of those eaten by the pelicans, and prey species similarity between the two is only 45% (Figure 10.6).

The fish diet of the Little Egret is markedly different to that of all other piscivores and this is probably a consequence of its different foraging behaviour. It forages in shallower water than the other herons (100 mm) and spends a greater proportion of its foraging time (58% vs. 14–44%) in areas with aquatic macrophytes. Aquatic plants provide shelter for fish from aerial predators and impede the foraging of surface divers. Little Egrets either run about actively among the plants, catching the fish they

Table 10.7. Fish diets of piscivorous birds summarised from Whitfield & Blaber (1978, 1979a, b). ++ =main prey type, + =occasional prey type

	Eastern White Pelican	White-breasted Cormorant	Reed Cormorant	Grey Heron	Goliath Heron	Great White Egret	Little Egret	African Fish Eagle	Caspian Tern	Pied Kingfisher
Acanthopagrus berda	+	++	++	++	++	++		+	++	
Ambassis productus									+	+
A. natalensis										+
Anguilla sp.	+									
Barbus paludinosus	++								+	
Clarias gariepinus	+	+						++	++	
Diplodus sargus		+								
Elops machnata										
Gerres acinaces		+								
Gilchristella aestuaria									+	+
Glossogobius giurus	+	+	+				++			
Hyporhamphus capensis									+	++
Atherina breviceps										+
Johnius dussumieri	+	++	+					+	++	
Mugilidae		++	+	++	++	++		++	+	++
Pomadasys commersonnii	+	+	+		+			+	++	+
Atherinomorus lacunosus									+	
Pseudocrenilabrus philander	++								+	+
Rhabdosargus holubi		+	+						+	+
R. sarba		++							+	+
Oreochromis mossambicus	++	++	++	++	++	++	++	+	++	++
Solea bleekeri		++					+			
Sphyraena sp.	+									
Redigobius dewaali							+			
Stolephorus holodon									+	
Terapon jarbua	+	+	+		+	++			+	++
Thryssa vitrirostris		++		++			+		++	+
Tylosurus leiura								+		

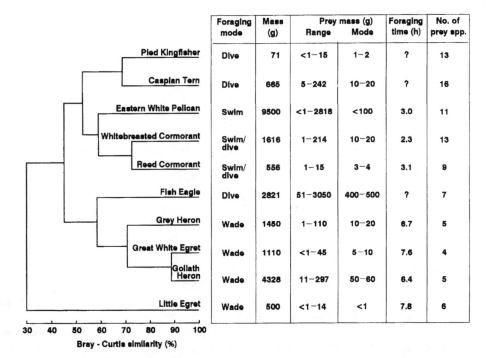

	Foraging mode	Mass (g)	Prey mass (g) Range	Prey mass (g) Mode	Foraging time (h)	No. of prey spp.
Pied Kingfisher	Dive	71	<1–15	1–2	?	13
Caspian Tern	Dive	665	5–242	10–20	?	16
Eastern White Pelican	Swim	9500	<1–2818	<100	3.0	11
Whitebreasted Cormorant	Swim/dive	1616	1–214	10–20	2.3	13
Reed Cormorant	Swim/dive	556	1–15	3–4	3.1	9
Fish Eagle	Dive	2821	51–3050	400–500	?	7
Grey Heron	Wade	1450	1–110	10–20	6.7	5
Great White Egret	Wade	1110	<1–45	5–10	7.6	4
Goliath Heron	Wade	4328	11–297	50–60	6.4	5
Little Egret	Wade	500	<1–14	<1	7.8	6

Bray - Curtis similarity (%)

Figure 10.6. **Dietary similarity (based on presence/absence of prey species), foraging mode, mass, prey mass, daily foraging time and** prey diversity of 10 piscivorous birds at Lake St Lucia. (Data from Whitfield & Blaber 1978, 1979a, b.)

disturb, or they dislodge fish by foot-trembling, a technique also used by Hamerkops. In this way they are able to exploit a spectrum of species unavailable to birds using other hunting techniques. Figure 10.6 analyses only the fish component of the diet. Numerically, however, fish make up only 19% of the diet of Little Egrets, and their diet is thus even more different from the other species than Figure 10.6 suggests. Other important dietary items for this species are crustaceans, gastropods and bivalves. Among the other piscivores, only Grey Herons, Great White Egrets, African Fish Eagles and Pied Kingfishers occasionally take prey other than fish, and for none of these species does the numerical contribution of other prey to the diet exceed 15%.

As with invertebrate-feeding waders, diets of piscivores are plastic and birds respond to changes in the relative or absolute abundances of different fish species. Blaber (1973) suggested that piscivore abundance in the West Kleinemond and Kasouga River estuaries was linked to the abundance of the Cape Stumpnose *Rhabdosargus holubi*; subsequently, Whitfield (1978) was able to show empirically that local piscivore densities at Lake St Lucia were strongly correlated ($p < 0.01$) with local fish abundance (all species combined).

Changes in the diets of African Fish Eagles and Whitebreasted Cormorants at Lake St Lucia have been demonstrably linked with changes in fish abundance. In 1975, Sharptooth Catfish *Clarias gariepinus* made up 5% of the diet of African Fish Eagles; this increased to 23% in 1976, concomitant with a major increase in the proportional representation of *Clarias* in gill-net catches. Whitebreasted Cormorants have one of the most diverse diets (13 spp.) of the Lake St Lucia piscivores. In the early summer of 1976 there was a massive increase in the abundance of the pelagic Orangemouth Glassnose *Thryssa vitrirostris* in the deeper areas of the Lake, and Whitebreasted Cormorants in this area fed almost exclusively on this species.

Further west, at the Swartvlei River estuary in the southern Cape, the diet of Whitebreasted Cormorants varies seasonally depending on movements of fish in and out of the estuary (Whitfield 1986). The Cape Silverside *Atherina breviceps* dominates the diet throughout the year, but the Estuarine Roundherring *Gilchristella aestuaria* more than quadruples in importance in early summer. This study also serves to highlight the risks of extrapolating details of diet between sites. *Atherina breviceps* and *G. aestuaria* are both abundant at Lake St Lucia, but neither species has been recorded in the diet of Whitebreasted Cormorants. *Gilchristella aestuaria* is also common at Kosi Bay, but is also not eaten by the cormorants (Jackson 1984).

Studies of bird diets at the Berg River, Swartkops River and St Lucia estuaries highlight three important points.

1 Bird abundance and diversity are influenced more by prey abundance and availability than by prey species diversity.
2 The principal axes of ecological segregation in foraging ecology are those of spatial dispersion of birds on macro- and micro-scales (as functions of habitat choice and foraging mode) and selection for prey of different sizes, rather than different species.
3 With a few exceptions, such as Openbilled Stork and Knot, estuarine birds are plastic in their diet choice; the presence of particular prey taxa is therefore not a good predictor of the composition of the bird community.

Feedback loops: responses of prey to avian predators

Unravelling the responses of predator and prey to one another has posed a central problem in behavioural ecology. Predators aggregate where prey are common, yet prey avoid (or are depleted in) areas where predators are common, and the two responses thus conflict (Sih 1984). Three potential patterns of spatial distribution are predictable. In a Type 1 situation, a predator response should dominate when prey are immobile. The definition of mobility is relative, but, by comparison with the mobility of birds, most estuarine benthic invertebrates should probably be classified as immobile. If this is correct, then invertebrate-feeders should respond to local prey abundance or biomass, as has been demonstrated at both the Swartkops and Berg River estuaries (Martin & Baird 1987; Kalejta & Hockey 1994). There is some evidence that the same Type 1 response is shown by piscivores to fish, in as much as piscivore abundance is correlated with fish abundance at Lake St Lucia (Whitfield 1978). This interaction may not, however, be as simple as Whitfield's data suggest.

In a Type 2 situation, the prey response should dominate when prey are relatively mobile and have a spatial refuge from predators. High mobility of both predators and prey characterises the interaction between piscivorous birds and fish. The extent to which estuarine fish have a spatial refuge from predators within estuaries, however, is unknown. Deep water may provide a spatial refuge, especially at high tide. Several surface-diving, swimming and wading piscivores opportunistically 'herd' shoals of fish into shallow water, where they are presumably easier to catch. This is clearly a predator-dominated response, but one that only happens occasionally. Its infrequency suggests that, at least for some of the time, shoaling fish do achieve a spatial refuge or reduce the strength of the predator response by scattering. The piscivore–fish interaction may be intermediate between a Type 2 and a Type 3 situation. In the latter, both predator and prey are mobile but no spatial refuge exists for prey. In this Type 3 situation (*sensu* Sih 1984) the predator and prey responses may cancel each other out, resulting in no clear pattern.

Prey responses to predators in South African estuaries have been little studied. In a series of paired cage-and-control experiments at the Berg River, Kalejta (1993b) found the abundance of nereids close to the mud surface in control plots to be consistently lower than in cages. However, she was unable to determine whether this was a prey depletion effect or whether the nereids had moved deeper into the substratum in response to predators.

The predator–prey 'arms race', whether behavioural or morphological, can be viewed as an oscillation in which the evolution of anti-predator behaviour in the prey is followed by a compensatory change in the behaviour of the predator. However, what appears to be an anti-predator behaviour may simply be a response to environmental conditions, and decreased predation is purely coincidental. For example, cold weather in north temperate estuaries produces changes in the behaviour of some invertebrates which reduce the hunting success of their shorebird predators (e.g. Goss-Custard 1969, 1970a, b; Evans 1976; Goss-Custard et al. 1978; Reading & McGrorty 1978; Evans & Dugan 1984). Whilst cold weather is unlikely to have the same effect in South African estuaries, hot and windy weather may cause invertebrates to bury deeper in the substratum to avoid desiccation.

Flocking (shoaling, herding) has frequently been shown to have predator-avoidance value for birds, fish and mammals, but is not a term normally applied to intertidal invertebrates. However, fiddler crabs (Uca spp.) feed in herds, and it has been suggested that this serves to minimise risk to individuals where predation by birds is intense (Zwarts 1985). Salmon and Hyatt (1983), by contrast, suggest that herding in Uca does not occur when predation pressure is high; this possible example of anti-predator 'flocking' therefore requires further investigation.

Prey removal and nutrient cycling

Energy demands of estuarine birds

The energy demands of individual birds vary seasonally, being influenced by factors such as breeding requirements, pre-migratory fattening, post-migratory recovery and moult. Additional, but less predictable, energy demands arise due to thermoregulation costs and, to an unknown extent, parasite loading.

Field metabolic rates of birds scale allometrically with body mass. Kersten and Piersma (1987) postulate that waders, which are major consumers on estuaries, have unusually high metabolic rates (both basal and field). However, values derived from their equation for existence metabolism differ little from those obtained using Nagy's (1987) equation for field metabolic rate based on doubly-labelled water studies. Indeed, Nagy's equation predicts higher energy expenditure, especially for small waders. This suggests that there is nothing 'unusual' about the metabolic rates of waders.

Whilst net energy demand scales allometrically, gross energy intake – a better measure of the impact of birds on their food resources – is influenced by assimilation efficiency. The assimilation efficiency of herbivorous birds (c. 58%) is lower than that of species which eat invertebrates or fish (c. 75%) (Velásquez et al. 1991). Thus, species at low trophic levels recycle proportionally more material to the environment than do species at high trophic levels.

The primary form in which nutrients are recycled by birds is guano. However, birds make additional contributions to the nutrient pool in the form of carcasses and feathers. The latter contribution is, however, fairly small. Based on figures in Summers (1977), the nutrient contribution from feathers to Langebaan Lagoon is only about 1.5% of the energy returned to the system in the form of guano.

Energy cycling by herbivores

Herbivores are a dominant component of the avifauna in few South African estuaries (Appendix 10.1). An exception is the Bot River estuary in the southwestern Cape: the hydrological conditions of this estuary are variable, as the mouth is often closed for extended periods. On average, herbivores make up 93.5% of the total waterbird biomass at this site (Heÿl & Currie 1985). Of this proportion, Redknobbed Coots alone make up 97.1%. Their numbers tend to be highest during autumn, but vary considerably from year to year depending on hydrological conditions. The two main food species for Redknobbed Coots in the Bot River estuary are the submerged aquatic macrophytes *Ruppia maritima* and, when salinities are low during periods of mouth closure, *Potamogeton pectinatus*.

Stewart & Bally (1985) found that the daily metabolised energy of captive coots was 516 kJ per bird: this is 70.2% of the field metabolic rate predicted from Nagy's (1987) equation. They extrapolated their consumption estimates by a factor of 1.34 to account for the increased energy cost of free existence (based on Murdock 1975), but should probably have used a factor closer to 1.42 (from Nagy 1987). Their results described below have been corrected to the latter factor.

The estimated daily consumption by a single coot is 90 g (dry weight) of plant food: of this, approximately 54 g (DW) is recycled as guano. The ratio of food to guano energy content is low (1.07) and, based on average values from Table 10.1 of Stewart and Bally (1985), corrected for the additional cost of free existence, each bird consumes

1278 kJ per day, of which 718 kJ is returned to the system as guano. This gives an assimilation efficiency of only 44%, compared with the 58% reported in their study. The original data from which the value of 58% was obtained are not presented. An assimilation efficiency of 44% falls within the range of assimilation efficiencies recorded for Brent Geese *Branta bernicla* (38%) and Wigeon *Anas penelope* (47%) feeding on leaves of *Zostera noltii* (calculated from Table 10.3 in Madsen 1988).

Between May 1980 and April 1981, a period when coot numbers were above average, these birds collectively consumed about 805 tonnes (DW) of aquatic macrophytes. This represents between 10.5 and 12.5% of the annual macrophyte production as estimated by Bally *et al.* (1985). Equivalent guano production is 483 tonnes, with an energy value of 6.42×10^9 kJ. If the reported assimilation efficiency of 58% is correct, the latter value would be reduced to 4.87×10^9 kJ.

The area of the Bot River estuary is 1359 ha (Heÿl & Currie 1985). The consumption figures above translate into an energy removal by coots of 23 043 kJ ha^{-1} per day. This is a high figure when compared with more saline and perennially open estuaries. At the Swartkops estuary there is negligible consumption by herbivores in either summer or winter. At the Berg River estuary, summer consumption is estimated to be 1958 kJ ha^{-1} d^{-1}, assuming no subtidal feeding by herbivores. During the winter, when the estuary is fresher, herbivore numbers increase and consumption rises to 3492 kJ ha^{-1} d^{-1}; this is still, however, only 15% of average consumption at the Bot River during periods when the estuary mouth is closed. When the mouth is open, most coots leave the estuary within 24 hours because the macrophyte beds are destroyed as water levels drop by about 2.5 metres. At Langebaan Lagoon, which has no input of river water, consumption is even lower. During summer, herbivores remove only 9 kJ ha^{-1} d^{-1}, and in winter, they take 210 kJ ha^{-1} d^{-1} (Velásquez *et al.* 1991).

At Swartvlei in the southern Cape, the average number of coots present is 1851 (Fairall 1981). Assuming a gross intake per bird of 1278 kJ d^{-1}, this translates to an energy removal by these birds of 3505 kJ ha^{-1} d^{-1}, the area of the estuary being 675 ha (Fairall 1981). The main food of coots at Swartvlei is *Potamogeton pectinatus*, and they consume approximately 3.6% of the annual production of this macrophyte.

At Lake St Lucia the water surface area varies between 22 500 and 41 700 ha. Assuming a gross energetic demand of four times Standard Metabolic Rate (calculated from Lasiewski & Dawson 1967), Berruti (1983) calculated that herbivore energy demand ranged from 1.08×10^6 to 9.73×10^6 kJ d^{-1}, with an average of 5.43×10^6 kJ d^{-1}. If the lake averages half full (32 100 ha), this translates to a consumption of only 169 kJ ha^{-1} d^{-1}. This is slightly less than the winter consumption by herbivores at Langebaan Lagoon.

Energy cycling by invertebrate-feeders

Several studies around the world have attempted to estimate or measure the removal rate of invertebrates by birds from estuaries. Estimates of removal rates range between 6 and 49% of annual production, and for some individual prey species, up to 90% mortality of the initial standing stock has been reported (Evans *et al.* 1979; Baird *et al.* 1985). Over a four-month period during summer, Curlew Sandpipers foraging on the mudflats of the Berg River removed 58% of the initial numbers and 77% of the initial biomass of nereids (Kalejta 1993*b*). Interestingly, this intense predation could not be detected with a predator-exclusion experiment because the mortality rate of nereids was counterbalanced by their high reproductive rate.

The proportion of production removed by estuarine invertebrate-feeders is broadly comparable with that removed by herbivores (from freshwater and estuarine environments), which have been estimated at 2 to 60% of annual macrophyte production (Nienhuis 1978; Prevost *et al.* 1978; Kiorboe 1980; Verhoeven 1980; Fairall 1981; Stewart & Bally 1985).

Velásquez *et al.* (1991) compared estimates of invertebrate removal from Langebaan Lagoon and two estuaries (Berg and Swartkops). Despite the fact that migrant waders are most abundant during the summer, prey removal by invertebrate-feeders at Langebaan Lagoon is greatest during the winter, due almost entirely to a seasonal influx of Greater Flamingos. At Langebaan Lagoon, the estimated average removal rate of invertebrates during summer (September–April) is equivalent to 4320 kJ ha^{-1} d^{-1}, rising to 5264 kJ ha^{-1} d^{-1} in winter. At the estuaries of the Berg and Swartkops Rivers removal rates are, on average, higher, but the seasonal difference in removal rates is greater at the Berg River than at the Swartkops River. The summer removal rates at the two sites are similar: 6982 kJ ha^{-1} d^{-1} at the Berg and 7633 kJ ha^{-1} d^{-1} at the Swartkops. The winter removal rate at the Berg River is 41% of summer removal, whereas at the Swartkops it is 75%.

Hockey *et al.* (1992) calculated midsummer removal rates by waders alone from the intertidal mudflats of the same three sites. This is the time when numbers of migrant waders are maximal and population energy intakes are thus higher than the seasonal average. The intake rates (in kJ ha^{-1} d^{-1}) were 12 316, 9725 and 5254 for the Berg River, Swartkops River and Langebaan Lagoon respectively. On average, these values for the two estuaries are 3.7 times the average consumption rates by birds at five British estuaries at the same time of year (Hockey *et al.* 1992).

Taking annual energy intake by invertebrate feeders at

the three South African sites, and assuming a 75% assimilation efficiency, energy returned to the system as guano ranges from 1160 kJ ha^{-1} d^{-1} at Langebaan Lagoon, to 1400 kJ ha^{-1} d^{-1} at the Berg River and 1745 kJ ha^{-1} d^{-1} at the Swartkops River. These values are substantially in excess of the amount of energy recycled by herbivores at these sites, except at the Berg River in winter, when energy recycled by herbivores is slightly more than double that recycled by invertebrate-feeders (Velásquez *et al.* 1991). The use of an assimilation efficiency of 75% is a compromise. Among avian carnivores, average assimilation efficiency ranges from 56% to 95% and even for a single prey species (Whimbrels eating fiddler crabs *Uca tangeri*) can vary between 55% and 76% (Zwarts & Blomert 1990).

At Lake St Lucia, invertebrate-feeders remove more energy from the system than do either herbivores or piscivores. Berruti (1983), using methodology described above, calculates a mean energy removal of 2.08 × 10^7 kJ d^{-1}. This is nearly four times the rate of energy removal by herbivores and 3.2 times the removal rate by piscivores at this site. Assuming a lake area of 32 100 ha (see above), daily removal by invertebrate-feeders is 648 kJ ha^{-1}, an order of magnitude less than the removal rates by these birds at the Berg and Swartkops estuaries and Langebaan Lagoon.

Energy cycling by piscivores

No direct measurements of energy removal or recycling by an estuarine piscivore assemblage have been made in South Africa. There are several reasons for this, including problems in tracking the short-term movements of piscivores in and out of estuaries.

Based on seasonal count data (several counts per month from the Berg and Swartkops Rivers and replicated biannual counts from Langebaan Lagoon), Velásquez *et al.* (1991) made an attempt to calculate removal rates by piscivores. In doing so, they assumed that all prey were caught subtidally and used only subtidal areas in their calculations. However, most piscivores also forage intertidally at high tide and roost on intertidal flats at low tide. They thus recycle energy to the entire system, and values below have been corrected to account for this.

Piscivores in South African estuaries follow an opposite pattern of seasonal energy removal to herbivores: consumption peaks during the summer months, due in the main to an influx of Palearctic-breeding terns. The total daily energy intake by piscivores at Langebaan Lagoon and the Swartkops River estuary is much less than the intake by invertebrate-feeders. During summer, when consumption by piscivores is maximal, they account for only 4% and 13% of energy consumption by invertebrate-feeders at Langebaan Lagoon and the Swartkops River respectively. Corresponding figures for the winter are 0.7% and 13%. Their proportional contribution to energy

Table 10.8. Average annual energy removal by aquatic foraging birds from estuaries/coastal lagoons and the proportional contributions made by members of three dietary categories. Values are corrected to total estuarine area (intertidal and subtidal)

	Berg River estuary	Langebaan Lagoon	Swartkops estuary	Lake St Lucia[a]
Total removal by birds (kJ ha^{-1} d^{-1})	7343	3287	4177	1019
% herbivores	25	2	0	17
% invertebrate feeders	55	95	88	63
% piscivores	20	3	12	20

[a] Assuming a lake area of 32 100 ha.

removal at the Berg River is greater. During summer, their rate of energy removal is 30% that of invertebrate-feeders, while in winter this figure rises to 68%. A substantial proportion of the winter intake is accounted for by Eastern White Pelicans and cormorants.

In terms of recycled energy (averaged over subtidal and intertidal areas), the contribution made by piscivores is less than that of invertebrate-feeders. Averaged over the year, piscivores recycle 374 kJ ha^{-1} d^{-1} at the Berg River, 121 kJ ha^{-1} d^{-1} at the Swartkops River and only 23 kJ ha^{-1} d^{-1} at Langebaan Lagoon. These values are almost certainly underestimates as they take no account of influxes of piscivores that have been feeding elsewhere. At the Berg River estuary, for example, very large numbers of terns and cormorants enter the estuary at night to roost – these numbers are not reflected in daytime counts and hence in calculations of energy cycling, although considerable defaecation does occur at these roosts and roosting birds thus act as net energy importers to the system.

At Lake St Lucia, Berruti (1983) calculated energy removal by piscivores to average 6.5 × 10^6 kJ d^{-1}. Correcting for a lake area of 32 100 ha (see above), this removal rate is equivalent to 202 kJ ha^{-1} d^{-1}. This is a very low rate of removal when compared with the Berg and Swartkops estuaries and with Langebaan Lagoon.

Importance of different foraging categories to energy cycling in estuaries

Table 10.8 summarises total energy intake by birds in three categories at four sites ranging geographically from the Berg River estuary in the west to Lake St Lucia in the east. At all four sites the major role in energy cycling is played by invertebrate-feeders, which account for 55% to 95% of energy consumed. Herbivores account for a maximum of 25% of energy consumption (at the Berg River), and the piscivore contribution to energy removal peaks at 20% at both the Berg River and Lake St Lucia.

From this analysis, certain tentative conclusions can be drawn.

1   There is no clear geographical pattern in the contribution made by different foraging categories to estuarine energy flow.

2   In most estuaries, energy flow through the bird community is dominated by the contribution of invertebrate-feeders. These birds thus act as an important energy sink in estuaries.

3   Herbivory by birds is important only in those estuaries with substantial submerged macrophyte production (where energy flow through herbivores may exceed that through other birds). Production of submerged macrophytes tends to be maximal in estuaries whose mouths are closed for extended periods. The amount of energy recycled by individual herbivores is greater than that recycled by birds feeding at higher trophic levels because of the former's relatively low assimilation efficiency.

4   The impact of piscivory in estuaries is less easily quantified than the impacts of herbivorous and invertebrate-feeding birds. This is because of mass movements of piscivores, especially cormorants and terns, in and out of estuaries, as well as short-term and seasonal immigration and emigration of fish. The level of piscivory at individual estuaries is probably linked in part to the proximity of breeding sites for colonial piscivores.

5   In estuaries, as distinct from lagoons, there is a tendency for energy removal by birds to decrease from the west to the east. This is at variance with Kalejta & Hockey's (1991) prediction that (invertebrate) production should be highest in the warmer estuaries of the east. It does, however, parallel a pattern of decreasing intertidal primary production (on rocky shores) from west to east (Bustamante et al. 1995).

Pastures of plenty or a delicate equilibrium?

Are estuarine birds food-limited?

The total sustainable annual offtake by estuarine consumers is roughly equivalent to the total net annual production of food resources. Yet, in the few cases where estimates of consumption relative to production have been made, consumption by birds has been found to be a surprisingly low proportion of annual production. Kalejta (1992) estimated that the migrant wader population at the Berg River estuary removes about 26% of annual invertebrate production. At Langebaan Lagoon, invertebrate-feeders consume 24% of annual invertebrate production (Underhill 1987a). At

the Swartkops River estuary, consumption of invertebrate production by shorebirds is estimated to be only about 6% (Baird et al. 1985). Redknobbed Coots, which make up 97% of the biomass of herbivores at the Bot river estuary, consume about 10% of the annual production of submerged macrophytes in the estuary, but their numbers decrease dramatically after the estuary is opened and the macrophyte stock all but disappears (Bally et al. 1985; Stewart & Bally 1985).

These figures may suggest that estuarine birds are not food-limited (e.g. Puttick 1980; Underhill 1987a), but realistic conclusions cannot be drawn without a thorough knowledge of the availability of prey to birds. The amount of food available to birds in relation to total biomass and production is limited by several factors, including natural mortality, predation from other sources, the space and time available in which to feed, and the behavioural rhythms and responses of the prey. The behaviour of invertebrate prey is influenced, not only by the availability of their own food, but by tidal ebb and flow, photoperiod, oxygen, temperature and the presence of their avian predators (Goss-Custard 1980; Pienkowski 1981; Zwarts & Dirksen 1989). Studies which have attempted to measure instantaneous prey availability in relation to abundance indicate that this proportion is very small (e.g. Myers et al. 1980; Zwarts & Blomert 1992) but can also be very variable between sites (Piersma et al. 1993). The situation at the Swartkops estuary is an extreme example: mudprawns on or at the surface over a full low tide cycle represent on average less than 0.01% of the numbers in the sediment (Martin 1991).

Without detailed knowledge of the biomass of available prey, estimates of the amount of prey removed from an estuary give scant indication of the proximate factors that limit bird numbers on estuaries. The best perspective may be gained through examining distributional, behavioural and energetic evidence for competition amongst foraging birds.

Distributional evidence for competition

The densities of estuarine birds are almost invariably correlated with the abundance, biomass or production of their most important prey species (e.g. Goss-Custard et al. 1977; Bryant 1979; Martin & Baird 1987; Hockey et al. 1992; Kalejta & Hockey 1994). These relationships exist on both temporal and spatial scales. At Lake St Lucia, the relative densities of piscivorous birds and their prey are correlated on a seasonal basis ($r_{11} = 0.743$, $p < 0.01$: Whitfield 1978). Temporal changes in bird numbers in relation to prey biomass, particularly where non-migratory species are concerned, provide convincing evidence that the number of birds present is closely linked to temporal changes in estuarine carrying capacity. We define the carrying capacity for birds as being the level of prey consumption which, when combined with rates of prey mortality from other sources, does not exceed prey production on a long-term basis (Piersma 1987; Hockey *et al.* 1992).

Spatial relationships between bird density and prey numbers or biomass imply the avoidance, although not necessarily the elimination, of competition. At the Swarkops estuary, where nearly all waders share a single main prey species, *Upogebia*, their combined densities are correlated with mudprawn biomass across different sections of the estuary ($r_8 = 0.79$, $p < 0.01$; Martin 1991). Grey Plovers and Whimbrels are the commonest migrant birds at the Swartkops estuary. During winter, adults migrate north to breed in the Palearctic, but immature birds remain behind in South Africa. In winter, these remaining birds concentrate their foraging activity in the areas of highest *Upogebia* biomass. As adults return the following spring, less prey-rich areas are occupied sequentially as bird densities on the estuary increase (Turpie 1994). Similar patterns of sequential habitat occupancy, and of bird numbers tracking prey distribution patterns, have been found at the Berg River estuary. The densities of both Curlew Sandpipers, the most common species, and Grey Plovers, are correlated with the density and biomass, respectively, of their preferred prey, *Ceratonereis* polychaetes (Kalejta & Hockey 1994). Curlew Sandpipers are not strongly size-selective predators, are facultative tactile/visual foragers, and concentrate where prey are most abundant. Grey Plovers are strongly size-selective obligate visual predators and concentrate in areas where large polychaetes are common rather than in areas where absolute polychaete density is maximal. These differences in preferred foraging conditions are reflected in the birds' diets. The small *C. keiskamma* is eaten most often by Curlew Sandpipers and the larger *C. erythraeensis* by Grey Plovers (Kalejta & Hockey 1994).

At the Berg River estuary, migrant waders do not spread themselves evenly across the estuary upon their arrival, concentrating first at one or two mudflats and then spreading out to occupy the remainder as their numbers on the estuary increase. Migrant waders also have a profound impact on the distribution of resident waders at this estuary (Kalejta 1991). Resident waders occupy the most profitable sites (those favoured by returning migrants) during winter and early spring (when they also breed), and move to the least preferred mudflats as the densities of migrants rise during summer. Not only does the distribution of resident waders change during summer, but some 20% of them leave the estuary entirely, possibly as a result of competition with migrants (Velásquez *et al.* 1991). At this time of year the availability of alternative wetland habitats is at its lowest. At the Swartkops estuary, where *Upogebia* dominates the diets of most invertebrate feeders, a similar pattern of sequential mudflat occupancy is evident. Returning Whimbrels and Grey Plovers settle in the most prey-rich areas of the estuary, spreading to other areas as

numbers increase (Turpie & Hockey 1996). By midsummer there is a strong correlation between bird density and *Upogebia* biomass across the estuary.

Behavioural and energetic evidence for competition

Food limitation for estuarine birds may be manifest in the form of competition between them; such competition may arise through either exploitation or interference. Exploitative competition is best reflected by the birds in terms of changes in their energy intake rates and foraging effort. Interference competition, on the other hand, is more directly reflected in the form of overt aggressive behaviour.

The spatial requirements of foragers place an upper limit on their potential foraging densities, regardless of prey density. Tactile foragers, such as Curlew Sandpipers and flamingos, are able to form more cohesive flocks than visual foragers, such as plovers and herons. Flock-foraging is advantageous in terms of reduced individual predation risk and vigilance effort and also allows for efficient location of food patches (Abramson 1979; Myers 1980; Clark & Mangel 1984; Buchanan *et al.* 1988). Visual foragers, which need more foraging space in order to locate prey, relinquish these advantages in order to minimise interference from their neighbours. Both tactile and visual foragers are potentially subject to exploitative competition, but visual foragers are more susceptible to interference competition (Goss-Custard 1976), and consequently, are more aggressive than tactile foragers in defence of personal space (Burger *et al.* 1979). Whereas aggression is often suppressed amongst tactile foragers at high densities (Puttick 1981; Kalejta 1991), it is more pronounced among visual foragers, some of which also defend feeding territories in order to reduce the level of interference from conspecifics (Myers *et al.* 1979). The foraging behaviour of visual predators thus ultimately limits the impact they can have on their prey populations: in theory, they can become space-limited before becoming food-limited. For tactile foragers the reverse is likely to be true. Common Redshanks *Tringa totanus*, which are facultative visual/tactile foragers, switch from visual to tactile foraging when densities become high enough that interference competition depresses their prey capture rate (Goss-Custard 1976).

At the Berg River estuary the intake rates of Grey Plovers are negatively correlated with the density of conspecifics. The foraging success of tactilely-foraging Curlew Sandpipers, on the other hand, is positively correlated with their foraging density (Hockey *et al.* 1992), reflecting benefits accruing to tactile flock-foragers in locating and exploiting food-rich patches. However, the existence of a positive relationship between foraging success and density cannot be interpreted in terms of food limitation or otherwise. Rather, it implies that the principal prey items of Curlew Sandpipers are patchily distributed.

Both Grey Plovers and Whimbrels forage visually at the Swartkops River estuary. Seasonal changes in the energy intake rates of both species are negatively correlated with their foraging densities (Turpie 1994) and birds attempt to maximise their instantaneous energy intake rates by increasing foraging speed when intake rates fall. Both of these observations provide evidence for the existence of intraspecific interference competition. The availability of both prawns and crabs in the estuary is highest during the mid- to late summer months (Els 1982; Martin 1991), but it is at this time that bird densities are maximal and energy intake rates are low relative to those achieved in spring and early summer (Turpie 1994).

If the structure of estuarine bird communities is moulded by competition, it could be predicted that the role of interspecific competition would be greater (at the community level) than that of intraspecific competition. However, interspecific effects are more difficult to demonstrate empirically because of the increased number of variables (species) involved. At the Swartkops estuary, interspecific competition probably occurs between Grey Plovers and Whimbrels. The lines of evidence supporting this are firstly, that intraspecific competition certainly exists among both species (implying that food resources are limiting), and secondly, that the two species have almost identical prey resource spectra. This situation supports the predictions of Optimal Foraging Theory in which resource limitation should lead to niche expansion and hence, in a situation where prey species richness is low, to high niche overlap.

Although intertidally-foraging waterbirds are limited in their foraging time, it is not uncommon to see them roosting or loafing during the low tide period. For example, Whimbrels at the Swartkops estuary feed for a maximum of only 64% of low tide exposure time. This occurs because of a 'digestive bottleneck' (Zwarts & Dirksen 1990; Kersten & Visser 1996), whereby birds have to stop foraging in order to digest their stomach contents. Large birds, which eat larger prey, tend to forage for a smaller proportion of the available time than do small birds (Engelmoer et al. 1984).

Apart from the winter months, when few migrants are present, Grey Plovers at the Swartkops estuary defend fixed feeding territories against conspecifics. The establishment of territories in spring, as bird numbers increase, suggests that foraging space may be limiting for Grey Plovers, and provides further evidence for the existence of intraspecific interference competition. Furthermore, these territories shrink in size as the number of Grey Plovers on the estuary increases. At least one quarter of Grey Plovers do not establish foraging territories, and are forced into marginal foraging areas where they achieve a lower energy intake rate than do territory holders. This intake rate also decreases

more rapidly than that of territory holders during the summer (Turpie 1995a). Whimbrels may be subject to greater levels of intraspecific competition than Grey Plovers, as they rapidly abandon territorial behaviour when numbers on the estuary increase in late spring. Instead, Whimbrels engage in non-territorial encounters with foraging neighbours, the frequency of which increases over the summer season as their densities increase and their energy intake rates decrease.

Although there is, respectively, good and inferential evidence for intra- and interspecific competition occurring among estuarine waterbirds in South Africa, our understanding of the effects of this competition, particularly interspecific competition, is very limited. Even among the well-studied large waders of the Swartkops estuary, we still cannot, even *a posteriori*, use this information to explain community structure. The primary value of demonstrating the existence of such competition is the inference that food resources on these estuaries must be limiting, at least for invertebrate feeders, implying that populations of these birds are close to carrying capacity. An important caveat should be added here. Most detailed ornithological research in South African estuaries has been carried out at sites where bird densities are high. Some large estuaries, such as Knysna Lagoon, support relatively low densities of birds: research is needed as to why this should be so.

Failure to migrate – further evidence for food limitation?
The conclusion that estuarine bird populations in South Africa are close to carrying capacity, at least at some sites, is supported by an unusual phenomenon that occurred at Langebaan Lagoon in 1980. Adult Knots failed to migrate north to their Palearctic breeding grounds, despite having moulted into breeding plumage, and remained at Langebaan throughout the austral winter (Hockey & Douie 1995). The most plausible explanation for such a phenomenon is that the birds had failed to accumulate sufficient energy reserves to start the journey north. Knots are dietary specialists on their non-breeding grounds, eating mostly bivalve prey (Cramp & Simmons 1983), which may explain why other species were not affected in the same way. Birds with even minor injuries which may affect their foraging efficiency also fail to migrate: one adult Grey Plover which failed to migrate from the Swartkops estuary had a broken upper mandible, and, although it was able to feed, was probably not capable of gaining enough weight for migration (J.K.T., personal observation).

The juveniles of many long-distance migrants, despite having made their first southward migration successfully, do not migrate to the breeding grounds in their first year, remaining in South Africa during the winter and following summer (e.g. Martin & Baird 1988). Several studies have shown that the intake rates of juvenile waders are less than those of adults. In South Africa this has been demonstrated during the pre-migration period (March) for both Whimbrels (Turpie & Hockey 1993) and Curlew Sandpipers (Hockey et al. 1998). Juvenile Whimbrels move faster than adults while foraging and make more prey capture attempts per unit time. However, their energy intake per 24 h is only 74% that of adults. Among Curlew Sandpipers, foraging effort differed little between foraging adults and juveniles, but prey capture rates of adults were between 44% and 100% greater than those of juveniles, the discrepancy being maximal when prey density was high. Given these disparities, even if juveniles started fattening before adults, delays at stop-over sites en route may result in their arriving on the breeding grounds too late to complete the breeding cycle. If late arrival alone makes the probability of successful breeding by young birds very low, mortality during migration does not have to be high for the behaviour to be selected against. There are some exceptions to these generalisations: for example, Little Stints do migrate north (and breed) at one year old (Cramp & Simmons 1983). Among Dunlins Calidris alpina, juveniles which spend the non-breeding season close to the breeding grounds are more likely to return to the breeding grounds at one year old than are individuals faced with a longer migration (Soikkeli 1967; Cramp & Simmons 1983).

Competition and the broad-scale distribution of migrant waterbirds

The broadscale patterns of movement and distribution of migrants are thought by many to be shaped by competition for resources (e.g. Cox 1968, 1985; Von Haartman 1968; Pienkowski & Evans 1985). This is particularly evident in the case of ecologically similar species pairs, such as Curlew Sandpiper and Dunlin, which segregate on both their breeding and non-breeding grounds, often forming a leap-frog pattern whereby the more northerly breeder migrates further south. Rather more enigmatic is the variability in migration distance that occurs within species: many species which breed in the high latitudes of the northern hemisphere migrate to non-breeding sites as distant from one another as southwestern Europe and South Africa. Why some birds should undertake journeys thousands of kilometres longer than those of their conspecifics remains a subject of much debate. Some argue that the birds wintering furthest to the south are competitively inferior individuals which have been unable to remain at sites closer to the breeding grounds (Pienkowski & Evans 1985). Long-term data on year-to-year variation in wader numbers at Langebaan Lagoon do not support the hypothesis that southern estuaries act as 'overflow' areas (Underhill 1987a). Furthermore, there is no behavioural evidence to suggest that migrants compete more intensely at northerly non-breeding sites – waders appear to be as

aggressive at the southern extremes of their ranges as in the north. Increasing evidence points to the fact that the relatively high risk in migrating longer distances may be balanced by advantages which increase survivorship on the non-breeding grounds.

Higher ambient temperatures and lower levels of predation by birds of prey are two important advantages of migrating further south. Many estuarine waterbirds die of starvation during the cold winter months in the northern hemisphere (Meininger et al. 1991), and large numbers fall prey to raptors and owls which converge on these estuaries during winter (Whitfield et al. 1986). Recent findings also suggest that the energy intake rates of waders are higher in the tropics and at the southern extremes of their ranges than in the northern hemisphere (Turpie 1994) even though their daily energy demands are lower than those of birds remaining at north temperate latitudes at this time of year because of higher ambient temperatures. Summers et al. (1989) calculated that the annual food consumption of Turnstones breeding in Siberia and migrating to South Africa was 10% less than that of Turnstones breeding in Greenland and migrating to western Europe. Considering all of the above, waterbirds which migrate to South Africa are probably as well off, in terms of fitness, as their conspecifics which undertake lower risk migrations to higher risk non-breeding sites further north. This scenario suggests a form of the 'ideal free distribution' (sensu Fretwell & Lucas 1969), whereby equilibrium is maintained by broadscale competition for resources, rather than an unstable system governed by latitudinally-determined competition intensity. This supports the contention of Hockey et al. (1992) that non-breeding wader populations are close to estuarine carrying capacity over a wide latitudinal range, which in turn has important consequences for their conservation.

Conservation of estuarine waterbirds

South Africa's role in the conservation of estuarine waterbirds

The level of endemism to South Africa amongst estuarine waterbird species is low, with only 10 species being endemic or near-endemic to southern Africa (Appendix 10.1). Moreover, none of these ten species is highly dependent on estuaries. This phenomenon can be attributed to the widespread availability of estuarine habitats on a global scale in contrast to, for example, the Karoo biome, where the level of bird species endemism is high (Clancey 1986). However, South Africa's estuarine avifauna includes 22 of the 102 mainland-breeding Red Data Book species, several of which are highly dependent on estuaries (Brooke 1984; Appendix 10.1). This large number of Red Data species occurring at South African estuaries is not a pattern unique to the region: it is a global phenomenon which has been brought about by the steady loss of wetland habitats in general (Finlayson & Moser 1991).

Wetland degradation worldwide has led to population decreases of many waterfowl species, particularly over the past few decades (e.g. Myers 1983; Myers *et al.* 1987). The rate at which coastal wetlands are lost is directly related to human population pressure; in California, over two-thirds of the coastal wetlands have been developed (Myers *et al.* 1987), and in 1984 in South Korea, some 345 500 ha of intertidal land were scheduled for reclamation (Piersma 1985). In other instances, decreases in waterbird populations have been brought about from less obvious causes than reclamation and development, such as the spread of invasive cord grass *Spartina anglica* in British estuaries (Goss-Custard & Moser 1988).

Historical data against which to assess population changes at South African estuaries are few. Of 16 wader species in southern Africa with predominantly coastal distributions, two (Ringed Plover and Curlew) are known to have decreased in abundance this century. The conservation status of inland waders (including terrestrial species) is less encouraging: 12 out of 32 species are known to have decreased in abundance this century (Hockey & Douie 1995). While changes in the status of inland species can be largely attributed to human modification of the environment, changes in the status of coastal species until now have been mostly a consequence of changes elsewhere within their range, such as loss or degradation of breeding habitat, rather than due to degradation of local estuarine habitats. There have been some documented changes in patterns of distribution and abundance of resident estuarine species other than waders. During this century Great White Egrets have extended their range westwards, as have African Spoonbills and Glossy Ibises (Hockey *et al.* 1989; Martin 1991). In these cases, the spread is thought to be due primarily to the proliferation of artificial water bodies inland, rather than to any intrinsic change in the nature of estuaries (Hockey *et al.* 1989). In the Western Cape alone, 31 estuarine bird species are thought to have increased in abundance for this reason. By contrast, only ten estuarine species are thought to have decreased in abundance in the same region (Hockey *et al.* 1989). In the Eastern Cape, Pygmy Geese and Yellowbilled Storks have decreased in abundance. Reduction in numbers of the former was due to the draining of freshwater vleis adjacent to the Swartkops estuary (Martin & Baird 1987). The reason for the decrease in Yellowbilled Stork numbers is unknown, but human disturbance is a likely cause. In South Africa, one estuarine bird species, the African Skimmer, is known to have become nationally extinct during the twentieth century (Brooke 1984).

The rate of degradation of estuarine habitats has led northern hemisphere researchers to investigate the resilience of waterbird populations to these changes. The key question that needs to be answered in this regard is whether estuarine bird populations are at carrying capacity. Consider the arguments presented in the preceding section. If estuarine birds are food-limited at some times of year in South Africa, implying that their numbers approach the carrying capacity of local estuaries, this means that the loss of habitat will almost certainly lead to a loss of birds. These threats are not restricted to South Africa but are believed to exist along the entire east Atlantic seaboard (e.g. Goss-Custard 1985; Smit & Piersma 1989).

A significant proportion of the waterbird numbers at South African estuaries in summer are Palearctic-breeding migrant waders. Whereas the importance of conserving local Red Data and endemic species is clear, the importance of South Africa's role in the conservation of non-breeding migrants needs to be clarified. Despite their proportionally large contribution to South Africa's estuarine avifauna, the numbers of migrant waders at South African estuaries are small in comparison to the numbers that inhabit the expansive coastal wetlands of the northern hemisphere. In Europe, estuaries and other coastal wetlands which are orders of magnitude larger than South Africa's largest coastal wetlands are not uncommon, and at least 38 sites along the Atlantic coast of Europe support over 20 000 waders (Smit & Piersma 1989). Even these sites pale by comparison with several sites along the West African coast, the most impressive of which, the Banc d'Arguin in Mauritania, supports over 2 million waders (Smit & Piersma 1989). Of the migratory waders which visit South Africa, populations of only two species, Curlew Sandpiper and Little Stint, exceed 5% of either the combined population of western Europe and the West African coast or the West African populations alone (data from Cooper & Hockey 1981; Hagemeijer & Marteijn 1993; Rose & Scott 1994).

These small proportions may raise some doubts about the importance of directing conservation measures at these migrants in South Africa, particularly in the light of the 1% criterion of the Ramsar Convention, whereby individual sites are considered to be of importance for waders when they contain more than 1% of a flyway population of any one species (Smit & Piersma 1989). Very few sites in South Africa would qualify as being of international importance on these grounds. Furthermore, if migrants reaching South Africa are competitively inferior individuals as has been suggested (e.g. Pienkowski & Evans 1985), then, for the most part, populations visiting South Africa become even less important in an international context. If, on the other hand, these birds are as 'fit' as their more northerly counterparts, as argued in the preceding section, then their conservation is justified purely in terms of maintaining intraspecific genetic diversity.

The low numbers of migratory birds found in South African estuaries compared with further north can be explained by the relatively limited availability of estuarine habitat. Despite their low total numbers, however, the densities of waterbirds on South African estuaries are higher than those at sites further north. Densities of migrant waders on the intertidal flats of the Berg River estuary exceed those recorded at any other coastal wetland in the east Atlantic (Hockey et al. 1992), and the foraging densities of Whimbrels at the Swartkops estuary are higher than recorded anywhere else in their Afrotropical non-breeding range (Turpie 1994). This pattern can be explained by the fact that peak numbers of migrants in South Africa coincide with the summer production period of their invertebrate prey, whereas birds migrating to non-breeding sites in the northern hemisphere encounter the opposite conditions (Hockey et al. 1992). Even given equal production, sites with synchronised production and predation peaks have a higher carrying capacity than do estuaries in which production and predation are asynchronous (Hockey et al. 1992). The carrying capacity per unit area of estuaries thus increases from north to south for boreal-breeding migrants. Consequently, and of considerable importance in conservation terms, many more Palearctic migrants can be conserved per unit area at estuaries in South Africa than further north. An important corollary to this is that the loss of even small areas of estuarine habitat in South Africa will impact a greater number of birds than loss of an equivalent area further north.

The effective conservation of migrants depends on the existence of reserve networks which incorporate a variety of breeding, non-breeding and staging sites. Because many birds undertake such long-distance migrations, reserve networks can only be developed through international cooperation. At present, no recognised migrant reserve network for wetland birds exists along the Palearctic–Afrotropical flyways as exists along the Nearctic–Neotropical migration routes (Hunter et al. 1991). However, progress towards this ideal has recently been made. In June 1995, 54 of 64 range states signed the final act of the African–Eurasian Agreement on the Conservation of Migratory Waterbirds (the Bonn Convention: Moser & van Vessem 1995).

Top priority estuaries and a core network for waterbirds

Opportunities for conservation are limited, and necessitate the identification of priority areas and of core networks for the minimal representation of South African estuarine biodiversity. The majority of South African estuaries are small and only 42 estuaries (including Langebaan Lagoon) support more than 500 non-passerine waterbirds during summer. Six South African estuaries support more than 20000 non-passerine waterbirds, with numbers at

Table 10.9. The top ten South African estuaries for waterbirds, selected on the basis of ranked values for each of four single-criteria indices (species richness – SR, total number of birds – TOT, conservation value index – CVI and conservation status index – CSI) for a total of 42 estuaries. Ranks for each index are given in italics

Estuary	SR		TOT		CVI[a]		CSI[b]	
1 St Lucia	62	3	38307	2	1393	1	155	2
2 Berg	63	2	45594	1	744	2	143	3
3 Richard's Bay	68	1	12732	11	337	3	161	1
4 Langebaan	43	7	37330	3	329	4	100	8
5 Orange	48	6	21514	6	239	6	126	4
6 Olifants	51	5	15561	8	124	10	116	7
7 Rietvlei	51	5	13414	9	146	8	122	5
8 Verlorenvlei	54	4	6760	13	171	7	122	6
9 Bot River	43	7	23471	5	118	12	95	10
10 Wilderness	43	7	13359	10	267	5	81	12

[a] Calculated as the sum of the percentage of the total South African coastal population of each species at each site.
[b] Calculated by assigning each species a score, ranging from lowest-scoring nonbreeding migrant species to highest scoring Red Data species.
Source: From Turpie (1995b).

Langebaan Lagoon, Lake St Lucia and the Berg River estuary exceeding 35000 (Turpie 1995b). In designing a reserve network for the conservation of South Africa's estuarine biota, this somewhat skewed distribution of bird numbers on South African estuaries facilitates the incorporation of a large proportion of the estuarine avifauna within relatively few sites. Nevertheless, the choice of a core reserve network for birds requires careful consideration of more than bird numbers alone.

Turpie (1995b) prioritised South African estuaries (including Langebaan Lagoon) for waterbirds, based on summer counts of 88 non-passerine species for which reliable count data exist from around the South African coast. The 42 estuaries with more than 500 birds were ranked in terms of their total bird numbers, species richness, an index of conservation value, based on rarity, and an index of species conservation status. Rather than combining these criteria in a complex index, the four separate series of rankings were integrated subjectively in order to produce a final ranking of the ten most important sites for estuarine waterbirds (Table 10.9). It is important to note, however, that this list is dominated by west coast sites, and that a prioritisation exercise does not take into account the fact that two or more top-ranking sites may support similar waterbird communities. A more representative reserve network may be achieved by conserving top sites within biogeographical zones (e.g. Hockey & Branch 1994). However, this approach would be complicated for estuarine waterbirds in South Africa because of the uncertainty as to whether much of the south coast can be recognised as a discrete biogeographic unit.

Turpie (1995b) circumvented this problem by using an iterative technique known as 'complementarity' analysis,

Table 10.10. A set of sites, identified by complementarity analysis, required for the conservation of 87 estuarine waterbird species in South Africa, based on complementarity analysis

	Number of species targeted for conservation	Cumulative number of species conserved
1 St. Lucia [a]	41	41
2 Langebaan [a]	16	57
3 Wilderness [a]	11	68
4 Berg [a]	5	73
5 Orange [a]	3	76
6 Verlorenvlei [a]	5	81
7 Richards Bay [a]	3	84
8 Kleinrivier	2	86
9 Swartkops	1	87

[a] Estuaries already afforded some level of protection.
Source: From Turpie (1995*b*).

which selects a set of sites that conserves all species. This technique has been widely used in reserve selection procedures, although it has been pointed out by Underhill (1994) that these algorithms can yield suboptimal results. In Turpie's (1995*b*) study, a total of nine sites were selected (refined *a posteriori* using the results of the prioritisation procedure) which conserved 87 of the 88 species in at least one site at which they met certain abundance criteria (Table 10.10). It is noteworthy that the three most important sites in this core reserve network are geographically disparate, with one each on the west, south and east coasts. Six of these nine estuaries/lagoons are already afforded some degree of protection, two at the level of a National Park (Table 10.10), whilst the Berg River, Kleinrivier and Swartkops River estuaries are in urgent need of official conservation status and action in order to establish a core network for estuarine waterbirds in South Africa. It is important to realise, however, that the mere designation of an estuary as a nature reserve or equivalent is inadequate to ensure its future conservation. Land use and water management in the catchment are key determinants of the ecological well being of estuaries; estuary management and conservation thus require an holistic approach. The key value in undertaking an analysis such as that of Turpie (1995*b*) is to identify those estuaries for which effective conservation management of the catchment is of vital importance in conserving South Africa's avian diversity.

rectly, influence sediment stability (Daborn *et al.* 1993). Major roles in energy cycling are played by those species which are part of aquatic, rather than terrestrial, food chains. Regardless of the trophic level at which birds operate as consumers, they target the most abundant prey species and thus probably play a role in maintaining the diversity of other taxa through the removal of competitive dominants. There are distributional, behavioural and energetic lines of evidence supporting the hypothesis that (at least) invertebrate-feeders are close to carrying capacity in South African estuaries. The greatest energy flow through birds is through these invertebrate-feeders and they therefore play integral roles in estuarine community energetics both as an energy sink and in recycling about 25% of the energy they consume.

Estuarine bird species richness increases from the cool temperate west to the subtropical east, but energy turnover through birds decreases along the same axis, suggesting, as on rocky shores, that estuarine productivity is maximal in the west.

Most South African estuaries are small, and only 42 support 500 or more non-passerines in summer. The greatest numbers are found at large estuaries/lagoons with extensive areas of shallow water and mudflats/saltmarshes. It is at these sites, such as Langebaan Lagoon and the Swartkops estuary, where the proportional role of birds in estuarine communities is thought to be greatest.

Relative to coastal wetlands further north in the east Atlantic, the carrying capacity of South African estuaries for invertebrate-feeders is high, reflected in high densities of foraging birds. From a conservation perspective, therefore, many more birds can be conserved per unit estuary area in South Africa than further north.

Most of the top-ranked estuaries for birds in South Africa have some form of legal protection. The key omission from this network of reserves is the Berg River estuary in the Western Cape. This site supports more non-passerine waterbirds than any other estuary/coastal lagoon in South Africa and, on all criteria examined, ranks as one of the top four most important estuaries for birds in the country. Apart from a current application for registration as a Ramsar site, the Berg River estuary has no official conservation status. It is also imminently threatened by both upstream water abstraction and development in its lower reaches.

Conclusions: the roles and status of birds in South African estuaries

Birds contribute to the biological functioning of estuaries both as consumers and as recyclers, and can even, indi-

Acknowledgements

We thank George Branch, Peter Evans, Paul Martin, Roy Siegfried and Les Underhill for comments on an earlier draft.

Appendix 10.1. Estuarine bird species of South Africa (names from Hockey 1994); mean mass (Dunning 1993, Maclean 1993); level of estuarine dependence; migratory status; summer totals from counts (Hockey 1993; Martin & Baird 1987; Ryan & Cooper 1985; Ryan et al. 1986, 1988; Tree & Martin; 1993; Underhill & Cooper 1984); diet and habitat (Maclean 1993); functional use of estuaries. Dependence (Dep): 1, not dependent; 2, semi-dependent; 3, highly dependent. Migratory status (Mig): R, resident, I, intra-African migrant, M, intercontinental migrant. Diet (X, main dietary category; x, minor dietary category): A, aquatic vegetation, B, aquatic invertebrates, C, fish and/or amphibians, D = terrestrial seeds or fruit, E = terrestrial/aerial invertebrates, F = other vertebrates/carrion; Habitat: W = open water, M = mudflats, F = floodplain, S = saltmarsh, E = tall emergent vegetation; Use: Br = breeding, F = feeding, R = roosting. + = endemic/near-endemic to southern Africa; * = Red Data Book species (Brooke 1984).

		Weight (g)	Dep	Mig	Summer total[a]	Diet						Habitat					Use		
						A	B	C	D	E	F	W	M	F	S	E	Br	F	R
Order Podicipediformes																			
Family Podicipedidae																			
Great Crested Grebe	Podiceps cristatus	595	2	R,I	319	x	x	X				x					x		
Blacknecked Grebe	P. nigricollis	298	2	R,I	2975		x	X										x	
Dabchick	Tachybaptus ruficollis	147	1	R	543		x	X				x					x		
Order Pelecaniformes																			
Family Pelecanidae																			
Eastern White Pelican	Pelecanus onocrotalus	9500	2	R,I	1024		x	X				x					x	x	
Pinkbacked Pelican*	P. rufescens	5450	3	R	177		x	X				x					x	x	x
Family Phalacrocoracidae																			
Whitebreasted Cormorant	Phalacrocorax carbo	1616	1	R	1788		x	X				x		x			x	x	x
Cape Cormorant+	P. capensis	1230	1	R	8182		x	X				x						x	x
Reed Cormorant	P. africanus	556	1	R	1637		x	X				x		x			x	x	x
Crowned Cormorant+	P. coronatus	760	1	R	3		x	X			x	x						x	x
Family Anhingidae																			
Darter	Anhinga melanogaster	1508	1	R	470		x	X				x					x	x	x
Order Ciconiiformes																			
Family Ardeidae																			
Grey Heron	Ardea cinerea	1450	1	R	608		x	X		x	x	x	x	x	x	x	x	x	x
Blackheaded Heron	A. melanocephala	1135	1	R	148		x	X		X	x		x	x	x	x	x	x	x
Goliath Heron	A. goliath	4328	2	R	119		x	X		x	x	x	x	x	x	x	x	x	x
Purple Heron	A. purpurea	920	2	R	95			X		x	x		x	x	x	x	x	x	x
Great White Egret	Casmerodius albus	1110	2	R	374		x	X		x	x	x	x	x	x	x	x	x	x
Little Egret	Egretta garzetta	500	2	R	1047		x	X	x	x	x	x	x	x	x	x	x	x	x
Yellowbilled Egret	E. intermedia	500	1	R,I	35			X	x	x			x		x	x	x	x	x
Black Egret	E. ardesiaca	313	1	R	25		x	X					x			x	x	x	x
Cattle Egret	Bubulcus ibis	345	1	R	616		x	x	x	x				x		x	x	x	x
Squacco Heron	Ardeola ralloides	299	1	R	10		x	x	x	x			x	x		x	x	x	x
Greenbacked Heron	Butorides striatus	214	2	R	8		x	x	x	x			x	x		x	x	x	x
Rufousbellied Heron*	B. rufiventris	?	1	–	1			x		x				x		x		x	x
Blackcrowned Night Heron	Nycticorax nycticorax	535	1	R	70		x	x	x	x			x	x	x	x	x	x	x
Whitebacked Night Heron*	Gorsachius leuconotus	?	2	R	–			x	x	x			x	x		x	x	x	x
Little Bittern*	Ixobrychus minutus	110	2	R	3		x	x	x	x						x	x	x	x
Dwarf Bittern*	I. sturmii	142	1	R	—			x		x						x		x	x
Bittern*	Botaurus stellaris	1231	1	R,I	—		x	x	x	x						x		x	x
Family Scopidae																			
Hamerkop	Scopus umbretta	422	1	R	43		x	X		x	x		x	x		x	x	x	x
Family Ciconiidae																			
Black Stork*	Ciconia nigra	3000	2	R	4		x	X		x	x		x	x				x	x
Woollynecked Stork*	C. episcopus	2185	2	R	3		x	x		x	x		x	x					x
Openbilled Stork*	Anastomus lamelligerus	1151	1	–	—		X	x		x	x			x				x	x
Saddlebilled Stork*	Ephippiorhynchus senegalensis	6163	1	R	11		x	X	x	x	x	x	x	x			x	x	x
Yellowbilled Stork*	Mycteria ibis	2167	2	I	222		x	X	x	x	x		x	x			x	x	x

	Weight (g)	Dep	Mig	Summer total[a]	Diet A	B	C	D	E	F	Habitat W	M	F	S	E	Use Br	F	R
Order Phoenicopteriformes																		
Family Plataleidae																		
Sacred Ibis — *Threskiornis aethiopicus*	1498	1	R	395		x	x	x	x	x	x	x	x	x	x	x	x	x
Glossy Ibis — *Plegadis falcinellus*	634	1	R,I	168		x	x	x	x			x	x	x	x	x	x	x
African Spoonbill — *Platalea alba*	1604	2	R,I	977		x	x			x	x	x	x	x	x	x	x	x
Family Phoenicopteridae																		
Greater Flamingo* — *Phoenicopterus ruber*	2714	3	I	22 015	x						x	x	x	x		x		x
Lesser Flamingo* — *P. minor*	1625	3	I	7 735	x	x					x	x	x	x			x	x
Order Anseriformes																		
Family Anatidae																		
Whitefaced Duck — *Dendrocygna viduata*	739	1	R,I	378		x					x		x			x	x	x
Fulvous Duck — *D. bicolor*	728	1	R,I	54		x		x	x		x		x			x	x	x
Egyptian Goose — *Alopochen aegyptiacus*	2110	1	R	2 103	x	x		x			x	x	x	x		x	x	x
South African Shelduck+ — *Tadorna cana*	1236	2	I	1 316	x			x	x		x	x	x			x	x	x
Yellowbilled Duck — *Anas undulata*	894	2	R,I	5 852	x	x		x	x		x	x	x			x	x	x
Cape Teal — *A. capensis*	400	2	R	2 010	x	x	x				x		x	x		x	x	x
Hottentot Teal — *A. hottentotta*	291	1	R	311	x	x					x		x	x		x	x	x
Redbilled Teal — *A. erythrorhyncha*	568	2	R,I	2 179	x	x	x		x		x		x	x	x	x	x	x
Cape Shoveller+ — *A. smithii*	615	2	R,I	3 032	x	x	x		x			x	x	x	x	x	x	x
Pygmy Goose* — *Nettapus auritus*	277	2	R	8	x	x					x		x			x	x	x
Knobbilled Duck+ — *Sarkidiornis melanotos*	1661	1	R,I	—	x			x			x		x			x	x	x
Spurwinged Goose — *Plectropterus gambensis*	6250	1	R,I	513	x	x		x			x	x	x			x	x	x
Order Falconiformes																		
Family Accipitridae																		
Yellowbilled Kite — *Milvus parasitus*	667	1	I	—						x	x	x	x				x	x
Palmnut Vulture* — *Gypohierax angolensis*	1609	3	I	—			x	x		x	x	x	x				x	x
African Fish Eagle — *Haliaeetus vocifer*	2821	2	R	108			x		x	x	x		x	x		x	x	x
African Marsh Harrier — *Circus ranivorus*	518	2	R	42			x		x	x		x	x			x	x	x
Black Harrier+ — *C. maurus*	?	1	R	—			x		x	x		x	x			x	x	x
Family Pandionidae																		
Osprey — *Pandion haliaetus*	1486	2	I,M	10			x			x	x		x	x			x	x
Family Falconidae																		
Peregrine Falcon — *Falco peregrinus*	594	1	R,M	—					x	x						x	x	x
Lanner Falcon — *F. biarmicus*	587	1	R,I	—					x	x						x	x	x
Order Gruiformes																		
Family Gruidae																		
Southern Crowned Crane — *Balearica regulorum*	3600	1	R	6			x	x	x				x					x
Family Rallidae																		
African Rail — *Rallus caerulescens*	164	1	R	4	x	x	x	x	x				x	x	x	x	x	x
Black Crake — *Amaurornis flavirostris*	91	1	R	1	x	x	x		x			x	x	x	x	x	x	x
Redchested Flufftail — *Sarothrura rufa*	32	1	R	3	x	x							x			x	x	x
Purple Gallinule — *Porphyrio porphyrio*	543	1	R	68	x	x	x		x	x		x	x	x	x	x	x	x
Moorhen — *Gallinula chloropus*	247	1	R	68	x	x	x		x		x		x	x	x	x	x	x
Redknobbed Coot — *Fulica cristata*	738	2	R,I	46 172	x	x	x	x	x		x		x	x	x	x	x	x
Family Heliornithidae																		

Order Charadriiformes

	Scientific name			Status	Count	Distribution columns (presence ×)
Family Jacanidae						
African Jacana	*Actophilornis africanus*	189	1	R,I	2	× × (… × × × × × ×)
Lesser Jacana*	*Microparra capensis*	41	1	R,I	—	× × (… × × × × × ×)
Family Rostratulidae						
Painted Snipe	*Rostratula benghalensis*	118	1	R	2	× (… × × × × ×)
Family Haematopodidae						
African Black Oystercatcher[+]	*Haematopus moquini*	699	1	R	263	× (… × × × × × × ×)
Family Charadriidae						
Ringed Plover	*Charadrius hiaticula*	64	2	M	4037	× × … × ×
Whitefronted Plover	*C. marginatus*	49	1	R	1536	× × … × × ×
Chestnutbanded Plover*	*C. pallidus*	35	1	R,I	423	× × … × × ×
Kittlitz's Plover	*C. pecuarius*	43	1	R,I	1532	× × … × × × ×
Threebanded Plover	*C. tricollaris*	34	1	R	373	× × … × × × ×
Mongolian Plover	*C. mongolus*	57	3	M	152	× × … × ×
Sand Plover	*C. leschenaultii*	106	3	M	36	× × … × ×
Grey Plover	*Pluvialis squatarola*	216	3	M	6734	× × … × × × ×
Blacksmith Plover	*Vanellus armatus*	158	1	R	1247	× × … × × × ×
Wattled Plover	*V. senegallus*	237	1	R	15	× × … × ×
Family Scolopacidae						
Turnstone	*Arenaria interpres*	99	2	M	2500	× × … × ×
Terek Sandpiper	*Xenus cinereus*	74	3	M	563	× × … × × ×
Common Sandpiper	*Actitis hypoleucos*	44	1	M	1246	× × … × × ×
Wood Sandpiper	*Tringa glareola*	60	1	M	690	× × … × × × ×
Marsh Sandpiper	*T. stagnatilis*	75	2	M	802	× × … × × × ×
Greenshank	*T. nebularia*	191	2	M	3476	× × … × × × ×
Knot	*Calidris canutus*	133	3	M	3252	× × … × × × ×
Curlew Sandpiper	*C. ferruginea*	57	2	M	74516	× × … × × × ×
Little Stint	*C. minuta*	24	2	M	20220	× × … × × × ×
Sanderling	*C. alba*	56	1	M	5147	× × … × × ×
Ruff	*Philomachus pugnax*	142	1	M	10380	× × … × × × ×
Ethiopian Snipe	*Gallinago nigripennis*	110	1	—	122	× × … × × × ×
Bartailed Godwit	*Limosa lapponica*	292	3	M	346	× × … × × ×
Curlew	*Numenius arquata*	725	3	M	331	× × … × × × ×
Whimbrel	*N. phaeopus*	422	2	M	1981	× × … × × × ×
Family Recurvirostridae						
Avocet	*Recurvirostra avosetta*	319	1	R,I	5782	× × … × × × ×
Blackwinged Stilt	*Himantopus himantopus*	161	1	R	1870	× × … × × × ×
Family Burhinidae						
Water Dikkop	*Burhinus vermiculatus*	304	1	R	108	× × … ×
Family Glareolidae						
Redwinged Pratincole*	*Glareola pratincola*	66	2	I	131	× … × ×
Family Laridae						
Kelp Gull	*Larus dominicanus*	924	1	R	5739	× × … × × × ×
Greyheaded Gull	*L. cirrocephalus*	280	1	R	2210	× × … × × × ×
Hartlaub's Gull[+]	*L. hartlaubii*	292	1	R	4871	× × … × × × ×
Caspian Tern*	*Hydroprogne caspia*	655	3	R,I	437	× × … × × × ×
Swift Tern	*Sterna bergii*	345	1	—	1987	× × … × × × ×
Lesser Crested Tern	*S. bengalensis*	200	2	—	176	× … × × × ×
Sandwich Tern	*S. sandvicensis*	267	2	M	2914	× × … × × × ×
Common Tern	*S. hirundo*	124	2	M	55339	× × … × × × ×
Damara Tern*[+]	*S. balaenarum*	60	2	—	53	× × … × × × ×
Little Tern	*S. albifrons*	57	3	M	1133	× × … × × × ×
Whiskered Tern	*Chlidonias hybridus*	72	1	—	28	× × … × ×
Whitewinged Tern	*C. leucopterus*	57	1	M	670	× × … ×

	Weight (g)	Dep	Mig	Summer total*	Diet A	Diet B	Diet C	Diet D	Diet E	Diet F	Hab W	Hab M	Hab F	Hab S	Hab E	Use Br	Use F	Use R
Order Cuculiformes																		
Family Cuculidae																		
Burchell's Coucal _Centropus burchellii_	170	1	R	—			x		X	x						x	x	x
Order Strigiformes																		
Family Strubidae																		
Marsh Owl _Asio capensis_	315	1	R	—			x		x	X					x	x	x	x
Order Caprimulgiformes																		
Family Caprimulgidae																		
Natal Nightjar _Caprimulgus natalensis_	87	2	R	—					X		x		x			x	x	x
Mozambique Nightjar _C. fossii_	60	1	R,I	—					X		x		x			x	x	x
Order Alcediniformes																		
Family Alcedinidae																		
Pied Kingfisher _Ceryle rudis_	71	2	R	506	x	x	x		x		x		x	x		x	x	x
Giant Kingfisher _Megaceryle maxima_	363	1	R	47	x	x	x				x		x	x		x	x	x
Halfcollared Kingfisher _Alcedo semitorquata_	39	1	R	1	x	x	x		x		x				x		x	x
Malachite Kingfisher _A. cristata_	18	1	R	28	x	x	x		x		x		x		x	x	x	x
Mangrove Kingfisher _Halcyon senegaloides*_	62	3	—	—		x	x		x	x		x			x		x	x
Order Passeriformes																		
Family Hirundinidae																		
European Swallow _Hirundo rustica_	18	1	M	—					X		x	x	x	x	x		x	x
Whitethroated Swallow _H. albigularis_	21	1	—	—					X		x	x	x	x	x	x	x	x
Wiretailed Swallow _H. smithii_	13	1	R	—					X		x		x	x	x		x	x
Greater Striped Swallow _H. cucullata_	27	1	—	—					X		x	x	x		x		x	x
Lesser Striped Swallow _H. abyssinica_	17	1	—	—					X		x	x	x		x		x	x
Greyrumped Swallow _Pseudhirundo griseopyga_	10	1	R,I	—					X		x		x		x	x	x	x
Sand Martin _Riparia riparia_	14	1	M	—					X		x		x		x		x	x
Brownthroated Martin _R. paludicola_	13	1	R,I	—					X		x		x	x	x		x	x
Banded Martin _R. cincta_	25	1	—	—					X		x		x	x	x		x	x
Family Sylviidae																		
African Marsh Warbler _Acrocephalus baeticatus_	10	1	—	—					X						x		x	x
Cape Reed Warbler _A. gracilirostris_	15	1	R	—	x				X				x		x		x	x
African Sedge Warbler _Bradypterus baboecala_	14	1	R	—			x		X				x	x			x	x
Broadtailed Warbler _Schoenicola brevirostris*_	17	1	R	—					X						x		x	x
Grassbird _Sphenoeacus afer_	31	1	R	—					X				x	x			x	x
Fantailed Cisticola _Cisticola juncidis_	9	1	R	—					X				x	x			x	x
Redfaced Cisticola _C. erythrops_	15	1	R	—					X				x		x		x	x
Levaillant's Cisticola _C. tinniens_	11	1	R	—					X				x		x		x	x
Family Motacillidae																		
African Pied Wagtail _Motacilla aguimp_	27	1	R	59					X			x	x			x	x	x
Cape Wagtail _M. capensis_	21	2	R	1235		x	x		X			x	x			x	x	x
Yellow Wagtail _M. flava_	17	1	M	—					X			x	x				x	x
Orangethroated Longclaw _Macronyx capensis+_	46	1	R	—					X				x			x	x	x
Pinkthroated Longclaw _M. ameliae*_	?	3	R	—				x	X				x			x	x	x

Family / Common name	Scientific name	Count		Status	
Family Sturnidae					
European Starling	*Sturnus vulgaris*	76	1	R	—
Indian Myna	*Acridotheres tristis*	110	1	R	—
Pied Starling	*Spreo bicolor*	99	1	R	—
Wattled Starling	*Creatophora cinerea*	68	1	I	—
Family Ploceidae					
Thickbilled Weaver	*Amblyospiza albifrons*	44	1	R	—
Spectacled Weaver	*Ploceus ocularis*	25	1	R	—
Spottedbacked Weaver	*P. cucullatus*	35	1	R,I	—
Cape Weaver	*P. capensis*	43	1	R,I	—
Southern Masked Weaver	*P. velatus*	27	1	R,I	—
Lesser Masked Weaver	*P. intermedius*	21	1	R	—
Golden Weaver	*P. xanthops*	41	1	R	—
Yellow Weaver	*P. subaureus*	27	1	R	—
Southern Brownthroated Weaver	*P. xanthopterus*	22	1	R	—
Red Bishop	*Euplectes orix*	22	1	R,I	—
Yellowrumped Widow	*E. capensis*	31	1	R,I	—
Redshouldered Widow	*E. axillaris*	26	1	R,I	—
Whitewinged Widow	*E. albonotatus*	19	1	R,I	—
Family Estrildidae					
Common Waxbill	*Estrilda astrild*	8	1	R	—
Family Viduidae					
Pintailed Whydah	*Vidua macroura*	15	1	R,I	—

a Total summer counts are minimum population estimates only.

References

Abramson, M. (1979). Vigilance as a factor influencing flock formation among Curlews *Numenius arquata*. *Ibis*, **121**, 213–16.

Baird, D., Evans, P.R, Milne, H. & Pienkowski, M.W. (1985). Utilization by shorebirds of benthic invertebrate production in intertidal areas. *Oceanography and Marine Biology Annual Review*, **23**, 573–97.

Bally, R., McQuaid, C.D. & Pierce, S.M. (1985). Primary production of the Bot River estuary, South Africa. *Transactions of the Royal Society of South Africa*, **45**, 333–45.

Bennett, B.A. (1994). The fish community of the Berg River estuary and an assessment of the likely effects of reduced freshwater inflows. *South African Journal of Zoology*, **29**, 118–25.

Berruti, A. (1980). Status and review of waterbirds breeding at Lake St Lucia. *Lammergeyer*, **28**, 1–19.

Berruti, A. (1983). The biomass, energy consumption and breeding of waterbirds relative to hydrological conditions at Lake St Lucia. *Ostrich*, **54**, 65–82.

Blaber, S.J.M. (1973). Population size and mortality of the marine teleost *Rhabdosargus holubi* (Pisces: Sparidae) in a closed estuary. *Marine Biology*, **21**, 219–25.

Brooke, R.K. (1984). *South African Red Data Book – Birds*. South African National Science Programmes Report No. 97. Pretoria: CSIR.

Brown, A.C. & Jarman, N. (1978). Coastal marine habitats. In *Biogeography and ecology of southern Africa*, ed. M.J.A. Werger, pp. 1239–77. The Hague: Dr W. Junk.

Bryant, D.M. (1979). Effects of prey density and site character on estuary usage by overwintering waders. *Estuarine, Coastal and Marine Science*, **9**, 369–84.

Buchanan, J.B., Schick, C.T., Brennan, L.A. & Herman, S.G. (1988). Merlin predation on wintering Dunlins: hunting success and Dunlin escape tactics. *Wilson Bulletin*, **100**, 108–18.

Burger, J., Hahn, D.C. & Chase, J. (1979). Aggressive interactions in mixed-species flocks of migrating shorebirds. *Animal Behaviour*, **27**, 459–69.

Bustamante, R.H., Branch, G.M., Eekhout, S., Robertson, B., Zoutendyk, P.,

Schleyer, M., Dye, A., Hanekom, N., Keats, D., Jurd, M. & McQuaid, C.D. (1995). Gradients of intertidal primary productivity around the coast of South Africa and their relationships with consumer biomass. *Oecologia*, **102**, 189–201.

Clancey, P.A. (1986). Endemicity in the southern African avifauna. *Durban Museum Novitates*, **13**, 245–84.

Clark, C. & Mangel, M. (1984). Foraging and flocking strategies: information in an uncertain environment. *American Naturalist*, **109**, 626–41.

Cooper, J. & Hockey, P.A.R. (1981). *The atlas and site register of South African coastal birds*, Part two. *The Atlas*. Report to the South African National Committee for Oceanographic Research.

Cox, G.W. (1968). The role of competition in the evolution of migration. *Evolution*, **22**, 180–92.

Cox, G.W. (1985). The evolution of migration systems between temperate and tropical regions of the New World. *American Naturalist*, **126**, 451–574.

Cramp, S. & Simmons, K.E.L. (eds.) (1983). *The birds of the Western Palearctic*, Vol. 3. Oxford: Oxford University Press.

Daborn, G.R., Amos, C.L., Brylinsky, M., Christian, H., Drapeau, G., Faas, R.W., Grant, J., Long, B., Paterson, D.M., Perillo, G.M.E. & Piccolo, M.C. (1993). An ecological cascade effect: migratory birds affect stability of intertidal sediments. *Limnology and Oceanography*, **38**, 225–31.

Day, J.H. (1981a). The estuarine flora. In *Estuarine ecology with particular reference to Southern Africa*, ed. J.H. Day, pp. 77–99. Cape Town: A.A. Balkema.

Day, J.H. (ed.) (1981b). *Estuarine ecology with particular reference to Southern Africa*. Cape Town: A.A. Balkema.

Dodson, S.I. & Egger, D.L. (1980). Selective feeding by Red Phalaropes on zooplankton of Arctic ponds. *Ecology*, **61**, 755–63.

Dunning, J.B. (ed.) (1993). *Handbook of avian body masses*. Boca Raton, Florida: CRC Press.

Els, S. (1982). *Distribution and abundance of two crab species on the Swartkops estuary saltmarshes and the energetics of the Sesarma catenata population*. MSc thesis, University of Port Elizabeth.

Emanuel, B.P., Bustamante, R.H., Branch,

G.M., Eekhout, S. & Odendaal, F.J. (1992). A zoogeographic and functional approach to the selection of marine reserves on the west coast of South Africa. *South African Journal of Marine Science*, **12**, 341–54.

Engelmoer, M., Piersma, T., Altenburg, W. & Mes, R. (1984). The Banc d'Arguin (Mauritania). In *Coastal waders and wildfowl in winter*, ed. P.R. Evans, J.D. Goss-Custard & W.G. Hale, pp. 272–93. Cambridge: Cambridge University Press.

Evans, P.R. (1976). Energy balance and optimal foraging strategies in shorebirds: some implications for their distributions and movements in the non-breeding season. *Ardea*, **64**, 117–39.

Evans, P.R. & Dugan, P.J. (1984). Coastal birds: numbers in relation to food resources. In *Coastal waders and wildfowl in winter*, ed. P.R. Evans, J.D. Goss-Custard & W.G. Hale, pp. 8–28. Cambridge: Cambridge University Press.

Evans, P.R., Herdson, D.M., Knights, P.J. & Pienkowski, M.W. (1979). Short-term effects of reclamation of parts of Seal Sands, Teesmouth, on wintering waders and shelducks. *Oecologia*, **41**, 183–206.

Fairall, N. (1981). A study of the bioenergetics of the red-knobbed coot *Fulica cristata* on a South African estuarine lake. *South African Journal of Wildlife Research*, **11**, 1–4.

Fasola, M. & Canova, L. (1993). Diel activity of resident and immigrant waterbirds at Lake Turkana, Kenya. *Ibis*, **135**, 442–50.

Finlayson, M. & Moser, M. (1991). *Wetlands*. Oxford: Facts on file.

Fretwell, S.D. & Lucas, H.L. (1969). On territorial behaviour and other factors influencing habitat distribution in birds. I. Theoretical development. *Acta Biotheoretica*, **19**, 16–36.

Fry, C.H., Keith, S. & Urban, E.K. (1988). *The birds of Africa*, Vol. 3. London: Academic Press.

Gerritsen, A.F.C. & Meiboom, A. (1986). The role of touch in prey density estimation by *Calidris alba*. *Netherlands Journal of Zoology*, **36**, 530–62.

Gerritsen, A.F.C., Van Heezik, Y.M. & Swennen, C. (1983). Chemoreception in two further *Calidris* species (C.

maritima and *C. canutus*). *Netherlands Journal of Zoology*, **33**, 485–96.

Goss-Custard, J.D. (1969). The winter feeding ecology of the Redshank *Tringa totanus*. *Ibis*, **111**, 338–56.

Goss-Custard, J.D. (1970a). Factors affecting the diet and feeding rate of the Redshank (*Tringa totanus*). In *Animal populations in relation to their food resources*, ed. A. Watson, pp. 101–10. Oxford: Blackwell.

Goss-Custard, J.D. (1970b). The responses of Redshank (*Tringa totanus* (L.)) to spatial variations in the density of their prey. *Journal of Animal Ecology*, **39**, 91–113.

Goss-Custard, J.D. (1976). Variation in the dispersion of Redshank, *Tringa totanus*, on their winter feeding grounds. *Ibis*, **118**, 257–63.

Goss-Custard, J.D. (1980). Competition for food and interference among waders. *Ardea*, **68**, 31–52.

Goss-Custard, J.D. (1985). Foraging behaviour of wading birds and the carrying capacity of estuaries. In *Behavioural ecology: ecological consequences of adaptive behaviour*, ed. R.M. Sibly & R.H. Smith, pp. 169–88. Oxford: Blackwell Scientific Publications.

Goss-Custard, J.D. & Moser, M.E. (1988). Rates of change in the numbers of Dunlin, *Calidris alpina*, wintering in British estuaries in relation to the spread of *Spartina anglica*. *Journal of Applied Ecology*, **25**, 95–109.

Goss-Custard, J.D., Kay, D.G. & Blundell, R.M. (1977). The density of migratory and overwintering Redshank, *Tringa totanus* (L.) and Curlew, *Numenius arquata* (L.), in relation to the density of their prey in south-east England. *Estuarine, Coastal and Marine Science*, **5**, 497–510.

Goss-Custard, J.D., Jenyon, R.A., Jones, R.E., Newbery, P.E. & Williams, R.le B. (1978). The ecology of The Wash, II. Seasonal variations in the feeding conditions of wading birds (Charadrii). *Journal of Applied Ecology*, **14**, 701–19.

Hagemeijer, W. & Marteijn, E.C.L. (1993). Waders. In *Coastal waterbirds in Gabon, winter 1992*, ed. F.J. Schepers & E.C.L. Marteijn, pp. 135–71. WIWO Report No. 41. Zeist.

Hanekom, N., Baird, D. & Erasmus, T. (1988). A quantitative study to assess standing biomasses of macrobenthos in soft substrata of the Swartkops estuary, South Africa. *South African Journal of Marine Science*, **6**, 163–74.

Harrison, P. (1983). *Seabirds: an identification guide*. Beckenham, UK: Croom Helm.

Heppleston, P.B. (1970). Anatomical observations on the bill of the Oystercatcher (*Haematopus ostralegus occidentalis*) in relation to foraging behaviour. *Journal of Zoology (London)*, **161**, 519–24.

Heÿl, C.W. & Currie, M.H. (1985). Variations in the use of the Bot River estuary by water-birds. *Transactions of the Royal Society of South Africa*, **45**, 397–417.

Hockey, P.A.R. (1993). *Potential impacts of water abstraction on the birds of the lower Berg River wetlands*. Unpublished report, Percy FitzPatrick Institute of African Ornithology, University of Cape Town.

Hockey, P.A.R. (1994). *Birds of southern Africa: checklist and alternative names*. Cape Town: Struik.

Hockey, P.A.R. & Branch, G.M. (1994). Conserving marine biodiversity on the African coast: implications of a terrestrial perspective. *Aquatic Conservation: Marine and Freshwater Ecosystems*, **4**, 345–62.

Hockey, P.A.R. & Douie, C. (1995). *Waders of southern Africa*. Cape Town: Struik Winchester.

Hockey, P.A.R., Navarro, R.A., Kalejta, B. & Velasquez, C.R. (1992). The riddle of the sands: why are shorebird densities so high in southern estuaries? *American Naturalist*, **140**, 961–79.

Hockey, P.A.R., Turpie, J.K. & Velásquez, C.R. (1998). What selection pressures have driven the evolution of deferred northward migration by juvenile waders? *Journal of Avian Biology*, **29** (in press).

Hockey, P.A.R., Siegfried, W.R., Crowe, A.A. & Cooper, J. (1983). Ecological structure and energy requirements of the sandy beach avifauna of southern Africa. In *Sandy beaches as ecosystems*, ed. A. McLachlan & T. Erasmus, pp. 507–21. The Hague: Dr W. Junk.

Hockey, P.A.R. & Steele, W.K. (1990). Intraspecific kleptoparasitism and foraging efficiency as constraints on food selection by Kelp Gulls *Larus dominicanus*. In *Behavioural mechanisms of food selection*, ed. R.N. Hughes, pp. 679–706. Berlin: Springer-Verlag.

Hockey, P.A.R. & Underhill, L.G. (1984). Diet of the African Black Oystercatcher *Haematopus moquini* on rocky shores: spatial, temporal and sex-related variation. *South African Journal of Zoology*, **19**, 1–11.

Hockey, P.A.R., Underhill, L.G., Neatherway, M. & Ryan, P.G. (1989). *Atlas of the birds of the southwestern Cape*. Cape Town: Cape Bird Club.

Howard, R.K. & Lowe, K.W. (1984). Predation by birds as a factor influencing the demography of an intertidal shrimp. *Journal of Experimental Marine Biology and Ecology*, **74**, 35–52.

Hunter, L., Canevari, P., Myers, J.P. & Payne, L.X. (1991). Shorebird and wetland conservation in the western hemisphere. In *Conserving migratory birds*, ed. T. Salathé, pp. 279–90. ICBP Technical publication No. 12. Cambridge: ICBP.

Jackson, S. (1984). Predation by Pied Kingfishers and Whitebreasted Cormorants on fish in the Kosi estuary system. *Ostrich*, **55**, 113–32.

Kalejta, B. (1991). *Aspects of the ecology of migrant shorebirds (Aves: Charadrii) at the Berg River estuary, South Africa*. PhD thesis, University of Cape Town.

Kalejta, B. (1992). Time budgets and predatory impact of waders at the Berg River estuary, South Africa. *Ardea*, **80**, 327–42.

Kalejta, B. (1993a). Diets of shorebirds at the Berg River estuary, South Africa: spatial and temporal variation. *Ostrich*, **64**, 123–33.

Kalejta, B. (1993b). Intense predation cannot always be detected experimentally: a case study of shorebird predation on nereid polychaetes in South Africa. *Netherlands Journal of Sea Research*, **31**, 385–93.

Kalejta, B. & Hockey, P.A.R. (1991). Distribution, abundance and productivity of benthic invertebrates at the Berg River estuary, South Africa. *Estuarine, Coastal and Shelf Science*, **33**, 175–91.

Kalejta, B. & Hockey, P.A.R. (1994). Distribution of shorebirds at the Berg River estuary, South Africa, in relation to foraging mode, food supply and environmental features. *Ibis*, **136**, 233–9.

Kersten, M. & Piersma, T. (1987). High levels of energy expenditure in shorebirds: metabolic adaptations to an energetically expensive way of life. *Ardea*, **75**, 175–87.

Kersten, M. & Visser, W. (1996). The rate of food processing in the Oystercatcher: food intake and energy expenditure constrained by a digestive bottleneck. *Functional Ecology*, **10**, 440–8.

Kiorboe, T. (1980). Distribution and production of submerged macrophytes in Tipper Ground (Ringkobing Fjord, Denmark), and the impact of waterfowl grazing. *Journal of Applied Ecology*, **17**, 675–87.

Lasiewski, R.C. & Dawson, W.R. (1967). A re-examination of the relation between standard metabolic rate and body weight in birds. *Condor*, **68**, 13–23.

Maclean, G.L. (1993). *Roberts' birds of southern Africa*. Cape Town: John Voelcker Bird Book Fund.

Macnae, W. (1957). The ecology of plants and animals in the intertidal regions of the Zwartkops estuary near Port Elizabeth, South Africa. Part 2. *Journal of Ecology*, **45**, 361–85.

Madsen, J. (1988). Autumn feeding ecology of herbivorous wildfowl in the Danish Waddensea. *Danish Review of Game Biology*, **13**, 1–34.

Marais, J.F.K. & Baird, D. (1980). Seasonal abundance, distribution and catch per unit effort of fishes in the Swartkops estuary. *South African Journal of Zoology*, **15**, 66–71.

Martin, A.P. (1991). *Feeding ecology of birds on the Swartkops estuary, South Africa*. PhD thesis, University of Port Elizabeth.

Martin, A.P. & Baird, D. (1987). Seasonal abundance and distribution of birds on the Swartkops estuary. *Ostrich*, **58**, 122–34.

Martin, A.P. & Baird, D. (1988). Lemming cycles – which Palearctic migrants are affected? *Bird Study*, **35**, 143–5.

Martin, A.P. & Randall, R.M. (1987). Numbers of waterbirds at a commercial saltpan, and suggestions for management. *South African Journal of Wildlife Research*, **17**, 75–81.

Meininger, P.L., Blomert, A.-M. & Marteijn, E.C.L. (1991). Watervogelsterfte in het Deltagebied, ZW-Nederland gedurende de drie koude winters van 1985, 1986 en 1987. *Limosa*, **64**, 89–182.

Millard, N.A.H. & Broekhuysen, G.J. (1970). The ecology of South African estuaries, Part X: St Lucia: a second report. *Zoologica Africana*, **5**, 277–308.

Moser, M.E. & Van Vessum, J. (1995). Finally ... agreement on waterbirds! *IWRB News*, **14**, 11.

Murdock, L.C. (1975). *Physiology and bioenergetics of the American Coot, Fulica americana*. MA thesis, California State University, Fullerton, California.

Myers, J.P. (1980). Territoriality and flocking by Buff-breasted Sandpipers: variations in non-breeding dispersion. *Condor*, **82**, 241–50.

Myers, J.P. (1983). Conservation of migrating shorebirds: staging areas, geographical bottlenecks and regional movements. *American Birds*, **37**, 23–5.

Myers, J.P., Connors, P.G. & Pitelka, F.A. (1979). Territoriality in non-breeding shorebirds. *Avian Biology*, **2**, 231–46.

Myers, J.P., Williams, S.L. & Pitelka, F.A. (1980). An experimental analysis of prey availability for Sanderlings *Calidris alba* Pallas feeding on sandy beach crustaceans. *Canadian Journal of Zoology*, **58**, 1564–74.

Myers, J.P., Morrison, R.I.G., Antas, P.Z., Harrington, B.A., Lovejoy, T.E., Sallaberry, M., Senner, S.E. & Tarak, A. (1987). Conservation strategy for migratory species. *American Scientist*, **75**, 18–26.

Nagy, K.A. (1987). Field metabolic rate and food requirement scaling in mammals and birds. *Ecological Monographs*, **57**, 111–28.

Nienhuis, P.H. (1978). An ecosystem study in Lake Grevelingen, a former estuary in the SW Netherlands. *Kieler Meeresforschungen, Sonderheft*, **4**, 247–55.

Pearce, E.A. & Smith, C.G. (1984). *The world weather guide*. London: Hutchinson.

Pienkowski, M.W. (1981). How foraging plovers cope with environmental effects on invertebrate behaviour and availability. In *Feeding and survival strategies of estuarine organisms*, ed. N.V. Jones & W.J. Wolff, pp. 179–92. New York: Plenum Press.

Pienkowski, M.W. (1983). Changes in the foraging pattern of plovers in relation to environmental factors. *Animal Behaviour*, **31**, 244–64.

Pienkowski, M.W. & Evans, P.R. (1985). The role of migration in the population dynamics of birds. In *Behavioural ecology: ecological consequences of adaptive behaviour*, ed. R.M. Sibly & R.H. Smith, pp. 331–52. Oxford: Blackwell Scientific Publications.

Piersma, T. (1985). Abundance of waders in the Nakdong estuary, South Korea, in September 1984. *Wader Study Group Bulletin*, **44**, 21–5.

Piersma, T. (1987). Production of intertidal benthic animals and limitations to their predation by shorebirds: an heuristic model. *Marine Ecology Progress Series*, **38**, 187–96.

Piersma, T., De Goeij, P. & Tulp, I. (1993). An evaluation of intertidal feeding habitats from a shorebird perspective: towards relevant comparisons between temperate and tropical mudflats. *Netherlands Journal of Sea Research*, **31**, 503–12.

Prevost, M.B., Johnson, A.S. & Landers, J.L. (1978). Production and utilization of waterfowl foods in brackish impoundments in South Carolina. *Proceedings of the Annual Conference of the Southeastern Association of Fish & Wildlife Agencies*, **32**, 60–70.

Prochazka, K. (1994). *Habitat partitioning in shallow-water cryptic ichthyofaunal communities in the western and southwestern Cape, South Africa*. PhD thesis, University of Cape Town.

Puttick, G.M. (1980). Energy budgets of Curlew Sandpipers at Langebaan Lagoon, South Africa. *Estuarine, Coastal and Marine Science*, **11**, 207–15.

Puttick, G.M. (1981). Sex-related differences in foraging behaviour of Curlew Sandpipers. *Ornis Scandinavica*, **12**, 13–17.

Reading, C.J. & McGrorty, S. (1978). Seasonal variations in the burying depth of *Macoma balthica* (L.) and its accessibility to wading birds. *Estuarine, Coastal and Marine Science*, **6**, 135–44.

Rose, P.M. & Scott, D.A. (1994). *Waterfowl population estimates*. IWRB Publication No. 29. Slimbridge: IWRB.

Ryan, P.G. & Cooper, J. (1985). Waders (Charadrii) and other coastal birds of the northwestern Cape Province, South Africa. *Bontebok*, **4**, 1–8.

Ryan, P.G., Cooper, J., Hockey, P.A.R. & Berruti, A. (1986). Waders (Charadrii) and other water birds on the coast and adjacent wetlands of Natal, 1980–1981. *Lammergeyer*, **36**, 1–33.

Ryan, P.G., Underhill, L.G., Cooper, J. & Waltner, M. (1988). Waders (Charadrii) and other waterbirds on the coast, adjacent wetlands and offshore islands of the southwestern Cape Province, South Africa. *Bontebok*, **6**, 1–19.

Salmon, M. & Hyatt, G.W. (1983). Spatial and temporal aspects of reproduction in North Carolina fiddler crabs (*Uca pugilator* Bosc.). *Journal of Experimental Marine Biology and Ecology*, **70**, 21–43.

Schneider, D.C. (1978). Equalization of prey numbers by migratory shorebirds. *Nature*, **271**, 353–4.

Siegfried, W.R. (1981). The estuarine avifauna of southern Africa. In *Estuarine ecology with particular reference to Southern Africa*, ed. J.H. Day, pp. 223–50. Cape Town: A.A. Balkema.

Sih, A. (1984). The behavioral response race between predator and prey. *American Naturalist*, **123**, 143–50.

Sinclair, J.C., Hockey, P.A.R. & Tarborton, W.R. (1993). *Sasol birds of southern Africa*. Cape Town: Struik.

Smit, C.J. & Piersma, T. (1989). Numbers, midwinter distribution, and migration of wader populations using the East Atlantic flyway. In *Flyways and reserve networks for waterbirds*, ed. H. Boyd & J.-Y. Pirot, pp. 24–63. IWRB Special Publication No. 9. Slimbridge: IWRB.

Smith, P.C. (1975). *A study of the winter feeding ecology and behaviour of the Bar-tailed Godwit* (Limosa lapponica). PhD thesis, University of Durham.

Soikkeli, M. (1967). Breeding cycle and population dynamics in the Dunlin (*Calidris alpina*). *Annales Zoologici Fennici*, **4**, 158–98.

Steele, W.K. & Hockey, P.A.R. (1995). Factors influencing the rate and success of intraspecific kleptoparasitism among Kelp Gulls (*Larus dominicanus*). *Auk*, **112**, 847–59.

Stephenson, T.A. & Stephenson, A. (1972). *Life between tidemarks on rocky shores*. San Francisco: W.H. Freeman.

Stewart, B.A. & Bally, R. (1985). The ecological role of the Red-knobbed Coot *Fulica cristata* Gmelin at the Bot River estuary, South Africa: a preliminary investigation. *Transactions of the Royal Society of South Africa*, **45**, 419–26.

Summers, R.W. (1977). Distribution, abundance and energy relationships of waders (Aves: Charadrii) at Langebaan Lagoon. *Transactions of the Royal Society of South Africa*, **42**, 483–95.

Summers, R.W., Underhill, L.G., Clinning, C.F. & Nicoll, M. (1989). Populations, migrations, biometrics and moult of the Turnstone *Arenaria i. interpres* on the east Atlantic coastline, with special reference to the Siberian population. *Ardea*, **77**, 145–68.

Tree, A.J. & Martin, A.P. (1993). Report on the Jan/Feb 1993 CWAC counts in the eastern Cape. *Bee-eater*, **44**, 18–26.

Turpie, J.K. (1994). *Comparative foraging ecology of two broad-ranging migrants, Grey Plover* Pluvialis squatarola *and Whimbrel* Numenius phaeopus *(Aves: Charadrii) in tropical and temperate latitudes of the western Indian Ocean*. PhD thesis, University of Cape Town.

Turpie, J.K. (1995a). Nonbreeding territoriality: causes and consequences of seasonal and individual variation in Grey Plover *Pluvialis squatarola* behaviour. *Journal of Animal Ecology*, **64**, 429–38.

Turpie, J.K. (1995b). Prioritizing South African estuaries for conservation: a practical example using waterbirds. *Biological Conservation*, **74**, 175–85.

Turpie, J.K. & Hockey, P.A.R. (1993). Comparative diurnal and nocturnal foraging behaviour and energy intake of premigratory Grey Plovers *Pluvialis squatarola* and Whimbrels *Numenius phaeopus* in South Africa. *Ibis*, **135**, 156–65.

Turpie, J.K. & Hockey, P.A.R. (1996). Foraging ecology and seasonal energy budgets of Grey Plovers *Pluvialis squatarola* and Whimbrels *Numenius phaeopus* at the southern tip of Africa. *Ardea*, **84**, 57–74.

Underhill, L.G. (1987a). Waders (Charadrii) and other waterbirds at Langebaan Lagoon, South Africa, 1975–1986. *Ostrich*, **58**, 145–55.

Underhill, L.G. (1987b). Changes in the age structure of Curlew Sandpiper populations at Langebaan Lagoon, in relation to lemming cycles in Siberia. *Transactions of the Royal Society of South Africa*, **46**, 209–14.

Underhill, L.G. (1994). Optimal and suboptimal reserve selection algorithms. *Biological Conservation*, **70**, 85–7.

Underhill, L.G. & Cooper, J. (1984). *Counts of waterbirds at southern African coastal wetlands*. Unpublished report, Western Cape Wader Study Group and Percy FitzPatrick Institute of African Ornithology, Cape Town.

Underhill, L.G., Waltner, M. & Summers, R.W. (1989). Three-year cycles in breeding productivity of Knot *Calidris canutus* wintering in southern Africa suggest Taimyr Peninsula provenance. *Bird Study*, **36**, 83–7.

Van Heezik, Y.M., Gerritsen, A.F.C. & Swennen, C. (1983). The influence of chemoreception on the foraging behaviour of two species of sandpiper, *Calidris alba* and *Calidris alpina*.

Netherlands Journal of Sea Research, **17**, 47–56.

Velásquez, C.R. & Hockey, P.A.R. (1992). The importance of supratidal foraging habitats for waders at a south temperate estuary. *Ardea*, **80**, 243–53.

Velásquez, C.R., Kalejta, B. & Turner, E. (1990). The Berg River estuary: an important wetland for Caspian Terns *Sterna caspia* in South Africa. *Marine Ornithology*, **18**, 65–8.

Velásquez, C.R., Kalejta, B. & Hockey, P.A.R. (1991). Seasonal abundance, habitat selection and energy consumption by waterbirds at the Berg River estuary, South Africa. *Ostrich*, **62**, 109–23.

Verhoeven, J.T.A. (1980). The ecology of *Ruppia*-dominated communities in western Europe. III: Aspects of production, consumption and decomposition. *Aquatic Botany*, **8**, 209–53.

Von Haartman, L. (1968). The evolution of resident versus migratory habit in birds. Some considerations. *Ornis Fennica*, **45**, 1–7.

Whitfield, A.K. (1978). Relationship between fish and piscivorous bird densities at Lake St Lucia. *South African Journal of Science*, **74**, 478.

Whitfield, A.K. (1986). Predation by Whitebreasted Cormorants on fishes in a southern Cape estuarine system. *Ostrich*, **57**, 248–9.

Whitfield, A.K. & Blaber, S.J.M. (1978). Feeding ecology of piscivorous birds at Lake St Lucia, Part 1: diving birds. *Ostrich*, **49**, 185–98.

Whitfield, A.K. & Blaber, S.J.M. (1979a). Feeding ecology of wading birds at Lake St Lucia, Part 2: wading birds. *Ostrich*, **50**, 1–9.

Whitfield, A.K. & Blaber, S.J.M. (1979b). Feeding ecology of piscivorous birds at Lake St Lucia, Part 3: swimming birds. *Ostrich*, **50**, 10–20.

Whitfield, D.P., Evans, A.D. & Whitfield, P.A. (1986). The impact of raptor predation on wintering waders. *Proceedings of the International Ornithological Congress*, **19**, 674–87.

Winter, P.E.D. (1979). *Studies on the distribution, seasonal abundance and diversity of the Swartkops estuary ichthyofauna*. MSc thesis, University of Port Elizabeth.

Zwarts, L. (1985). The winter exploitation of fiddler crabs *Uca tangeri* by waders in Guinea-Bissau. *Ardea*, **73**, 3–12.

Zwarts, L. & Blomert, A.-M. (1990).

Selectivity of Whimbrels feeding on fiddler crabs explained by component-specific digestibilities. *Ardea*, **78**, 193–208.

Zwarts, L. & Blomert, A.-M. (1992). Why Knots *Calidris canutus* take medium-sized *Macoma balthica* when six prey species are available. *Marine Ecology Progress Series*, **83**, 113–28.

Zwarts, L. & Dirksen, S. (1989). Feeding behaviour of Whimbrel and anti-predator behaviour of its main prey, the fiddler crab *Uca tangeri*. In *Report of the Dutch–Mauritanian project, Banc d'Arguin 1985–1986*, ed. B.J. Ens, T. Piersma, W.J. Wolff & L. Zwarts, pp. 205–7. WIWO Report 25. The Netherlands: Texel.

Zwarts, L. & Dirksen, S. (1990). Digestive bottleneck limits the increase in food intake of Whimbrels preparing for spring migration from the Banc d'Arguin, Mauritania. *Ardea*, **78**, 257–78.

11 Estuaries as ecosystems: a functional and comparative analysis

Dan Baird

Keurbooms River estuary

Introduction

The scientific base of estuarine ecology in South Africa has increased dramatically over the past two decades. South African ecologists have gathered numerous observations and published prolifically on various aspects of the dynamics of plant and animal populations and communities in a large number of estuaries (see Whitfield 1995). Much of the results emanated from these studies are incorporated in this volume. Although some of the research was reductionist in nature, the data generated were most useful and contributed significantly to the holistic and more integrative approach to the study of estuaries as identifiable ecosystems, which found its application during the mid-1980s in South Africa (Allanson 1992).

The holistic study of ecosystems requires the synthesis of information into a format representing the system in some form or another. Since the flow of energy or material in a system, and exchanges with adjacent systems is a universal and fundamental property of an ecosystem, biological oceanographers have recently given greater emphasis to the quantification of stocks and flows of material in marine ecosystems (Fasham 1985; Hannon 1985; Wulff *et al.* 1989, Heymans & Baird 1995). Quantitative descriptions of energy or material flows provide significant insights into the fundamental structure and function of an ecosystem, and outline the efficiencies with which such currencies are transferred, transformed and assimilated within the system (Allen 1985; Baird & Ulanowicz 1989). These presentations, moreover, provide a starting point for ecosystem flow analysis, structural comparison, and simulation modelling, as well as providing an heuristic tool to help shape and formulate future research.

Estuaries are complex ecosystems and it is unlikely that all the components and interrelationships will ever be fully analysed and understood. However, sufficient information is available on a few South African estuaries for which flow networks can be constructed and analysed. This chapter thus specifically aims to present a statement on the status of whole ecosystem studies on estuaries in South Africa, with the following objectives.

1 To present detailed networks of the flux of carbon (a surrogate for energy) between the living and non-living compartments of selected estuaries.

2 To analyse these flow models by means of network analysis which includes input–output analysis, trophic and cycling analysis and the calculation of whole ecosystem properties such as Total System Throughput, Development Capacity, and System Overhead.

3 To compare some functional processes and global ecosystem attributes, derived from network analysis and other means, of the South African estuaries with each other, as well as with similar systems on a global basis.

4 To examine temporal changes of whole system behaviour, affected by anthropogenic influences, of a South African estuary.

5 To review recent developments in predictive ecosystem modelling.

Despite the fact that a large amount of information has been generated on a number of estuaries in South Africa, data on the standing stocks (biomass) of the living and non-living components, and the rate of exchange between various components of individual systems are remarkably sparse. It is not the intention to discuss single functional processes (e.g. predation rates by birds) in this chapter, as they are addressed elsewhere in the volume, but rather to focus on whole system studies. The three estuaries examined here are the Swartkops, Kromme and Bot River estuaries, and their ecological attributes are compared with the Ythan estuary in Scotland, the Ems estuary in the Netherlands, and the Chesapeake Bay estuary on the east coast of the USA. These three estuaries from South Africa represent an ordered triad ranging from closed, to mildly impacted, to a relatively pristine system.

Study sites and database

The Swartkops estuary (32° 57′ S, 25° 38′ E), has been studied by numerous workers over the past 20 years, and their results are summarised by Baird et al. (1986) and Baird (1988) and Baird and Ulanowicz (1993). Results of ecological studies on the Kromme estuary (34° 09′ S, 24° 51′ E) were obtained from numerous sources, and summarised by Bickerton and Pierce (1988), Heymans and Baird (1995), Baird and Heymans (1996), and Baird et al. (1992). Similarly, the Bot River estuary (34° 21′ S, 19° 04′ E) was studied in detail by, for example, Bally et al. (1985), Bally and McQuaid (1985), Branch et al. (1985), Bennett and Branch (1990), De Decker and Bally (1985), Coetzee (1985), Heÿl and Currie (1985), Koop et al. (1983), Stewart and Bally (1985), and Bennett (1989).

Many of the estuaries along the South African coast are subject to similar environmental and anthropological influences. One of these common influences is a lack of freshwater inflow, or a reduction in freshwater inputs, due to either low rainfall in the catchment, or to the construction of storage dams. The effect of reduced fresh water inflow is manifested in two ways: firstly, in the closure of the mouth from the sea due to reduced scouring action of floods, and secondly, where the mouth remains open, the development of a virtual homogeneous salinity regime in the estuary. The Bot River estuary is a typical example of a 'closed' estuary which remains cut off from the sea by a dune barrier for months or even years, and is only breached when the water in the estuary reaches unacceptably high levels, and mass mortalities of marine and estuarine fish occur (Bally 1985; Fromme 1985). When closed, the salinity fluctuates from low values ($<$10) to hypersaline conditions ($>$35) (Bally & McQuaid 1985). The Kromme estuary receives a very small amount of freshwater, leading to a homogenous salinity structure with no salinity gradient from mouth to the upper reaches. This estuary is, however, relatively pristine in the sense that no industrial and very little domestic and agricultural effluents are discharged into the river and estuary (Baird & Peryra-Lago 1992).

The Swartkops River, on the other hand, flows through

Table 11.1. *Some properties of the Swartkops, Kromme and Bot River estuaries*

Attribute	Swartkops	Kromme	Bot River
Area (km^2)	4	3	13.6
Temperature range (°C)	13–26	13–28	12–20
Salinity	10–35	33–35	1–42
No. of model compartment	15	16	15
Net phytoplankton production (mg C m^{-2} d^{-1})	313	28	95
Net macrophyte production (g C m^{-2} d^{-1})	1.54	2.086	8.142
Total standing stock (g C m^{-2})	398	213	965
Total production (primary and secondary, g C m^{-2} d^{-1})	3.981	2.547	8.42
Net NPP efficiency (%)[a]	38	9	12
Detritivory: herbivory ratio	1.5:1	7:1	3:1

[a] NPP efficiency gives fraction of plant production (phytoplankton, macrophytes, benthic algae) directly grazed by herbivores.

a highly urbanised and industrial region, with many anthropogenic inputs varying from untreated sewage to industrial effluents. The estuary is thus impacted to some degree, although it still supports rich and diverse plant and animal communities. It is heavily utilised for recreational purposes, e.g. yachting, board sailing, swimming, fishing, etc., while developments (housing, roads, etc.) on the floodplain pose an ever increasing threat to the ecological functioning of the system. The inlet is, however, always open to the sea, and the estuary receives freshwater continuously, and exhibits a typical salinity gradient from mouth to the head.

The three estuaries considered here span the spectrum of the kind of systems typically found along the coast, namely from semi-permanently closed, to relatively pristine but freshwater deprived, to nutrient-rich and somewhat impacted systems. These systems are a representative sample of South African estuaries, although it must be noted that many others may occupy intermediate positions which do differ to some degree with any of the three mentioned above. The physical, chemical and biological characteristics of all three systems are well documented (see key references mentioned above) while some pertinent characteristics of each estuary are given in Table 11.1.

Ecosystem flow models and trophic structure

The available data on the various components and the interactions between the compartments of each system are depicted in Figures 11.1, 11.2 and 11.3. The flow models of the Swartkops and Bot estuaries consist of 15 compartments, and the Kromme estuary of 16. The flow models of the Kromme and Bot River estuaries each have three

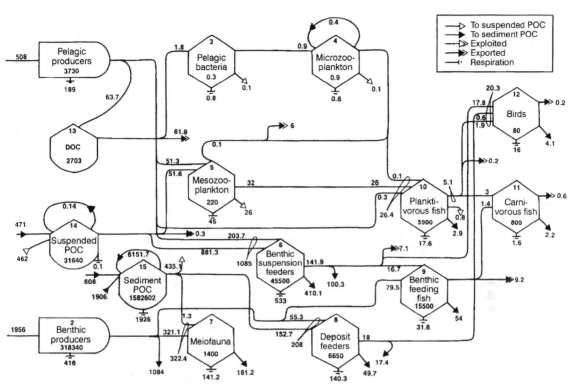

Figure 11.1. **Carbon flow network of the Swartkops estuary. Biomass given in mg C m^{-2} and flows in mg C m^{-2} d^{-1}. Feedback loops in com- partments 4, 14 and 15 indicate feeding of some species on others within the same compartments.**

primary producer modules, whereas for the Swartkops estuary benthic microalgae and halophytes were combined and thus have only two primary producer compartments. Few birds are present in the Kromme estuary, and the fish component was not subdivided in the Bot River estuary as in the other two systems. All three networks have similar configuration, structure, and number of compartments, which allow comparison of these systems with each other, and with other systems with a similar number of compartments.

The networks illustrate the standing stocks of the living and three non-living compartments (suspended and particulate organic carbon and dissolved organic carbon (DOC) in mg C m^{-2} and the flows between the compartments in mg C m^{-2} d^{-1}. For primary producers, phytoplankton, benthic microalgae and macrophytes, gross primary production (GPP) was assumed equal to the sum of net primary production (NPP) and respiration (R). Exudates, as a fraction of NPP, from plants during photosynthesis were channelled to the dissolved organic carbon pool (DOC). The GPP values were considered to be the inputs into each system.

For heterotrophic organisms the carbon uptake by each living compartment was balanced by production, respiration and excretion. System outputs consisted of respira-

tion, exports in the form of suspended material, burial and export of unutilised macrophytic material, and emigration of fish and birds. Each system was also assumed to be in a steady state, that is, annually averaged inputs were balanced by the corresponding outputs for each compartment. [Data incorporated in the Kromme model were collected from 1988 to 1992.]

Each of the flow networks contains primary producer compartments, such as macrophytes, benthic microalgae (mainly pennate diatoms) and phytoplankton. Submerged aquatic macrophytes, mainly *Ruppia maritima* which is replaced by *Potamogeton pectinatus* in the upper low salinity reaches, and dense strands of emergent macrophytes (mainly *Phragmites australis*) dominate as primary producers in the Bot River estuary. In the Kromme estuary the submerged macrophytes *Zostera capensis* and *Caulerpa falciformis* predominate, whereas in the Swartkops estuary emergent halophytes such as *Spartina maritima* are the major primary producers. Phytoplankton biomass and production is higher in the Swartkops estuary than in the Kromme and Bot River estuaries (see Table 11.1).

The networks contain compartments of free-living bacteria (<2 μm), microzooplankton (2–200 μm) and mesozooplankton (>200 μm), invertebrate detritivores

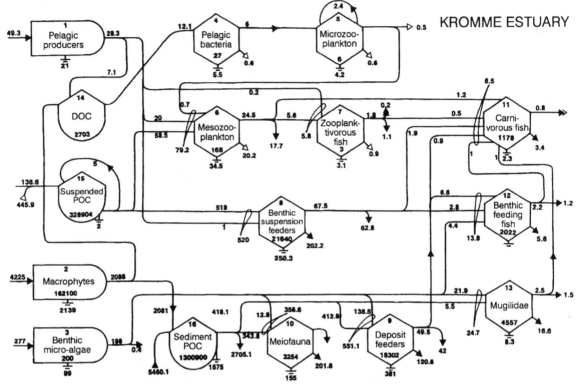

Figure 11.2. **Carbon flow network of the Kromme estuary. Biomass given in mg C m^{-2} and flows in mg C m^{-2} d^{-1}. Feedback loops in compartments 5 and 15 indicate feeding of some species on others within the same compartments.**

and suspension feeders, meiofauna, fish and birds. Very few birds occur in the Kromme estuary, whereas both the Swartkops and Bot River estuaries support large numbers of birds which play a significant role in the feeding dynamics of these systems. Structurally the systems are very similar, although the species composition of the various communities may differ substantially between the systems. Essentially the same trophic guilds, e.g. suspension feeders, deposit feeders (or detritivores), etc., occur in the three systems, and appear to be universal components of virtually all estuarine systems (McLusky 1981). There are, however, differences in the relative abundance of the individual species, and in the role some communities play in the food webs of different systems. For example, the birds in the Swartkops estuary, mainly Palearctic migrants (Martin & Baird 1987; Hockey & Turpin, Chapter 10), feed predominantly on benthic macrofauna (compartment 6, Figure 11.1). In the Bot River estuary, however, the dominant species, the Red-knobbed coot, is herbivorous and feeds mainly on *Ruppia maritima* and *Potamogeton pectinatus*. Similarly, the fish communities differ in diversity and relative abundance, varying from 22 200 mg C m^{-2} in the Swartkops, to 7757 mg C m^{-2} in the Kromme, to only 62 mg C m^{-2} in the Bot River estuary.

Global measures of system organisation

The comparison of the characteristics of species, populations, natural communities, and even ecosystems, has been an important tool for ecologists for more than 100 years. The comparison of attributes of, for example, ecosystems, has been inspirational in the formulation of hypotheses and concepts such as succession, natural selection and evolution, and contributed to the development of the principles of community ecology (Downing 1991). The interest in comparative ecosystems ecology and in the comparison of system attributes appears to be growing (Strayer 1991). This approach requires the synthesis and generalisation of data on systems, the identification of properties that are distinctive at the ecosystem level of organisation but not at lower levels, and the parallel development of supportive theoretical and methodological frameworks.

A suitable approach to intersystem comparison lies in the comparison of descriptive networks (as illustrated in Figures 11.1, 11.2 and 11.3) of flow of energy or material between the various components in different systems. The volume *Flows of material and energy in marine ecosystems*, edited by Fasham (1984), emphasised the value of this

BOT RIVER ESTUARY

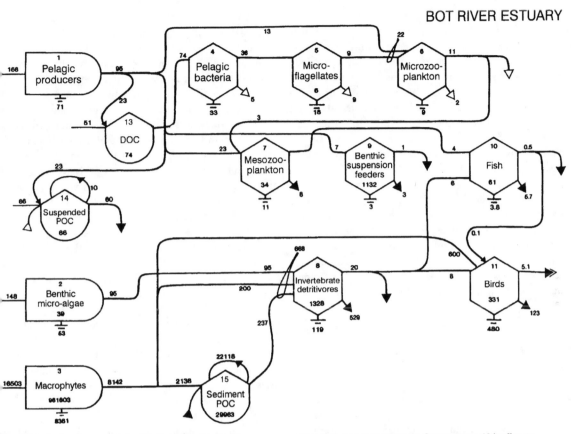

Figure 11.3. **Carbon flow network of the Bot River estuary. Biomass given in mg C m⁻² and flows in mg C m⁻² d⁻¹. Feeback loops in compartments 14 and 15 indicate feeding of some species on others** within the same compartments. Compartment 12 (sediment bacteria) is merged with compartment 15 (sediment POC) in diagram.

approach, and a number of recent publications illustrated the usefulness of comparing systems on a geographical scale (Baird *et al.* 1991; Baird & Ulanowicz 1993), and of changes in a system over time (Asmus and McKellar 1989; Baird & Ulanowicz 1989, Ducklow *et al.* 1989; Field *et al.* 1989*b*; Warwick & Radford 1989).

Most of these studies use a set of algorithms, known collectively as network analysis, which are designed to evaluate the structure of multi-compartmental models (or 'networks') of ecosystem. Network analysis of ecosystem models yields numerous analyses, such as the structure and methods of cycling discussed above, and allows the calculation of global system properties such as development capacity, throughput and ascendancy. In this section the flow networks of the three estuaries are examined by network analysis which includes input–output analysis (Leontief 1951; Hannon 1973), throughflows, storages and information theory (Ulanowicz 1986). Detailed reference to the underlying theoretical concepts and methodology of network analysis is given by Ulanowicz and Kay (1991), Ulanowicz (1986) and Kay *et al.* (1989).

The following system properties were used to compare

South African estuaries with each other, and with some elsewhere in the world: (1) Total System Throughput (*T*), reflecting the size or activity of the system in terms of the sum of flows through all the individual compartments; (2) Ascendancy (*A*) which represents both the size and organisation of flows (or mutual information), i.e. the product of *T* and the average mutual information of flows in a system. Ulanowicz (1986) postulated that as a system grows and develops towards maturity over time, the ascendancy of the system should increase; (3) Development Capacity (*C*) which represents the uppermost limit of ascendancy, and which measures the potential for a particular network to develop; (4) System Overhead (*O*), numerically represented by *C*−*A*, and is generated by all hierarchical, thermodynamic, environmental and resource related inter-determinancies, that keep the system ascendancy from reaching *C*, its theoretical maximum (Ulanowicz 1995). Odum (1969) listed 24 attributes to characterise more 'mature' systems, while Ulanowicz (1986) suggested that ascendancy encapsulates the concept of 'maturity' to a large degree. Mann *et al.* (1989) proposed the dimensionless *A:C* ratio as a suitable index of a system's development status, whilst Baird *et al.* (1991) used the *internal indices* of ascendancy and development capacity, A_i and C_i, functions of internal exchanges only (thus excluding exogenous transfers into and out of the system) in their analysis and comparison of six marine ecosystems. They (Baird *et al.* 1991) argued that a highly organised system internalises most of its activity so that it can function relatively independent of 'outside' supplies and demands. There is, however, a lively debate among systems ecologists at present on goal functions in ecology and appropriate measures of ecosystem stability and maturity (cf. Christensen 1995; Müller & Leupelt 1997).

However, the calculation of all the above indices is sensitive to the structure of the model network (e.g. number of compartments) and its degree of aggregation. Here the flow networks are cast in models of the same number of components in which the degree of aggregation among the living components is essentially the same. The same currency (carbon) has also been used.

Trophic structure and cycle analysis

The descriptive flow models of each of the systems contain a considerable amount of information on the dynamic structure of individual estuaries in terms of the flows between the various components, biomass, and energy pathways. These models are at best, however, simplified representations of complex ecosystems, but nevertheless provide the database and information necessary for more sophisticated analyses which were developed and applied during the past two decades.

(a) Trophic structure of estuaries
The rate of primary production and standing crop varies greatly between the systems. Phytoplankton production in the Kromme estuary is particularly low (28 mg C m^{-2}d^{-1}), compared with rates of 255 and 95 mg C m^{-2}d^{-1} in the Swartkops and Bot River estuaries respectively. It is also lower than in other estuaries such as Palmiet River estuary (34° 21′ S, 19° 00′ E; 190 mg C m^{-2}d^{-1}), and the Sundays River estuary (34° 43′ S, 25° 51′ E) (for comparative detail on primary producers see Adams, Bate and O'Callaghan, Chapter 5). The low phytoplankton production in the Kromme estuary could be ascribed to the low concentrations of dissolved inorganic nitrogen and phosphorus in the water column of the system (Baird & Pereyra-Lago 1992), and low rates of freshwater inflow.

Of interest is the exceptionally high standing crop (951 g C m^{-2}) and total production rate (8.14 g C m^{-2}d^{-1}) of macrophytes in the Bot River estuary compared with the others (see Table 11.1 and Figures 11.1–11.3). The difference

in system activity is also reflected in the values of total systems throughput (defined as the sum of flows through all compartments and which is a measure of the 'activity' of the system (Field *et al.* 1989a), as shown in Table 11.3. However, 99% of the Bot River estuary's standing crop and 91% of its production is ascribed to the submerged and emerged macrophyte components in the estuary. These values are lower in the Swartkops and Kromme estuaries where the standing crop of macrophytes accounts for 80% and 76% respectively of the total biomass, and macrophyte production about 38% and 82% respectively of the total production in the two estuaries. Another point of interest is the ratio of pelagic phytoplankton production to benthic primary production, and the fate of primary plant production in these estuarine ecosystems. The ratio of phytoplankton to benthic plant production varies from 1:5 in the Swartkops estuary to the very high ratios of 1:80 and 1:87 in the Kromme and Bot River estuaries respectively (see Table 11.1).

The relatively low standing stock of macrozooplankton in the latter two estuaries may also be as a result of low phytoplankton production. The direct utilisation of plants by herbivores, as the net NPP efficiency ratios (see Table 11.1), shows low values for the Kromme and Bot River estuaries (9% and 12% respectively), but a higher ratio of about 38% for the Swartkops estuary indicating greater grazing activity in the latter system. There also appears to be an inverse relationship between the net NPP efficiency and the detrivory:herbivory ratio which is low in the Swartkops estuary (1.5:1), but higher in the Kromme (7:1) and Bot River estuary (3:1), which means that detritivory appears to be of greater importance in systems where the direct utilisation of primary producers is low. This phenomenon has also been observed in the Ems estuary (the Netherlands) where the net NPP efficiency is high (98%) and the detritivory:herbivory ratio low (0.5:1). The Ythan estuary in Scotland, a mildly impacted system, on the other hand, has a low NPP efficiency (10.5%), and a high detritivory:herbivory ratio of 10:1. In coastal seas, including upwelling systems, where pelagic primary producers form the base of the food web, the net NPP efficiency ratio ranges between 40%, 82% and 87% in the Southern Benguela and Peruvian upwelling systems, and the Baltic Sea, respectively; whereas the detritivory:herbivory ratios in these systems are inversely low at 0.01:1, 0.3:1, and 1.5:1 respectively (see Baird *et al.* 1991).

(b) Structure and magnitude of cycling in estuaries
A universal and important phenomenon of all ecosystems is the recycling of matter and energy, and occurs when flows cycle among the compartments of the system.

The positive feedback of matter and energy is a critical process which determines the overall structure of systems,

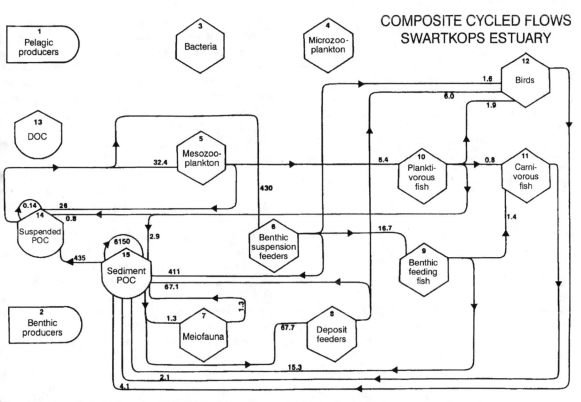

Figure 11.4. **The cycling structure of carbon in the Swartkops estuary. Units of flow and compartment numbers are the same as in Figure**

11.1. **Feedback loops in compartments 14 and 15 indicate cycling within the compartment between POC and bacteria.**

and which contributes to their autonomous behaviour (Ulanowicz 1986). Kay (1984) furthermore argued that the organisation of the flow structure in ecosystems reflects not only the rate of energy dissipation, but also the conservation of material through the cycling process within the system. By studying the structure and magnitude of cycling it is possible to identify the number of cycles and the routes of recycling in a system. An individual cycle represents a unique pathway which begins and ends in the same compartment and in which no system compartment appears more than once. A collection of cycles having the same smallest transfer is called a nexus (Baird & Ulanowicz 1993). It has also been suggested that the greater diversity of nexuses containing longer cycles implies that there is less 'stress' on the system and that the relative amount of cycling increases as an ecosystem becomes more stressed (Ulanowicz 1983; Baird & Ulanowicz 1993).

The Finn Cycling Index (FCI: Finn 1976) gives the proportion of the flow in a system that is recycled compared with the total system throughput, T. The FCI is thus equal to Tc/T, where Tc is the amount of system activity devoted to cycling. FCI also measures the retentiveness of a system in that the higher the index, the greater the proportion of material recycled and the more mature the system *sensu*

Odum (1969). Three aspects of cycling are discussed here, namely (i) the number of cycles and their distributions against cycle lengths, (ii) the Finn Cycling Index, and (iii) the average path length for each system.

The methodology for analysing the structure of cycling present in a network is described by Ulanowicz (1983, 1986) and Kay et al. (1989). Networks of the cycled flows for each system are illustrated in Figures 11.4, 11.5 and 11.6. The cycled flows (in mg C m^{-2} d^{-1} in the diagrams) reflect the amount of compartmental throughput that returns to that compartment after travelling the indicated loop. The number of cycles per nexus and the total number of cycles in each system are given in Table 11.2a and the amount of cycled flow over various path lengths in Table 11.2b. The number of cycles for each system is given in Table 11.3.

The Bot River estuary exhibits surprisingly few cycles, only five, compared with 15 and 19 for the Swartkops and Kromme estuaries respectively. Of importance, however, is the distribution of cycles per nexus and the amount of cycled flow over various path lengths as shown in Table 11.2a and b. The results show that all the cycling in the Swartkops and Bot estuaries occurs through single cycle nexuses (Table 11.2a). Table 11.2b shows that most of the cycling involves short and fast loops in all three systems

Table 11.2 Results from the cycle analyses of the three estuaries

(a) Distribution (%) cycles per nexus

Cycles per nexus	Swartkops	Kromme	Bot River
1	100	89	100
2	0	11	0
3	0	0	0
4	0	0	0

(b) Percentage of cycled flow through loops of various path lengths.

Path length	Swartkops	Kromme	Bot River
1	80	62	0
2	2.2	17.9	98
3	16.5	19.7	1.9
4	1	0.5	0.1
5	0.3	0.1	0.01
Total cycled flow (mg C m^{-2} d^{-1})	7679	4378	5473

Table 11.3 Indices of system organisation derived from network analysis

Index	Swartkops	Kromme	Bot River
Total System Throughput (T, mgCm^{-2}d^{-1})	17 541	16 897	15 234
Finn Cycling Index (FCI, %)	43.8	26	36
Average Path Length (APL)	3.95	2.38	3.22
Number of Cycles	15	19	5
Production, Biomass ratio (day^{-1})	0.1	0.08	0.11
Development Capacity (C, mgC, bits)	62 652	59 422	42 454
Ascendancy (A, mgC, bits)	17 565	20 022	20 703
System Overhead (O, mgC, bits)	22 517	30 183	21 751
Relative Ascendancy (A/C, %)	28	33.7	48.8
Internal Capacity (C_i, mgC, bits)	32 359	29 880	18 132
Internal Ascendancy (A_i, mg bits)	9 842	8 597	10 451
Internal Relative Ascendancy (A_i/C_i, %)	30.4	28.8	57.6
Flow Diversity (C/T)	3.6	3.5	2.8

involving mainly bacteria and sediment POC, but particularly so in the Bot River estuary where 98% of the cycled flow is involved with the decomposition process on the sediment. In the Swartkops and Kromme estuaries 17.8% and 20.3% of cycling takes place over longer and slower loops, but only a very small fraction (3%) in the Bot River involves longer loops. A simple cycling structure may thus be characteristic of 'closed' systems, where the pulsing effect of tides and the resuspension and redistribution of material is less than in 'open' estuaries. The simple structure may also be a result of stress, such as hypersalinity which is characteristic of this system (Bally & McQuaid 1985). FCI values are highest for the Swartkops estuary at 43.8%, lowest from the Kromme estuary at 26%, with the Bot River estuary recycling about 36% of its total throughput (see Table 11.3).

Another aspect of the cycling process is the Average Path Length (APL) which measures the average number of transfers a unit of flux will experience from its entry into the system until it leaves it, thus effectively describing the average length of the food web. This index is derived from APL = $(T - Z)/Z$ where Z is the sum of all exogenous inputs and T the total system throughput (Field et al. 1989a). It is commonly held that the food chain (or APL) is short in upwelling areas of the ocean, whereas the APL should be longer in, for example, estuaries where energy is passed along a greater number of steps from primary producers to top predators. Recent studies confirmed this notion. Baird et al. (1991) reported APL values of 2.24 and 2.54 for the Peruvian and Southern Benguela upwelling regions respectively, whereas APL values for most estuaries were all greater than 3. For example, this index was calculated as 3.61 for the Chesapeake Bay on the east coast of the USA, 3.42 for the Ems estuary in the Netherlands, 3.95 and 3.22 for the Swartkops and Bot estuaries respectively, and 2.38 for the Kromme estuary. This means that a unit of carbon will be transferred at least one step more in the Swartkops and Bot River estuaries where material and energy thus have a longer residence time than in the Kromme estuary. The shorter APL in the Kromme may well be as a result of the high rate of detrivory in combination with the high rate of energy transfer activity at the lower trophic levels (Baird & Heymans 1996).

When considering the cycle distributions, FCI and APL values, it would appear that the mildly polluted Swartkops estuary and the 'closed' Bot River estuary show cycle distribution different from the 'pristine' Kromme estuary. The former two recycle relatively large proportions (43.8% and 36%) of their total flows through short cycles, and have relatively high APL values. A large proportion of their internal production is thus retained within the systems, and the recycling takes place over a few short circuits in each. The Kromme estuary shows a more even distribution of cycled flows through different path lengths, but has lower APL and FCI values than the other two systems, and thus occupies an intermediate position between 'open' and 'closed' systems.

The topologies of the composite cycling structure of the Swartkops and Kromme are similar, whereas in the Bot River estuary fewer compartments participate in cycling (see Figures 11.4–11.6). Most of the cycling in the Bot River estuary relates to sediment POC and bacteria, while none of the pelagic compartments (e.g. mesozooplankton) appears to be involved. A common feature of estuaries is the resuspension of sedimentary particulate matter, which is then cycled through the suspended POC compartment to other components such as suspension feeders and mesozooplankton. This process is characteristic of 'open' estuaries with relatively fast tidal currents such as the

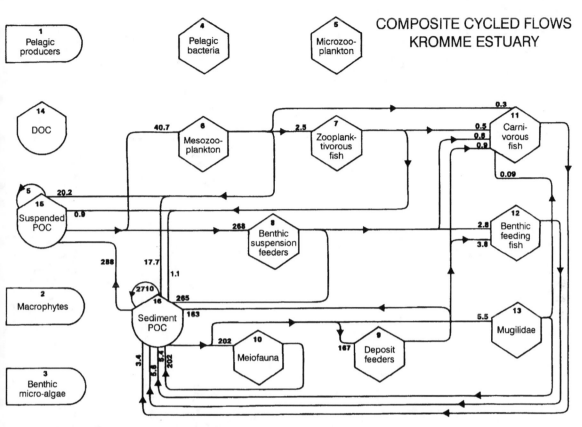

Figure 11.5. **The cycling structure of carbon in the Kromme estuary.** Units of flow and compartment numbers are the same as in Figure

11.1. Feedback loops in compartments 15 and 16 indicate cycling within the compartment between POC and bacteria.

Swartkops and Kromme estuaries (Baird & Ulanowicz 1993). Resuspension is virtually absent in the Bot River estuary, which explains the absence of cycled flows through the suspended POC compartment.

System properties

The results from the analyses of the networks of the three estuary systems are given in Table 11.3. The T (total system throughput), C (development capacity) and A (ascendancy) values of all three systems are within the same order of magnitude. The $A{:}C$ and $A_i{:}C_i$ ratios for the Bot River estuary are higher (see Table 3) than for the other two systems. The internal relative ascendancy ($A_i{:}C_i$) ratio of the Bot River estuary increases by almost 10%, relative to the $A{:}C$ ratio, when only internal fluxes are considered, indicating that this system can maintain its activity without too much dependence on external inputs. The Kromme estuary, on the other hand, shows a decline of 5% in the $A_i{:}C_i$ to $A{:}C$ ratios, pointing to a greater dependency on external connections to adjacent systems.

In comparing three South African systems, it would appear that when only the FCI values are considered, then the Swartkops estuary is the more mature according to Odum's (1969) dictum that the aggregate amount of cycling is an indication of system development. If only the $A{:}C$ ratios are considered, the Bot River estuary would be considered the more mature; thus contrasting inferences. When the cycling structure is considered then we find that the high FCI of the Bot River estuary belies a very simple cycle flow network consisting of only five cycles with most of the activity taking place over short and fast cycles. The Swartkops estuary shows similar cycling features, although it has a more complex cycling topology than the Bot River estuary. The Kromme estuary, on the other hand, presently has a more complex cycling structure with greater proportion of carbon cycled over longer and slower loops (see Table 11.2b). It is also interesting to note that the flow diversity index, C/T, a measure of the number of inter-actions in the food web and the evenness of those flows, are lowest in the Bot, and highest in the Swartkops estuary (see Table 11.3).

In a detailed comparative study of six marine ecosys-tems, Baird *et al.* (1991) concluded that the developmental

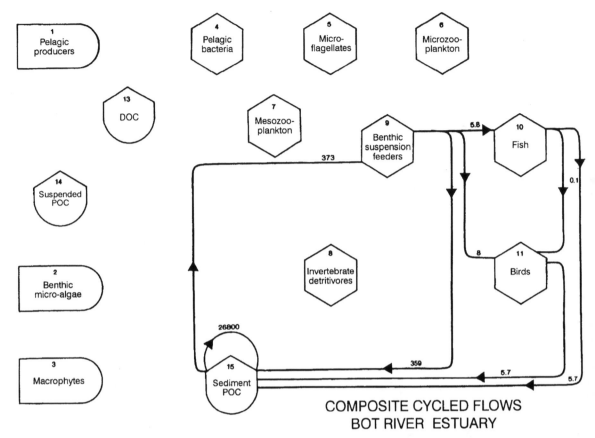

COMPOSITE CYCLED FLOWS
BOT RIVER ESTUARY

Figure 11.6. **The cycling structure of carbon in the Bot River estuary. Units of flow and compartment numbers are the same as in Figure 11.1. Feeback loop in compartment 15 indicates cycling within the compartment between POC and bacteria.**

(or 'maturity') status of a system can be assessed not only from a suite of indices, but also that only dimensionless indices such as $A_i{:}C_i$ ratio and FCI may be used for comparative purposes. Real system values, such as T, C and A, may not be useful for inter-system comparisons, but they are applicable when the same system is compared over time (for example, see Baird & Ulanowicz 1989; Field *et al.* 1989*b*; Baird & Heymans 1996).

Baird *et al.* (1991) have shown in their analyses an inverse relationship between FCI and the $A_i{:}C_i$ ratio. Their results have shown upwelling regions as mature systems having high *A:C* ratios, but simple recycling structures and low FCI values, compared with, for example, the Swartkops estuary with a low *A:C* ratio, but a high FCI. The Swartkops estuary is mildly polluted and thus presumably more stressed, than, for example, the Kromme estuary which exhibits a lower FCI but a slightly higher *A:C* ratio. The Bot River estuary appears to be a contradiction in terms. It retains and cycles a relatively large proposition (36%) of its system throughput, and has a high *A:C* ratio (49%) when compared with the *A:C* ratio of the Swartkops (28%), Kromme (34%) Ythan (34%), and Ems (38%) estuaries,

and the Chesapeake Bay (38%), but lower than that of the Southern Benguela (51%) and Peruvian (48%) upwelling systems, and the Baltic Sea (51%) (Baird *et al.* 1991; Baird & Ulanowicz 1993; the present study). The Bot River estuary occupies an intermediate position between typically closed estuaries, which are opened on a regular basis, and coastal lakes which are permanently cut off from the sea. It may well be that high 'maturity' and recycling indices are typical of 'closed' but periodically opened estuarine systems, benefiting from continuous freshwater and nutrient inputs and episodic marine stimulation. Since many South African estuaries are 'closed', more information on the functioning of these systems is clearly needed.

Temporal changes in ecosystem behaviour: a case study

Ecosystems are not static but are in a continuous process of evolution from one state to another. These changes take place over geological time, and long-term succession and development and are thus rarely documented. However, terrestrial and aquatic systems also undergo short-term changes brought about by seasonal climatic changes which affects standing stocks, production rates and community composition. In some instances longer term fluctuations in, for example, rainfall patterns may also affect ecosystems in one way or another. In addition, human action may also alter the biotic make-up and functional process of a system through the removal or addition of species, chemical pollution, nutrient imbalances, physical changes to the landscape, etc. Often these changes, over short and longer time scales, are subtle and not immediately recognisable, and are hardly comprehensively documented. The question of how to quantify these changes is a vexing one and can only be resolved if comparable data sets on, for example, trends in biomass fluctuations exist and some analytical methods are available to standardise procedures.

Network analysis allows the comparison of different systems on a geographical scale, and the examination of the same system on temporal scales of days, months, seasons or years. For example, Field et al. (1989b) examined the daily succession of a plankton community following an upwelling event, Warwick and Radford (1989) assessed seasonal changes in an estuarine benthic community, Ducklow et al. (1989) the annual N cycle in open ocean plankton systems, and Baird and Ulanowicz (1989) and Baird et al. (1995) the seasonal dynamics of carbon and nitrogen respectively in the mesohaline Chesapeake Bay system.

Although temporal changes in South African estuarine systems do occur, virtually no comprehensive data sets documenting system level changes exist. An exception is the Kromme River estuary for which information exists on biomass and rates of flow between the various living and non-living components over a period of 15 years, 1979 to 1994. The first set of data was collected prior to the construction of a second large impoundment, the Mpofu Dam, in the catchment in 1984, and the second set of information collected during the years 1988–92.

The impoundment severely reduced the mean annual runoff from the catchment into the estuary from 117×10^6 m^3 for the period 1924–80 (Bickerton & Pierce 1988) to an average of about 1.3×10^6 m^3 since 1984 (after the construction of the dam) (Baird et al. 1992). The reduced freshwater input resulted in a virtual homogenous axial salinity regime in the estuary, with salinities varying from 35 in the mouth to 30 at the head. Hypersaline conditions (salinities >35) occasionally occurred in the upper reaches during summer months when evaporation rates are maximal. The normal salinity gradient, a characteristic feature of open estuaries, has been absent since 1984, and has occurred only sporadically during minor flooding

conditions for short periods of time (Heymans & Baird 1995). In this section the data and results collected prior to the construction of the dam in 1984 are compared with more recent information to illustrate changes which may have occurred at the ecosystem level, using network analysis techniques.

Baird and Heymans (1996) postulated that in the absence of any other physico-chemical impacts (CSIR 1991; Baird & Pereyra-Lago 1992; Baird et al. 1992), except reduced freshwater inflows, any observed changes that may have taken place could be ascribed to the altered salinity regime of the estuary. Two detailed flow diagrams were constructed representing the two time periods respectively, each consisting of 27 compartments and illustrated in Figures 11.7 and 11.8. A detailed account of the network model of the latter period (1989–92) is given by Heymans and Baird (1995), in which information on the species composition of the various compartments is given, as well as the flow of energy through each compartment and their associated species. The pre-1984 network was constructed on the same basis, but with different values for biomass and flows as derived from the literature.

Results from the analyses by Baird and Heymans (1996) showed that several attributes of the system and of individual compartments changed significantly. The total standing biomass of the estuary increased from 563 g C m^{-2} in 1984 to 648 g C m^{-2} in 1992, an increase of about 15%. Although the rate of primary production has increased slightly (from 5.20 to 5.52 g C m^{-2} d^{-1}), significant changes took place in the standing stock of the producers. Phytoplankton biomass decreased from 3.7 g C m^{-2} prior to 1984 to 0.4 g C m^{-2} in 1992, whilst the biomass of submerged macrophytes (particularly Zostera capensis and Caulerpa falciformis) increased from 59.9 g C m^{-2} to 125.7 g C m^{-2} (Adams 1991) (see Figures 11.7 and 11.8). Reasons for these changes are that dissolved inorganic nutrients in fresh water inflows support phytoplankton production and biomass, whereas a decrease in turbidity and current velocities, and a more stable sediment and salinity environment contributed to higher macrophyte production (Hanekom and Baird 1988, Adams and Talbot 1992, Adams 1994). It would thus appear that systems with substantial and sustained freshwater inflow have, in general, higher pelagic phytoplankton production as in the Swartkops estuary, whereas little freshwater inflow and reduced salinity gradients favour the growth of submerged macrophytes.

The standing stocks of the heterotrophic components (compartments 5–24 in Figures 11.7 and 11.8) doubled from 22.6 g C m^{-2} prior to 1984 to 48.8 g C m^{-2} in the early 1990s. However, not all of the heterotrophic components increased: in fact, some showed a noticeable decrease in standing stock. Of note is the fact that water column com-

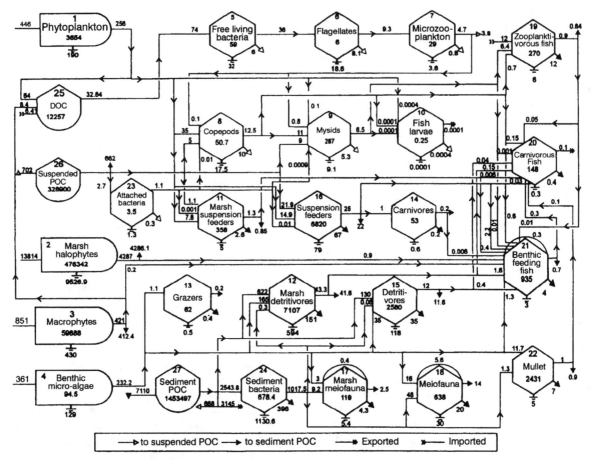

Figure 11.7. **Carbon flow model of the Kromme estuary representing the pre-1984 period. Biomass given in mg C m⁻² and flows in mg C m⁻² d⁻¹.**

munities (compartments 5–9) decreased, with copepods (comp. 8) and mysids (comp. 9) in particular declining dramatically (see Figures 11.7 and 11.8). The decline in copepods could be attributed to the reduction in phytoplankton, their primary food source, whilst mysids declined due to a reduction in their main prey, copepods. The collective production rate of all living compartments, however, decreased slightly from 6.87 to 6.79 g C m⁻² d⁻¹, whilst the annual P:B ratio, a system index, decreased by about 14% from 4.46 to 3.82.

Most of the increase in biomass occurred in marsh suspension feeders (comp. 11), mainly in the mudprawn *Upogebia africana*, marsh detritivores (comp. 12), mainly *Paratylodiplax algoensis*, *P. edwardsii* and *Sesarma catenata*, and suspension feeders in the non-marsh region of the estuary (mainly *Lamya capensis*, *Loripes clausus*, *Macoma litoralis*, and *U. africana* (comp. 16). There was also a noticeable increase in the two species of carnivorous gastropods, *Hydatina physis* and *Natica genuana* (comp. 14), as well as in the standing stocks of several fish components.

Zooplanktivorous fishes (mainly *Gilchristella aestuarius* and *Atherina breviceps*) increased from 270 mg C m⁻² in 1984 to 715 mg C m⁻² in 1992, benthic feeding fish (mainly *Lithognathus lithognathus and Rhabdosargus globiceps*) and *R. holubi* (an epiphytic browser), and species of the family Gobiidae (comp. 21) increased from 935 to 2249 mg C m⁻². Species of the family Mugilidae (mullets, comp. 22) also increased from 2431 to 3100 mg C m⁻² during the intervening period (see Figures 11.7 and 11.8).

Of interest is the decrease in the direct consumption of primary producers (plants) by herbivores. During the pre-1984 period about 8.6% of NPP was consumed by herbivores, whereas only 4.1% was utilised by 1992. This decline in consumption efficiency is mainly attributable to the reduction in phytoplankton production and consumption by zooplankton. Macrophyte consumption is low (Marais 1984), thus little of the production is directly utilised and most of it appears to be reduced to detritus and channelled through the detritus food web (Hanekom & Baird 1988). There was an increase in the suspension feeders and detri-

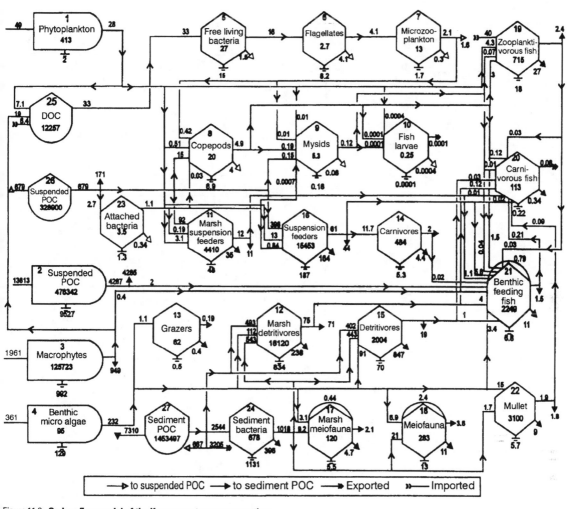

Figure 11.8. **Carbon flow model of the Kromme estuary representing the period 1988–92. Biomass given in mg C m^{-2} d^{-1}.**

tivores and thus an increase in detritus consumption, which is reflected in the detrivory:herbivory ratio which increased from 8:1 in the pre-1984 years to 15:1 in 1992.

The carbon flow models of the Kromme estuary, depicted in Figures 11.7 and 11.8 and representing the pre- and post-1984 periods respectively, indicate that the configuration of the flow structure has remained essentially the same. No large changes in species composition or the presence of 'new' species have been observed since 1984. Changes in biomass in several components and in the rates of flows in the system between the various compartments were, however, clearly observed, which could have had an effect on the trophic and cycling structures, and on system properties; these possibilities are reviewed below.

The trophic efficiency (i.e. the efficiency of energy and material transfer in an ecosystem) can be obtained by the mapping of complex networks of trophic transfers into an abstract, linear food chain consisting of a number of discrete but finite steps or trophic levels, the so-called 'Lindeman Spine'. This process is analogous to principal component analysis which allows the aggregation of a multi-species assemblage into a number of smaller principal components (Field *et al.* 1989a). The Lindeman Spine illustrates the net amount each level receives from the previous one, as well as the amount it creates through respiration, exports, detritus for recycling, and the net production transmitted to the next level. The non-living organic detrital pool and autotrophs are usually merged and assigned to the first trophic level.

The flow networks of the two time periods under consideration were mapped into Lindeman Spines by using the programme NETWRK (Ulanowicz & Kay 1991), together with their associated routes of detritus recycling and illustrated in Figures 11.9a and b. Detrital returns from all trophic levels are shown as well as the loop of

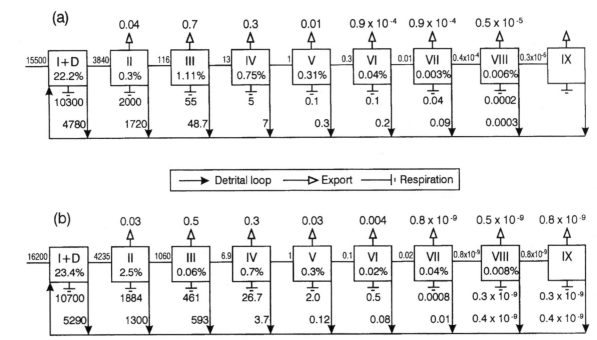

Figure 11.9. **The Lindeman Spine for (a) the pre-1984 period and (b) the years 1988–92 representing the aggregation of the flow networks of the Kromme estuary (Figures 11.7 and 11.8) into concatenated chains of transfers for the two periods, respectively. Primary** producers and detritus are merged into the first trophic level I+D). Percentage values in boxes refer to the trophic efficiency at each level.

detrital returns from plants (I) which are merged with the detrital pool (D) to represent the first trophic level. Respiration and export losses from each level are shown as well as the flow of energy (in mg C m^{-2} d^{-1}) from one level to the next. The trophic efficiencies, defined as that fraction of the total input into a level that is passed on to the next, are given as percentages in the boxes. Each spine consists of nine levels, but negligible transfers occur at the sixth and higher trophic levels. The trophic level efficiencies generally decrease with ascending level, although the efficiencies at the fifth level in the pre-1984 and post-1984 spines are unexpectedly higher than the preceding ones. This means that the system is quite effective in delivering resources to the higher trophic components, despite the relatively low efficiencies over the first few transfers (Baird & Ulanowicz 1993).

Results from the Lindeman Spine analysis indicate reduced transfer efficiencies at the first two levels during the period 1988–92 (25%) than for the same levels for the pre-1984 period (48%: see Figure 11.9a and b). These values are low when compared to other systems such as the Swartkops (45%), the Ems (77%) and the Ythan (53%) estuaries (cf. Baird & Ulanowicz 1993; Baird & Heymans 1996). The lower efficiency value at the lower trophic levels in the Kromme estuary points to a less effective direct utilisation of primary producer products at present than during earlier years (pre-1984). The increase in the detritivory:herbivory ratio (from 8:1 to 15.5:1), however, indicates that the increase in trophic efficiency at the second level (see Figures 11.9a and b) is attributed to an increase in the uptake of detrital material as opposed to grazing on plants. The geometric mean of the trophic efficiencies in the pre-1984 period was 4.5, nearby double that of 2.8 calculated for recent years, and indicates a drastic decrease in the overall efficiency of energy utilisation and transfer in the system.

Although the FCI and the APL in the Kromme estuary remained relatively constant over the past 15 years (i.e. at about 11% for the FCI and 1.8 for the APL for the 27 compartment models: see Baird & Heymans 1996), subtle but significant changes occurred in cycling structure. The number of cycles remained constant at 100 in the pre-1984 period and 99 during the years 1988–92, and there was a fairly even distribution of cycles per nexus during both periods. However, a surprisingly large nexus of 16 cycles was present in the pre-1984 period in which the weak arc (i.e. representing the slowest flow between two connecting compartments in the cycling structure) existed between attached bacteria and copepods. This nexus was not present in the 1988–92 cycle structure and can be attributed to the much reduced abundance of phytoplankton and copepods and their subsequent diminished role in the dynamics of the food web (Baird & Heymans 1996). A shift in the amount of flow cycled through various path lengths also occurred during the intervening period. During the pre-1984 period more than 55% of cycled flow occurred via longer and slower loops of path lengths of 3 and more. By 1992, 85% of the recycling took place over short path lengths involving only two compartments, pointing to a high rate of decomposition and detritus production. It would appear that due to reduced water column productivity (e.g. phytoplankton and zooplankton) and the commitant increase in benthic primary production, the Kromme estuary has become more dependent on the recycling of unutilised macroalgae material, enhancing the flow of energy through the detritus food web.

Changes have also been observed in the global indices of system organisation. There was an increase in the total system throughput (or system activity T) by about 7% from pre-1984 to 1992, about a 13% increase in development capacity, 8% in ascendancy and 23% in system overhead (or redundancy). Despite an increase in C and A, the relatively large increase of 23% in system overhead indicates that the cost to the system to function the way it does has increased since 1984. There was, however, a decline in both the relative ascendancy $A{:}C$ and the normalised internal $A_i{:}C_i$ ratios. The $A{:}C$ ratio declined from 48% to 46% and the $A_i{:}C_i$ ratio from 40% to 38% since 1984, and it can be inferred that the system as a whole declined somewhat in terms of its maturity and internal organisation. This is also reflected in the decline of the P:B ratio (from 4.5 to 3.8) which is considered to be an index of the functional status of an ecosystem.

This study on the temporal evolution of the Kromme estuarine system has shown that the level of organisation and maturity of the system has declined since the inflow of fresh water was severely reduced in the early 1980s. Changes were observed not only at the systems level, but also strongly at the individual component level. Fluctuations of individual species are also more likely to be observed than fluctuations at the system level, which may be subtle and not easily detected. Baird and Heymans (1996) concluded that the Kromme system has, probably irreversibly, settled in a new equilibrium state at a different level of organisation.

Simulation modelling of estuarine processes

The concept of modelling ecosystem behaviour originated in the 1940s when quantitative explanations of plankton production were sought through food requirements at different trophic levels (Laevatsu & Favorite 1981). The modelling of physical and biological oceanographic processes has since then gained substantial interest, and there is an abundant and growing literature on simula-

tion modelling. There are two approaches to the modelling of biological systems, namely the reductionist and the holistic. Reductionist modelling simulates distinct processes in isolation from the whole system, whereas the system approach attempts to simulate whole system behaviour, and may be represented mathematically by a number of coupled process models. The goal of simulation modelling is to mimic the behaviour of processes or systems, to test hypotheses, and to make realistic predictions of the future state of that system under arbitrary environmental conditions. Numerous system models have been constructed which gave insights into the behaviour of particular ecosystems. Although we have learned much from reductionist methods about how processes and subsystems interact, the predictive capabilities of whole system models, and even of coupled process models, fall short of expectations, especially in the area where models are needed most, namely in making predictions for management. Reasons for errant predictions are that simulation predictions may involve ranges of variables which lie outside the range of measurements of the system, the inherent variability of biological parameters, the hierarchical nature of ecosystems, the adaptability of parts, and the capacity for self-organisation that natural ecosystems possess (Mann et al. 1989). For example, Dugdale (1975) has shown how difficult it becomes to simulate systems in which the cast of species changes rapidly and arbitrarily, and Lorenz (1969) demonstrated how non-linearities in system equations often give rise to progressive amplification of errors in the data used to initiate the simulations.

Examples of simulation models are plentiful and include excellent attempts by, for example, Jansson et al. (1984) on the Baltic Sea, Kremer and Nixon (1978) on Narrangansett Bay, a coastal marine ecosystem, Stigebrandt and Wulff (1987) on the dynamics of nutrients and oxygen in the Baltic Sea, Summer et al. (1980) on the coupling of an estuarine subsystem and the sea, and Moloney et al. (1986) on the effect of sedimentation and microbial degeneration in a plankton community. Extensive studies on the Bristol Channel and the Severn estuary in the UK led to the construction of the GEMBASE (General Ecosystem Model for the Bristol Channel and Severn Estuary) Systems model at the Plymouth Marine Laboratory (Radford & Uncles 1980), simulating, among other applications, the potential impact of a barrage in the Severn estuary on the estuarine ecosystem.

One of the most comprehensive simulation modelling exercises undertaken for an estuarine ecosystem is that of Baretta and Ruardij (1988) on the Ems estuary in the Netherlands. They divided the estuary along its longitudinal axis into five compartments, linked together with a transport model simulating the exchange of water, dissolved and particulate matter between them. Each of the compartments was modelled by means of three similar interlinked sub-models simulating biological processes in the pelagic, benthic and epibenthic subsystem. By analysing and simulating carbon flow in the networks of the various compartments and relevant subsystems, Baretta and Ruardij (1988) showed the whole systems model to be dynamically stable and that abiotic variables influenced the predictive capabilities of the model, and raised the question whether the reductionist approach to subsystem modelling is applicable to whole system modelling. Although the predictive outputs of the model were somewhat suspect, Baretta and Ruardij (1988) concluded that the model was nevertheless most useful in exploring the possible consequences of man-made stresses on the system.

Simulation modelling of South African estuaries is still in its infancy. Although some processes have been modelled, e.g. carbon and inorganic nutrient flux across the estuary–sea boundary (Baird et al. 1987; Baird & Winter 1990, 1991), closure and breaching of an estuarine mouth and accompanying changes in salinity and tidal fluxes (Slinger 1996a), mathematical modelling of estuarine macrophyte dynamics (Busse et al. 1994), and the use of a one-dimensional hydrodynamic model to simulate water levels, current velocities and transport processes, system modelling has to date received little attention. A noticeable exception is the attempt by Slinger (1996a, b) to develop a generic modelling system to provide predictions of the physical, chemical and/or biological responses of South African estuaries to variable rates of freshwater inflow.

Slinger (1996b) used five models relating to the determination of the freshwater requirements of estuaries, and by linking these models was able to develop a protocol for prediction and subsequent management applications. The models used to develop an estuarine modelling system were: (i) the MIKE 11 modelling system (Danish Hydraulics Institute 1992) which includes one-dimensional hydrodynamic, transport dispersion, and water quality modules with outputs in the form of time series of water levels, current velocities and salinity concentrations along the length of an estuary; (ii) a plant–estuarine decision support system which provides information on the response of estuarine plant communities (subtidal, intertidal and supratidal macrophytes, phytoplankton) to variable freshwater inputs; (iii) an estuarine systems model designed to provide medium to long-term time histories of the physical dynamic aspects of an estuary (height of the sill at the mouth, water levels, mean salinities, stratification and circulation state) under different freshwater inflow scenarios, and to evaluate the efficacy of management policies involving freshwater releases and mouth breachings; (iv) an estuarine ecosystem evaluation model which provides indices of ecosystem performance under different inflow and mouth conditions by simulating the response of the mudprawn (Upogebia africana) in terms of biomass and the population size structure, as well as the success of fish recruitment; (v) a dynamic vegetation model which yields time histories of the biomass and production rates of selected estuarine macrophytes (Zostera capensis, Ruppia cirrhosa and Phragmites australis) in relation to different freshwater inflow conditions; and (vi) a faunal prediction module which comprises a formal procedure for acquiring expert opinions on the potential response of various estuarine faunal communities and/or species.

The underlying logic of the model linkages was that if the physical and chemical responses of the estuary could be predicted in relation to different freshwater inflow scenarios, then the resultant responses of the various biotic components of the estuary are also predictable. To test the applicability of the linked modelling system and explore the limitations and ranges of applicability of the individual models, two estuaries, namely the Kromme estuary (a large, permanently open system) and the Great Brak estuary (a small, temporarily open system) were subjected to the analyses as case studies. Both rivers had been impounded by the building of a dam quite close to the limit of tidal effect.

A variety of freshwater inflow scenarios were formulated, and Slinger (1996b) provided detailed predicted responses of each of the modules to these scenarios. For example, the response of Kromme estuary to inflow rates ranging from the natural runoff situation, to situations resembling freshwater inflows of 10%, 20% and 40% of the natural mean annual runoff (MAR), to situations of no freshwater inflows for up to three years were simulated, yielding interesting and valuable output results. The estuarine systems model successfully predicted the response of the physical environment under different inflow scenarios, particularly the axial salinity differences. An important output was that fish recruitment will be severely affected should freshwater inflows drop below 10% of the MAR owing to sharp reductions in axial salinity differences. Results from the dynamic vegetation model were successfully validated against an expert plant estuarine decision support system and the prediction of the response of Zostera to various inflow regimes appeared particularly acceptable. The invertebrate production index, however, did not prove to be a sensitive indicator of ecosystem functioning, owing to the fact that hypersalinities were not well predicted, nor were combined temperature–salinity conditions successfully modelled.

For the Great Brak case study, a variety of freshwater inflow scenarios and mouth breaching actions were formulated. These encompassed inflows characteristic of the natural situation, a pre-impoundment situation and wet and dry periods under the post-impoundment scenario. The responses of various estuarine components were then predicted using the different models (Slinger 1996b).

Significant results included the simulation of the closure of the mouth, and the subsequent infilling and breaching of the estuary as predicted by the estuarine systems model. The provision of concurrent time histories of mouth condition, water levels, salinities, tidal fluxes and stratification–circulation conditions enabled simulations and predictions of the medium to long term responses of biotic components in a temporarily open system under various flow conditions. The plant estuarine decision support system and the dynamic vegetation model concurred in their predictions, indicating increasing dominance by salt-tolerant species and the displacement of *Phragmites* beds by *Zostera* under the post-impoundment dry period scenario. The predictions of the responses of the fish and the estuarine mudprawn confirmed the deleterious consequences of the post-dam dry period, indicating that *Upogebia africana* population densities would be about 30% of the natural situation and that fish recruitment would be severely curtailed. These scenarios assumed continuous releases rather than pulse releases, which can also be tested. The predictions, therefore, do not reflect the optimal use of the available freshwater under the post-dam dry period.

Thus, while the estuarine modelling system has many limitations and cannot hope to simulate a whole ecosystem response, it is anticipated that further biological simulation will be stimulated by the provision of appropriate information on the altered abiotic conditions in response to changes in freshwater inflow (Slinger 1996b).

General conclusions

A review of available information on estuaries showed that, with the exception of three systems, insufficient data are available to construct representative and quantified flow models and to analyse these on a holistic basis. Our knowledge on the behaviour of estuarine ecosystems in response to physical perturbations and variability thus remains relatively sparse. However, this is a universal problem and not confined only to South Africa. The question of the characterisation of ecosystems under stress remains largely unanswered, although considerable progress has been made in that field over the past twenty years.

Ecosystems are complex entities, and they undergo changes over short and long time scales in response to external perturbations such as rapid temperature change, sudden or chronic nutrient addition, organic loading, etc. There appears to be, however, no clear consensus as to how to quantify the effects of stress on ecosystems under all circumstances. Odum's (1969) twenty-four ecosystem attributes appear to correlate with community development, including ratios such as biomass to production which increases in relatively undisturbed systems. One of the most widely measured indices of system response to perturbations, or 'stress' is species diversity. This index is usually derived, however, for specific sub-sets of communities such as benthic invertebrates, zooplankton, etc. present in an ecosystem, and is seldom calculated for the whole community from bacteria to vertebrates. Thus diversity indices represent only small subsystems which may or may not behave in the same way as the whole system.

Platt et al. (1981) are of the opinion that information on ecosystem processes is more likely to provide answers to questions on community dynamics. Thus, the flow structure and magnitude of flows in an ecosystem are more likely to reflect signs of impact than do individual compartments. This concept is clearly illustrated in the Kromme estuary by Baird and Heymans (1996), who illustrated relatively large fluctuations in the abundance and productivity of some species and communities over time, but only subtle changes at the ecosystem level. It would appear that changes at the species, community and system level were brought about through a reduction in freshwater inflow and a subsequent change in the overall salinity regime in the estuary. At the system level the structure of cycling changed, but not the magnitude. With the disappearance of the more complex and longer recycling pathways the cycling structure became simpler, which is, according to Odum (1969), indicative of a system in an early development phase and of one under stress. Other system attributes such as the P:B, relative A:C and normalised A_i:C_i ratios declined over time (see Table 11.3) also pointing to the possibility that the system is now more stressed to some degree or other than it was before the advent of reduced freshwater inflows.

In their comparison of marine ecosystems on a global scale Baird et al. (1991) and Baird and Ulanowicz (1993) concluded that a high FCI may not be a characteristic of mature systems, but, when viewed in conjunction with the cycling structure, rather an indication of stress. The Swartkops estuary, for example, has in comparison with other estuarine systems a high FCI, a simple cycling structure, and a low P:B value, indicating a system under stress. The closed Bot River estuary exhibits a similar trend in the value of these system properties, also indicative of stress on the system. It is interesting to note that the stress imposed on these two systems are different in origin: the Swartkops estuary is impacted by chemical and sewage pollution, where as the Bot River estuary's stress is probably mediated more by physical processes, i.e. closing of the mouth, reduced salinities) than by chemical pollution. Clearly the characterisation and assessment of the magnitude of stress on estuarine ecosystems should include descriptions of both the dynamic properties of the system and the abiotic forcing functions.

This chapter has attempted to synthesise information of estuarine ecosystem in a holistic manner and although the dynamics of whole systems are not yet fully understood, significant progress in this field has taken place over the past two or three decades. Analytical and mathematical procedures and methodology have been developed at an international level and have been successfully applied to local systems, including estuaries, in South Africa. The comparison between different systems from different geographic regions in South Africa has highlighted differences in functional processes, such as cycling, and in system properties such as the magnitude of cycling, growth and development. A study on the Kromme estuary over time showed clearly the response of not only individual species or trophic related groups of species to changes in the freshwater inflow regime, but also of the system as a whole.

References

Adams, J. (1991). *The distribution of estuarine macrophytes in relation to fresh water in a number of eastern Cape estuaries.* MSc thesis, University of Port Elizabeth.

Adams, J. (1994). *The importance of freshwater to the survival of estuarine plants.* PhD thesis, University of Port Elizabeth.

Adams, J. & Talbot, M.M.B. (1992). The influence of river impoundments on the estuarine sea grass, *Fostera capensis* Setchel. *Botanica Marina*, **35**, 69–75.

Allanson, B.R. (1992). *An assessment of the SANCOR Estuaries Research Programme from 1980 to 1989.* Committee for Marine Science Occasional Report. No. 1. Pretoria: Foundation for Research Development.

Allen, P.M. (1985). Ecology, thermodynamics and self-organisation: towards a new industry of complexity, In *Ecosystem theory for biological oceanography*, ed. R.E. Ulanowicz & T. Platt. *Canadian Bulletin of Fisheries and Aquatic Sciences*, **213**, 139–62.

Asmus, M.L. & McKellar, H.N. (1989). Network analysis of the North Inlet salt marsh ecosystem. In *Network analysis in marine ecology: methods and applications*, ed. F. Wulff, J.G. Field & K.H. Mann. *Coastal and Estuarine Studies*, **32**, 220–31. Heidelberg: Springer-Verlag.

Baird, D. (1988). Synthesis of ecological research in the Swartkops Estuary. In *The Swartkops Estuary: Proceedings of a symposium held on 14 and 15 September, 1987 at the University of Port Elizabeth*, ed. D. Baird, J.F.K. Marais & A.P. Martin.

South African National Scientific Programmes Report No. 156, 41–56.

Baird, D., Hanekom, N.M. & Grindley, J.R. (1986). Estuaries of the Cape. In *Estuaries of the Cape, Part II,: Synopsis of available information on individual systems*, eds. A.E.F. Heydorn & J.R. Grindley, Report No. 23, Swartkops. *CSIR Research Stellenbosch. Report* **422**, 1–82.

Baird, D., & Heymans, J.J. (1996). Assessment of ecosystem changes in response to fresh water inflow in the Krom River estuary, St Francis Bay, South Africa: a Network Analysis approach. *Water SA.*, **22**, 307–18.

Baird, D., Mcglade, J.J. & Ulanowicz, R.E. (1991). The comparative ecology of six marine ecosystems. *Philosophical Transactions of the Royal Society, London* B, **333**, 15–29.

Baird, D., Marais, J.F.K., & Bate, G.C. (1992). An environmental analysis for the Krom River area to assist in the preparation of a structure plan. *Institute for Coastal Research, University of Port Elizabeth, South Africa. Report No. C*, **16**, 1–56.

Baird, D., Marais, J.F.K. & Martin, A.P. (eds.) (1988). The Swartkops Estuary. Proceedings of a symposium held on 14 and 15 September, 1987 at the University of Port Elizabeth. *South African National Scientific Programmes Report No.* **156**, 1–107.

Baird, D. & Pereyra-Lago, R. (1992). Nutrient status and water quality assessment of the Marina Glades canal system. Krom Estuary, St Francis Bay. *Water SA*, **18**, 37–42.

Baird, D. & Ulanowicz, R.E. (1989). The

seasonal dynamics of the Chesapeake Bay ecosystem. *Ecological Monographs*, **59**, 329–64.

Baird, D. & Ulanowicz, R.E. (1993). Comparative study on the trophic structure, cycling and ecosystem properties of four tidal estuaries, *Marine Ecology Progress Series*, **99**, 221–37.

Baird, D., Ulanowicz, R.E. & Boynton, W.R. (1995). Seasonal nitrogen dynamics in Chesapeake Bay: a network approach. *Estuarine, Coastal and Shelf Science*, **41**, 137–62.

Baird, D. & Winter, P.E.D. (1990). Annual flux of dissolved nutrients through a well-mixed estuary. In *Estuary water quality management: monitoring, modelling and research*, ed. W. Michaelis. *Coastal and Estuarine Studies*, **36**, 335–40. Heidelberg: Springer-Verlag.

Baird, D. & Winter, P.E.D. (1991). The exchange of phosphate between the Swartkops estuary and Algoa Bay, South Africa. *South African Journal of Science*, **87**, 192–7.

Baird, D., Winter, P.E.D. & Wendt, G. (1987). The flux of particulate material through a well-mixed estuary. *Continental Shelf Research*, **7**, 1399–1403.

Bally, R. (1985). Historical records of the Bot River estuarine system. *Transactions of the Royal Society of South Africa*, **45**, 323–45.

Bally, R, & McQuaid, C.D. (1985). Physical and chemical characteristics of the water of the Bot River estuary, South Africa. *Transactions of the Royal Society of South Africa*, **45**, 317–22.

Bally, R., and McQuaid, C.D. & Pierce, S.M. (1985). Primary productivity of the Bot

River estuary, South Africa. *Transactions of the Royal Society of South Africa*, **45**, 333–46.

Baretta, J. & Ruardij, P. (eds.) (1988). Tidal flat estuaries: simulation and analysis of the Ems Estuary. *Ecological Studies*, **71**, 1–353. Heidelberg: Springer-Verlag.

Bennett, B.A. (1989). The diets of fish in three south-western Cape estuarine systems. *South African Journal of Zoology*, **24**, 163–77.

Bennett, B.A. & Branch, G.M. (1990). Relationships between production and consumption of prey species by resident fish in the Bot, a cool temperate South African estuary. *Estuarine, Coastal and Shelf Science*, **31**, 139–55.

Bickerton, I.B. & Pierce, S.M. (1988). Estuaries of the Cape. In *Synopsis of available information on individual systems, Part II*. eds. A.E.F. Heydorn, P.D. Morant, Rep. No. 33, Krom (CMS 45), Seekoei (CMS 46) and Kabeljous (CMS 47). *Stellenbosch CSIR Research, Report* **432**, 1–109.

Branch, G.M., Bally, R., Bennett, B.A., De Decker, H.P., Fromme, G.A.W., Heyl, C.W. & Willis, J.P. (1985). Synopsis of the impact of artificially opening the mouth of the Bot River estuary: implications for management. *Transaction of the Royal Society of South Africa*, **45**, 465–83.

Busse, J., Hearn, J.W. & Adams, J. (1994). Mathematical modelling of estuarine macrophytes. *Paper presented at a symposium on Aquatic Ecosystems: Ecology, Conservation and Management*, 13–16 July 1994, Port Elizabeth.

Christensen, V. (1995). Ecosystem maturity – towards quantification. *Ecological Modelling*, **77**, 3–32.

Coetzee, D.J. (1985). Zooplankton and some environmental conditions in the Bot River estuary. *Transactions of the Royal Society of South Africa*, **45**, 363–77.

CSIR (1991). Sedimentation in the Krom estuary: data report. *CSIR Report No. EMA-D 9108*. Stellenbosch: CSIR.

Danish Hydraulics Institute (1992) Mike 11. A microcomputer based modelling system for rivers and channels. *Reference Manual Danish Hydraulics Institute, Holstrom, Denmark.*

De Decker, H.P. & Bally, R. (1985). The benthic macrofauna of the Bot River estuary, South Africa, with a note on its melofauna. *Transactions of the Royal Society of South Africa*, **45**, 379–97.

Downing, J.A. (1991). Comparing apples with oranges: methods of interecosystem comparison. In *Comparative analysis of ecosystems: patterns, mechanisms and theories*, ed. J. Cole, G. Lovett & S. Findlay, pp. 24–45. New York: Springer-Verlag.

Ducklow, H.W., Fasham, M.J.R. & Vézina, A.F. (1989). Derivation and analysis of flow networks for open ocean plankton systems. In *Network analysis in marine ecology: methods and applications*, ed. F. Wulff, J.G. Field & K.H. Mann. *Coastal and Estuarine Studies*, **32**, 159–205.

Dugdale, R.C. (1975). Biological modelling. In *Modelling of marine systems*, ed. J.C.J. Nihoul, pp. 187–205. Amsterdam: Elsevier.

Fasham, M.J.R. (ed.) (1984). *Flows of energy and material in marine ecosystems: theory and practice*. New York: Plenum.

Fasham, M.J.R. (1985). Flow analysis of materials in the marine euphobic zone. In *Ecosystem theory for biological oceanography*, ed. R.E. Ulanowz & T. Platt. *Canadian Bulletin of Fisheries and Aquatic Sciences*, **213**, 139–62.

Field, J.G., Moloney, C.L. & Attwood, C.G. (1989a). A detailed guide to network analysis. In *Network analysis in marine ecology: methods and applications*, ed. F. Wulff, J.G. Field & K.H. Mann. *Coastal and Estuarine Studies*, **32**, 15–61. Heidelberg: Springer-Verlag.

Field, J.G., Wulff, F. & Mann, K.H. (1989b). The need to analyse ecological networks. In *Network analysis in marine ecology: methods and applications*, ed. F. Wulff, J.G. Field & K.H. Mann. *Coastal and Estuarine Studies*, **32**, 3–12. Heidelberg: Springer-Verlag.

Finn, J.T. (1976). Measures of ecosystem structure and function derived from analysis of flows. *Journal of Theoretical Biology*, **56**, 363–84.

Fromme, G.A.W. (1985). The hydrology of the Bot River estuary. *Transactions of the Royal Society of South Africa*, **45**, 305–16.

Hanekom, N. & Baird, D. (1988). Distribution and variations in the seasonal biomass of eelgrass *Zostera capensis* in the Krom Estuary, St Francis Bay, South Africa. *South African Journal of Marine Science*, **7**, 51–9.

Hannon, B. (1973). The structure of ecosystems. *Journal of Theoretical Biology*, **41**, 535–46.

Hannon, B. (1985). Ecosystem flow analysis. In *Ecosystem theory for*

biological oceanography, ed. R.E. Ulanowicz & T. Platt. *Canadian Bulletin of Fisheries and Aquatic Sciences*, **213**, 139–62.

Heÿl, C.W. & Currie, M.H. (1985). Variations in the use of the Bot River estuary by water-birds. *Transactions of the Royal Society of South Africa*, **45**, 397–417.

Heymans, J.J. & Baird, D. (1995). Energy flow in the Krom estuarine ecosystem, St Francis Bay, South Africa. *Estuarine, Coastal and Shelf Science*, **41**, 39–50.

Jansson, B.-O., Wilust, W. & Wulff, F. (1984). Coupling the sub-systems – the Baltic Sea as a case study. In *Flows of energy and materials in marine ecosystems*, ed. M.J. Fasham, pp. 549–95. New York: Plenum.

Kay, J.J. (1984). *Self-organization in living systems*. PhD thesis, University of Waterloo, Ontario, Canada.

Kay, J.J., Graham, L.A. & Ulanowicz, R.E. (1989). A detailed guide to network analysis. In *Network analysis in marine ecology: methods and applications*, ed. F. Wulff, J.G. Field & K.H. Mann. *Coastal and Estuarine Studies*, **32**, 15–61.

Koop, K., Bally, R. & McQuaid, C.D. (1983). The ecology of South African estuaries. Part XII. The Bot River, a closed estuary in the south-western Cape. *South African Journal of Zoology*, **18**, 1–10.

Kremer, J.N. & Nixon, S.W. (1978). *A coastal marine ecosystem: simulation and analysis*, pp. 1–217, Heidelberg: Springer-Verlag.

Laevatsu, T. & Favorite, F. (1981). Holistic simulation models of shelf-sea ecosystems. In *Analysis of marine ecosystems*, ed. A.R. Longhurst, pp. 701–27. London: Academic Press.

Leontief, W. (1951). *The structure of the American economy, 1919–1939*, Second edition. New York: Oxford University Press.

Lorenz, E.N. (1969). The predictability of a flow which possesses many scales of motion. *Tellus*, **21**, 29–307.

McLusky, D.S. (1981). *The estuarine ecosystem*. Glasgow: Blackie.

Mann, K.H., Field, J.G. & Wulff, F. (1989). Network analysis in marine ecology: an assessment. In *Network analysis in marine ecology: methods and applications*, ed. F. Wulff, J.G. Field & K.H. Mann, *Coastal and Estuarine Studies*, **32**, 260–82.

Marais, J.F.K. (1984). Feeding ecology of major carnivorous fish from four eastern Cape estuaries. *South African Journal of Zoology*, **19**, 210–23.

Martin, A.P., & Baird, D. (1987). Seasonal abundance and distribution of birds on the Swartkops Estuary, Port Elizabeth. *Ostrich*, **38**, 122–34.

Moloney, C.L., Bergh, M.D., Field, J.G. & Newell, R.C. (1986). The effect of sedimentation and microbial nitrogen regeneration in a plankton community: a simulation investigation. *Journal of Plankton Research*, **8**, 427–45.

Müller, F. & Leupelt, M. (eds.) (1997). *Ecotargets, goal functions, and orientors*. Heidelberg: Springer-Verlag.

Odum, E.P. (1969). The strategy of ecosystem development. *Science*, **164**, 262–70.

Platt, T., Mann, K.H. & Ulanowicz, R.E. (1981). Mathematical models in biological oceanography. *UNESCO Monographs on Oceanographic Methodology 7*. Paris: UNESCO Press.

Radford, P.J. & Uncles, R.J. (1980). Ecosystem models and the prediction of ecological effects. In *An environmental appraisal of tidal power stations with particular reference to the Severn Barrage*, ed. T.L. Shaw, pp. 109–27. London: Pitman.

Slinger, J. (1996a). A co-ordinated research programme on decision support for the conservation and management of estuaries. Project No. K5/577/0/1 undertaken for the Water Research Commission. *Pretoria Final Report of the Predictive Capability sub-project*.

Slinger, J. (1996b). Modelling the physical dynamics of estuaries for management purposes. PhD thesis, University of Natal, Pietermaritzburg.

Stewart B.A. & Bally, R. (1985). The ecological role of the red-knobbed coot *Fulica cristata* Gmelin at the Bot River estuary, South Africa: a preliminary investigation. *Transactions of the Royal Society of South Africa*, **45**, 419–26.

Stigebrandt, A. & Wulff, F. (1987). A model for the dynamics of nutrients and oxygen in the Baltic proper. *Journal of Marine Research*, **45**, 729–59.

Strayer, D.L. (1991). Comparative ecology and undiscovered public knowledge. In *Comparative analysis of ecosystem: patterns, mechanisms and theories*, ed. J. Cole, G. Lovett & S. Findlay, pp. 3–6. New York: Springer-Verlag.

Summer, J.K., McKellar, H.N., Dave, R.F. & Kithcens, W.M. (1980). A simulation model of estuarine subsystem coupling and carbon exchange with the sea. II. North Inlet model structure, output and validation. *Ecological Modelling*, **11**, 101–40.

Ulanowicz, R.E. (1983). Identifying the structure of cycling in ecosystems. *Mathematical Biosciences*, **65**, 219–37.

Ulanowicz, R.E. (1986). *Growth and development: ecosystems phenomenology*. New York: Springer-Verlag.

Ulanowicz, R.E. (1995). Network growth and development: ascendancy. In *Complex ecology: the part–whole relation in ecosystems*, ed. B.C. Patten, S.E. Jorgensen & S.I. Averbach, pp. 643–55. New Jersey: Prentice Hall.

Ulanowicz, R.F. & Kay, J.J. (1991). A package for the analysis of ecosystem flow networks. *Environmental Software*, **6**, 131–42.

Warwick, R.M. & Radford, P.J. (1989). Analysis of the flow network in an estuarine benthic community. In *Network analysis in marine ecology: methods and applications*, ed. F. Wulff, J.G. Field & K.H. Mann. *Coastal and Estuarine Studies*, **32**, 220–31. Heidelberg: Springer-Verlag.

Whitfield, A.K. (1995). Available information on individual South African estuarine systems. *Water Research Commission Report* No. 577/1/95, pp. 1–204.

Wulff, F., Field, J.L.G. & Mann, K.H. (eds.) (1989). Network analysis in marine ecology: methods and applications. *Coastal and Estuarine Studies*, **32**, 220–231. Heidelberg: Springer-Verlag.

12 Influence of Man and management of South African estuaries

Patrick Morant and Nevil Quinn

Bulungula River estuary

Introduction

Worldwide the demands of a burgeoning population are placing an ever increasing pressure on the resources of the planet. South Africa is no exception. Much of southern Africa is semi-arid or is prone to extremes of drought and flood. Against this background South Africa is faced with the task of feeding and improving the living conditions of its people. Already major inter-basin water transfer schemes have been implemented to meet the needs of Gauteng, South Africa's industrial heartland, and of greater Cape Town. Under such circumstances water allocated for environmental purposes is viewed by many as water wasted. As estuaries constitute the 'last in line' in catchments this has serious implications for their long-term well-being.

Besides the effects of activities in their catchments there is direct pressure on the estuaries themselves. The 3000 km long South African coastline is rugged, has very few truly sheltered embayments and is dominated by a high energy wave regime and strong winds throughout much of the year. These conditions have led to coastal development pressure being focused on estuaries. Since most of South Africa's major industries are located in the interior this pressure mainly has taken the form of residential and recreational developments. In the last two decades, however, a change in this situation has begun notably with the industrialisation of Richards Bay and Saldanha Bay.

Management of South African estuaries has been, in the main, undertaken on a piecemeal basis usually driven by sectoral interests such as property owners, anglers and yachtsmen. It is possible that the apparently benign nature of residential and recreational development along the coast created a false sense of security and complacency about the state of the coastal environment, including the estuaries. Consequently, until recently, no attempt has

been made to manage estuaries as integral components of catchments nor as key components of the coastal zone. In general they have been treated as islands in which a few components, usually fishes and bait organisms, have received attention from the managing authorities.

In addition, South Africa has, and will continue to have, a federal form of government. This has led to the fragmentation of responsibility for the management of the environment and, as a consequence, has compounded the problems discussed above.

The state of South African estuaries

During the 1970s concern grew regarding the conditions of estuaries in South Africa. Water abstraction, dam construction, soil erosion and pollution were clearly affecting more and more estuaries. Few remained in a state approximating the pristine, particularly in KwaZulu-Natal (Heydorn 1972, 1973). In response to this concern about the degradation of the estuaries of KwaZulu-Natal the Natal Town and Regional Planning Commission initiated a study to compile information and assess the condition of the estuaries of the province. Only 20 of the estuaries evaluated could be considered to be in a good condition (Begg 1978). Siltation was seen as the greatest threat to estuaries primarily as a result of intensive sugarcane cultivation in the catchments. Flow diversion and the construction of dams and weirs were also recognised as problems (Table 12.1).

A similar review of estuaries of the coast between the Orange and Great Kei Rivers was commissioned in 1979

Table 12.1. Summary of the various types and likely consequences of man-induced modifications on the estuarine environment in KwaZulu-Natal

Feature	Expected environmental effect	Probable results in terms of estuarine productivity	
		Adverse	Beneficial
Incoming silt	Losses in storage capacity; losses of water area; increased turbidity	Decreased production due to decreased light penetration; exclusion of visual feeders; suffocation of benthic fauna; decreased carrying capacity	Reed encroachment; mud flat formation
Dredging	Substrate disruption; turbidity; wetland disruption; increased tidal exchange		Deepened areas offer refuge; increased passive transport of juvenile into estuary
Disruption of riverine vegetation	Erosion and silt transport	As above	As above
Disruption of wetlands	Silt transport; reduced winter flow; destruction of sources of detritus	As above, with particularly severe decrease in productivity due to elimination of detritus sources	None
Drainage diversion	Reduced flow in estuary giving rise to mouth closure; or else increased water velocities giving rise to siltation.	Impeded immigration; decreased production at all levels.	None
Dam construction	Modification of flow, giving rise to mouth closure; or else increased water velocities giving rise to siltation.	Impeded immigration; decreased production at all levels.	None
Dam construction	Modification of flow, giving rise to mouth closure.	As above; including diminished beach nourishment; barriers to movement.	Silt interception.
Road and rail construction	Reduced tidal exchange; reduced scour; water area losses; decrease in depth.	Decreased production at all levels.	None
Weirs and causeways Breaching	Altered water levels; altered salinity; altered circulation.	Lowered carrying capacity; decreased utilisation; barriers to movement.	None
Pollution	Deterioration in water quality; increased biological demand for oxygen; presence of toxic compounds.	Modification of species composition and abundance; development of sludge communities; poor survival.	Increased birdlife; limited enhancement of fertility.

Source: From Begg (1978).

(Heydorn & Tinley 1980). Both Begg, and Heydorn and Tinley drew attention to the need for policies to ensure the sustainable management of the coastal zone and provided preliminary guidelines for land use planning and management in the coastal zone. Day and Grindley (1981) provided further insight into management needs in South Africa, assessing estuary management problems and reviewing appropriate legislation, reiterating the need for sound management and the implementation of procedures such as environmental impact assessments.

These concerns were again highlighted in a subsequent review of KwaZulu-Natal estuaries culminating in the publication of policy proposals for estuaries of the region (Begg 1984). Site-specific assessment of the condition of certain estuaries (Blaber *et al.* 1984) ensured that concerns remained in the public eye, leading to a national assessment of the status of South African estuaries (Heydorn, (ed.) 1986).

This was the first published attempt to provide a complete overview of the state of the estuaries of South Africa: it did not, however, cover the estuaries of the Transkei, then a so-called 'independent homeland'. The study was undertaken as a result of an initiative of the Department

of Water Affairs in response to increasingly voiced concerns about the effect of water extraction from rivers on the downstream environment, especially estuaries. The Department was seeking an acceptable balance between the water demands inland for agricultural and domestic purposes and the freshwater requirements of estuaries, if the latter were to be kept in a viable ecological (and hence also economic) condition.

Four key questions were to be addressed.

- On what criteria should the environmental importance of an estuary be assessed?
- What is the relative environmental importance of the different estuarine systems?
- What are the effects of variation in freshwater input on the biotic and abiotic characteristics of estuarine systems?
- What freshwater input pattern is required to obtain and/or maintain an agreed state of the estuarine system concerned?

From these four questions, but particularly from the last one, emerged the need for determining an '*agreed state'* for each of the estuaries of the Western and Eastern Cape and Kwazulu-Natal.

The following criteria were applied in the designation of these estuaries for the purpose of the study.

> Good, Fair or Poor (present condition) designations were used according to (1) the intensity and characteristics of developments around the estuary and (2) the extent of agricultural and urban development in the catchment. It was found that if the catchment was not too greatly disturbed as regards serious sources of organic and toxic wastes, erosion, and river regulation, or any of these in combination, the ecological condition of the estuary could generally be described as 'Good'. 'Fair' would imply a degree of noticeable ecological degradation in the catchment, as a result of e.g. agricultural or urban development, or river degradation resulting from severe environmental change in areas contiguous to the estuary. A 'Poor' estuary would be one which has been either physically altered, e.g. by canalisation (for example, the Soutrivier Canal at Cape Town), or Sezela in Kwazulu-Natal before rehabilitation. The recommended estuarine state to be pursued is defined according to three categories:

Category 1: *conserve in present state* was assigned to estuaries which were either considered to be pristine and/or representative of a coastal region and well preserved. Land use patterns in their catchment should be carefully considered in view of the possible impacts on the estuarine system.

Category 2: *conserve but permit controlled development* was assigned to those estuaries where limited development had already taken place but which were considered to be in a good enough state to be conserved with further development strictly controlled at a low intensity.

Category 3: *develop but according to environmentally acceptable guidelines* was assigned to estuaries where the surrounding areas, including the catchment, were developed, are being developed or are considered suitable for development. These developments must be according to environmentally acceptable guidelines.

The assessment panels made it clear that the findings of the study should be seen as guides to, rather than definitive statements on, the condition of the estuaries evaluated. They recommended that, as more knowledge of the estuaries becomes available, the assessments should

Table 12.2. The estuaries of the Western and Eastern Cape Provinces and KwaZulu-Natal: percentage distribution of estuaries in the condition categories

(a) Western and Eastern Cape Provinces

Size	No. of estuaries	Present condition (%)			
		Good	Fair	Poor	Unscored
Large	35	6	83	11	0
Small	118	30	41	22	7
Large and small combined	153	24	50	20	6

(b) KwaZulu-Natal

Size	No. of estuaries	Present condition (%)			
		Good	Fair	Poor	Unscored
Large	6	67	16,5	16,5	0
Small	66	24	49	27	0
Large and small combined	72	28	46	26	0

Source: From Heydorn (1986).

be reviewed and revised. This recommendation applied equally to the 'condition' and to the 'required state'.

The 'required state' assigned to the estuaries was seen as a goal for the managing authorities controlling the destinies of these systems. The assigned required state was considered to be the minimum level to which management should be directed. Heydorn (1986) stated that the required state should be incorporated within the context of regional planning for the areas in which the estuaries occur. The evaluation panels stressed that the management of estuaries should be undertaken on a regional and not a site-specific basis. They further stated that

> Whatever is done, it will not be possible to manage any given estuary according to the desires of all sectors of the population. A lack of a clear policy for estuaries on a regional basis leads to *ad hoc* management, often to the detriment of the system concerned, under pressure from vocal (minority) special-interest groups.

The results of the assessments are summarised in Table 12.2*a*, Western and Eastern Cape Provinces and Table 12.2*b*, Kwazulu-Natal. The overall distribution of estuaries in the condition categories is remarkably similar for the three provinces. However, when the large and small estuaries are analysed separately, it is clear that Kwazulu-Natal is proportionally much better endowed with large estuaries in good condition, mainly in Zululand, whereas the pattern for the smaller estuaries does not differ significantly between the three provinces. This indicates either that the small estuaries of Kwazulu-Natal are in a

Table 12.3. Current condition of South African estuaries

	Condition			
Region	**Excellent** estuary in a near pristine state (negligible human impact)	**Good** no major negative anthropogenic influences on either the estuary or the catchment (low impact)	**Fair** noticeable degree of ecological degradation in the catchment and/or estuary (moderate impact)	**Poor** major ecological degradation arising from a combination of anthropogenic influences (high impact)
Cool temperate	1 (10%)	2 (20%)	2 (20%)	5 (50%)
Warm temperate	34 (28%)	52 (43%)	21 (17%)	13 (11%)
Subtropical	39 (33%)	22 (19%)	36 (31%)	20 (17%)
Total	74 (30%)	76 (31%)	59 (24%)	38 (15%)

Source: From Whitfield (1995).

better condition or, conversely, that those of the Western and Eastern Cape are in a much poorer condition than is generally thought. As pointed out earlier, the condition of catchments in South Africa is relevant in this context.

Heydorn (1986) emphasised that while the 'required state' of an estuary should not be subject to constant modification, the actual condition can change very rapidly as a result of factors such as the construction of a major impoundment in a catchment, developments in the immediate environment or at the mouth of an estuary, or an infrequent episodic event such as Cyclone Domoina. However, the slow degradation of catchments and hence of estuaries is ongoing and almost imperceptible degradation of an estuary can also take place as a result of unco-ordinated minor developments in its vicinity.

This expert panel assessment of South African estuaries has not been repeated, and consequently there is no way of knowing how reliable the approach has been. If the panel assessment is repeated the procedure should be modified to allow all panels to be present on each occasion. The panels for each province should be refereed by the panels of the other provinces sitting as observers. This would permit greater consistency of evaluation. There is little doubt, however, that it can be undertaken rapidly and cost-effectively.

In general the Heydorn (1986) assessment has stood the test of time. The most obvious anomaly is the required state (3) (see page 291) assigned to the Orange River mouth which has since been granted Wetland of International Importance status in terms of the Ramsar Convention. In fact the Orange River mouth is one of the very few international Ramsar sites, being shared by South Africa and Namibia. A number of other Western and Eastern Cape estuaries would probably be upgraded to a required state of 2, i.e. conserve but permit controlled development. These include the Kromme, Klein, Goukou and Cintsa.

The KwaZulu-Natal assessment was more conservative in that the majority of estuaries were assigned a required state of 2. The Tugela's condition was rated as 'poor' and

the required state was 3. In view of the current interest in the Tugela estuary as a result of the potential impact of the planned inter-basin transfer of water from its headwaters, it would be of interest to know whether the panel would have revised its evaluation, for example by assigning a required state of 2.

Ramm (1988) developed a community degradation index which highlighted the degree to which the diversity of some estuarine communities has become reduced. Application of this index to 62 estuaries in KwaZulu-Natal revealed that only 22% fell into the categories of 'relatively undegraded' and 'slightly degraded' whereas 78% were moderately, strongly or severely degraded (Ramm 1990).

The results of the most recent assessment of the condition of South African estuaries (Whitfield 1995) are shown in Table 12.3. Although this assessment finds that 60% of South African estuaries are in good or excellent condition, it should be noted that this is primarily attributable to the large number of estuaries in an excellent/good condition which occurred in the former Transkei region (now part of the Eastern Cape Province) and were therefore not considered in previous assessments. For comparative purposes, Whitfield's (1995) assessment of Kwazulu-Natal estuaries yields 48% in a fair condition, 26% in a poor condition, 25% in a good condition and only one estuary out of a total of 73 considered to be in an excellent condition.

Major impacts on estuaries

South African estuaries are subject to a range of impacts such as catchment degradation caused by commercial and subsistence farming, impoundments and water abstraction, invasion of catchments by woody alien plants, pollution, artificial breaching, urban encroachment and harbour development. However, of these environmental insults probably the single most important is the reduction of freshwater inflow into estuaries as a result of the construction of dams and the direct abstraction of water.

The reduction in freshwater supply to South African estuaries

Prior to the European settlement of South Africa the estuaries were subject to the vagaries of a climate characterised by erratic rainfall, floods and droughts. The ever increasing demand for water caused by industrialisation and a rapidly growing population has exacerbated these effects. The flow and flood regimes of many rivers have been altered by impoundment and abstraction with the result that the periods of low flow have become extended.

Many, if not the majority, of South African estuaries became isolated from the sea by the formation of a sand berm across the mouth during periods of little or no fluvial input. Such an estuary will remain closed until the berm is breached by a large flood or the estuary basin fills and the berm is overtopped. Abstraction of water leads to reduced fluvial input to an estuary thus extending the periods of closure. In many cases this period can exceed three years.

Permanently open estuaries such as the Kromme, Boesmans and Kariega may develop reverse salinity gradients: that is, the upper reaches may become hypersaline, in the absence of fluvial input (Bickerton & Pierce 1988).

The frequency and duration of estuary mouth openings are major factors affecting the estuarine biota, particularly the juveniles of marine fishes which enter estuaries which serve as nurseries (Whitfield 1994a). Other than fishes, studies of the effect of restricted fluvial inflow and mouth closure have been limited to those of Adams et al. (1992), Adams & Talbot (1992) and Adams & Bate (1994a, b, c, d) on macrophyte vegetation (Chapter 5), and Wooldridge (1991, 1994) on crustacean larvae and zooplankton e.g. *Pseudodiaptomus hessei* (Chapter 7).

Earlier reviews of research needs identified diminished water supply to estuaries as a potentially serious problem (Begg 1978; Noble & Hemens 1978). Recent droughts in parts of the subcontinent have reinforced the concern about the effects of reduced freshwater inflow into estuaries (Reddering 1988a; Whitfield & Bruton 1989; Whitfield & Wooldridge 1994; Allanson & Read 1995). In an assessment of research needs relating to the surface water resources of South Africa (Cousens et al. 1988), the establishment of minimum flow requirements for estuaries and the development of models to assess fluvial contribution to estuaries were explicitly identified and ranked as research priorities.

The basis for the concern regarding diminished freshwater supply to estuaries is related to the continually rising demand for water, given a spatially and temporally inequitable supply. It is estimated that South Africa's water resources (excluding any potential contribution by desalination or recycling) will be fully utilised early in the 21st century (Davies et al. 1995). Sixty-five per cent of southern Africa receives less than 500 mm of rainfall annually and 21% receives less than 200 mm, while the average annual rainfall for the whole region (497 mm) is considerably less than the world average of 860 mm (Department of Water Affairs 1986). The growing demand for assured water supplies has necessitated the construction of large storage dams, and in some cases vast inter-basin transfer schemes. Numerous small agricultural dams, as well as barrages and weirs have been built to sustain stock-watering and irrigation requirements. Instead of being available as streamflow to estuaries, water is stored and subject to consumptive losses, including high evaporation. It is estimated that as little as 8% of the mean annual runoff reaches the coastal zone (Department of Water Affairs 1986). The invasion of catchments by woody alien plants, particularly in the southern and southwestern regions of the Western Cape province, has led to decreases of up to 30% in water yield (Le Maitre et al. 1996) further exacerbating the impact of abstraction and impoundments.

More significant than the loss of additional streamflow (volume) due to impoundment, is the alteration of the pattern of supply (volume and flow rate). Impoundments have the effect of attenuating floodpeaks. The extent to which this occurs depends on the capacity of the reservoir and the characteristics of the particular flood event. As Reddering (1988b) has noted, the equilibrium between successive periods of deposition by wave action and scouring by episodic events can be readily disturbed by modifications in river discharge.

If the scouring potential of flood events is reduced by impoundment attenuation, it follows that an estuary may open to the sea at a reduced frequency, thereby limiting the opportunity for key processes necessary for maintaining the estuarine habitat to occur. Begg (1978) has argued that closure of the estuary mouth in KwaZulu-Natal estuaries has become the artificial norm, and furthermore that this often occurs at the most critical time for inshore spawning of estuarine-dependent fish species.

Environmental management authorities in South Africa have been slow to accept that estuaries should be given equal priority for water resource allocation. Generally until the last ten years an almost universally held view has been that water supplied to estuaries is water wasted. This is epitomised by Burman (1970) in his book *Waters of the Western Cape* in which he states

> We need not talk of reusing water, whilst eighty per cent of our rivers run to waste

and

> Even the average annual runoff of the Palmiet is calculated at 46 400 000 000 gallons (209 000 ML). As the forty-odd

dams in the valley altogether hold no more than
5 800 000 000 gallons (26 400 ML), it is obvious that the
Palmiet River can make a big contribution to the future of
the Western Cape . . .

Public awareness of the importance of freshwater
inflow into estuaries began with the problems occurring
at Lake St Lucia. In the late 1950s and early 1960s, as a consequence of reduced inflow under drought conditions,
the salinity in the Lake rose to the extent that public
concern resulted in the matter being raised in Parliament.
The St Lucia Commission of Enquiry, to investigate the
'alleged threat to plant and animal life in St Lucia Lake', was
established in response to this concern (Kriel 1966; Crass
1982).

First attempts at quantifying estuary freshwater requirements
The concept of allocating a proportion of runoff to
environmental management was first introduced to the
engineering and water resource development community
in 1983 (Roberts 1983). In projecting future national water
demand, Roberts made allowance for the requirements
of estuaries, wetlands and nature conservation, suggesting a figure amounting to 11% of the estimated total
water requirements of all sectors in the year 2000.
Acknowledging that the initial estimate was simplistic,
Roberts (1983) urged engineers and scientists to undertake
research to refine his estimate (King & Tharme 1993).

The first regional assessment of estuary freshwater
requirements was produced three years later (Jezewski &
Roberts 1986). These estimates were based on two separate
components: a flooding requirement and an evaporative
requirement. The former was considered necessary to
open temporarily closed estuaries, to flood wetlands and to
flush out accumulated sediment, while the latter was considered important to counter the loss due to evaporation,
thereby preventing the occurrence of hypersalinity in the
estuary (Whitfield & Wooldridge 1994). Estimation of the
flooding requirement was almost four times the evaporative requirement and was considered to be less accurate
than the estimate of the evaporative requirement, leading
the Department of Water Affairs (1986) to recommend that
future research would have to concentrate on refining
these estimates in particular.

Rather than simply being annual allocations of freshwater to estuaries, these estimates were compiled to
provide better predictions of runoff available for utilisation and were incorporated in the review of South African
water resources (Department of Water Affairs 1986).
According to the Department of Water Affairs, the then
total freshwater requirement of estuaries and lakes
(Table 12.4) amounted to 5% of the virgin MAR of rivers,
and was anticipated to represent as much as 15% of the

Table 12.4. Estimated water requirements of estuaries and lakes

Coastal region	Requirements (million m³ y⁻¹)					
	Evaporative		Flooding		Total	
	Quantity	%	Quantity	%	Quantity	%
Cape	150	37	1080	64	1230	58
KwaZulu-Natal	260	63	620	36	880	42
Total (RSA)	410	100	1700	100	2110	100
Transkei	190		470		660	
Ciskei	6		14		20	
Grand Total	606		2184		2790	

Source: From Department of Water Affairs (1986).

utilisable resource. The total amount of water required
for environmental management (including nature
conservation) was estimated to reach 2954×10^6 m³ per
year by the year 2000, comprising approximately 13% of
the total demand, and representing almost 10 times the
amount originally estimated by the Commission on
Water Affairs in 1970.

In recognising estuaries as valid 'users' of water the
Department of Water Affairs (1986) also raised the interesting contention that not all estuaries are equally important, arguing that

> it may be found that certain estuaries have little ecological
> value and enjoy low priority when water is scarce, whilst
> others would be regarded as being so important that they
> would be allocated water in almost any circumstances.

Furthermore the Department of Water Affairs (1986)
recognised that the requirements presented in Table 12.4
are in competition with other demands, suggesting that it
may not be possible to meet the management demand of
each estuary and that 'benefits would have to be weighed carefully against water use in each case.'

Improved estimates of estuary freshwater requirements
During the early 1990s the Department of Water Affairs
and Forestry reaffirmed its commitment to establishing
the water requirements of the environment (Department
of Water Affairs and Forestry 1991) and commissioned a
number of studies to determine more accurately the fresh
requirements of certain estuaries (e.g. CSIR 1992a, b, c). The
principal methodology utilised in these approaches was to
convene a workshop of interested parties, and particularly experts, in various aspects of estuary management. Through a discussion of issues, participants would
aim to converge on a preliminary estimate, providing
recommendations in the form shown in Table 12.5.

During this period pioneering research on the
response of estuarine macrophytes to freshwater was

Table 12.5. Summary of the conclusions and recommendations of a workshop to determine the freshwater requirements of three KwaZulu-Natal estuaries

	Mdloti Estuary	Lovu Estuary	Mkomazi Estuary
Classification	Blind estuary	Periodically blind estuary	Permanently open system
Management objectives	1 Seasonally open mouth (approx. 25% open) 2 Flushed regularly	1 Seasonally open mouth (approx. 80% open) 2 Flushed regularly	Permanently open mouth (approx. 100% open)
Key environmental factors	1 Maintenance of character (nursery habitat) 2 Socio-economics (water quality is important)	1 Maintenance of character (nursery habitat) 2 Socio-economics (water quality is important)	1 Maintenance of character: importance in regional context is unknown. 2 Water level (upstream salinities)
First estimate of the freshwater requirement	10 to 20 million $m^3 y^{-1}$	35 to 45 million $m^3 y^{-1}$	220 to 330 million $m^3 y^{-1}$
Confidence level	65%	55%	50%
Further requirements	1 IEM 2 Baseline survey 3 Basic monitoring programme	1 IEM 2 Baseline survey 3 Basic monitoring programme	1 Baseline survey 2 Basic monitoring programme 3 Research into regional importance

Source: From CSIR (1992*a*).

being undertaken (Adams 1992; Adams *et al.* 1992; Adams & Talbot 1992), culminating in an assessment of the importance of freshwater in the maintenance of estuarine plants (Adams 1994) and a decision support system for determining the freshwater requirements of estuarine plants (Adams & Bate 1994*a*) (Chapter 5).

Current South African approaches to the management of freshwater inflow

The current approach to the management of freshwater inflow to South African estuaries is best illustrated by an examination of the Great Brak estuary, arguably the most successful example of estuary management in South Africa. The Great Brak estuary is located approximately halfway between Mossel Bay and George on the southern coast of the Western Cape. The estuary has an area of approximately 79 ha while the total catchment of the Great Brak River extends to 200 km², with an average annual runoff of approximately 37×10^6 m³ per annum (CSIR 1990). The Wolwedans Dam was completed in 1989, and was constructed with the main purpose of supplying water to the Mossgas petrochemical plant, and to supplement the growing water demand for the Mossel Bay region. Prior to construction, the water resources of the Great Brak were exploited on a modest scale for agricultural, industrial and domestic purposes: currently 65% of the mean annual runoff is impounded (1990).

According to Morant (1983), even before the construction of the Wolwedans Dam, normal flow to the estuary had been significantly reduced by afforestation, the damming of small tributaries for irrigation, and the building of larger dams such as the Ernest Robertson Dam. For example, along the Varings River there are six irrigation dams (Morant 1983). Afforestation was initiated as early as 1911, and approximately one-third of the catchment is currently afforested. The cumulative effects of changing land

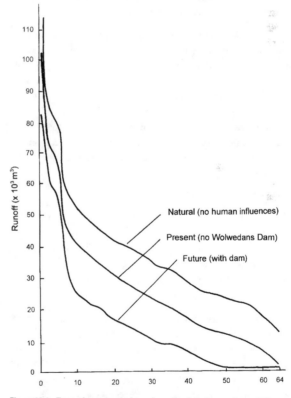

Figure 12.1. Exceedance curve based on simulated annual runoff for the period 1920 to 1983. The three curves indicate the relative effects of land use change as well as the attenuation of the Wolwedans Dam (CSIR 1990).

use and streamflow attenuation by dam construction are shown in Figure 12.1 (CSIR 1990).

Subsequent to the announcement that the Wolwedans Dam was going to be constructed the Department of Water Affairs established a steering committee, the Great Brak Environment Committee (GEC), with the specific aim of investigating the effect of the dam on the Great Brak estuary, and to establish an effective estuarine management plan for optimal use of the water allocated for estuary purposes (CSIR 1990). In providing an allocation to the estuary a preliminary report by the Department of Water Affairs (Department of Water Affairs 1988/89), indicated that

- an amount of 1×10^6 m^3 per year would be sufficient for the estuary;

- any further needs would be met during normal and above normal (wet) years when occasional overflow of the dam would occur;

- during dry years only the evaporative losses of the estuary would need to be met; and

- the capacity of the outlet works would provide the possibility of flushing the mouth of the estuary from time to time (CSIR 1990).

The subsequent investigation comprised studies of estuarine hydrodynamics, water quality and sedimentation as well as elements of the estuarine ecosystem including the distribution of vegetation and benthic fauna. The study on estuarine hydrodynamics concluded that the estuary mouth would be closed more often and for longer periods, resulting in higher water levels in the estuary which would be subject to higher evaporative and seepage loss (CSIR 1990). The ecological study (CSIR 1990) concluded that:

- the ecosystem is in a fairly stable condition;

- an open mouth is required to maintain estuarine biodiversity;

- salinity in the estuary should be maintained between 7 and 40;

- water levels in excess of 1.22 m (MSL) for extended periods should be avoided.

Water quality was expected to decline as a result of decreased dilution and reduced flushing, leading to eutrophication and periodic oxygen depletion. However, this assessment was based on the fact that the town of Great Brak lacked an effective sewerage system and consequently, it was recommended that a new sewerage system be constructed. A socio-economic assessment was also undertaken (CSIR 1990), and anticipated socio-economic impacts of the dam were investigated by means of a questionnaire circulated to residents.

The primary recommendation of the CSIR (1990) was that

the present estuarine environment be maintained by substituting active management for the natural processes operative prior to the construction of the dam. This implies

Table 12.6. Checklist and scoring criteria for determining whether management action is necessary

Criteria	Score Yes	Score No
Is the mouth open?	2	0
Is the estuary water level less than +1.22 msl?	2	0
Is there a bad smell and / or excessive algal growth in the water?	0	1
Is the E. coli level less than 1000?	2	0
Is the salinity level more than 7 and less than 40?	2	0
Are fish dying or under stress e.g. gaping at the surface for air?	0	2
Is it February?	−1	0
Is it June?	−1	0
Is it November?	−2	0

Source: From CSIR (1990).

that the natural effect of unregulated river flow should be replaced by controlled water releases, together with the excavation of the beach berm, to open the estuary mouth when required.

It was also recommended that in order to determine when water releases or breaching would be necessary a monitoring programme should be established. Monitoring should be undertaken on a monthly basis during the off-season period, but weekly during the holiday season. At each sampling interval a checklist would be filled in and scored according to the criteria shown in Table 12.6. A rule-based model in the form of a flowchart would then be used to determine what management action should be taken (CSIR 1990).

A monitoring programme was also recommended, with the purpose of evaluating the proposed management approach. Key aspects of the monitoring programme included: continuous monitoring of river inflow and water level in the estuary; an annual bathymetric survey of the lower estuary for at least the initial 10 years; and monitoring of plant communities as well as species diversity and abundance of the fauna along baseline transects on an annual basis, for a similar period. This monitoring programme has been running for several years now and three major monitoring reports have been produced (CSIR 1992c, 1993, 1994).

Although many countries have recently begun to recognise the water requirements of the natural environment, the emphasis has been on the determination and allocation of instream flow requirements for rivers and associated wetland and riparian environments. Over the last few decades considerable effort has been focused on the development of methodologies to determine the instream freshwater requirements of these systems, and consequently a wide range of techniques for impact assessment and flow requirement assessment is now available (Bovee

1982; Gore & Nestler 1988; Milhous *et al.* 1989; King & Tharme 1993). In contrast, very little effort has been directed at the development of similar techniques for determining the water requirement of estuaries.

Shortcomings of previous South African approaches
As noted by others (Whitfield & Bruton 1989; Whitfield & Wooldridge 1994), the regional allocations determined by Jezewski and Roberts (1986) did not take into account the numerous biotic–abiotic relationships within estuaries, focusing instead on an estimate of the flooding and evaporative requirements. Indeed, these estimates were not intended as allocations, but merely approximations which would provide water managers with a more realistic estimate of the utilisable runoff from catchments. Thus, while the determination of the first estimates was rooted in recognition of the water needs of the environment, it was not an assessment of the *ecological* water requirements of estuaries. In addition, the estimates of Jezewski and Roberts (1986) did not provide an indication in the seasonal or monthly distribution of this annual allocation (Whitfield & Bruton 1989; Whitfield & Wooldridge 1994).

Recent approaches (CSIR 1992*a*, *b*, *c*, 1996), reflect the ecological water requirements of estuaries more explicitly as they are determined by consensus of specialists, whereby each specialist is an authority representing some component of the ecosystem such as fish or birds. However, the approach is based on a collective and intuitive expert assessment rather than a scientifically defensible methodology, and consequently very low confidence assessments accompany these estimates. Furthermore, these estimates are usually annual totals and do not specify monthly flow regimes or ranges in acceptable variability. Because explicit ecological goals are not specified there is no way of knowing in the long term whether these estimates are realistic, and while the experts always stress the need for monitoring, it is seldom undertaken by government authorities. Lack of monitoring and investment in basic research by government departments has led to a growing dissatisfaction among estuarine scientists and a reluctance to participate in estuarine flow requirement workshops. Scientists are concerned that their initial estimates are accepted as a 'quick fix' solution tantamount to 'rubber stamping' without providing them the opportunity to re-evaluate estimates in conjunction with a structured research and monitoring programme.

Although the Great Brak estuary provides an example of where a relatively comprehensive monitoring strategy has been established, it should be noted that the freshwater requirements of this estuary were allocated *prior* to any impact assessment, as was the decision to construct the dam. Nevertheless the management system is currently the only one of this nature in place and is widely regarded as being successful. The advantages and limitations of this management approach have been discussed in other chapters of this volume.

Artificial breaching of estuary mouths
Many of the impacts of urban encroachment on estuaries are self-evident. These include permanent alterations of the shoreline by structures, construction of jetties and launching ramps, human disturbance of wildlife, diffuse and point source pollution. However, in South Africa an almost unique by-product of poor urban planning is the need to breach estuary mouths to lower water levels to prevent the inundation of houses.

Many South African estuaries close when there is little or no fluvial input. When river flow resumes the water dams up behind the bar and may rise to a point when flooding occurs. Artificial breaching is undertaken to safeguard properties built below the normal breaching level of the particular system. Thus the entire functioning of an ecosystem can be jeopardised by township planning undertaken in ignorance of the natural envelope of variability of an estuary.

The level to which the water is permitted to rise, before artificial breaching is undertaken, is often a contentious issue: conservationists want it to be the highest level possible whereas property owners require a margin of safety. At two such locations a reference mark has been determined to formalise the breaching procedure. At Swartvlei in the southern region of the Western Cape the breaching level has been set at 2 metres above mean sea level (Whitfield *et al.* 1983). A safety factor has been incorporated into the breaching guidelines: under high risk conditions, i.e. when the upper lake (Swartvlei) level is already high and when there are continuous heavy rains and a rising lake level, consideration can be given to opening the mouth before the +2 m MSL level is reached. A similar set of guidelines has been established for Klein River estuary (Hermanus Lagoon) in the southern Western Cape province where the breaching level has been set at +2.1 m MSL (CSIR 1988).

Artificial breaching may have serious long-term impacts upon the sediment dynamics and biota of an estuary. Artificial opening of an estuary mouth when the water level is below that when breaching naturally occurs results in a reduced scour potential. In the long term this leads to accumulation of sediments in the estuary mouth, thereby compounding the original problem. De Decker (1989) cites Waldron (1986) who reports that marine sand has penetrated the mouth of Kleinriviervlei (Hermanus Lagoon) to a distance of 4 kilometres since 1938. Between 1948 and 1988 Kleinriviervlei has been breached artificially at least forty times, on occasion twice in one year.

The effect of artificial breaching on the biota may be severe. Nowhere is this better illustrated than at the Bot River estuary. The Bot River estuary is a large blind estuary on the Western Cape south coast. Its connection with the sea is maintained by artificial breaching; natural openings of the mouth occur rarely, if at all (Koop 1982). For over 80 years the mouth has been artificially breached every few years, initially to restore populations of marine fish and thus improve fishing, and more recently to prevent floodwaters from threatening properties on the banks (Branch et al. 1985). Following a breaching the estuary remains open for 2–4 months. In summer when the river inflow is minimal or ceases altogether the mouth closes. Initially hypersaline conditions may prevail as a result of evaporation: Koop (1982) reports salinities reaching 40. With the onset of the winter rains the estuary fills and the salinity declines to as low as 4–5.

These rapidly changing extremes of salinity and water level have severe consequences for the biota. The physical impact of breaching on the Bot River estuary is dramatic: in a few hours the water level drops from 2.8 m above mean sea level to 0.2 m above MSL, exposing 40% of the basin to desiccation. The extensive beds of *Ruppia maritima* L. are stranded and eventually die. Since macrophytes contribute 72% of the total annual primary production (Bally et al. 1985) the impact on the overall ecosystem can only be massive. Since most of the benthic invertebrates in the system are associated with macrophyte beds, they too are largely eliminated, such that reductions in invertebrate biomass of 84–92% in weed beds, and a 54% decline in overall invertebrate biomass, occur (De Decker & Bally 1985). The impact of such massive perturbations on the diversity of the benthic biota is demonstrated by De Decker and Bally (1985), who reported the presence of only 27 benthic species in the Bot River estuary. Of these only six are consistently abundant and contribute 95% of the benthic invertebrate biomass. This contrasts with 319 species in the permanently open Knysna estuary (Day et al. 1952; Day 1967), and the seasonally open Klein River estuary with 134 species (Scott et al. 1952).

Breachings have a similarly massive impact on the avifauna. The Bot River estuary may support up to 36000 mainly herbivorous Redknobbed Coots *Fulica cristata* as well as some thousands of other waterfowl species. Following the two breachings in August and October 1981 the avian biomass of the estuary fell from a peak of 40 000 kg to only 700 kg (Heÿl & Currie 1985).

The affected biota is slow to recover from the impact of breaching: 14 months after breaching, macrophyte beds had become re-established over only 40% of the area that had been previously colonised (Branch et al. 1985). Similarly, the waterbird populations may take up to three years to reach pre-breaching levels (Heÿl & Currie 1985).

The ichthyofauna is the only component of the biota of this system which benefits from the opening of the estuary to the sea. Thirty-two species of fishes have been recorded from the estuary: these can be divided into three categories depending upon whether they breed in the sea, in estuaries or in freshwater. Twenty-one of the species breed only in the sea and the majority of these can only leave the estuary when the mouth is open. These fishes, therefore, are unable to sustain themselves reproductively while confined in the closed estuary. They move into the estuary as juveniles and, once in, may take from two to five years to reach sexual maturity (Bennett et al. 1985) Artificial opening thus allows a considerable proportion of the local ichthyofauna to have access to the estuary. As a result, the practice of breaching enables the estuary to serve as a valuable nursery for fishes of many species.

Management of the Bot River estuary has been subject to conflicting interests for many years, resulting in a completely *ad hoc* approach to breaching. High water levels are favoured by yachtsmen, birdwatchers and recreational swimmers, whereas low water levels are favoured by anglers and property owners. Branch et al. (1985) reviewed various management options and concluded that the estuary should be breached every 3–4 years or when the salinity fell below 6. Below a salinity value of 6 estuarine fishes are at risk. While this solution manages to satisfy most of the interested parties most of the time, it still does not avoid the effects of the massive perturbation resulting from artificial breaching. Two points of interest arise from the approach of Branch et al. (1985).

- They did not recommend an investigation into the re-establishment of the natural mouth which appears to be located immediately west of Hawston. As late as August 1917 the mouth was located in this position (see Figure 5 in van Heerden 1985). Should it be possible to re-establish the old mouth channel the Bot River estuary might be able to function naturally without the need for artificial breaching.

- Their site-specific approach is typical of that which has bedevilled environmental management in South Africa for so long. Clearly, if restoration of the natural mouth is not possible, the Bot River system cannot be managed to satisfy all the conflicting interests. Had Branch et al. (1985) assessed the management of the Bot River estuary in the context of the immediate region they might have decided that, despite the effect on marine and estuarine fishes, the best option would be to let the estuary change into a freshwater coastal lake. In the region, along approximately 40 km of coastline, there are three small (Palmiet, Kleinmond and Onrus) and two large (Bot and Klein (Hermanus

Lagoon)) estuarine systems. These five systems are capable, if managed on a holistic basis, of providing a diversity of wetland/estuarine environments. The Palmiet and Kleinmond are small 'blackwater' (acid, peat-stained) systems, the Onrus is a small reed (*Phragmites*) fringed system infrequently connected to the sea, the Klein River estuary is maintained as an estuary and the Bot (if left to its own devices) would become a coastal lake.

A similar situation exists in KwaZulu-Natal where estuary mouths may be breached for a variety of reasons. A policy has been established to guide response to calls for the artificial breaching of estuaries (Natal Parks Board 1989).

Sedimentation

Sediments of terrestrial origin

Soil erosion poses a major threat to estuaries, particularly those of KwaZulu-Natal and those of the Ciskei and Transkei regions of the Eastern Cape Province. For example, Begg (1978) states that

Siltation has had a greater detrimental effect on Natal's estuaries than any other single factor. At least 45 out of 73 estuaries show deterioration through incoming silt and sand.

The soil erosion results from overgrazing, poor farming techniques (both commercial and subsistence), destruction of wetlands in catchments, and removal of river- and streambank vegetation.

During marine geological studies of the continental shelf of KwaZulu-Natal, Martin (1987) estimated that the modern rates of sediment supply are 12 to 22 times greater than the geological average. The Tugela catchment is estimated to yield 10.5 million tonnes of sediment per annum (Orme 1974). This is equivalent to 375 tonnes per km^2 per year from a catchment of approximately 28 000 km^2. By comparison, the 1.33 million km^2 catchment of the Zambezi yields 75 tonnes per km^2 per year.

The consequences of this high sediment yield on estuaries are manifold. The nature of the benthic communities will change from those that favour a sandy substratum to those more at home in mud. The estuary basin will become shallower, particularly in the case of seasonally-open systems, thereby affecting the temperature regime of the water column. Shallow water will favour encroachment by reeds such as *Phragmites*, resulting in reduced water surface area. In the sugar cane growing regions of KwaZulu-Natal and the pineapple growing areas of the Eastern Cape, leached fertilizers and other agrochemicals compound the problems caused by siltation.

Although it is the ultimate fate of small beach-barrier estuaries to be infilled with sediment, the rate at which it is occurring in the eastern parts of South Africa is unacceptable. The remedy is education. Residents of catchments have to be educated as to the value of wetlands, streambank vegetation, careful tillage, proper stocking levels and so forth. Unfortunately education is time-consuming. The integrated catchment management approach to be embodied in the new Water Act may go some way towards an improvement in the quality of catchments besides serving as a mechanism for more equitable freshwater allocation.

To date there has been one attempt to restore a catchment in order to improve the condition of its estuary. Begg (1978) proposed that the Siyai Lagoon (*sic*) be selected for a pilot study in catchment restoration. He recommended that it should be a coordinated, multi-disciplinary exercise involving scientists, engineers and sugar cane farmers. The Siyai was selected because it was a system with a sufficiently small catchment to bring the project within the bounds of economic and practical means. It was also selected because of the knowledge acquired by Dr Ian Garland (the farmer whose land surrounds the lagoon) in terms of moderating river flow, and restoring natural vegetation along the watercourses, in sponge areas and marshes. With the cooperation of the few farmers higher in the catchment, and that of the South African Sugar Association, much could be learnt from the project. Hopefully it would demonstrate that the damage done is not irreversible, that such a system can recover, and that the technique used could be applied elsewhere in KwaZulu-Natal.

Sediments of marine origin

Large volumes of marine sediments are transported along the South African coastline by the littoral drift. At any point along the coast, between 250 000 m^3 and 1 200 000 m^3 of sediment are estimated to be transported per year (Coppoolse *et al.* 1994). The littoral drift is the source of the sediment which forms sandbars across the mouths of estuaries or flood-tide deltas within them. Reduction of fluvial input by impoundments and abstraction results in estuaries being closed more frequently and for longer periods or, in the case of those that remain open, suffer massive influxes of marine sand which may limit recreational opportunities, e.g. the Nahoon (Wiseman *et al.* 1993), or render navigation hazardous (Bickerton & Pierce 1988).

The Nahoon Dam in the Eastern Cape province, with a capacity of 22.1 million m^3 holds a volume roughly equivalent to two-thirds the mean annual runoff (34×10^6 m^3) of the Nahoon River. In addition an important characteristic is that years with below average runoff are more common than years with above average runoff. Wiseman *et al.* (1993) present simulated runoff data which show that in only 17 years in 55 does the runoff exceed the mean annual runoff (MAR). The effect of the variable runoff, and the dam which can impound a significant proposition of the MAR, is to reduce the frequency with which large floods scour the estuary thus permitting a greater accumulation of sand in the estuary between floods.

It is not only large formal impoundments that have an impact on estuaries. There are no such dams on the two rivers feeding the small Seekoei estuary on the Western Cape south coast. However, during a recent drought the estuary virtually dried up resulting in an almost complete mortality of estuarine biota, conspicuously fishes (P. Morant, personal observations). It is possible that the drought was sufficiently severe that desiccation was inevitable but there is little doubt that the numerous earthen farm dams on the seasonal streams in the catchment had a significant impact on the volume water reaching the estuary. Farm dams have been shown to have a severe effect in catchments with low runoffs, particularly in dry years (Maaren & Moolman 1986). Even in a relatively large catchment such as that of the Berg River in the relatively well-watered Western Cape it has been estimated that up to 40% of the total mean annual runoff is retained in farm dams on seasonal tributaries. It is critical, therefore, that these small dams and their effect on runoff are not overlooked in calculating the volume of water available to estuaries. The effect upon the biota of the influx of marine sediments into estuaries has been little studied. However, the general effect will be to favour those species which prefer coarser sediments over those preferring muds and fine sediments.

Pollution

Unlike most countries with extensive coastlines, South Africa's industries are concentrated in the interior, primarily as a result of the discovery of gold on the Witwatersrand at the end of the nineteenth century. As a result of this accident of history and location, until recently, with the exception of Durban Bay, industrial pressure on the coastal zone in general and estuaries, in particular, was relatively slight. Lately the pace of industrialisation of the coastal zone has quickened: Richards Bay is set to rival Durban as an industrial centre and port and Saldanha Bay in the Western Cape is also being rapidly industrialised. It is interesting that although the Saldanha Bay–Langebaan system is not an estuary, it shares many features with South African estuaries (see Chapters 5 and 8), and as a result Day (1981*a*) included it in his 'Summaries of Current Knowledge'.

Of those South African estuaries which are recipients of industrial (mainly point sources), domestic and agricul-

tural pollutants (mainly diffuse sources), few have ceased to have any estuarine function at all. Such estuaries have been canalised, e.g. Umbilo and Mhlatuzana in Durban Bay, Bakens and Papkuils in Algoa Bay (Port Elizabeth) and the Black in Table Bay (Cape Town), and serve as little more than urban drains.

The greatest pollution pressure on estuaries occurs in KwaZulu-Natal as a result of agrochemicals leaching from sugar cane fields; faecal contamination from unserviced informal settlements in catchments, especially in the vicinity of Durban and the northern section of the South Coast; and effluents from industries such as sugar mills.

It should be stated, however, that, with the exception of the canalised systems, none of the estuaries is beyond redemption. For example, in 1978 the Sezela was described as 'the most grossly polluted lagoon in Natal . . . a foul-smelling, filthy black cesspool' (Begg 1978). Untreated sugar mill wastes had been discharged directly into the lagoon for many years and the impacts of this continuous discharge were obvious. No living aquatic fauna were collected in 15 trawls conducted during surveys in 1982 (Begg 1984).

A rehabilitation programme was initiated in 1981 consisting of: the construction of a high capacity activated sludge plant to treat all the mill effluents; a breaching regime requiring the estuary mouth to be opened monthly to flush accumulated materials from the lagoon; and the re-establishment of fluvial input. The success of the programme was demonstrated by 1986 when 15 species of fishes were recorded in the lagoon (Ramm *et al.* 1987). Significantly, ten of these were species dependent on estuaries for the completion of their life cycles.

Currently the Water Act 54 of 1956 is by far the most important statute relating to the control of water resources in South Africa. However, it should be noted that, although all three of the main use categories recognised in the Act – namely use for agricultural, urban and industrial purposes – do give rise to water pollution, control measures are directed primarily at industrial use. The reason for this is that industrial use, through the production of effluents as defined in the Act, to a large extent causes pollution through point sources which are more susceptible to control than diffuse sources of pollution (Fuggle & Rabie 1992). The implication for many estuaries is that diffuse sources of pollution, such as informal settlements and agriculture, will largely remain uncontrolled. Coupled with increased freshwater abstraction the potential for eutrophication of many estuaries, particularly those that close seasonally, is high (see Chapter 4).

From a management perspective the quality of water in estuaries has a Cinderella status and tends to be managed on a case-by-case basis by the Department of Water Affairs and Forestry. Whereas Receiving Water Quality Objectives

(RWQO) have been established for freshwater and marine systems (Department of Water Affairs and Forestry 1993, 1995) there are no RWQO for estuaries. In part this is due to the dynamic nature of estuaries which may range from fresh to hypersaline. For those estuaries which do not have industries in their catchments the RWQO for freshwater systems, if implemented, provide the best opportunity to maintain influent water quality at the highest possible level. In addition, the integrated catchment management approach embodied in the new Water Act will require that the needs of all users will have to be addressed in the allocation of water resources. This allocation process will address both the quantity and the quality of the water available. Environmental impact assessments should be mandatory before any effluent discharges into estuaries are permitted.

Encroachment

Urban encroachment

Communication between nature conservation agencies and provincial and metropolitan planning authorities in South Africa has always been poor. A consequence of this is the encirclement of many estuaries by built environments. The planning authorities are responding to the demand for seaside recreational developments on a rugged coastline with few sheltered embayments and exposed to a high wave energy regime. Estuaries have naturally become the focus of coastal development. Nowhere is this more clearly demonstrated than along the south Coast of KwaZulu-Natal where almost every system is surrounded by resort development (Begg 1978).

On the other hand, while the conservation agencies have been slow to integrate estuary management 'vertically' into catchments, they appear to have failed completely to integrate estuary management 'horizontally' into the adjacent coastal zone. Consequently, many estuaries have become almost completely isolated from the coastal zone by urbanisation without any attempt to assess the impact on the terrestrial biota which may have made use of the estuarine environment. Such development, for example, reduces the utility of estuaries to many birds since there is no undisturbed shoreline to provide peaceful refuge; had The Island in the Great Brak estuary (Morant 1983) remained undeveloped, it is almost certain that the estuary would have supported a much larger and richer avifauna than it does at present. In the southern and western regions of the Western Cape Province, Cape clawless otters *Aonyx capensis* exploit the intertidal and subtidal zone but require freshwater to maintain their fur; thus they must have safe access to rivers (Kruuk 1995). Should development encircle an estuary mouth otters may no longer be able to use it for access to the sea since

they would have to run the gauntlet of lights, boats, domestic dogs, etc. The 'horizontal' role that estuaries play in the coastal zone urgently requires investigation.

Structures

Urbanisation brings with it associated infrastructure such as roads and railways. These require bridges across rivers and estuaries which were (and often still are) almost always built with no thought to environmental requirements. The minimum and cheapest structure to convey the road or railway over a river while ensuring safety during a design flood, e.g. 1-in-100 years is usually the objective. The result generally is a multispan bridge with long approach embankments. A typical example is the bridge spanning the Uilkraals estuary in the Western Cape. The 220 m long bridge comprises a 120 m causeway and a 100 m multispan concrete bridge. It is notable that since its construction the bloodworm *Arenicola loveni* has disappeared from the estuary upstream of the bridge (Heydorn & Bickerton 1982).

The most extensive series of bridges over estuaries in South Africa are those spanning the estuaries of the KwaZulu-Natal South Coast. The south coast railway spans every estuary, usually very close to the sea, for the entire distance from Durban to the Zotsha estuary some 150 km to the south (Begg 1978). In addition the N2 arterial highway and ancillary roads span the same estuaries. Since no baseline studies were ever undertaken of these estuaries prior to bridge construction it is difficult to assess the scale of impact that has resulted. There is little doubt, however, that the bridges have reduced scour potential and have resulted in the retention of sediment with all the attendant consequences.

Other structures such as launching ramps, jetties, sea-walls and groynes have been built with little regard to the consequences. In an attempt to increase awareness of the effects of various structures and development actions on estuaries and the coastal zone, the Council for the Environment (1989) produced *Guidelines for coastal land-use*. The Department of Environmental Affairs and Tourism through the Coastal Management Advisory Programme (CMAP), now CoastCARE, has run seminars and short courses for provincial and municipal authorities to introduce them to the concepts contained in the *Guidelines*. It is hoped that this will be the beginning of more effective management of the coastal zone and estuaries.

Mariculture

The paucity of sheltered bays for mariculture in South Africa has led to the consideration of estuaries for the culture of bivalves and fish: notably the Knysna estuary, in which a successful oyster culture company was established in 1960. The potential for conflict between existing interests and the incoming mariculture industry is high (Dankers 1993). Various interest groups, users and environmentalists claim rights over space, water quality, biomass, landscapes, special ecosystems and endangered species. As urban areas, industries, tourism, harbour activities, agriculture, mariculture and fisheries expand, these resources become increasingly limited. Congestion effects lead to losses of economic efficiency or amenities in the use of natural assets and, in some cases, to their destruction. Coastal aquatic ecosystems are particularly affected.

The increase in environmental problems with respect to mariculture results from both endogenous and exogenous causes (Bailly & Paquotte 1996). The first is attributable to the development of mariculture that goes beyond the capacity of estuarine ecosystems to process: in particular, the wastes produced by such operations. The second is due to the competition of rapidly growing activities in the coastal zone. For these reasons it is vital that all proposals to establish mariculture operations in estuaries be subjected to a full environmental impact assessment conducted according to the Integrated Environmental Management (IEM) procedure endorsed by the Department of Environment Affairs and Tourism (Department of Environment Affairs 1992).

The IEM procedure provides the best method to ensure resolution of (potential) conflicts and for the allocation of resources. Only after a full investigation of the costs and benefits, social, economic and environmental, should a mariculture operation be permitted in an estuary not previously used for mariculture.

Environmental policy developments in South Africa

International Conventions

Environmental issues have achieved greater prominence in international affairs in recent years: the 1992 United Nations Conference on Environment and Development ('the Rio Earth Summit') and associated Rio Declaration being a reflection of this. One of the major highlights of the Earth Summit was the opening for signature of the Convention on Biological Diversity. The objectives of the convention are

> the conservation of biological diversity, the sustainable use of its components and the fair and equitable sharing of the benefits arising out of the utilization of genetic resources.
> (Convention on Biological Diversity 1994).

More specifically, Article 10 of the Convention requires that contracting parties

> integrate consideration of the conservation and sustainable use of biological resources into national decision making and adopt measures relating to the use of biological resources to avoid or minimise adverse impacts on biological diversity.

Contracting Parties are also required to implement environmental impact procedures for proposed projects which may have significant adverse effects on biological diversity, with a view to avoiding or minimising such effects.

Although the signing of the Convention on Biodiversity by South Africa in 1994 exacts a greater commitment to the sustainable utilisation of natural resources such as estuaries, such a commitment was made over 25 years ago with the signing of the Ramsar Convention by the South African Government in 1975. The Ramsar Convention (Convention on Wetlands of International Importance Especially as Wetland Habitat, 1971) was the first of the modern global nature conservation conventions, with the primary purpose of preventing the decline of wetland habitats, which includes estuaries and lagoons, and ensuring that they remain ecologically viable (Matthews 1993). The signing of the convention by South Africa committed the country to safeguarding and promoting the 'wise use' of the nation's wetlands and estuaries. With the publication of the World Conservation Strategy in 1980 (IUCN 1980), conservation or 'wise use' came to be defined as

> the management of human use of the biosphere so that it may yield the greatest sustainable benefit to present generations while maintaining its potential to meet the needs and aspirations of future generations.

Legal framework

Environmental law in South Africa has developed haphazardly in response to environmental needs and pressures. Historically, in South Africa the notion of environment has often been identified with the preservation of wildlife and the conservation of nature. In contrast, the new (1996) Constitution for the first time includes a direct reference to the environment and lays down as one of the fundamental rights of citizens, that (Article 29)

> Every person shall have the right to an environment which is not detrimental to his or her health or well-being.

Hence the Constitution emphasises the *human* aspect of the environment rather than being concerned only with the impact of humans on nature.

The Constitution sets out the overall legislative framework, and makes reference to the environment in the Bill of Rights where Section 24 states that

Every person has the right –

(a) to an environment that is not harmful to their health or well-being, and

(b) to have the environment protected, for the benefit of present and future generations, through reasonable legislative and other measures that –

   (i) prevent pollution and ecological degradation;

   (ii) promote conservation; and

   (iii) secure ecologically sustainable development and use of natural resources while promoting justifiable economic and social development.

This section is likely to influence judicial decision-making by requiring that due weight be given to environmental considerations, traditionally undervalued by the courts.

South Africa has a plethora of legislation dealing with the environment. Nevertheless, it is generally agreed that the environmental legislation has been thus far relatively ineffective. The main reasons for this are that responsibility for the formulation of legislation and enforcement thereof is divided amongst a large number of different authorities which generally are not adequately trained, funded or organised. Thus, the legislative framework related to the environment in South Africa is affected by the complex division of delegated legislative powers and functions between the national and the provincial levels.

Against this background of a strong environmental component in the new Constitution and the existing body of environmental law, South Africa has embarked on a process aimed at developing a new environmental policy. This has resulted in the publication of a Consultative National Environmental Policy Process discussion document *Towards a New Environmental Policy for South Africa* (CONNEPP 1996). In due course the discussion document and the comments on it will form the basis of a White Paper on the Environment. In parallel, the Department of Water Affairs and Forestry has initiated a similar process aimed at producing a new Water Act for the country. In addition to these two broad initiatives, a further document *The Philosophy and Practice of Integrated Catchment Management: Implications for Water Resource Management in South Africa* (Water Research Commission 1996) has been released for discussion. Whatever the outcome of these initiatives, it is clear that the decision-making process concerning environmental issues will be much more open and consultative than hitherto. It is hoped that they will also stimulate one more holistic approach to environmental management.

The draft Environmental Policy for South Africa stresses the importance of sustainable resource management, with specific objectives to improve biodiversity conservation, ensure the sound management of fragile ecosystems, and to integrate the management and sustainable development and utilisation of the marine coastal zone. A specific objective is to ensure the sustainable and rational utilisation, conservation and management of water resources based on ecosystem and community needs. The latter objective is a key theme of the Water Law Principles (Water Research Commission 1996) whereby

> the quantity, quality and reliability of water required to maintain the ecological functions on which humans depend should be reserved so that the human use of water does not individually or cumulatively compromise the long term sustainability of aquatic and associated ecosystems.

Indeed, it is intended this principle be specifically enshrined in the law in that

> the water required to meet people's basic needs and the needs of the environment should be identified as 'the Reserve', and should enjoy priority of use.

The onus, therefore, is upon scientists and managers to provide the information necessary to ensure that sufficient water is allocated to secure the future of South Africa's estuaries.

Management of estuaries

Definition of an estuary

One of the problems faced by environmental managers in South Africa is the difficulty of adequately defining an estuary. Ecologists may accept a degree of fuzziness in the boundaries of a given system. However, environmental managers have to operate within the legal framework of the country and, therefore, require a 'watertight' definition. Day (1980) posed the question, 'What is an estuary?' and South African environmental scientists have been wrestling with the problem ever since.

The South African coastline to the west of Cape Agulhas borders on the South Atlantic, and to the east on the Indian Ocean. The east coast waters are characterised by the warm waters of the southward-flowing Agulhas Current (Heydorn *et al.* 1978), those of the west coast by sporadic upwelling of cold, nutrient-rich waters typical of the Benguela Current (Shannon 1985). Along the southwest and south coasts extensive mixing of water masses occurs. The coastal climate of southern Africa (see Chapter 3) is modulated by oceanographic conditions: summer rainfall occurs along the east coast, bimodal rainfall along the south coast, winter rainfall along the southwest coast, and semi-arid conditions along the west coast. Another characteristic of the rainfall regime is the marked inter-annual variability (Tyson 1986).

A consequence of the variable rainfall regime is that few estuaries in South Africa fit conveniently into the northern hemisphere definition of such systems in having a perennial inflow of freshwater and tidal action. Day (1980) attempted to address this problem by modifying Pritchard's (1967) definition of an estuary. According to Pritchard:

> An estuary is a semi-enclosed coastal body of water which has a free connection with the open sea and within which seawater is measurably diluted with fresh water derived from land drainage.

Day rightly indicated the difficulties with the application of Pritchard's definition to South African estuaries: '. . . a free connection with the open sea . . .' excludes estuaries which are temporarily isolated from the sea during the dry season and '. . . diluted with freshwater . . .' excludes estuaries which become hypersaline when evaporation exceeds fresh water inflow. In southern Africa and other semi-arid countries there are many 'blind' estuaries which are closed by sandbars for shorter or longer periods. Day (1980), therefore, proposed the following definition:

> An estuary is a partially enclosed coastal body of water which is either permanently or periodically open to the sea and within which there is a measurable variation of salinity due to the mixture of seawater with fresh water derived from land drainage.

The compilers of the South African National report to the United Nations Conference on Environment and Development held in Rio de Janeiro in June 1992 (CSIR 1992d) recognised the inadequacy of Day's definition with respect to South African estuaries and proposed the following:

> In South Africa an estuary is considered to be that portion of a river system which has, or can from time to time have, contact with the sea. Hence, during floods an estuary can become a river mouth with no seawater entering the formerly estuarine area. Conversely, when there is little or no fluvial input an estuary can be isolated from the sea by a sandbar and become a lagoon which may become fresh, or hypersaline, or even completely dry.

From a management point of view there is a need to have a definition of an estuary which is *legally* unambiguous rather than one that only describes the ecological function of such systems. This issue is particularly per-

tinent in South Africa at present since the entire corpus of Water Law is being reviewed (Water Research Commission 1996). At a workshop held in Port Elizabeth on 27 September 1996 members of the Consortium for Estuarine Research and Management (CERM) agreed that, from a Water Law perspective, estuaries should be considered to be integral components of river systems. In other words, a catchment would encompass an entire river basin drainage from the watershed to the sea. This approach obviates the difficulty of determining and defining the location of the head of an estuary. The CERM workshop delegates recognised that the *spatial* extent of an estuary also needs legally unambiguous definition. It was proposed that a specific contour above Mean Sea Level should be used to define the limits of an estuary for management purposes. This contour was not defined but it was agreed that it should include water levels necessary to maintain high saltmarsh and other similar wetland vegetation and, possibly, the 1:50 year flood line. In the absence of vegetation and flood data for the majority of South African estuaries it was suggested that the $+3.0$ m MSL contour should be investigated.

The need for a pragmatic definition and approach is highlighted by the omission by Harrison *et al.* (1994a, b, 1995a, b) and Whitfield (1995) of many rivers which reach the sea, because they are of little or no value to fishes. Since the work of Harrison *et al.* and of Whitfield was undertaken on behalf of the Department of Environmental Affairs and Tourism and the Water Research Commission respectively, there is a real risk that their narrow ichthyofaunal perspective will lead the managing authorities to discount the environmental value of fish-poor systems. It should be borne in mind that the total area of estuaries in South Africa has been estimated to be between 500 and 600 km^2 (Heydorn 1989). This constitutes 0.05% of the area of the country. It is critical therefore, to ensure that all South Africa's estuaries receive adequate protection and are effectively and sustainably managed.

Management of estuaries

The management of estuaries requires that the decision-making process follows a formal protocol to guide thinking and to focus upon issues of concern. Two important components vital to the achievement of successful management of estuaries are reference frameworks and predictive tools. Reference frameworks are necessary for the ranking of estuaries in terms of their importance to various components of the biota, e.g. birds, fishes, vegetation, etc., or for the ranking of the estuaries themselves as complete systems. Predictive tools provide insight into the response of various components of estuaries to change, natural or anthropogenic. These predictive tools usually take the form of computer-based models. Finally, management decisions should be converted into an implementation programme that can be monitored cost-effectively.

Reference frameworks

Botanical importance rating

The botanical importance rating developed by Coetzee *et al.* (1997) (Chapter 5) is aimed at enabling a single numerical botanical score to be calculated for estuaries. The score is the arithmetic sum of four factors: plant community area cover, condition (degree of impact), association with the estuary and plant community richness. The principle behind the scoring was: the greater the area cover of a plant community, the fewer impacts associated with a community and the greater the number of communities, the higher the final score.

A cover score of either 20, 40, 60, 80 or 100 (essentially an expanded Braun-Blanquet cover classification: Poore 1955*a*, *b*) is assigned to each plant community: the higher the area cover, the higher the score. The multiplication factor is determined from the number of impacts affecting each community: the greater the number of impacts, the lower the multiplication factor. The cover score is multiplied by the multiplication factor to obtain a single score for each plant community. This number is then multiplied by the community importance value.

Plant communities are assigned an importance value according to their degree of association with the estuary. Because of their close association with the water, submerged macrophytes (*Zostera* and *Ruppia*) are regarded as the most important plant community. Intertidal salt-marshes are regarded as the second most important plant community. Reedswamps and sedge communities, ranked third in importance, act as natural biological filters, are important for bank stabilisation and contribute to the diversity of aquatic life. Supratidal marshes experiencing little tidal flooding are likely to be less important with respect to nutrient exchange than intertidal marshes or submerged macrophyte beds. Consequently supratidal salt marshes are assigned the lowest importance value.

Coetzee *et al.* (1997) applied their formula to 33 estuaries for which suitable botanical data were available. The botanical importance scores for the different estuaries were calculated using the formula described. The scores were normalised to make the differences easier to conceptualise. The estuary with the highest score was regarded as having a score of 100 and the rest were ranked as a percentage of the highest score. Mangroves were excluded from this study as they only occur in subtropical South African estuaries, which are not covered by the *Estuaries of the Cape* (now Western and Eastern Cape provinces) report series (Heydorn & Grindley 1981–1985; Heydorn & Morant 1986–1990; Morant 1993–) used as the

Table 12.7. The botanical importance rating for Western and Eastern Cape estuaries in two categories: Permanently open and Temporarily open

Permanently open			Temporarily open		
	Formula score	Normalised score		Formula score	Normalised score
Olifants	410	100	Kabeljous	300	100
Duiwenhoks	330	80	Great Brak	290	97
Gqunube	280	68	Rooiels	260	87
Goukou	265	65	Wildevoël	250	83
Keurbooms	250	61	Buffels Oos	240	80
Gamtoos	250	61	Seekoei	200	67
Breë	200	49	Hartenbos	170	57
Heuningnes	200	49	Groot Wes	160	53
Kowie	185	45	Onrus	150	50
Gourits	180	44	Quko	140	47
Kromme	175	43	Uilkraal	140	47
Swartkops	170	41	Elsies	120	40
Nahoon	170	41	Eerste	115	38
Sout	40	10	Piesang	90	30
			Qinira	90	30
			Silvermine	75	25
			Hout Bay	45	15
			Sir Lowrys Pass	30	10
			Lourens	30	10

Source: From Coetzee *et al.* (1997).

botanical data source. When application of the Coetzee *et al.* (1997) formula is extended to subtropical estuaries and mangrove communities, the method of score normalisation will require manipulation.

The results of the Coetzee *et al.* (1997) study accord with the known botanical condition of the estuaries of the Eastern and Western Cape Provinces both when all the estuaries are ranked together and when they are ranked in two classes, 'permanently open' and 'temporarily open' (Table 12.7). However, caution must be applied when using the results of the botanical importance rating: two low-scoring estuaries, Quko and Sout, are so because of a lack of diversity of vegetation communities and not because of any degradation.

Using waterbirds to determine the conservation priority of estuaries

Turpie (1995) explored methodologies and criteria for determining the priorities of estuaries for the conservation of waterbirds. Both single and multiple criteria indices were tested. The single criteria indices included those for diversity, abundance, rarity and conservation status. Two multiple criteria indices were also investigated, namely the Site Value Index, combining the conservation status, site endemism and population size indices; and a Priority × Diversity Index (Bolton & Specht 1983), which combines indices for site endemism and abundance.

Turpie (1995) tested the evaluation criteria using a data-

Table 12.8. The top 10 estuaries, determined subjectively on the basis of species richness (SR), conservation status (CSI), total numbers (TOT), and conservation value (CVI), Shannon-diversity (*H'*) is only considered in that it diminishes the value of high-scoring sites which have low Shannon-diversity

Estuary	Reason for ranking
1 St Lucia	Top 3 for all criteria; very high CVI
2 Berg	Top 3 for all criteria; high numbers
3 Richards Bay	Top 3 for most criteria, high CSI
4 Langebaan	Very high TOT and CVI
5 Orange	High ranks for all criteria
6 Olifants	Moderately high ranks for all criteria
7 Rietvlei	Moderately high ranks for all criteria
8 Verlorenvlei	High in most criteria except total numbers
9 Wilderness	High CVI
10 Bot River	Top-10 for SR and CSI, high TOT but low *H'*

Source: From Turpie (1995).

Table 12.9. The top three estuaries within each of the two main biogeographical zones for estuarine avifauna in South Africa (west and east) and in the overlap zone (south)

West	South	East
1 Berg	1 Wilderness	1 St Lucia
2 Langebaan	2 Botrivier	2 Richards Bay
3 Orange	3 Kleinrivier	3 Mgeni

Source: From Turpie 1995).

base comprising the summer population counts for 88 waterbird species at 42 sites (estuaries) along the coast of South Africa between the Orange River in the west and Kosi Bay in the east. Only sites which supported total waterbird populations exceeding 500 were used. With the exception of the Shannon diversity index (*H'*), all rankings were significantly correlated with one another. However, the degree of similarity between the results obtained from the various indices does not provide an indication of their usefulness.

A key finding of Turpie's (1995) study was that the final evaluation of sites should, ideally, involve a subjective assessment of the results of single-criterion rankings, according high priority to those sites which rank highly in terms of all of the criteria considered, and by assessing the merit of sites which feature only in terms of certain criteria. It has been argued (Klopatek *et al.* 1981) that different criteria should be considered both separately and together, but others (e.g. Järvinen 1985, Götmark *et al.* 1986) express doubt as to whether multivariate indices have any use at all in conservation work, and suggest that there is little point in constructing more 'overall' conservation indices. One of the most severe limitations of the use of multivariate indices is that the prime reason for a site's position in the hierarchy is hidden in a complex formula. The factors contributing to a site's importance need to be explicit to decision-makers.

The top 10 South African estuaries for waterbirds, evaluated subjectively according to the above criteria, are listed with reasons in Table 12.8.

Ranking procedures have limited value in terms assessing the priority of sites for conservation unless the biogeographical zonation of waterbirds is taken into account. Estuaries in different biogeographical zones cannot be compared legitimately, and it is important that reserves are sited within each (Rebelo & Siegfried 1992; Turpie &

Crowe 1994). In South Africa, waterbirds fall into three zones, with two main zones along the Atlantic and Indian Ocean coasts, and the southern coast of the Western Cape constituting a zone of overlap (Siegfried 1981): a more appropriate ranking of the top sites is presented in Table 12.9.

Estuarine Health Index

The Estuarine Health Index (EHI) is intended to provide the basis for an overall assessment of the state of South African estuaries (Cooper *et al.* 1992, 1993*a*, *b*, 1995; Harrison *et al.* 1994*a*, *b*, 1995*a*, *b*). These authors developed an Estuarine Health Index which has been applied to the estuaries of KwaZulu-Natal and is being applied to the estuaries of the Northern, Western and Eastern Cape Provinces. The purpose of the EHI methodology is to integrate physical, chemical, biological and aesthetic criteria into a measure of estuarine health (see also Chapter 4). In this it departs from the single biotic component approach of Coetzee *et al.* (1997) (vegetation) and Turpie (1995) (birds).

Inspection of the results of the application of the EHI suggests that the entire methodology requires extensive review and revision before rankings emanating from it can be treated with any confidence. A major problem appears to be the confusion of classification (geomorphology) with importance (fishes), with health (fishes) and with a composite health index (fishes, water chemistry, aesthetics, etc.).

The relevance of a geomorphological classification system to the *health* of an estuary appears to be limited. The geomorphology of the basin will affect the nature of the plant and animal communities occupying the estuary concerned but can have little effect upon the 'health' which will respond to much shorter-term processes, e.g. the fluvial regime. In fact the 'permanently open' and 'temporarily closed' categories used by Coetzee *et al.* (1997) are simpler and more functional to apply than the nine estuarine geomorphological types and their subcategories presented in the EHI.

The basis of the biological component of the EHI is a comparison of the 'expected' fish fauna with that actually 'observed'. Fishes were chosen because they 'are relatively easy to identify, most samples can be sorted and identified in the field and the general public can relate to statements about conditions of the fish community' (Cooper *et al.* 1995). The acceptability of fishes to the general public aside, the use of such highly mobile animals as biological indicators is fraught with problems. This criticism could also be directed with some justification to the use of other mobile organisms such as birds. However, birds can be sampled (counted) non-destructively and cost-effectively using skilled amateur observers. Also, Turpie (1995) confined her assessment to the *importance* of estuaries for birds, and did not attempt to extrapolate her findings to the *health* of the estuaries assessed.

The major assumption underlying the use of fishes for determining the EHI is that differences in the fish fauna are due to habitat degradation, i.e. a decline in the 'health' of an estuary. The problem with this in the South African context is that the natural envelope of variability of many estuaries is large, primarily as a result of the very variable rainfall regime. The proponents of the EHI appear to have completely overlooked the need to understand the hydrological and hydrodynamic regime of estuaries in order to assess the validity of their results.

Fishes are highly mobile and will respond to such changes by migrating to more suitable conditions. The 'healthy' fish fauna of an estuary will, therefore, vary considerably depending upon the time within the drought–flood cycle. Consequently, frequent sampling of estuaries will be required in order to establish adequate baselines against which the condition at any one time can be assessed. Clearly frequent sampling is expensive and puts into question the assertion by Cooper *et al.* (1995) that fishes can be sampled cost-effectively in comparison with other indicator organisms such as benthic macrofauna.

The EHI makes use of presence–absence data which gives disproportionate emphasis to rare species. The lack of a measure of relative abundance is a serious drawback. Degradation or other disturbances are likely to affect different species in different ways, and may increase or decrease the *abundance* of certain species without affecting the *number* of species present. This is particularly noticeable when one considers angling species which currently are heavily exploited in estuaries.

The most serious difficulty with the use of fishes to determine the biological health of South African estuaries is the inapplicability of the method to a number of systems. For example, two estuaries in Namaqualand, namely the Groen and the Spoeg, received a zero biological health rating (Harrison *et al.* 1994a) and yet have been identified as key components of a proposed national park. Similarly many of the estuaries within the Tsitsikamma Coastal National Park cannot be rated using a fish-based EHI but they are vital for the survival of the Cape clawless

otter *Aonyx capensis* which forages in the inter- and subtidal zones of the area; otters require freshwater in order to maintain the insulating properties of their fur (Kruuk 1995). Despite the absence of fishes these systems clearly remain vital components of the coastal environment and any method to assess the biological health of South African estuaries should utilise a range of organisms such that the absence of a particular group does not result in the inability to assess a particular estuary. There is a real risk that the narrow ichthyofaunal perspective of Harrison *et al.* (1994a, b, 1995a, b) could lead managing authorities to discount the environmental value of fish-poor systems.

The validity of including an assessment of water quality in the EHI is questionable. In most cases the data used are from single 'snapshot' surveys which cannot account for the variability of estuaries both seasonally and inter-annually. It is interesting to note that while the Department of Water Affairs and Forestry (DWA&F) has established Receiving Water Quality Objectives (RWQO) for freshwater and marine environments, it has not done so for estuaries (Department of Water Affairs and Forestry 1993, 1995). DWA&F accepts that estuaries are highly dynamic and that each estuary requires individual assessment before effluent discharges, if any, are permitted into it.

The final component of the EHI is the Aesthetic Health Index. Harrison *et al.* (1994a, 1995a) state that

> the basic premise of the Aesthetic Health Index is that an estuary which is totally unimpacted by man, reflecting a maximum degree of 'naturalness', is in a perfect or pristine state – and deviation from this state is indicative of degradation.

The problem arises as to how 'pristine' or a similar reference benchmark is defined. There is a danger that a Eurocentric perspective is adopted which assumes that prior to the European colonisation of southern Africa the environment was 'pristine'. However, this view overlooks the presence of humans, and their associated livestock, in the precolonial era. The term 'pristine', therefore, is misused by Harrison *et al.* (1994a, 1995a) since almost all environments have been affected by human activity to some degree (see also Chapter 13). In addition, the concept of a pristine environment often implies, albeit inadvertently, that it is unchanged or unchanging in the face of natural and anthropogenic forces.

The foregoing critique of the Estuarine Health Index highlights its many shortcomings and hence, the results presented in Cooper *et al.* (1992, 1993a, b) and Harrison *et al.* (1994a, b, 1995a, b) need to be treated with caution. It is clear that too much is expected of the fish-based EHI and that oversimplification actually leads to a loss of information.

Whitfield (1996) states that data from ichthyofaunal surveys require specialist interpretation and

> ... are usually beyond the comprehension of most coastal managers and planners. A method of condensing these data into a more readily usable format is essential if this information is to be used in any planning or management process.

However, management of an ecosystem or environment requires an understanding of its components and the processes controlling it. These cannot be reduced to an index or a few indices and placed in the hands of planners and managers with the expectation that the estuaries will retain their viability and that their future will be safeguarded.

Only two comprehensive attempts, therefore, have been made to assess the state of South Africa's estuaries. The methodologies adopted differed greatly as did the time and cost to undertake them. If the prime objective of such studies is the provision of a simple management tool there is little doubt that the expert panel approach (e.g. Heydorn 1986) has much to commend it. The development of an **ichthyofaunal importance rating** to complement Turpie's (1995) waterbird and Coetzee *et al.*'s (1997) estuarine vegetation work would be an immediately useful tool.

Predictive tools

Once the importance of an estuary as a whole, or to a component of its biota, e.g. birds, has been established there is a need for predictive tools to assist with the assessment of the effect of actions, both anthropogenic and natural, upon estuaries.

Hydrodynamic models

Currently in South Africa only the hydrodynamic model Mike-11 (Danish Hydraulic Institute 1992) is being applied in assessments of freshwater requirements of estuaries. Simulation of the physical characteristics of estuaries at the scale of days and months is readily achieved using one-dimensional hydrodynamic simulation programmes such as Mike-11. Such software packages generally require bathymetric cross-sections, water level recordings and river inflow as input data, and are capable of providing time histories of water levels, flow velocities and even salinity through modelling of hydrodynamic and transport–dispersion processes (Huizinga 1994).

A limitation of standard multi-dimensional hydrodynamic models is that they cannot accommodate rapid alterations in estuary mouth conditions. The effects of different mouth configurations on tidal flow through the mouth can be investigated using 1-D models but sharp alterations in mouth state cannot be accommodated during a simulation.

Estuarine systems model

Slinger (1996), realising the limitations of the 1-D model and the need for medium to long-term predictions for management, developed the estuarine systems model (see Chapter 3). This is a semi-empirical model which treats the estuary as a basin connected to the sea by a channel of variable height. Time histories of water levels, tidal fluxes, mean salinities, stratification, circulation state, freshwater and tidal flushing, and sill height at the mouth are simulated. The provision of concurrent information on mouth conditions and associated effects on salinity and water levels is a significant improvement for the management of South African estuaries as is the capability to simulate the process of mouth closure and breaching.

The model has been implemented for the Great Brak Estuary and the role of episodic wave events in causing closure of the mouth highlighted. The model demonstrates that reduced freshwater flows lead to an increase in the sensitivity of the system to marine forcing. This serves as a warning for the future of many freshwater deprived estuaries in South Africa.

Upogebia africana model

The mudprawn *Upogebia africana* is widely distributed in southern African estuaries from Langebaan Lagoon in the west to Inhambane, Mozambique in the east (Hanekom & Erasmus 1988). *Upogebia africana* is an important prey item for estuarine fish and is extensively exploited for bait by anglers. The mudprawn has an obligate marine phase in its life cycle: Zoea I larvae migrate from estuaries into the sea and pass through Zoea II–IV before returning to the estuary as postlarvae (Wooldridge 1991, 1994). The requirement for a marine phase to complete the life cycle requires that the estuary mouth is open at the right time and for sufficiently long to permit the return of postlarvae. This life cycle requirement can thus be used as a clear objective for the management of the estuary mouth.

The *Upogebia africana* model represents a first attempt to link processes operating at a physiological scale to those operating at a population scale, in an approach to the management of estuaries in South Africa.

In recent years a number of population models which explicitly account for attributes of population structure such as size and age have been developed. These models account for the fact that individuals at different ages, sizes, stages and locations may respond differently to environmental gradients and hence contribute differently to processes such as reproduction. In addition, at the level of ecosystem management, population attributes such as size class distribution are frequently used as indices of ecosystem health (CSIR 1990). For this reason, and the fact that the recruitment of a group of *U. africana* occurs as a discrete event, an approach was selected whereby the growth and mortality of individual cohorts comprising the total population are modelled (Quinn 1997).

Physiologically-based growth models of individual fish have been widely used and vigorously tested (Rice & Cochran 1984). These models estimate growth from consumption, based on an understanding of the physiology of the fish as a function of their body size and temperature. Under ideal temperature conditions they would grow at a maximum rate and the size of the organism would be determined directly from a relationship relating size and age. Under non-ideal conditions the growth rate is slower, represented in the model by a reduction in the rate of ageing. This concept of differential ageing gives rise to two terms: chronological age and physiological age. The former refers to the actual age of the organism while the latter is an expression of the net effective increase in maturity due to unfavourable conditions.

Quinn's (1997) *Upogebia africana* model extends the approaches described above by attempting to include the interactive effects of both temperature and salinity on the growth and development of an invertebrate i.e. it incorporates the cumulative effects of unfavourable environmental conditions on the rate of growth and final dimensions of an organism.

The approach has relied on several assumptions, the most significant of which probably concerns the translation of a laboratory-based understanding of the organism's response to changes in temperature and salinity to those which may occur in a natural system. Assumptions regarding recruitment of postlarvae are perhaps the weakest of all. Virtually nothing is known about mortality of zoeae in the marine environment and the conditions under which postlarvae return to the estuary environment have only recently been determined (Wooldridge & Loubser 1996). The model has, however, provided evidence that continual recruitment into populations of *U. africana* within an estuary is strongly dependent on the existing population. Should a population within an estuary be lost by mouth closure or a discrete flooding or pollution event, re-establishment by recolonisation from an adjacent estuary may occur although this is likely to be slow.

Despite the numerous assumptions and considerable degree of conjecture inherent in the current formulation of the model, preliminary testing (using data from the Great Brak estuary: CSIR 1990) did appear to be able to predict, at least in broad terms, patterns of response to the historical record of the mouth status.

The Fish Recruitment Index

One of the primary ecological values attributed to estuaries is the role they play in providing habitats for juvenile fish, often referred to as the 'nursery function' of estuaries. Quinn (1997) formulated a modelling approach which would provide an indication of the extent to which an estuary was fulfilling this role for a variety of fish species. The basis of the Fish Recruitment Index is the observation that recruitment of juvenile marine fish is limited in the absence of strong axial salinity gradients (Quinn et al. 1997). Understanding of recruitment processes of juvenile marine fish is currently very limited, and is hampered by the difficulties of larva identification and sampling. While it has been suggested that juvenile fish do not respond to the axial salinity difference *per se* (Whitfield 1994a), it is not known whether all species respond in the same way to the flow conditions which give rise to axial salinity differences. Until evidence to the contrary is proposed, equal response of juvenile marine fish to prevailing axial salinity differences is assumed, and as this assumption is associated with the greatest of uncertainty, it may be the major weakness of the approach. An additional source of error may be the assumptions regarding the recruitment preferences of the species included in the model.

Although Whitfield (1994a) has shown a relationship between axial salinity difference and recruitment, it is likely that fish are responding to riverine-based olfactory cues entering the marine environment, and not salinity or salinity gradients *per se* (Whitfield 1996). The latter suggests that the actual volume of water entering the marine environment may also be relevant. As a consequence, it has been suggested that the index be modified to account for differences in flow volumes, particularly in the case of permanently open estuaries.

Notwithstanding the considerable scope for testing and further development of the Fish Recruitment Index, initial application in the Great Brak estuary suggested that annual runoff to the estuary could be halved without an appreciable decrease in the Fish Recruitment Index, but with a potentially sharp decline thereafter. While the index represents recruitment opportunity rather than actual recruitment, the index does at least provide an indication of where ecological risk is likely to increase substantially.

Estuary habitat

Habitat is usually conceptualised as the link between organisms and the environment. Generally, it is considered to be the place which provides an individual organism with the resources for survival and procreation (food, water, cover, space, mates, etc.). Southwood (1977) relates habitat to the range of an individual's movements, stating that for any species habitat may be defined as '*the area accessible to the trivial movements of the food harvesting stages*', while others have adopted a species or population focus, defining habitat as '*any part of the earth where the species can live, either temporarily or permanently*' (Krebs 1985).

Quinn (1997), however, considers an alternative view of

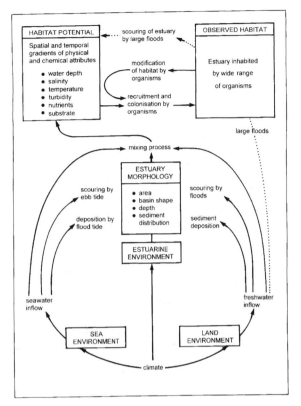

Figure 12.2. Relationship between potential habitat, habitat and the processes which sustain these habitats in the estuarine environment (Quinn 1997).

habitat, potential habitat, which recognises the primary environmental gradients giving rise to the 'space' which ultimately may be populated by individuals of a species, thereby transforming the 'space' into 'habitat'. Such an approach therefore places more emphasis on the identification of areas characterised by similarities in environmental gradients: the hypothesis being that areas which are subject to similar environmental gradients will have similar potential to support the same range of organisms. The most significant advantage of conceptualising habitat in this manner is that if sufficient is understood about the processes which maintain the primary habitat, the consequences of any changes in these processes can be translated into consequences for the organism inhabiting the primary habitat at that time. This approach therefore has obvious application in the management of estuaries.

The two primary processes maintaining estuarine environments are freshwater inflow and tidal exchange. The relationship between these processes and the primary or potential habitat that they create are illustrated in Figure 12.2. When occupied by an organism through recruitment or colonisation, potential habitat for a variety of organisms becomes the temporary habitat for a range of

organisms. The colonisation of the environment by vegetation and animals in turn modifies the habitat. For example, saltmarshes and mangroves produce large amounts of organic matter which enriches the substratum. The presence of vegetation such as submerged and emergent macrophytes reduces water flow velocities, thereby permitting the deposition of finer sediment particles and ultimately providing a greater area for colonisation by the plant. The stabilisation of shorelines and mudbanks by vegetation reduces scour in subsequent floods, thereby gradually altering the estuarine environment.

The recent rapid development of Geographic Information System (GIS) technology has provided a wealth of techniques which lend themselves to applications in environmental management. For example, Miller (1994) has illustrated the utility of GIS in mapping the distribution of species, identifying potential habitat for certain species as well as in mapping biodiversity. Pereira and Duckstein (1993) have combined GIS and multiple-criteria decision-making (MCDM) techniques to identify suitable habitat for the endangered Mount Graham red squirrel and Haines-Young et al. (1993) have presented a variety of applications of GIS in landscape ecology. While three-dimensional approaches have found wide application in mining, hydrology and landform analysis (Raper 1989), it appears that they have yet to be fully utilised in the modelling of ecological processes.

Habitat areas may differ from each other with respect to their suitability for certain species, as well as their durability (persistence) and changeability over time. Consequently if all the life requirements of the species are identified, 'habitat types within a mosaic could be classified as optimal, sub-optimal, marginal or non-inhabitable' (Kozakiewicz 1995). This concept is the basis of many of the spatially explicit habitat models reported in the literature. Thus, in an approach formulated by Pereira and Duckstein (1993) during a study of the habitat of the Mount Graham red Squirrel, expert opinion is used to determine scores representing the suitability of certain categories of components of habitat, such as elevation, slope, aspect, canopy cover, tree stem diameter and distance to clearings. Potential habitat predicted by the model and actual habitat were found to be significantly correlated.

Change in estuaries occurs at a variety of spatial and temporal scales. For example, some components of a benthic invertebrate's habitat such as substratum characteristics and vegetation cover may change over a long time scale, whereas others such as inundation or salinity may change over a shorter time scale. Thus at least two categories of change may be evident (Figure 12.3).

Because sediment and vegetation undergo change at less frequent intervals in estuaries, they are considered to

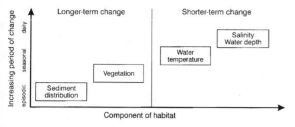

Figure 12.3. **Components of habitat and their rates of change (Quinn 1997).**

form the basis of potential habitat. In a study of the Great Brak estuary Quinn (1997) defined 21 unique combinations of sediment and vegetation. Since each area is defined only by the interaction of two variables it is relatively simple to score the preference of an organism for each of the areas. For example, of the 21 potential habitat areas in the Great Brak estuary, the benthic mudprawn *Upogebia africana* could occur in only seven. The potential habitat areas are also subject to fluctuating water levels and, as a consequence, will experience differing periods of inundation. The intertidal and submerged habitat available to benthic invertebrates such as *U. africana* consequently vary with the water level.

Whether an organism actually utilises the potential habitat or not, is considered to be dependent on other factors which fluctuate in the short term, such as whether or not the organisms' range of physiological tolerances (e.g. temperature and salinity) and other requirements are met in this potential habitat zone within the specified time interval.

The availability, extent and quality of the data on which this type of analysis is based often present difficulties. Bathymetric characteristics, as well as sediment and vegetation distribution are seldom mapped at the same scale with the same degree of accuracy, and consequently it is usually difficult to integrate these data sets. In the case of the Great Brak estuary, for example, sediment distribution was mapped on the basis of schematic diagrams and narrative descriptions.

The habitat model, therefore, has highlighted major deficiencies in the historical data and the data currently being collected. The implication is that data collection procedures and criteria will have to be revised.

There are obvious advantages to analysing and presenting spatial data in the manner proposed by Quinn (1997). For example, mapping these attributes at regular intervals permits rapid assessment of change: thus overlaying bathymetric maps compiled at different times will identify where deposition and erosion have occurred. Similarly, two vegetation maps would show where dieback or colonisation is occurring, providing a useful monitoring tool. More importantly, this approach allows

assessment of components in relation to other elements of environmental change. Variation in water level (exposure and inundation) and salinity are usually the primary factors influencing the distribution of vegetation in estuaries (Adams 1994), and often give rise to a distinct zonation (Day 1981b). As the tolerances of species to inundation are well known (Adams 1994), it is possible to predict the consequences of changing water levels for different vegetation communities. Indeed, if the approach described above provides a meaningful framework for the analysis of habitat potential in an estuary, there should be a strong correlation between the occurrence of vegetation and seasonal fluctuations in inundation and exposure, particularly given that water level appears to be a primary factor in determining habitat for vegetation (CSIR 1990).

A useful output of the approach which has been demonstrated, however, is the facility to evaluate the change in intertidal area, and species-specific habitat associated with each of the freshwater inflow scenarios. Intertidal areas are considered an important component of estuary habitat and are particularly important for particular groups of species such as waders. Estuarine mudbanks and sandbanks are also important for roosting birds; if the prevailing water level becomes too high these habitats are then not available for roosting. Precisely the same methodology can be used to evaluate the change in the availability of roosting habitat for birds as was used to determine change in intertidal area. Similarly the change in any defined set of environmental factors can be evaluated. For example, if a biologically significant vegetation community is known to flourish under a particular inundation regime, the consequences of changing the freshwater inflow and therefore the inundation regime can be explored.

One key assumption of the above approach in its current form should, however, be noted. In the application of the method to the Great Brak estuary, a constant water surface elevation was assumed. In reality the water surface slopes seawards, and thus at the same inflow rate the water may be at 0.71 m (MSL) at point A in the estuary and 0.68 m (MSL) at point B, closer to the estuary mouth. The implication of this is that the analysis of habitat should be compartmentalised into areas of similar water surface elevation. As hydrodynamic models compute water levels for each of the contiguous network of cells, these data are usually readily available. The analysis of habitat in the estuary would then occur by hydrodynamic cell, and would be summed to provide an overall value for the estuary.

The management process

In South Africa there has been a tendency for environmental policy formulation to become an end in itself. Begg

(1985) advanced policy proposals for a number of KwaZulu-Natal estuaries which, over a decade later, have yet to be implemented. Similarly, although the need for allocation of freshwater to estuaries has been acknowledged for several years, an integrated method for determining it has not been developed.

It follows, therefore, that laudable policy objectives such as the 'sustainable utilisation of estuaries' and the 'maintenance of biodiversity' will not be achieved in the absence of a strategy for their implementation. This strategy for implementation needs to be complemented by a supportive and integrated management framework which sets out a timetable for the action plan describing clearly the operational goals against which progress will be measured. The strategy should document the range of methods which will be employed to achieve these goals, and may include a mix of regulatory, non-regulatory and incentive based approaches.

Most importantly, successful policy also requires a 'toolbox' of decision-making and evaluatory techniques. The purpose of natural resource policy is to set the constraints and conditions of resource utilisation, and as can be expected, as the demand for natural capital such as water or land increases, so do the complexities of resource management.

The development of appropriate predictive capacity, located within a framework which acknowledges the failings of previous approaches to adaptive management (McLain & Lee 1996), will contribute significantly to defining a management approach which will demonstrate achievable ecological sustainability. Breen *et al.* (1994) have proposed a management approach for the natural environment of rivers which addresses many of these concerns (Figure 12.4). The concepts employed in this approach have been extended in formulating an approach appropriate for the management of estuaries.

Setting the vision

A vision describes the state in which society wishes estuaries to be, reflecting the value society places on estuaries and their functions. For management to be meaningful its purpose must be to achieve a desired and predetermined state of the natural environment (Breen, *et al.* 1994). The vision, in turn, forms the basis for policy. As the vision is a reflection of the wishes of society, it is important that setting the vision, as well as many other components of this management approach, occurs on the basis of public participation. Participation is a fundamental principle of the new environmental policy, and recognises that 'all interested and affected parties have a right to participate in environmental management and decision making' (Department of Environmental Affairs and Tourism 1996). This tenet is equally recognised in the new Water Law Principles

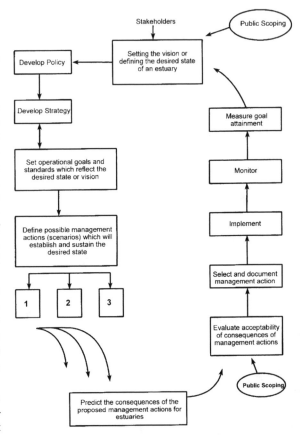

Figure 12.4. **A management framework for demonstrating ecological sustainability in estuaries (after Breen *et al.* 1994).**

which acknowledge the need to devolve management to a local level (Department of Water Affairs and Forestry 1996).

Setting operational goals

Goals such as 'maintaining ecological processes', 'preserving genetic diversity' or 'ensuring sustainable utilisation' cannot be considered operational goals as there are no direct means of assessing achievement. These goals must therefore be redefined in precise and meaningful terms (Breen *et al.* 1994). Setting operational goals has recently been expressed as a fundamental requirement of scientifically based ecosystem management (Christensen *et al.* 1996). The report of the Ecological Society of America Committee on the Scientific Basis for Ecosystem Management states that

> goals must be explicitly stated in terms of specific 'desired
> future trajectories' and 'desired future behaviours' for the
> ecosystem components and processes necessary for
> sustainability. Furthermore, these goals should be stated
> in terms that can be measured and monitored.
>
> (Christensen et al. 1996)

Defining and recording potential management actions

Management actions required to sustain the natural environment are usually based on wisdom in the absence of any structured decision-making framework. While decisions are considered to reflect the expertise of participants, the process, interpretations and assumptions which underlie these decisions are seldom recorded. As a consequence it is impossible to review the steps in decision-making to correct, improve or reverse decisions (Breen *et al.* 1994). Apart from giving attention to documenting the process of defining management actions, the actions themselves must be documented in a clear and unambiguous manner.

Predicting the consequences of management actions

Adaptive management is currently promoted as a desirable conceptual approach to ecosystem management (Christensen *et al.* 1996), and can be described as a structured system in which monitoring iteratively improves the knowledge base and helps to refine management plans (Ringold *et al.* 1996). Being able to predict the consequences of management actions will obviously limit the risk associated with the adaptive management approach. According to Christensen *et al.* (1996), models

> *can be useful in identifying particularly sensitive ecosystem components or in setting brackets around expectations for the behaviour of particular processes. They can be especially useful in identifying indices and indicators that provide a measure of the behaviour of a broad suite of ecosystem properties.*

Some would argue that for models to be useful management tools they *need* to identify indices or indicators which provide a realistic measure of ecosystem functioning. The approach proposed in this chapter argues that policy, goals and management actions should be specifically focused on the biological attributes of estuaries, and measured by one or more objective indices. This means that the consequences of potential action such as artificial breaching, or water release policies from proposed or existing impoundments, should be evaluated by indices which indicate the extent to which the proposal could affect the biological potential of the system.

Such an approach necessitates the development of predictive abilities at two levels. The hydrodynamic and physical consequences to an estuary, of an action such as artificial breaching or a water release from a dam, need to be determined; i.e. the effect on the physical environment, measured for example by changes in water level, temperature, salinity or access to the sea, needs to be established. The next stage of prediction requires the translation of changes in these environmental variables into conse-

quences for the biota. For example, how do changes in inflow alter habitat availability and consequently the dynamics of populations supported by these habitats; or alternatively a decrease in freshwater inflow together with high evaporation may result in salinities above the tolerances of organisms within the estuary.

Evaluate the acceptability of consequences of different management actions

Anthropogenically induced changes in the structure and functioning of natural systems alter the asset value ascribed to natural systems (Breen *et al.* 1994), and the response of society to the predicted change reflects the extent to which the change in asset value is acceptable. If the consequences are unacceptable, other management actions need to be considered and routed through the predictive process. This process is dependent on public participation and requires protocols for establishing trade-offs and documentation of the entire assessment process.

Monitoring and measuring goal attainment

Management is meaningless unless one is able to measure the extent to which management goals are being attained. Furthermore, monitoring needs to occur at a time scale that will permit the implementation of corrective action before undesirable changes in the state of the ecosystem occur. Implementation of monitoring programmes requires vision and commitment, the benefits of which are not always evident in the short term (Christensen *et al.* 1996).

Challenges

The prime challenge to those responsible for the management of South Africa's estuaries is to maintain their viability in the face of ever increasing pressures, particularly the demand for freshwater. This requires that the approach changes from the *ad hoc* 'tinkering', characteristic of so much of the so-called management of estuaries, to a more holistic approach such as that pioneered by the Department of Water Affairs and Forestry and the Water Research Commission.

In 1986, prior to embarking on assessment of the state of estuaries in the then provinces of Cape and Natal (Heydorn 1986), four key questions were posed.

- On what criteria should the environmental importance of an estuary be assessed?
- What is the relative environmental importance of the different estuarine systems?

- What are the effects of variation in freshwater input on the biotic and abiotic characteristics of estuarine systems?
- What freshwater input pattern is required to obtain and/or maintain an agreed state of the estuarine system concerned?

Just over a decade later, what progress has been made towards answering these questions? The criteria on which the importance of estuaries can be assessed remain the subject of debate. A consensus is forming that composite indices such as that used by Cooper *et al.* (1995) in the Estuarine Health Index are not universally applicable and can lead to a loss of information as a result of over-simplification. Similarly, a particular component of the biota, such as fishes, cannot be used to rank the importance of all South Africa's estuaries given their great diversity of form and function. The trend is towards considering a suite of plants and animals to provide rankings for particular components of the estuarine biota, e.g. vegetation (Coetzee *et al.* 1997), fishes (Whitfield 1994*a*, 1996) and birds (Turpie 1995).

There is a need to resolve differences in the methods used in these rankings. Two schools of thought have emerged. (i) Those who use the number of species and their abundance in deriving rankings. Turpie (1995) investigated this at length and pointed out that the concept of abundance depended on the overall population of a given species. This concept also underlies the vegetation cover classification used by Coetzee *et al.* (1997). (ii) Those who use presence–absence data, e.g. Cooper *et al.* (1995) and Whitfield (1994*a*, 1996), on fishes. A major problem with presence–absence data is that it over-emphasises the importance of rare species.

Invertebrates present a particular problem since their populations are difficult to quantify cost-effectively and their life cycles are not well understood, particularly with respect to the need for access to the open sea. Wooldridge (1991, 1994) has demonstrated the potential of the mud-prawn *Upogebia africana* as an indicator species, particularly with respect to the frequency that an estuary mouth is open. However, there is a requirement to determine the potential of a suite of estuarine invertebrates as indicators for a range of environmental conditions and to ensure that all the biogeographic regions are encompassed.

The relative importance of the different estuarine systems remains to be resolved. Any such ranking is unlikely to be acceptable on a whole system basis but can be effective when individual components of the biota are assessed. Turpie (1995) has undertaken the most exhaustive assessment of single- and multi-criteria indices applied to the ranking of estuaries for their importance to waterbirds. She states that

it is strongly advised that the prioritization of sites be based on results of single-criterion rankings. This elevates the subjective input to that of collating primary rankings rather than to the relative weighting of criteria in complex indices, thereby making the decision process more explicit.

(Turpie 1995)

Although Turpie's work was restricted to waterbirds it demonstrates the hazards of using composite indices for ranking estuaries in order of importance. It is suggested that a matrix method should be adopted so that the ratings for each component of the estuarine environment can be clearly seen. Thus some estuaries will be clearly seen to be important for a single component, e.g. fishes whereas others may be important for two or more components. Estuaries can be assessed in this way to provide ratings on a national, regional (political or biogeographical) or local scale. It is clear that, whatever ranking method is utilised, there is a need for informed, intelligent interpretation of the results by experts. It would be very dangerous to place any rating, particularly those based on composite indices, in the hands of an authority or manager who would apply it mechanically.

The challenge remains for the estuarine research and management community to develop mutually acceptable ranking systems for all components of the estuarine biota.

The effects of variation of freshwater input on the biotic and abiotic characteristics of estuaries have received some attention during the last decade, primarily stimulated by the need for the Department of Water Affairs and Forestry to make allocations for estuarine function. Reddering (1988*a*, *b*) has investigated the effect of reduced river discharge on selected estuaries. In contrast to Reddering's geomorphological approach, Slinger (1996) has developed a semi-empirical estuarine system which allows a variety of scenarios to be investigated cost-effectively in a short time.

Estuarine phytoplankton plays a critical role in supporting the zooplankton essential as prey for juvenile fishes using estuaries as nurseries. The importance of freshwater in sustaining estuarine phytoplankton needs much more intensive investigation. For example, Allanson and Read (1995) reported that the phytoplankton and zooplankton biomass levels in the freshwater-rich Great Fish River estuary are generally an order of magnitude greater than those of the freshwater-starved Kariega estuary.

The response of estuarine vegetation to freshwater inflow has been investigated by researchers at the University of Port Elizabeth (Adams 1992, 1994; Adams & Talbot 1992, Adams *et al.* 1992; Adams & Bate 1994*a*–*d*). The prime focus has been on the importance of freshwater to the survival of estuarine plants. The salinity tolerance envelope of a number of species has been investigated.

Invertebrates have received limited attention. In terms of management, the work of Wooldridge (1991, 1994) and Quinn (1997) on *Upogebia africana* shows the potential of invertebrates as indicators. During the Great Brak estuary study (CSIR 1990) benthic invertebrates were used as indicators of the state of the estuary and the success of the management policy. Species used were the marsh crab *Sesarma catenata*, the mudprawn *Upogebia africana* and the sand prawn *Callianassa kraussi*.

With respect to fishes the focus, primarily by Whitfield (1994a, b, 1996) has been on the importance of estuaries to fishes and the mechanisms by which juveniles enter estuaries. The relative importance of the role of axial gradients and olfactory cues needs to be elucidated.

The avifaunal studies of Turpie (1995) and Hockey and Turpie (Chapter 10, this volume), have been directed at assessing the importance of estuaries for Palearctic migrant waders and for the conservation of estuary-associated birds.

The resilience of individual estuarine organisms and of the estuaries themselves is poorly understood. Such an understanding is of great value when assessing the impact of natural or anthropogenic low flows, or droughts, on estuaries. In other words, what are the limits of the dynamic equilibrium under which an estuary can recover from perturbations? It is important to know what factors will induce irreversible change in any given type of estuary.

Freshwater input patterns to maintain estuaries in an acceptable condition have been subject to investigation in the course of determining the freshwater requirements of estuaries. The Department of Water Affairs and Forestry (DWA&F) has run workshops on the requirement of the Orange, Berg, Palmiet, Great Brak, Keurbooms, Tsitsikamma, Mgeni, Mdhloti, Lovu, Mkomazi and Tugela estuaries. The DWA&F favours the approach where a fixed volume of water is allocated to an estuary. This approach suggests an implicit unwillingness to recognise that the extraction of water from a river system has an impact and that the problem is to determine acceptable trade-offs between the impacts and the benefits (ecological, social, economic, etc.). The process promoted by the DWA&F is based on the instream flow requirement (IFR), i.e. there is some defined quantity of water which will maintain the estuary in a desired or agreed state. This conceals the fact that trade-offs are already made in defining the desired state. Furthermore, it prevents the consideration of the actual impacts on the system because effort is expended trying to define how much water an estuary or river needs in order to be maintained in an agreed state rather than assessing impacts.

Many researchers and managers find the DWA&F approach to be far too restrictive and favour the investigation of the *impact* of a range of release scenarios before selecting a particular flow regime. Essentially this scenario-based approach involves assessing the impacts of various development scenarios on the river and its estuary. The various scenarios can then be ranked according to the severity of their impact upon the components of the system such as the fauna, flora, state of the mouth and characteristic salinity variations. Each development scenario may not rank in the same order for the components chosen for the rating. Thereafter, examination of the full suite of information on what will happen to the different components under the various development scenarios permits trade-off decisions, both ecological and socio-economic, to be made. The important point is that only at the very last stage of the process is a value judgement made. It is separated from the quantifiable, scientific assessment of the impacts, the reasons for the rankings/ratings and decisions can be explicitly stated and detrimental effects are not hidden. The estuarine freshwater requirement is determined only at this final trade-off stage. This acceptable trade-off scenario automatically includes natural variations.

Critical to any adopted freshwater allocation scenario is the need to monitor the effectiveness of the water releases in achieving the set goals. The Great Brak estuary study has demonstrated convincingly the effectiveness of a well-designed monitoring programme in providing the information necessary for the adaptive management of an estuary. There is a need, therefore, to invest in the development of monitoring methods and criteria which can be implemented cost-effectively.

It is likely that integrated catchment management (ICM), to be incorporated in the new Water Act currently in draft form, will result in the impact assessment approach being adopted. The ICM methodology will require a goal-orientated approach more familiar to the business sector than the environmental management and conservation sectors (Bestbier *et al.* 1996). This will entail setting a vision with the participation of stakeholders and then translating it into goals which can be audited to determine the success of the management mechanisms adopted. There is a risk that the vision may be unrealistic. However, the process of translating it into goals will soon demonstrate the limitations imposed by the estuary and its catchment on the attainment of those goals. The key is an open and consultative process which will foster support for decisions taken. Ownership of the vision and the goals and the methods by which success is measured is critical to the future of all South African environments, not only estuaries.

One issue with respect to freshwater allocation is the

dubious value of using mean annual run-off (MAR) in the assessment of water availability. Leopold (1994) states that

> Serious errors in water management have been made by too heavy reliance on the average runoff value. A much more useful analysis is a frequency study of the whole array of available data so that the probability of various levels of excess or deficiency may be considered.

In a semi-arid country prone to droughts broken by floods it would be more realistic, therefore, to use a more conservative criterion such as the *median* runoff for water allocation calculations. In essence this is a precautionary approach which should help to engender a more responsible attitude to South Africa's limited water resources.

Until recently there has been an almost complete failure of estuary management authorities to consider estuaries as parts of catchments and of the adjacent coastal zone. As a result of the freshwater allocation studies estuary management has become integrated into catchments. However, the integration into the coastal zone remains neglected. It is hoped that the new initiative to develop a national coastal zone policy will remedy this problem. Little is known about the role estuaries play in the functioning of the immediately adjacent terrestrial environment.

Research into estuarine function will have to respond to the needs identified by the goal-setting process. To borrow business terminology, 'technology push' will have to be replaced by 'market pull'. In the main researchers have had the freedom to investigate any aspect of estuaries that interested them, whereas in the future they will have to respond to clearly enunciated needs emerging from the ICM process.

Overall much progress in the science underpinning estuary management has been made since Day and Grindley's (1981) review. Also the Integrated Environmental Management process has been developed and provides the necessary framework for decision-making (Fuggle & Rabie 1992). However, few initiatives have been fully implemented and fewer still have observed principles of integrated environmental management. Estuary management, indeed all environmental management, should operate within a framework that emphasises principles of sustainable development, biodiversity conservation, the precautionary principle, integrated management, self-regulation and sensitivity to local circumstances. Particular emphasis should be placed on the need for the various actors, agencies and levels of government to work together and seek consensus.

References

Adams, J.B. (1992). The importance of freshwater in the maintenance of estuarine plants. *The Naturalist*, **36**, 19–24.

Adams, J.B. (1994). *The importance of freshwater to the survival of estuarine plants*. PhD thesis, University of Port Elizabeth.

Adams, J.B. & Bate, G.C. (1994a). *The freshwater requirements of estuarine plants incorporating the development of an estuarine decision support system*. WRC Report No. 292/2/94. Pretoria: Water Research Commission.

Adams, J.B. & Bate, G.C. (1994b). The effect of salinity and inundation on the estuarine macrophyte *Sarcocornia perennis* (Mill.) A. J. Scott. *Aquatic Botany*, **47**, 341–8.

Adams, J.B. & Bate, G.C. (1994c). The ecological implications of tolerance to salinity by *Ruppia cirrhosa* (Petagna) Grande and *Zostera capensis* Setchell. *Botanica Marina*, **37**, 449–56.

Adams, J.B. & Bate, G.C. (1994d). The tolerance to desiccation of the submerged macrophytes *Ruppia cirrhosa* (Petagna) Grande and *Zostera capensis* Setchell. *Journal of Experimental Marine Biology and Ecology*, **183**, 53–62.

Adams, J.B., Knoop, W.T. & Bate, G.C. (1992). The distribution of estuarine macrophytes in relation to freshwater. *Botanica Marina*, **35**, 215–26.

Adams, J.B. & Talbot, M.M.B. (1992). The influence of river impoundment on the estuarine seagrass *Zostera capensis* Setchell. *Botanica Marina*, **35**, 69–75.

Allanson, B.R. & Read, G.H.L. (1995). Further comment on the response of eastern Cape Province estuaries to variable freshwater inflows. *South African Journal of Aquatic Sciences*, **21**, 56–70.

Bailly, D. & Paquotte, P. (1996). Aquaculture and environment interactions in the perspective of renewable resource management theory. *Coastal Management*, **24**, 251–69.

Bally, R., McQuaid, C.D. & Pierce, S.M. (1985). Primary productivity of the Bot River estuary, South Africa. *Transactions of the Royal Society of South Africa*, **45**, 333–45.

Begg, G.W. (1978). *The estuaries of Natal*. Natal Town and Regional Planning Commission Report 41, Pietermaritzburg.

Begg, G.W. (1984). *The estuaries of Natal Part 2*. Supplement to NTRP Report Vol. 41. Natal Town and Regional Planning Commission Report 55, Pietermaritzburg.

Begg, G.W. (1985). *Policy proposals for the estuaries of Natal*. Natal Town and Regional Planning Commission Report 43, Pietermaritzburg.

Bennett, B.A., Hamman, K.C.D., Branch, G.M. & Thorne, S.C. (1985). Changes in the fish fauna of the Bot River Estuary in relation to opening and closure of the estuary mouth. *Transactions of the Royal Society of South Africa*, **45**, 449–64.

Bestbier, R., Rogers, R., Blackmore, A., Kruger, J., Nel, L. & Biggs, H. (1996). *Guidelines for goal-orientated* conservation. Centre for Water in the Environment, Report No. 1/96. Johannesburg: University of Witwatersrand.

Bickerton, I.B. & Pierce, S.M. (1988). Report No. 33. Krom (CMS 45), Seekoei (CMS 46) and Kabeljous (CMS 47). In *Estuaries of the Cape. Part 2. Synopses of available information on individual systems*, ed. A.E.F. Heydorn & P.D. Morant. CSIR Research Report No. 432. Stellenbosch: CSIR.

Blaber, S.J.M., Hay, D.G., Cyrus, D.P. & Martin, T.J. (1984). The ecology of two degraded estuaries on the north coast of Natal, South Africa. *South African Journal of Zoology*, **19**, 224–40.

Bolton, M.P. & Specht, R.L. (1983). *A method for selecting conservation reserves.* Australian National Parks and Wildlife Service Occasional Paper No. 8.

Bovee, K.D. (1982). *A guide to stream habitat analysis using the instream incremental flow methodology.* Instream Flow Information Paper No. 12. Fort Collins: United States Department of the Interior, Fish and Wildlife Service.

Branch, G.M., Bally, R., Bennett, B.A., De Decker, H.P., Fromme, G.A.W., Heÿl, C.W. & Willis, J.P. (1985). Synopsis of the impact of artificially opening the mouth of the Bot River estuary: implications for management. *Transactions of the Royal Society of South Africa*, **45**, 465–83.

Breen, C.M., Quinn, N.W. & Deacon, A. (1994). *A description of the Kruger Park Rivers Research Programme (Second Phase).* Pretoria: Foundation for Research Development.

Burman, J. (1970). *Waters of the Western Cape.* Cape Town: Human & Rousseau.

Christensen, N.L., Bartuska, A.M., Brown, J.H., Carpenter, S., D'Antonio, C., Francis, R., Franklin, J.F., MacMahon, J.A., Noss, R.F., Parsons, D.J., Peterson, C.H., Turner, M.G. & Woodmansee, R.G. (1996). The report of the Ecological Society of America committee on the scientific basis for ecosystem management. *Ecological Applications*, **6**, 665–91.

Coetzee, J.C., Adams, J.B. & Bate, G.C. (1997). A botanical importance rating of selected Cape estuaries. *Water SA*, **23**, 81–93.

CONNEPP (1996). *Towards a new environmental policy for South Africa.* Discussion Document. Pretoria:

Department of Environmental Affairs and Tourism.

Convention on Biological Diversity (1994). Convention on Biological Diversity: Text and annexes. Switzerland, Geneva: The Interim Secretariat for the Convention on Biological Diversity.

Cooper, J.A.G., Harrison, T.D., Ramm, A.E.L. & Singh, R.A. (1992). *Biological health of South African estuaries. Part 1. Tugela to Mtamvuma 1991.* Report to the Department of Environment Affairs, Pretoria. Durban: CSIR.

Cooper, J.A.G., Harrison, T.D., Ramm, A.E.L. & Singh, R.A. (1993a). *Refinement, enhancement and application of the Estuarine Health Index to Natal's estuaries, Tugela – Mtamvuma. Executive Report.* Report to the Department of Environment Affairs, Pretoria. Durban: CSIR.

Cooper, J.A.G., Harrison, T.D., Ramm, A.E.L. & Singh, R.A. (1993b). *Refinement, enhancement and application of the Estuarine Health Index to Natal's estuaries, Tugela – Mtamvuma. Technical Report.* Report to the Department of Environment Affairs, Pretoria. Durban: CSIR.

Cooper, J.A.G., Ramm, A.E.L. & Harrison, T.D. (1995). The Estuarine Health Index: a new approach to scientific information transfer. *Ocean and Coastal Management*, **25**, 103–41.

Coppoolse, R.C., Schoonees, J.S. & Botes, W.A.M. (1994). Physical impacts of the disposal of dredger spoil at Richards Bay, South Africa. In *Proceedings of the 28th International Navigation Congress, Seville, 22–27 May 1994*, Section II-I, 73–85.

Council for the Environment (1989). *A policy for coastal zone management in the Republic of South Africa. Part 2. Guidelines for coastal land-use.* Pretoria: Council for the Environment.

Cousens, D.W.H., Braune, E. & Kruger, F.J. (1988). *Surface water resources of South Africa: research needs.* Pretoria: Water Research Commission.

Crass, R.S. (1982). Lake St. Lucia: a historical note on research and management effort. In *St Lucia Research Review*, ed. R.H. Taylor. Pietermaritzburg: Natal Parks Board.

CSIR (1988). *Dynamics of the Klein River with reference to the artificial breaching of the estuary mouth.* CSIR Contract Report EMA-C 8891. Stellenbosch: CSIR.

CSIR (1990). *Great Brak River: Estuary environmental study with reference to a management plan for the Wolwedans Dam and Great Brak River Mouth.* CSIR Report EMA-C9036. Stellenbosch: CSIR.

CSIR (1992a). *Freshwater requirements for Natal Estuaries. Part I: The relative importance of the coastal processes and estuarine dynamics of the Mgeni estuary with reference to identified key environmental parameters and the existing freshwater release policy.* CSIR Report EMAS-C92099. Stellenbosch: CSIR.

CSIR (1992b). *Freshwater requirements for Natal Estuaries. Part II: A first assessment of the freshwater requirements of three selected Natal estuaries: the Mdloti, Lovu and Mkomazi.* CSIR Report EMAS-C92101. Stellenbosch: CSIR.

CSIR (1992c). *The freshwater requirements of the Palmiet River mouth.* CSIR Contract Report C/SEA 8426. Stellenbosch: CSIR.

CSIR (1992d). *Building the foundation for sustainable development in South Africa.* National report to the United Nations Conference on Environment and Development (UNCED) held in Rio de Janeiro, June 1992. Pretoria: Department of Environment Affairs.

CSIR (1993). *Great Brak Estuary Management Programme Interim Report. Report on the monitoring results for the period April 1992 to March 1993.* CSIR Report EMAS-C93051. Stellenbosch: CSIR.

CSIR (1994). *Great Brak Estuary Management Programme: Report on the monitoring results for the period April 1993 to March 1994.* CSIR Report EMAS-C94013; DWAF Report V/K100/08/E006. Stellenbosch: CSIR.

CSIR (1996). *The effect of the 2015 run-off scenario on the Mvoti estuary.* CSIR Report ENV/S-C96038. Stellenbosch: CSIR.

Danish Hydraulic Institute (1992). *Mike-11. A microcomputer-based modelling system for rivers and channels.* Reference Manual (Version 3.01). Holstrom: DHI.

Dankers, N. (1993). Integrated estuarine management – obtaining a sustainable yield of bivalve resources while maintaining environmental quality. In *Bivalve Filter Feeders in Estuarine and Coastal Ecosystem Processes*, ed. R.F. Dame. NATO ASI Series. Series G: Ecological Sciences, **33**, 479–511.

Davies, B.R., O'Keeffe, J.H. & Snaddon, C.D. (eds.) (1993). *A synthesis of the ecological functioning, conservation and management of South African river* systems. WRC Report TT62/93. Pretoria: Water Research Commission.

Day, J.H. (1967). The biology of Knysna estuary, South Africa. In *Estuaries*, ed. G.H. Lauff, pp. 397–407. Washington, DC: American Association for the Advancement of Science.

Day, J.H. (1980). What is an estuary? *South African Journal of Science*, **76**, 198.

Day, J.H. (1981a). Summaries of current knowledge of 43 estuaries in southern Africa. In *Estuarine ecology with particular reference to Southern Africa*, ed. J.H. Day, pp. 251–329. Cape Town: A. A. Balkema.

Day, J.H. (1981b). The estuarine fauna. In *Estuarine ecology with particular reference to southern Africa*, ed. J.H. Day, pp. 147–86. Cape Town: A.A. Balkema.

Day, J.H. & Grindley, J.R. (1981). The management of estuaries. In *Estuarine ecology with particular reference to Southern Africa*, ed. J.H. Day, pp. 373–97. Cape Town: A.A. Balkema.

Day, J.H., Millard, N.A.H. & Harrison, A.D. (1952). The ecology of South African estuaries. Part 3: Knysna, a clear open estuary. *Transactions of the Royal Society of South Africa*, **33**, 367–413.

De Decker, H.P. (1989). Report No. 40. Klein (CSW 16). In *Estuaries of the Cape. Part 2. Synopses of available information on individual systems*, ed. A.E.F. Heydorn & P.D. Morant. CSIR Research Report No. 439. Stellenbosch: CSIR.

De Decker, H.P. & Bally, R. (1985). The benthic macrofauna of the Bot River estuary, South Africa with a note on its meiofauna. *Transactions of the Royal Society of South Africa*, **45**, 379–96.

Department of Environment Affairs (1992). *The Integrated Environmental Management procedure*. Pretoria: Government Printer.

Department of Environmental Affairs and Tourism (1996). *An environmental policy for South Africa: Green paper for public discussion*. Pretoria: Department of Environmental Affairs and Tourism.

Department of Water Affairs (1986). *Management of the water resources of the Republic of South Africa*. Pretoria: DWA.

Department of Water Affairs (1988/89). *Verslag oor die voorgestelde Mosselbaai Wolwedans-Staats-Waterskema*. Verslag van die Direkteur-Generaal: Waterwese aan die Minister van Waterwese. Pretoria: DWA.

Department of Water Affairs and Forestry

(1991). *Policy on water for the environment*. Pretoria: DWA&F.

Department of Water Affairs and Forestry (1993). *South African Water Quality Guidelines*. 5 volumes. Pretoria: DWA&F.

Department of Water Affairs and Forestry (1995). *South African Water Quality Guidelines for Coastal Marine Waters*, 4 volumes. Pretoria: DWA&F.

Department of Water Affairs and Forestry (1996). *Water Law Principles: Discussion Document*. Pretoria: DWA&F.

Fuggle, R.F. & Rabie, M.A. (eds.) (1992). *Environmental management in South Africa*. Cape Town: Juta and Co.

Gore, J.A. & Nestler, J.M. (1988). Instream flow studies in perspective. *Regulated rivers: Research and management*, **2**, 93–101.

Götmark, F., Åhlund, M. & Eriksson, M.O.G. (1986). Are indices reliable for assessing conservation value of natural areas? An avian case study. *Biological Conservation*, **38**, 55–73.

Haines-Young, R., Green, D.R. & Cousins, S.H. (eds.) (1993). *Landscape ecology and geographic information systems*. London: Taylor and Francis.

Hanekom, N. & Erasmus, T. (1988). Variations in size compositions of populations of *Upogebia africana* (Ortman) (Decapoda, Crustacea) within the Swartkops estuary and possible influencing factors. *South African Journal of Zoology*, **23**, 259–65.

Harrison, T.D., Cooper, J.A.G., Ramm, A.E.L. & Singh, R.A. (1994a). *Health of South African estuaries, Orange River – Buffels (Oos)*. Catchment and Coastal Environmental Programme, Technical Report. Durban: CSIR.

Harrison, T.D., Cooper, J.A.G., Ramm, A.E.L. & Singh, R.A. (1994b). *Health of South African estuaries, Orange River – Buffels (Oos)*. Catchment and Coastal Environmental Programme, Executive Report. Durban: CSIR.

Harrison, T.D., Cooper, J.A.G., Ramm, A.E.L. & Singh, R.A. (1995a). *Health of South African estuaries, Palmiet – Sout*. Catchment and Coastal Environmental Programme, Technical Report. Durban: CSIR.

Harrison, T.D., Cooper, J.A.G., Ramm, A.E.L. & Singh, R.A. (1995b). *Health of South African estuaries, Palmiet – Sout*. Catchment and Coastal Environmental Programme, Executive Report. Durban: CSIR.

Heydorn, A.E.F. (1972). South African

estuaries, their function and the threat to their existence. *Findiver*, **32**, 18–19.

Heydorn, A.E.F. (1973). South African estuaries: an economic asset or a national resource being squandered. *Natal Wildlife*, **14**, 10–15.

Heydorn, A.E.F. (ed.) (1986). *An assessment of the state of the estuaries of the Cape and Natal in 1985/86*. South African National Scientific Programmes Report, No. 130. Pretoria: CSIR.

Heydorn, A.E.F. (1989). The conservation status of southern African estuaries. In *Biotic diversity in southern Africa: concepts and conservation*, ed. B.J. Huntley, pp. 290–7. Cape Town: Oxford University Press.

Heydorn, A.E.F., Bang, N.D., Pearce, A.F., Flemming, B.W., Carter, R.A., Schleyer, M.H., Berry, P.F., Hughes, G.R., Bass, A.J., Wallace, J.H., Van der Elst, R.P., Crawford, R.J.M. & Shelton, P.A. (1978). Ecology of the Agulhas Current region: an assessment of biological responses to environmental parameters in the south-west Indian Ocean. *Transactions of the Royal Society of South Africa*, **43**, 151–90.

Heydorn, A.E.F. & Bickerton, I.B. (1982). Report No. 9. Uilkraals (CSW17). In *Estuaries of the Cape. Part 2. Synopses of available information on individual systems*, ed. A.E.F. Heydorn & J.R. Grindley. CSIR Research Report No. 408. Stellenbosch: CSIR.

Heydorn, A.E.F. & Grindley, J.R. (eds.) (1981–85). *Estuaries of the Cape. Part 2. Synopses of available information on individual systems*. CSIR Research Reports. Stellenbosch: CSIR.

Heydorn, A.E.F. & Morant, P.D. (eds.) (1986–90). *Estuaries of the Cape. Part 2. Synopses of available information on individual systems*. CSIR Research Reports. Stellenbosch: CSIR.

Heydorn, A.E.F. & Tinley, K.L. (1980). *Estuaries of the Cape. Part 1. Synopsis of the Cape Coast*. CSIR Research Report 380. Stellenbosch: CSIR.

Heÿl, C.W. & Currie, M.H. (1985). Variations in the use of the Bot River estuary by water-birds. *Transactions of the Royal Society of South Africa*, **45**, 397–417.

Huizinga, P. (1994). Recent advances in the understanding of estuary mouth dynamics. In *Proceedings of a Conference on Aquatic Ecosystems – Ecology, Conservation and Management*. Port Elizabeth.

IUCN (1980). *World Conservation Strategy: Living resource conservation for sustainable development*. Gland, Switzerland: International Union for Conservation of Nature and Natural Resources.

Järvinen, O. (1985). Conservation indices in land use planning: dim prospects for panacea. *Ornis Fennica*, **62**, 101–6.

Jezewski, W.A. & Roberts, C.P.R. (1986). *Estuarine and lake freshwater requirements*. Department of Water Affairs Technical Report No. TR129. Pretoria: Department of Water Affairs.

King, J.M. & Tharme, R.E. (1993). *Assessment of the instream flow incremental methodology, and initial development of alternative instream flow methodologies for South Africa*. WRC Report No. 295/1/94. Pretoria: Water Research Commission.

Klopatek, J.M., Kitchings, J.T., Olson, R.J., Kumar, K.O. and Mann, L.K. (1981). A hierarchical system for evaluating regional ecological resources. *Biological Conservation*, **20**, 271–90.

Koop, K. (1982). Report No. 18. Bot/Kleinmond System (CSW 13). In *Estuaries of the Cape. Part 2. Synopses of available information on individual systems*, ed. A.E.F. Heydorn & J.R. Grindley. CSIR Research Report No. 417. Stellenbosch: CSIR.

Kozakiewicz, M. (1995). Resource tracking in space and time. In *Mosaic landscapes and ecological processes*, ed. L. Hannson, L. Fahrig & G. Merriam. London: Chapman and Hall.

Krebs, C.J. (1985). *Ecology: the environmental analysis of distribution and abundance*. New York: Harper and Row.

Kriel, J.P. (1966). *Report of the commission of enquiry into the alleged threat to animal and plant life in St Lucia Lake*. Pretoria: Government Printer.

Kruuk, H. (1995). *Wild otters: predation and populations*. Oxford: Oxford University Press.

Le Maitre, D.C., Van Wilgen, B.W., Chapman, R.A. & McKelly, D.H. (1996). Invasive plants and water resources in the Western Cape Province, South Africa: modelling the consequences of a lack of management. *Journal of Applied Ecology*, **33**, 161–72.

Leopold, L.B. (1994). *A view of the river*. Cambridge, Mass.: Harvard University Press.

Maaren, H. & Moolman, J. (1986). The effects of farm dams on hydrology. In

Proceedings of the Second South African National Hydrology Symposium, Pietermaritzburg, 16–18 September 1985, ed. R.E. Schulze, pp. 428–41. Agricultural Catchments Research Unit, Report No. 22.

McLain, R.J. & Lee, R.G. (1996). Adaptive management: promises and pitfalls. *Environmental Management*, **20**, 437–48.

Martin, A.K. (1987). Comparison of sedimentation rates in the Natal Valley, south-west Indian Ocean, with modern sediment yields in east coast rivers of Southern Africa. *South African Journal of Science*, **83**, 716–24.

Matthews, G.V.T. (1993). *The Ramsar Convention on Wetlands: its history and development*. Gland, Switzerland: Ramsar Convention Bureau.

Milhous, R.T., Updike, M.A. & Schneider, D.M. (1989). *Physical habitat simulation system reference manual – Version II*. Instream Flow Information Paper No. 26. Fort Collins: United States Department of the Interior, Fish and Wildlife Service.

Miller, R.I. (1994). *Mapping the diversity of nature*. London: Chapman and Hall.

Morant, P.D. (1983). Report No. 20. Groot Brak (CMS 3). In *Estuaries of the Cape. Part 2. Synopses of available information on individual systems*, ed. A.E.F. Heydorn & J.R. Grindley. CSIR Research Report No. 419. Stellenbosch: CSIR.

Morant, P.D. (ed.) (1993–). *Estuaries of the Cape. Part 2. Synopses of available information on individual systems*. CSIR Research Reports. Stellenbosch: CSIR.

Natal Parks Board (1989). *Natal Parks Board Policy: Interference with the mouth of a lagoon or river (breaching)*. Board Minute: 6(a)(i). 30 June 1989.

Noble, R.G. & Hemens, J. (1978). *Inland water ecosystems in South Africa – a review of research needs*. South African National Scientific Programmes Report No. 34. Pretoria: CSIR.

Orme, A.R. (1974). *Estuarine sedimentation along the Natal coast, South Africa*. Office of Naval Research Technical Report No. 5. Washington, DC: Office of Naval Research.

Pereira, J.M.C. & Duckstein, L. (1993). A multiple criteria decision-making approach to GIS-based land suitability evaluation. *International Journal of Geographic Information Systems*, **7**, 407–24.

Poore, M.E.D. (1955a). The use of phytosociological methods in ecological investigations. I. The Braun-

Blanquet system. *Journal of Ecology*, **43**, 226–44.

Poore, M.E.D. (1955b). The use of phytosociological methods in ecological investigations. II. Practical issues involved in an attempt to apply the Braun-Blanquet system. *Journal of Ecology*, **43**, 245–69.

Pritchard, D.W. (1967). What is an estuary, physical viewpoint. In *Estuaries*, ed. G. Lauff. Washington, DC: American Association for the Advancement of Science.

Quinn, N.W. (1997). *An integrated modelling approach to the management of freshwater inflow to South African estuaries*. PhD thesis, University of Natal, Pietermaritzburg.

Quinn, N.W., Whitfield, A.K., Breen, C.M. & Hearne, J.W. (1998). An index for the management of South African estuaries for juvenile fish recruitment from the marine environment. *Fisheries Management and Ecology* (in press).

Ramm, A.E. (1988). The community degradation index: a new method for assessing the deterioration of aquatic habitats. *Water Research*, **20**, 293–301.

Ramm, A.E. (1990). Application of the community degradation index to South African Estuaries. *Water Research*, **24**, 383–9.

Ramm, A.E.L., Cerff, E.C. & Harrison, T.D. (1987). Documenting the recovery of a severely degraded coastal lagoon. *Journal of Shoreline Management*, **3**, 159–67.

Raper, J. (1989). *Three dimensional applications in Geographical Information Systems*. London: Taylor and Francis.

Rebelo, A.G. & Siegfried, W.R. (1992). Where should nature reserves be located in the Cape Floristic Region, South Africa? Models for the spatial configuration of a reserve network aimed at maximising the protection of floral diversity. *Conservation Biology*, **6**, 243–52.

Reddering, J.S.V. (1988a). Prediction of the effects of reduced river discharge on estuaries of the south-eastern Cape Province, South Africa. *South African Journal of Science*, **84**, 726–30.

Reddering, J.S.V. (1988b). Coastal and catchment basin controls on estuary morphology on the south-eastern Cape coast. *South African Journal of Science*, **84**, 154–7.

Rice, J.A. & Cochran, P.A. (1984).

Independent evaluation of a bioenergetics model for largemouth bass. *Ecology*, **65**, 732–9.

Ringold, P.L., Alegria, J., Czaplewski, R.L., Mulder, B.S., Tolle, T. & Burnett, K. (1996). Adaptive monitoring design for ecosystem management. *Ecological Applications*, **6**, 745–7.

Roberts, C.P.R. (1983). Environmental constraints on water resources development. In *Proceedings of the Seventh Quinquennial Convention of the South African Institute of Civil Engineers, Cape Town*.

Scott, K.M.F., Harrison, A.D. & Macnae, W. (1952). The ecology of South African estuaries. *Transactions of the Royal Society of South Africa*, **33**, 283–331.

Shannon, L.V. (1985). The Benguela ecosystem. 1. Evolution of the Benguela, physical features and processes. *Oceanography and Marine Biology Annual Review*, **23**, 105–82.

Siegfried, W.R. (1981). The estuarine avifauna of southern Africa. In *Estuarine ecology with particular reference to Southern Africa*, ed. J.H. Day, pp. 223–50. Cape Town: A.A. Balkema.

Slinger, J.H. (1996). *Modelling of the physical dynamics of estuaries for management purposes*. PhD thesis, University of Natal, Pietermaritzburg.

Southwood, T.R.E. (1977). Habitat, the templet for ecological strategies. *Journal of Animal Ecology*, **46**, 337–65.

Turpie, J.K. (1995). Prioritizing South African estuaries for conservation: a practical example using waterbirds. *Biological Conservation*, **74**, 175–85.

Turpie, J.K. & Crowe, T.M. (1994). Patterns of distribution and diversity of larger mammals in Africa. *South African Journal of Zoology*, **29**, 19–31.

Tyson, P.D. (1986). *Climatic change and variability in southern Africa*. Cape Town: Oxford University Press.

Van Heerden, I. (1985). Barrier/estuarine processes: Bot River Estuary – an interpretation of aerial photographs. *Transactions of the Royal Society of South Africa*, **45**, 239–51.

Waldron, M. (1986). The importance of water levels in the management of the Klein River estuary, Hermanus. MSc thesis, University of Cape Town.

Water Research Commission (1996). *Republic of South Africa. Water Law Review Process. The Philosophy and Practice of Integrated Catchment Management: Implications for Water Resource Management in South Africa*. Discussion Document. Pretoria, Department of Water Affairs and Forestry and Water Research Commission. WRC Report No. TT81/96. Pretoria: Water Research Commission.

Whitfield, A.K. (1994a). Abundance of larval and 0+ juvenile marine fishes in the lower reaches of three southern African estuaries with differing freshwater inputs. *Marine Ecology Progress Series*, **105**, 257–67.

Whitfield, A.K. (1994b). An estuary-associated classification for the fishes of southern Africa. *South African Journal of Science*, **90**, 411–17.

Whitfield, A.K. (1995). *Available scientific information on individual South African estuarine systems*. Report to the Water Research Commission by the Consortium for Estuarine Research and Management. WRC Report No. 577/1/95. Pretoria: Water Research Commission.

Whitfield, A.K. (1996). Fishes and the environmental status of South African estuaries. *Fisheries Management and Ecology*, **3**, 45–7.

Whitfield, A.K. & Bruton, M.N. (1989). Some biological implications of reduced freshwater inflow into Eastern Cape estuaries: a preliminary assessment. *South African Journal of Science*, **85**, 691–4.

Whitfield, A.K. & Wooldridge, T.H. (1994). Changes in freshwater supplies to southern South African estuaries: some theoretical and practical considerations. In *Changes in fluxes in estuaries*, ed. K.R. Dyer & R.J. Orth, pp. 41–50. International Symposium Series. Fredensborg, Denmark: Olsen and Olsen.

Whitfield, A.K., Allanson, B.R. & Heinecken, T.J.E. (1983). Report No. 22. Swartvlei (CMS 11). In *Estuaries of the Cape. Part 2. Synopses of available information on individual systems*, ed. A.E.F. Heydorn & J.R. Grindley. CSIR Research Report No. 421. Stellenbosch: CSIR.

Wiseman, K.A., Burns, M.E.R. & Vernon, C.J. (1993). Report No. 42. Nahoon (CSE44), Qunira (CSE45) and Gqunube (CSE46). In *Estuaries of the Cape, Part II: Synopses of available information on individual systems*, ed. P.D. Morant, CSIR Research Report No. 441. Stellenbosch: CSIR.

Wooldridge, T. (1991). Exchange of two species of decapod larvae across an estuarine mouth inlet and implications of anthropogenic changes in the frequency and duration of mouth closure. *South African Journal of Science*, **87**, 519–25.

Wooldridge, T.H. (1994). The effect of periodic inlet closure on recruitment in the estuarine mudprawn *Upogebia africana* (Ortmann). In *Changes in fluxes in estuaries*, ed. K.R. Dyer & R.J. Orth. Fredensborg, Denmark: Olsen and Olsen.

Wooldridge, T.H. & Loubser, H. (1996). Larval release rhythms and tidal exchange in the estuarine mudprawn *Upogebia africana*. *Hydrobiologia*, **337**, 113–21.

13 Perspectives

Brian Allanson, Dan Baird and Allan Heydorn

Great Fish River estuary

Introduction

The opportunity seldom arises to review and assess the progress of scientific endeavour within cognate fields of study. The chapters in this book document information on the functioning of estuaries over a wide spectrum of processes and on virtually all aspects of the estuarine ecosystem. They demonstrate the enormous advance made in this field of environmental science over the past twenty years in South Africa. And we hope that this scientific output and its management implications will be useful to others in different parts of the world.

Two concepts or themes of great importance emerge from the contents of this volume. The first encapsulates the human influence in its many guises of which the most immediately critical is the importance of adequate supplies of freshwater to estuaries, without which they stand to lose the characteristic features of such ecosystems. The ramifications of no or lesser freshwater inflow are manifold: affected are not only the various biota, in some way or another, but also physical and chemical processes and features of estuaries such as sediment movement and deposition, closing of inlets, mixing and other hydrodynamic properties, nutrient dynamics and flux rates. Less freshwater influences the nature and production of plant and animal communities, and so impacts upon the structure, magnitude and diversity of animal life in estuaries. The overriding message is clear: for estuaries to maintain their integrity as functional ecosystems, the nature and quality of freshwater is of extreme importance.

The second concept, namely that of ecosystem holism, is clearly articulated in many chapters of this volume. Whereas individual studies usually adopt a reductionist approach, the chapters have described and examined processes and biota as integral components and parts of the whole. The emphasis is on estuaries as ecosystems, and that they should be studied and managed in this way, thereby embracing an holistic approach to system ecology. In so doing we become aware of the complexity of ecosystems and their causal indirectness, and of the intricacies of the network of material and energy flow and cycling in estuaries.

The first concept

We have not thought it wise to debate the ultimate definition of an estuary! And while estuaries may defy vigorous definition, the general acceptance of their common facies is without question. They are the product of two adjacent ecosystems: the terrestrial watershed and the ocean.

The South African coastal area is greatly dissected by these interfaces between the river and the sea; from the steeply tilted hinterland subject to heavy summer rainfall of the eastern seaboard to those occurring on the less tilted west coast under arid conditions and which only become functional during rare events of exceptional precipitation. The estuaries are small in area, and although numerous (465 within a coastline of 3000 km: see Chapters 2 and 8), they cover only some 600 square kilometres of coastline, compared with approximately 10 000 km of coastline in the USA where estuaries cover an area of 107 722 km^2! – or some 11 km^2 per km of coastline. They have not, in general, attracted large settlements of people and their diverse activities; although the urban and industrial foci of Richards Bay and Durban in KwaZulu-Natal, the Buffalo and Swartkops Rivers in the Eastern Cape

Province, and Knysna and Saldhana in the Western Cape are important estuarine settlements. The decentralisation of industry from the once primary focus of the Witwatersrand, associated *inter alia* with increasing exports to the coastal rimland, is leading to an increase in the rate of development along the coast.

The archaeological evidence points quite clearly to sustained occupancy of the coastal zone throughout the late Pleistocene by individuals who were anatomically modern human beings (van Andel 1989). These people belonged to the Middle Stone Age culture and exploited the rich marine resources. Van Andel (1989) makes the point that the sea was, with one exceptional period, never very far away from their habitations.

The one exception referred to is the presumed long hiatus between the Middle Stone Age and the Holocene Later Stone Age, when the archaeological record afforded by the existing caves and associated sites was interrupted. This might be attributable to the migration of the population who were living in these coastal caves onto what is now a submerged coastal plain, but which was exposed during the glacial maximum of 25 000–15 000 years ago. The wide central coastal plain may have seen an extension of the fynbos landscape and with it a greater reliance by the population on the plant foods of this resource.

Notwithstanding the circumstantial nature of much of the archaeological evidence for this period, van Andel (1989) would support the exploitation of the coastal seas and presumably the estuaries by human beings for long periods of time. He notes that human occupancy of the southern African coast appeared earlier than anywhere else, although he does admit that this hypothesis 'appears less well supported at the moment than it should be in view of its great importance'.

The later occupancy by the hunter-gatherers of the 'Strandloper' and San people, and their subsequent displacement by pastoralists from the North East and colonialists from Europe, set the scene for the modern occupancy of this diverse and sensitive region. This modern occupancy brought with it a technology which, when superimposed upon the natural cycles of drought and flood perturbations, tended to exacerbate the rate at which changes occurred. Prime among these is the loss of estuarine wetland. The Restore America's Estuarine Programme reports that, while 10 years ago every new person added to the State of Maryland accounted for the loss of ⅓ acre of land, now, every new person causes the loss of ⅔ of an acre into housing developments, shopping malls, business parks and new roads.

We have no equivalent numerical assessment of this impact in South Africa, but the demands such growth are having upon the river and its estuary are unrelenting. It has been demonstrated, for example, that although there has

been considerable infilling by sediment transfer during the Holocene, the last 60–80 year period has seen a substantial increase in the rate of sedimentation and decreases in river flow principally of anthropogenic origin. Thus, in the absence of river floods or, more commonly, their attenuation, channel volume is reduced by the accumulation of marine sands. If unalleviated by scour this leads to a reduction in tidal prism volume until the estuary mouth closes.

Persistent reduction of river flow as a result of drought, water abstraction and gross interference with the hydraulic structure of estuaries over the past 60 years has resulted in some instances in the destruction of the ecosystem. Examples of such extremes are the Seekoei estuary near Jeffreys Bay which until recently was a dry sand basin, and the Mtwalume River estuary in KwaZulu-Natal in which sedimentation has become so severe that it is represented by a small lagoon. The elevated floodplain is exploited by man.

In management terms this is the ***first effect of river flow reduction!***

A possible dynamic classification of our estuaries depends largely upon discharge volume relative to tidal prism volume which ranges from 0.01 to 0.09 in those which are well mixed, 0.1 to 0.5 with partial stratification in middle reaches, and 0.6 to >1 in highly stratified estuaries, for example the Palmiet, Mgeni and Tugela rivers. Nevertheless, a major obstacle to the solution of many estuarine hydrodynamic problems has been the generally poor understanding of turbulent mixing processes in stratified fluids. Chapter 3 has resolved a number of these and contributed to The Estuarine System Model (Chapters 11 and 12) which overcomes, in some measure, the limitation of one-dimensional hydrodynamic models.

This model, developed by Slinger (1996), is used as the basic building block of a linked or integrative model which Jill Slinger and her colleagues developed between 1994 and 1996, under the auspices of the Consortium for Estuarine Research and Management. Six models which include (a) the Estuarine system model as the basic unit, (b) Mike 11 hydrodynamic and transport-dispersion model, (c) Mike 11 water quality model, (d) Plant estuarine decision support system, (e) Estuarine ecosystem model and (f) Dynamic vegetation model are linked primarily by the flow of appropriately summarised data from each model (Slinger 1996) for specific estuaries.

The overriding consideration Slinger (1996) draws attention to, is the verification and quantification of the effects of water deprivation upon mouth closure and resistence to scouring floods, or the change to marine dominance. Likewise it has been possible to predict and quantify a number of biotic responses, notably the change in macrophyte associations and, where the mouth closes, the loss of those burrowing anomuran prawns which depend upon tidal transport to the sea to complete their life cycles. The management implications of this integrative model have still to be appreciated.

While the chemistry of the water column and sediments is moderately well understood, the investigation of biogeochemical processes has attracted a number of interesting studies which illustrate the role of microbial, macrophyte and faunal modulation. However, no fixed pattern of nutrient export has been established. Some systems are sinks of nutrients while others are exporters. And even this is not a fixed property of the intertidal wetlands: they may switch the flux of nitrogen and or phosphorus depending upon their concentrations in the inundating water, which in turn is dependent upon the strength or weakness of river inflow and the state of the mouth of the estuary.

Our ignorance is, however, substantial when we ask questions about the impact of changes of river flow upon the conservative behaviour of essential elements and the consequential changes in silica:nitrogen or nitrogen: phosphorus ratios which are being delivered to the coastal seas. It is hard to imagine that with a coastline of some 3000 km dissected by 465 river systems, changes in these ratios as emphasised by Justic *et al.* (1995) for North American and European rivers would not have similar effects, namely an increase in eutrophication potential consequent upon silica limitation and the encouragement of non-siliceous algal phytoplankton (see Skreslet 1985 for review). A component of what may be further involved is the coupling between the benthic and pelagic domains in estuaries about which we have little understanding. Examination of the rate and magnitude of the exchange of macro-elements (e.g. N, P, Si, etc.) between the sediment and overlying waters, the crucial role of micro- and meio-fauna in regeneration and cycling, and the contribution of these to the water column productivity could be a fruitful avenue of research.

Questions which immediately spring to mind in this regard are: (1) To what extent is the increasing incidence of toxic red tides along the west coast a consequence of such changes? (2) Have chemical changes in the Great Berg, Oliphants, Orange, Mgeni and Tugela rivers begun the slow but inevitable drift towards eutrophication of the coastal seas, and superimposed in some way upon the natural oceanographic events? This we consider is an urgent and stimulating area of enquiry, requiring the closest liaison between the various branches of physical oceanography, aquatic chemistry and river hydrology.

The analysis of estuarine structure and function has been carried out at three levels: physical, chemical and biological, the latter involving (a) single species ecology (b) faunal and floral community ecology incorporating interactions with both physical and chemical environment, and (c) ecosystem description, which takes the form of (i)

the flux of carbon between living and non-living compartments and (ii) analysis of the flow models by means of network analysis. Each has contributed materially to the present holistic view of our estuarine systems.

While it has been established that the estuarine plants and animals are ecologically robust, research has also established the degree to which the estuarine ecosystem is dependent upon the physical environment, although considerable biological modulation is evident. This is particularly so in the change from macrophytes to phytoplankton communities when turbidity increases occur in an estuary. As turbidity is invariably associated with decreased salinity and an increase in river-borne nutrients, phytoplankton biomass increases. Turbulence in the shallow estuaries, e.g. the Great Fish River, is sufficiently marked to ensure that even under quite high light attenuation circulation within the water column ensures that the algal cells receive sufficient flux.

This switching between a macrophyte-based to a phytoplankton-based economy as elaborated by Allanson and Read (1995) is common in many southeast coast estuaries as river flow either increases or decreases. The zooplankton community responds to changes in river flow by alteration in species composition and abundance, as has been elegantly shown in the Sundays River estuary by Tris Wooldridge and his students.

There is a frequently quoted paradigm that estuaries, and by implication South African estuaries, are very productive systems. While the evidence in the Americas may be overwhelming, the productivity of our estuaries remains to be systematically measured. The studies of Christie (1981) in the Langebaan lagoon embayment are the first estimate we have of the productivity of various intertidal plant compartments.

The wide-ranging review by Janine Adams and her colleagues (Chapter 5) has established the structure, productivity and sensitivity to salinity changes of many of the important components of the estuarine plant community. The phytosociological structure of the saltmarshes of South African estuaries has been well researched, particularly in the Western and Eastern Cape Provinces, but it is their role in the biochemical economics of these systems which requires greater understanding. Fortunately the work reported in this volume on the mangrove estuaries is enriched by an extensive international literature and to which the studies of the botanists in KwaZulu-Natal have contributed meaningfully.

The finite role of the intertidal marshes and the epipsammic, epiphytic or benthic microalgae as energy and elemental sources should become a productive research field. In our view their contribution towards estuarine processes remains equivocal until more serious work is initiated. Of particular import is the contribution of C-4 and

C-3 plants in estuarine ecosystems. Are we correct in accepting that estuaries are detritus-based, which in essence means that a very large proportion of the primary production, particularly of the intertidal flats, salt-marshes and mangroves is not directly utilised by grazing herbivores but manipulated initially by the microbial community?

In some measure this has given rise to a rather lax attitude towards the determination of the precise nature or origin of the detritus. Fortunately new investigations have shown that the detritus derived from the most immediately available source, e.g. *Zostera capensis* is not a major component (if used at all) in the diet of, for example, burrowing animals. Plants possessing the C-4 photosynthetic pathway such as *Z. capensis* have $\delta^{13}C$ values ranging from 9 to 19‰. The increased $\delta^{13}C$ negativity of the C-3 plants such as *Sarcocornia* and *Chenolea* have values of -24 to -34‰ which matches the $\delta^{13}C$ negativity of razor clams and mud-prawns.

Recently Riera and Richard (1996) have shown that, along a trophic gradient from the head of the Marennes-Oleron estuary (France) to the open sea, gradients of $\delta^{13}C$ of particulate organic carbon changed from riverine -29.2 to -27.4‰ to sea littoral of -21.8 to -19.5‰. The isotopic composition of the oyster *Crassostrea gigas* followed a similar trend, indicating a predominant contribution of particulate organic matter from terrestrial detritus to the diet of *C. gigas* inhabiting the upper reaches of the estuary.

Similar studies by Jerling and Wooldridge (1995) on the $\delta^{13}C$ ratios of mesozooplankton in the Sundays River estuary have suggested that the high negative values of $\delta^{13}C$ of -28.3 to -32.0‰ indicate carbon source depleted of ^{13}C. To a degree they found matching levels of negativity in the <20 μm particle suspensoids in the estuary. This fine fraction is composed nanoflagellates and nanodino-flagellates as well as unidentified material.

These examples of the way in which energy pathways are being defined are removing the deadening hand of assumption from the field of estuarine trophic studies, and point to a re-evaluation of the significance of primary production within the diverse compartments, and where and how they are utilised.

It is but a short step from this structural level to the role of estuaries in the life cycle of a number of important marine fish. The interaction between the estuary and the surf zone has been unequivocally established by the innovative studies of Alan Whitfield (see Chapter 9). And in this regard it is important to recognise that, in contrast to the shallow, turbid coastal seas of the Indo-Pacific, the coastal seas of South Africa are deep and turbulent. Only the estuaries provide the shallow turbid conditions required by many juvenile stages of fish and other invertebrate taxa. This is particularly true in KwaZulu-Natal estuaries and

underlines the urgent need to introduce strategies to maintain tidal flux and re-establish it where realistic to do so in degraded systems.

The results from and interpretations of the diverse array of projects were brought together in the broader context of zoogeography, specifically the Indo-Pacific region, by Stephen Blaber in 1981. He has argued convincingly that the remarkable close affinity of the fishes of the southeast coast of Africa (63% of the species in St Lucia have been recorded in southeast Asian waters) is due very largely to the properties which the estuaries share with the extensive seas of the Indo-Pacific. They are shallow, turbid warm systems with varying salinities. And as the coastal seas of Southern Africa are far removed from this characterisation, the estuaries offer environmental conditions suitable for the juveniles of many species and as feeding grounds for adults.

Obviously there is subtraction as these estuarine qualities are replaced southwards and westwards. Nevertheless, the quantitative approach adopted in the Swartkops River estuary and adjacent seas and further developed in the Swartvlei system has provided unequivocal support for the following conclusions.

- Most fish found in estuaries are dependent upon a link between the estuary and the sea to complete their life cycle. This link is particularly important in early summer when most ichthyoplankton interchange takes place.

- Thus barriers to this migration such as sandbars will prevent larvae of estuarine spawners from reaching the sea, and conversely the larvae and postlarvae of marine species from entering the estuary.

In management terms this is **the second effect of reducing river inflow to estuaries!**

The quiet water and habitat diversity, in estuaries and embayments where tidal exchange is maintained, supports a rich array of benthic invertebrates. Much of this biodiversity is dependent upon stable physical conditions and, in particular, salinities and transparency with a concomitant increase in eelgrass meadows where salinities are near that of seawater. This is in contrast to the zooplankton community which has greater diversity and abundance where salinities and turbidities are under riverine control.

While the Langebaan marine embayment is not influenced by river inflow of any magnitude but rather by an exceptional resource of fresh groundwater inflow, it has, nevertheless, provided the stimulus for penetrative studies of the ecological roles of five key species, including the sand prawn *Callianassa kraussi*, the globular mud snail *Assiminea globosus*, the bloodworm *Arenicola loveni*, and the dwarf cushion star, *Patriella exigua* in the economy of the tidal lagoon. In general these animals enhance bacterial numbers and biomass, and reduce the density of diatoms and meiofauna. Sand prawns are prodigious burrowers, oxygenating the sediments through this activity. They are the only abundant animal which serves this function in the lagoon. In all tidal systems along the south and southeast coast this species tends to be restricted to the sandy portions of the estuary, being replaced in those areas where muddy sand occurs by the thallassinid burrowing prawn *Upogebia africana*, which serves a similar function.

The review presented by Casper de Villiers and Alan Hodgson and Anthony Forbes in Chapter 8 has highlighted the importance of biological modulation in ecosystems which were generally thought to be under the exclusive control of physical forcing, and has described the structure of the multi-dimensional niche of a number of species. The application of these findings to those estuaries and embayments which possess large intertidal floodplains normally under the influence of tidal action, e.g. of the estuaries of the Swartkops, Kariega, Kromme, Keiskamma and Knysna rivers, and the marine embayment Langebaan lagoon, has increased our understanding of their ecological properties, although as pointed out earlier the ecological quality of Langebaan lagoon on the west coast is not dependent upon surface river flow.

This raises something of a conundrum as it points to at least two quite distinct ecological requirements. Firstly, the richness of the benthos is almost entirely dependent upon stable salinity near to that of seawater and the quiet exchange of tidal flow, so clearly enunciated by John Day in 1958, and upon which a rich and varied bird life depends. Secondly, a nektonic and planktonic component is required which is defined by the axial gradient of salinity and turbidity, and upon which fish and prawn larvae depend for food and shelter. Clearly this conundrum is resolved by accepting that the possession and persistence of axial gradients within the estuary are the very fibre of their being – destroy the gradients and the estuarine character fades!

The Second Concept

Thus far we have assessed in some measure a reductionist approach to estuarine ecology. And there is no doubt that the rich resources of data and interpretation which this approach has generated have increased the overall success of those studies which have adopted the newer holistic and integrative descriptions of estuarine ecosystems. This direction in both global and South African studies is reviewed in Chapter 11 by Dan Baird, and demonstrates the absolute necessity of long and meaningful databases without which the application of network analysis is not possible. Fortunately three estuarine systems broadly representative of the array of estuarine types commonly found along the south and southeastern seaboard matched this requirement.

The contents of the various energy compartments within the Bot River, Kromme River and Swartkops River estuaries were sufficiently well known that flow networks could be constructed. From this level a trophic structure was erected within which the detrivory:herbivory ratio was assessed, and also the inverse relation to net primary production (NPP). In the Bot River estuary detrivory (3:1) is of greater importance than the direct utilisation of primary producers. This is in contrast to the Ems River estuary (Netherlands) where the NPP efficiency is high (98%) and the detrivitory:herbivory ratio is low (0.5:1).

A logical development of these findings is to examine the structure and magnitude of cycling in estuaries, given that recycling of matter is of universal importance in all ecosystems. The objective is to identify the number of cycles and routes of recycling in the system. Application of this analysis allowed Dan Baird and his colleagues to show conclusively that building a dam in the Kromme River above the ebb and flow caused the cycling within the system to change. An immediate effect was the disappearance of complex recycling pathways, Thus, with the loss of phytoplankton–zooplankton pathways, the cycling structure becomes simpler, indicating a system under stress. The value of network analysis is that it allows quantification of intuitive interpretations which hitherto have been so troublesome to express in any acceptable manner.

Environmental impacts

The significance of all this work lies in what it tells us about the interactive role of tidal and river flow. A balance between these two components (which varies from estuary to estuary) is essential if the full spectrum of 'estuarine' conditions is to pertain. And while there are estuaries which have survived the loss of river inflow, they have in effect become arms of the sea offering the quiet water conditions referred to earlier. The more general experience has shown that reduction of freshwater and tidal flow leads at the one extreme to hypersalinity, the loss of saltmarsh and riparian plants, phytoplankton and faunal diversity. This degradation continues until the estuary withers away. At the other extreme, where the mouth closes largely as a direct result of oceanographic events but river flow is sustained, a barrier lagoon or coastal lake is formed, in which the trophic dynamics is rearranged.

Many of the KwaZulu-Natal estuaries belong here. And while the formation of barrier lagoon systems has gone on throughout the Holocene, the elevated lagoons in which many of the smaller rivers end are directly the result of terrigenous accumulation as a consequence of poor land management practice in their catchments during the past 100 years.

Abood and Metzger (1996) have emphasised that any holistic approach to impact assessment in shallow-water estuarine habitats must recognise that the ecology of these areas has been shaped by their history of earlier impacts. While the estuaries of South Africa are not subject to the environmental extremes of North America and Europe, they experience episodic flooding which invariably lowers the salinity to <1 for up to 4 or 5 days and in some instances as long as 10 days. Fortunately the flood hydrographs of the coastal streams and rivers are short-lived so that under open mouth conditions the tidal prism is quickly restored. Superimposed upon this scenario is the spectre of sea level rise through global warming. The implication this has for the estuaries of the world is that they will become increasingly dominated by the sea, and possibly extend the period of tidal influence in those systems which have a tendency to close. This coupled with concomitant reduction in river flow will further underpin marine dominance.

We need, therefore, an approach, as Abood and Metzger (1996) have emphasised, which is based upon long data series and so allows the ability to discriminate among human impacts and natural variation. This, they argue, cannot be achieved with a snapshot of present conditions. It is in this context that South African estuarine environmental studies have been most successful in that they have and continue to record changes which are occurring through time – certainly over a period, so far, of 20 years.

Within the array of environmental studies those of the Estuarine and Coastal Research Unit (CSIR) series 'Estuaries of the Cape', and the Town and Regional Planning Commission of Natal's 'Estuaries of Natal' reports are of major importance. These remarkable works initiated by Alan Heydorn and John Grindley are encyclopaedic in content and, together with *Comparative Ecology of Natal's Smaller Estuaries*, Begg (1984), are rich sources of biological and environmental information about a diverse and sensitive coastline. The development of this work to encompass the estuaries of the Eastern and Western Provinces by Trevor Harrison in the Coastal and Environmental Programme of the CSIR based in Durban will provide the necessary yardstick against which to measure change.

The authors of Chapter 12, as modern practitioners, have deliberated in detail upon specific methodologies which have evolved as a result of the very real need to get to grips with those issues which, if not abated, will bring about the demise of these complex interfaces at the coastal edge.

In common with experience in the United States of America, and in particular the Save America's Estuaries Campaign, the major threats to the estuary habitat are:

1   Population growth by natural increase and migration and its attendant requirements.

2   Reduction in freshwater inflow through the continual building of dams, large and small, to meet this burgeoning growth.

3   Poor agricultural land practices and failures in urban planning resulting in an excessive sediment transfer to the basin of the estuary.

4   Diffuse and point sources of pollution; stormwater inflow, industrial and sewage discharge.

5   Apathy – the failure in understanding that there is a problem or that the communities and local authorities can do something about it. Our new Constitution requires that all citizens have the right to clean water and a healthy environment: But only the community can deliver this right!

The growth of our knowledge, and the insight we have gained on the structure and function of estuaries as ecosystems are reflected in this book. They have, we hope, demonstrated that estuaries should be managed as a whole, including the watershed and oceanic environment. The recent essays by Hopkinson and Vallino (1995) on the interrelationship between human activity in the watershed of estuaries, and of Odum *et al.* (1995) on the pulsing effects of tides on small time scales, of episodic floods on larger scales and the large spectrum of biological oscillations are inherent to the life of estuaries, and complement this review.

Of further significance has been the initiative and work of Professor Michael Bruton of the Environmental Education Trust, Two Oceans Aquarium, Cape Town. His insight into the manner in which the Ramsar Convention could be expanded to include criteria of importance to fish and fisheries, but not necessarily of importance to birds, when designating Ramsar sites will lead to the conservation of estuarine wetlands which may otherwise have been ignored.

Rogers and Bestbier (1997) in a particularly thought-provoking review of current literature, and an assessment of the demands required in attempting a definition of the desired state of riverine systems in South Africa have stressed the urgent need to change from the old 'Balance of Nature' paradigm in which the focus is 'on species conservation through population control in closed stable systems to one which emphasises temporal and spatial heterogeneity across the scales of ecological systems'.

It is of course quite impossible to imagine any estuarine

study being based upon the balance of nature paradigm – estuaries are excellent examples of open systems in which temporal and spatial heterogeneity are very much the order of the day! We submit, however, that it is this intrinsic property which makes their management by those who have not progressed beyond the balance of nature syndrome so very difficult.

The challenge of the present and future is the appreciation that this interface between land and sea is in flux – fragmented management, and fragmented research are past. But, notwithstanding, as ecologists and environmental scientists we are aware of the missing component of the equation: the political will of those who govern to initiate change!

References

Abood, K.A. & Metzger, S.G. (1996). Comparing impacts to shallow-water habitats through time and space. *Estuaries*, **19**, 220–8.

Allanson, B.R. & Read, G.H.L. (1995). Further comment on the response of south east coast estuaries to variable freshwater flows. *Southern African Journal of Aquatic Sciences*, **38**, 56–70.

Begg, G.W. (1984). *The comparative ecology of Natal's smaller estuaries*. Natal Town and Regional Planning Report 62, Pietermaritzburg.

Blaber, S.J.M. (1981). The zoogeographical affinities of estuarine fishes in south-east Africa. *South African Journal of Science*, **77**, 305–7.

Christie, N.D. (1981). Primary production in Langebaan Lagoon. In *Estuarine ecology with particular reference to southern Africa*, ed. J.H. Day, pp. 101–15. Cape Town: A.A. Balkema.

Day, J.H. (1958). The biology of Langebaan lagoon: a study of the effect of shelter from wave action. *Transactions of the Royal Society of South Africa*, **35**, 475–547.

Hopkinson, C.S. Jr & Vallino, J.J. (1995). The relationship among man's activities in watersheds and estuaries: a model of runoff effects on patterns of estuarine community metabolism. *Estuaries*, **4**, 598–621.

Jerling, H.L. & Wooldridge, T.H. (1995). Relatively negative δ^{13}C ratios of mesozooplankton in the Sundays River estuary, comments on potential carbon sources. *Southern African Journal of Aquatic Sciences*, **21**, 71–7.

Justic, D., Rabalais, N.N., Turner, R.E. & Dortch, Q. (1995). Changes in nutrient structure of river-dominated coastal waters: stoichiometric nutrient balance and its consequences. *Estuarine, Coastal and Shelf Science*, **40**, 339–56.

Odum, W.E., Odum, E.P. & Odum, H.T. (1995). Nature's pulsing paradigm. *Estuaries*, **4**, 547–55.

Riera, P. & Richard, P. (1996). Isotopic determination of food sources of *Crassostrea gigas* along a trophic gradient in the estuarine bay of

Marennes-Oleron. *Estuarine, Coastal and Shelf Science*, **42**, 347–60.

Rogers, K.H. & Bestbier, R. (1997). *Development of a protocol for the definition of the desired state of riverine systems in South Africa*. Department of Environmental Affairs and Tourism, Pretoria.

Slinger, J.H. (1996). *Modelling the physical dynamics of estuaries for management purposes*. PhD thesis, University of Natal, Pietermaritzburg.

Skreslet, S. (1985). Freshwater outflow in relation to space and time dimensions of complex ecological interactions in coastal waters. In *The role of freshwater outflow in coastal ecosystems*, ed. S. Skreslet, pp. 3–12. Berlin: Springer-Verlag.

van Andel, T.H. (1989). Late Pleistocene sea levels and the human exploitation of the shore and shelf of Southern South Africa. *Journal of Field Archaeology*, **16**, 133–51.

Systematic index

This index includes species mentioned in the text and tables. Those bird names which appear only in Appendix 10.1 are not repeated here.

General index